TO THE STUDENT:
HOW TO USE A STUDENT'S COMPANION
TO FUNDAMENTALS OF PHYSICS

A Student's Companion to Fundamentals of Physics is designed to be used closely with the text *FUNDAMENTALS OF PHYSICS* by Halliday, Resnick, and Walker. Chapters in the Companion correspond to chapters in the text and should be read along with the text. Each of the Companion chapters is divided into 5 sections: Important Concepts, Overview, Hints for Questions, Hints for Problems and Quiz.

The Important Concepts section provides a list of concepts that are introduced in the chapter. In order to understand the material of the chapter you should understand these concepts. Some definitions are given in *A Student's Companion*; for other definitions you should refer to the text. Boxes beside the concepts allow you to check a concept when it has been studied and you think you understand it well. You can then go back over the list and pick up the concepts that need further study.

The Overview section gives a summary of the chapter, including the important equations. Use this section to look up equations you need while doing the homework and also to review for exams. The summary is organized according to the sections of the chapter in the text. Most paragraphs deal with a single ideal each. A box beside the paragraph allows you to mark the paragraphs you understand, so you can devote more time to those that require more work.

Use the Hints for Questions and Hints for Problems section when you are doing your homework. Hints are given for all of the odd-numbered questions and about one-third of the odd-numbered problems at the ends of chapters.

The Quiz sections give you a chance to test your understanding of the chapter. Each quiz contains about 10 multiple-choice questions, mostly on qualitative aspects of the material in the chapter. Take the quiz when you have finished reading the chapter, check your answers against the answers given at the end of the quiz, and then go back and review pertinent material if you missed any of the questions.

Reading *A Student's Companion* is NOT a substitute for reading the text. Some derivations and applications are outlined in the Companion but they are necessarily shortened. The text contains much more detail and much fuller explanations. Study the text well, preferably before class, then use the Companion to remind yourself of the material you have studied. If it fails to jog your memory, restudy the appropriate portion of the text.

Acknowledgements   Many good people at John Wiley & Sons helped put together *A Student's Companion*. Thanks go to Stuart Johnson for suggesting the overall structure, which differs greatly from previous editions. Geraldine Osnato was instrumental in seeing the project through production and for handling a myriad of day-to-day chores associated with writing and publishing the Companion. I am grateful to them all. Very special thanks goes to Mary Ellen Christman, whose support and encouragement seem to know no bound.

J. Richard Christman
Professor Emeritus
U.S. Coast Guard Academy
New London, CT 06320

P

A Student's Companion

J. Richard Christman
Professor Emeritus
U. S. Coast Guard Academy

FUNDAMENTALS OF PHYSICS

Seventh Edition

David Halliday
University of Pittsburgh

Robert Resnick
Rensselaer Polytechnic Institute

Jearl Walker
Cleveland State University

WILEY

John Wiley & Sons, Inc.

Cover Photo: ©Jeff Hunter/The Image Bank/Getty Images

To order books or for customer service call 1-800-CALL-WILEY (225-5945).

ISBN 0-471-47062-7

Printed in the United States of America

10 9 8 7 6 5 4 3 2 1

Printed and bound by Courier Kendallville, Inc.

TABLE OF CONTENTS

Chapter 1
MEASUREMENT

Physics is an experimental science and relies strongly on accurate measurements of physical quantities. All measurements are comparisons, either direct or indirect, with standards. This means that for every quantity you must not only have a qualitative understanding of what the quantity represents but also an understanding of how it is measured. A length measurement is a familiar example. You should know that the length of an object represents its extent in space and also that length might be measured by comparison with a meter stick, say, whose length is accurately known in terms of the SI standard for the meter. Make a point of understanding both aspects of each new quantity as it is introduced.

Important Concepts

- ☐ unit
- ☐ standard
- ☐ base quantity
  (base unit, base standard)
- ☐ International System of Units

- ☐ conversion factor
- ☐ meter
- ☐ second
- ☐ kilogram
- ☐ atomic mass unit

Overview

1–2 Measuring Things

☐ A **unit** is a well-defined quantity with which other quantities are compared in a measurement. Examples: a unit for length is the meter, a unit for time is the second, and a unit for mass is the kilogram.

☐ Some units are defined in terms of others. For example, a unit for speed is the meter per second. Others are **base units** and are defined in terms of **standards**. Ideally, a standard should be accessible and invariable.

☐ A system of units consists of a unit for each physical quantity, organized so that all can be derived from a small number of independent base units.

1–3 The International System of Units

☐ This system is called the SI system (previously, the metric system).

☐ The three International System <u>base units</u> used in mechanics are:

> length: meter (abbreviation: m)
> time: second (abbreviation: s)
> mass: kilogram (abbreviation: kg)

☐ SI prefixes are used to represent powers of ten. The following are used the most:

Prefix	Power of Ten	Symbol
kilo:	10^3	k
mega:	10^6	M
centi:	10^{-2}	c
milli:	10^{-3}	m
micro:	10^{-6}	μ
nano:	10^{-9}	n
pico:	10^{-12}	p

Memorize them. When evaluating an algebraic expression, substitute the value using the appropriate power of ten. That is, for example, if a length is given as $25\,\mu$m, substitute 25×10^{-6} m. One catch: the SI unit for mass is the <u>kilogram</u>. Thus, a mass of 25 kg is substituted directly, while a mass of 25 g is substituted as 25×10^{-3} kg.

1–4 Changing Units

☐ There are usually several common units for each physical quantity. For example, length can be measured in meters, feet, yards, miles, light years, and other units.

☐ A quantity given in one unit is converted to another by multiplying by a **conversion factor**. Some conversion factors are listed in Appendix D of the text.

☐ Carefully study Section 1–4 to see how the value of a quantity given in one unit is converted to another unit.

1–5 Length

☐ The SI standard for the meter is the distance traveled by light during a time interval of $1/299,792,458$ s. This makes the speed of light exactly $299,792,458$ m/s.

☐ Table 1–3 gives some lengths. Note the wide range of values.

1–6 Time

☐ The SI standard for the second is the time taken for exactly $9,192,631,770$ vibrations of a certain light emitted by cesium-133 atoms.

☐ Table 1–4 lists some time intervals. Note the wide range of values.

1–7 Mass

☐ The SI standard for the kilogram is the mass of a platinum-iridium cylinder carefully stored at the International Bureau of Weights and Measures near Paris, France.

☐ A second mass unit is the **atomic mass unit**, abbreviated u and defined so the mass of a carbon-12 atom is exactly 12 u. $1\,u = 1.6605402 \times 10^{-27}$ kg.

☐ Table 1–5 lists some masses. Note the wide range of values.

Hints for Problems

3 (a) Multiply 1.0 km by the number of meters per kilometer (1.0×10^3 m/km) and the number of microns per meter ($1.0 \times 10^6 \, \mu$/m).

(b) Multiply 1.0 km by the number of meters per kilometer, by the number of feet per meter (3.218), and by the number of yards per foot (0.333).

$\left[\text{Ans: (a) } 10^9 \, \mu\text{m; (b) } 10^{-4}; \text{ (c) } 9.1 \times 10^5 \, \mu\text{m} \right]$

9 The volume of water is the area of the land times the depth of the water if it does not seep into the ground. You must convert 26 km^2 to acres and 2.0 in. to feet. For the first conversion use 1 km^2 = 1×10^6 m^2, 1 m^2 = $(3.281 \text{ ft})^2$, and 1 acre = 43 560 ft^2. See Appendix D.

$\left[\text{Ans: } 1.1 \times 10^3 \text{ acre-feet} \right]$

11 Use 1 fortnight = 7 d, 1 d = 24 h, 1 h = 3600 s, and 1 s = $1 \times 10^6 \, \mu$s.

$\left[\text{Ans: } 1.21 \times 10^{12} \, \mu\text{s} \right]$

17 (a) The ratio of the interval on clock B to the interval on clock A (600 s) is the same as the ratio of the interval between events 2 and 4 on clock B to the same interval on clock A.

(b) The ratio of the interval on clock C to the interval on clock B is the same as the ratio of the interval between events 1 and 3 on clock C to the same interval on clock B.

(c) Measure the time of the event (when clock A reads 400 s) from the time of event 2 on the diagram. The interval is 400 s − 312 s on clock A. Calculate the interval on clock B and add 125 s.

(d) Measure the time of the event (when clock C reads 15.0 s) from the time of event 1 on the diagram. The interval is 15.0 s − 92 s on clock C. Calculate the interval on clock B and add 25 s.

$\left[\text{Ans: (a) 495 s; (b) 141 s; (c) 198 s; (d) −245 s} \right]$

21 If M is the mass of Earth and m is the average mass of an atom, then the umber of atoms in Earth is $N = M/m$. M and m must have the same unit. Convert 5.98×10^{24} kg to atomic mass units or 40 u to kilograms. Use 1 u = 1.661×10^{-27} kg.

$\left[\text{Ans: } 9.0 \times 10^{49} \text{ atoms} \right]$

27 Convert both 25 wapentakes and 11 barns to square meters. Take 1 hide to be 110 acres (an average value) and multiply by 100 families and 4047 m^2/acre. Multiply 11 barns by 1×10^{-28} m^2. Calculate the ratio.

$\left[\text{Ans: } \approx 1 \times 10^{36} \right]$

31 (a) Convert 3.0 acres to square perches. Use 1 acre = 160 perch^2. Now add 100 perch^2 and convert the total to roods. Use 1 rood = 40 perch^2.

(b) First convert 1 perch to meters. Use 1 perch = 16.5 ft and 1 ft = 0.3048 m (see Appendix D). Now convert 1 perch^2 to square meters. Just square the last result. Finally use the result to convert the answer to part (a) to square meters.

$\left[\text{Ans: (a) 14.5 roods; (b) } 1.47 \times 10^4 \text{ m}^2 \right]$

35 (a) Use 1 ft = 0.3048 m, 1 gal = 231 in.^{3}, and 1 in.^{3} = 1.639×10^{-2} L.

(b) Convert liters to cubic meters. Use 1 L = 1×10^{-3} m^3.

(c) The volume of paint on a surface is equal to the product of the surface area and the thickness of the paint.

[Ans: (a) $11.3 \, \text{m}^3/\text{L}$; (b) $1.13 \times 10^4 \, \text{m}^{-1}$; (c) $2.17 \times 10^{-3} \, \text{gal/ft}^2$; (d) number of gallons to cover a square foot if spread uniformly]

39 (a) Convert 9 cubits to meters. For the lower limit use 1 cubit = 43 cm and for the upper limit use 1 cubit = 53 cm. also use 1 cm = 0.01 m.

(b) Use 1 m = 1000 mm to convert meters to millimeters.

(c) If D is the radius of the cylinder and L is its length, its volume is $\pi D^2 L/4$. You must convert the diameter, given in cubits, to meters.

[Ans: (a) 3.9 m, 4.8 m; (b) 3.9×10^3 mm, 4.8×10^3 mm; (c) $2.2 \, \text{m}^3$, $4.2 \, \text{m}^3$]

43 Do an experiment to find the average time in seconds between your exhaled breaths. Measure the time for, say, 10 breaths and divide by 10. Now calculate the number of seconds in a day. It is 60 s/min times 60 min/h times 24 h/d. The number of debugs in a day is this number divided by the duration of a debug in seconds.

[Ans: 2×10^4 to 4×10^4 debugs]

45 Multiply the three dimensions to find the volume of the receptacle in cubic centimeters. Use $1 \, \text{cm}^3 = 1.00 \times 10^{-3} \, \text{L}$ to convert to liters and 16 standard bottles = 11.356 L to convert to standard bottles. Find the largest integer number of nebuchadnezzars with total volume less than that of the receptacle and calculate the remainder. Then proceed to the next largest bottle and find the largest integer number of balthazars with total volume less than that of the remainder and compute the new remainder. Proceed in this way through the list of bottles, from the largest to the smallest.

[Ans: (a) 3 nebuchadnezzars, 1 methuselah; (b) 0.37 standard bottle; (c) 0.26 L]

49 The total mass M of the collection is the product of the number N of hydrogen atoms and the mass m of a single atom. Solve $M = Nm$ for N. You must convert kilograms to atomic mass units or atomic mass units to kilograms so that M and m have the same units. Use $1.0 \, \text{u} = 1.661 \times 10^{-27} \, \text{kg}$.

[Ans: 6.0×19^{26} atoms]

59 (a) Use equivalent triangles to show that the ratio of the diameters is the same as the ratio of the distances from Earth.

(b) Since the volume of a sphere is proportional to the cube of its diameter, the ratio of the volumes is equal to the cube of the ratio of the diameters cubed.

(c) The angle should be roughly half a degree. Use a little trigonometry to show that the diameter of the Moon is given by $d = 2r \sin \theta/2$, where θ is the angle.

[Ans: (a) 400; (b) 6.4×10^7; (c) actual diameter is about 3.4×10^6 m]

Quiz

Some questions might have more than one correct answer.

1. In the SI system of units which of the following quantities does NOT have a derived unit?
   A. weight
   B. speed
   C. acceleration
   D. force
   E. mass

2. Which of the following is NOT a unit of mass?
   A. gram
   B. kilogram
   C. pound
   D. slug
   E. newton·s²/m

3. A kilosecond is about
   A. 1 min
   B. 17 min
   C. 1 h
   D. 5 h
   E. 1 d

4. The area of the surface of a sphere is proportional to
   A. its radius
   B. the square of its radius
   C. the cube of its radius
   D. the square root of its radius
   E. none of the above

5. If all sides of a cube are doubled the volume of the cube will be
   A. twice as great as before
   B. four times as great as before
   C. six times as great as before
   D. eight times as great as before
   E. the same as before

6. Volume has the dimensions of
   A. length
   B. length squared
   C. length cubed
   D. the square root of length
   E. the cube root of length

7. The *order of magnitude* of a quantity is

    A. the power of ten associated with its value
    B. its size
    C. its ranking in a list of similar quantities
    D. the maximum possible value
    E. the minimum possible value

8. The area of a circle with radius R is given by

    A. $2\pi R$
    B. $2\pi R^2$
    C. $\pi R^2/4$
    D. πR^2
    E. $2R^2$

9. The volume of a sphere with radius R is given by

    A. πR^2
    B. πR^3
    C. $4\pi R^3$
    D. $4\pi R^2$
    E. $(4\pi/3)R^3$

10. The surface area of a sphere of radius R is given by

    A. πR^2
    B. πR^3
    C. $4\pi R^3$
    D. $4\pi R^2$
    E. $(4\pi/3)R^3$

Answers: (1) E; (2) C; (3) B; (4) B; (5) D; (6) C; (7) A; (8) D;l (9) E; (10) D

Chapter 2
MOTION ALONG A STRAIGHT LINE

This chapter introduces you to some of the concepts used to describe motion. Pay particular attention to their definitions and to the relationships between them.

Important Concepts

- [ ] particle
- [ ] coordinate axis
- [ ] origin
- [ ] coordinate
- [ ] displacement
- [ ] average velocity
- [ ] average speed
- [ ] (instantaneous) velocity

- [ ] (instantaneous) speed
- [ ] average acceleration
- [ ] (instantaneous) acceleration
- [ ] motion with constant acceleration
- [ ] free-fall acceleration
- [ ] free-fall motion

Overview

2–2 Motion

- [ ] In this section of the text, objects are treated as particles. A **particle** has no extent in space and has no internal parts that can move relative to each other. It may have other properties, such as mass.

- [ ] An extended object can be treated as a particle if all points in it move along parallel lines and it retains its shape. This means it cannot rotate and cannot deform. If an extended object can be treated as a particle, we may pick one point on the object and follow its motion. The position of a crate, for example, means the position of the point on the crate we have chosen to follow, perhaps one of its corners.

2–3 Position and Displacement

- [ ] The motion of a particle in one dimension can be described by giving its **coordinate** x as a function of time t. Draw a **coordinate axis** along the line of motion of the particle and select one point on the axis to be the **origin**. The distance from the origin to the particle is the magnitude of the coordinate. The coordinate is positive if it is on the side of the origin designated positive and negative if it is on the side designated negative.

- [ ] A value of the coordinate x specifies a *point* on the x axis. It has no extension in space.

☐ You must carefully distinguish between an *instant* of time and an *interval* of time. The symbol t represents an instant and has no extension. On the other hand, an interval extends from some initial time to some final time: *two* instants of time are required to describe it. Note that a value of the time may be positive or negative, depending on whether the instant is after or before the instant designated as $t = 0$.

☐ A **displacement** is a difference in two coordinates. If a particle goes from x_1 to x_2 during some interval of time, its displacement during that interval is $\Delta x = x_2 - x_1$. Notice that the *initial* coordinate is subtracted from the *final* coordinate. This definition is valid no matter what the signs of x_1 and x_2.

☐ Two particles that start at x_1 and end at x_2 have the same displacement no matter what their motions. The magnitude of the displacement during any time interval is different from the distance traveled during the interval if the direction of motion reverses during the interval.

2–4 Average Velocity and Average Speed

☐ If a particle goes from x_1 at time t_1 to x_2 at time t_2, its **average velocity** v_{avg} in the interval from t_1 to t_2 is given by

$$v_{\text{avg}} = \frac{x_2 - x_1}{t_2 - t_1} = \frac{\Delta x}{\Delta t},$$

where $\Delta x = x_2 - x_1$ and $\Delta t = t_2 - t_1$.

☐ If you are given the function $x(t)$ and are asked for the average velocity in some interval from t_1 to t_2, first evaluate the function for $t = t_1$ to find x_1, then evaluate the function for $t = t_2$ to find x_2 and finally substitute the values into the defining equation.

☐ On a graph of x versus t the average velocity over the interval from t_1 to t_2 is given by the slope of the line from t_1, x_1 to t_2, x_2. On the graph on the right $t_1 = 2.0\,\text{s}$ and $t_2 = 8.0\,\text{s}$. The average velocity in this interval is the slope of the dotted line.

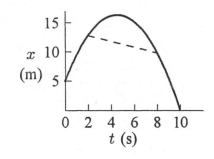

☐ A downward sloping line (from left to right) has a negative slope and indicates a negative average velocity. An upward sloping line has a positive slope and indicates a positive velocity.

☐ Carefully distinguish between average velocity and **average speed**. The average speed over a time interval Δt is defined by

$$s_{\text{avg}} = \frac{d}{\Delta t},$$

where d is distance traveled in the interval.

2–5 Instantaneous Velocity and Speed

☐ The **instantaneous velocity** is the velocity at an instant of time (not over an interval). It is defined as the limiting value of the average velocity in an interval as the interval shrinks to zero. The term "velocity" means instantaneous velocity.

☐ A positive velocity means that the particle is traveling in the positive x direction. Similarly, a negative velocity means that the particle is traveling in the negative x direction.

☐ If the function $x(t)$ is known, the velocity at any time t_1 is found by differentiating it with respect to t and evaluating the result for $t = t_1$. The velocity at any time t is given by

$$v(t) = \frac{dx(t)}{dt} .$$

☐ On a graph of x versus t, the velocity at any time t_1 is the slope of the straight line that is tangent to the curve at the point corresponding to $t = t_1$. On the graph on the right the slope of the slanted dotted line gives the velocity at time $= 7.0$ s.

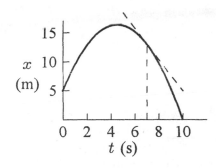

☐ The **instantaneous speed** (or just plain **speed**) of a particle is the magnitude of its velocity.

2–6   Acceleration

☐ If the velocity of a particle changes from v_1 at time t_1 to v_2 at a later time t_2, then its **average acceleration** a_{avg} over the interval from t_1 to t_2 is given by

$$a_{avg} = \frac{v_2 - v_1}{t_2 - t_1} = \frac{\Delta v}{\Delta t} ,$$

where $\Delta v = v_2 - v_1$ and $\Delta t = t_2 - t_1$. Notice that the velocity at the *beginning* of the interval is subtracted from the velocity at the *end*. Also notice that *instantaneous* velocities appear in the definition.

☐ If the function $x(t)$ is given, first differentiate it with respect to t to find the velocity as a function of time, then evaluate the result for times t_1 and t_2 to find values for v_1 and v_2. Substitute the values into the defining equation for the average acceleration.

☐ On a graph of v versus t, the average acceleration over the interval from t_1 to t_2 is the slope of the line from t_1, v_1 to t_2, v_2.

☐ The **instantaneous acceleration** gives the acceleration of the particle at an instant of time, rather than over an interval. It is the limiting value of the average acceleration as the interval shrinks to zero. "Acceleration" and "instantaneous acceleration" mean the same thing.

☐ If the function $v(t)$ is known, the instantaneous acceleration is found by differentiating it with respect to t. If the function $x(t)$ is known, the instantaneous acceleration is found by differentiating it twice with respect to t:

$$a(t) = \frac{dv}{dt} = \frac{d^2x}{dt^2} .$$

☐ On a graph of v versus t, the instantaneous acceleration at any time t_1 is the slope of the straight line that is tangent to the curve at $t = t_1$.

☐ Note that a positive acceleration does not necessarily mean the particle speed is increasing and a negative acceleration does not necessarily mean the particle speed is decreasing. The speed increases if the velocity and acceleration have the same sign and decreases if they have opposite signs, no matter what the signs.

☐ At the instant a particle is momentarily at rest its acceleration is not necessarily 0. If, at an instant, the velocity is zero but the acceleration is not, then in the next instant the velocity is not zero and the particle is moving.

2–7 Constant Acceleration: A Special Case

☐ If a particle is moving along the x axis with constant acceleration a, its coordinate and velocity are given as functions of time t by

$$x(t) = x_0 + v_0 t + \tfrac{1}{2}at^2 \qquad \text{and} \qquad v(t) = v_0 + at\,,$$

where x_0 is its coordinate at $t = 0$ and v_0 is its velocity at $t = 0$. Notice that the second equation is the derivative of the first.

☐ Eq. 2–16 of the text is also extremely useful. It is

$$v^2 = v_0^2 + 2a(x - x_0)\,.$$

It gives the velocity as a function of the coordinate.

☐ Problems involving motion with constant acceleration are worked by solving these equations for the unknowns. Usually two events are described in the problem statement. Select the time to be zero at one of the events. x_0 is the coordinate of the particle and v_0 is its velocity then. The other event occurs at some time t. x is the coordinate and v is the velocity then. The other quantity that enters is the acceleration a. Of the six quantities that occur in the equations, four are usually given or implied and two are unknown.

2–8 Another Look at Constant Acceleration

☐ Integration can be used to derive the equations for motion with constant acceleration.

2–9 Free-Fall Acceleration

☐ Every object near the surface of Earth, if acted on only by the gravitational force of the Earth, has the same acceleration, called the **free-fall acceleration** and denoted by g. It is downward, toward Earth. The value of g is independent of the mass or shape of the object. It varies slightly from place to place on the Earth.

☐ In the absence of air resistance, a ball (or any other object) thrown upward has the same acceleration at all times: during its upward flight, during its downward fall, and even at the very top of its trajectory.

☐ The constant-acceleration equations are revised somewhat to deal with free-fall motion. A vertical y axis is drawn with the upward direction being positive and the equations are written

$$y(t) = y_0 + v_0 t - \tfrac{1}{2}gt^2\,, \qquad v(t) = v_0 - gt\,, \qquad \text{and} \qquad v^2 = v_0^2 - 2g(y - y_0)\,.$$

They can be obtained from the equations given previously for motion along an x axis by substituting y for x and $-g$ for a.

☐ Problems dealing with free fall are solved in exactly the same way as other problems dealing with constant acceleration.

2–10   Graphical Integration in Motion Analysis

☐ An integral can be interpreted as the area under the curve of the integrand versus the integration variable, from the lower limit of the integral to its upper limit. Thus the velocity is the area under an acceleration versus time graph and the displacement is the area under a velocity versus time graph.

☐ Often the curve can be sectioned so the area under each section can be easily computed. The region under a section might be a rectangle or triangle, for example.

☐ Be careful about signs here. Those portions for which the integrand is negative contribute negative amount to the integral and their area must be taken to be negative. You must calculate the negative and positive contributions separately.

Hints for Questions

1 (a) The average velocity is the displacement divided by the time. Since the times are all the same you need to look at the displacement from the initial position to the final position.
(b) Now you must look at the lengths of the paths.
[Ans: (a) all tie; (b) 4, tie of 1 and 2, then 3]

3 (a) and (b) See if the velocity is positive or negative at the point. If it is positive the particle is traveling in the positive direction on the axis and if it is negative the particle is moving in the negative direction.
(c) The direction of motion reverses at points where the velocity is zero. This is not always true but for this graph the sign of the velocity is different on the two sides of a zero.
(d) The acceleration is the slope of the velocity versus time graph.
[Ans: (a) positive direction; (b) negative direction; (c) 3 and 5; (d) 2 and 6 tie, then 3 and 5 tie, then 1 and 4 tie (zero)]

5 (a) Since the acceleration is constant the average velocity over an interval is equal to the instantaneous velocity at the halfway point in time. The cream tangerine is slowing as it rises.
(b) The widow height is equal to the a product of the average velocity and the time of passage.
(c) The cream tangerine is in free fall.
(d) The acceleration of the cream tangerine is the change in its velocity during passage divided by the time of passage.
[Ans: (a) 3, 2, 1; (b) 1, 2, 3; (c) all tie; (d) 1, 2, 3]

7 Imagine what would happen if the speed of the blue car is greater than the speed of the red car and they going in the same direction. If the speed of the blue car is less than the speed of the red car, could the blue car go faster and still avoid a collision?
[Ans: 60 km/h, not zero]

9 Write the equations for the coordinates of the cars as functions of time. When the red car passes the blue car their coordinates are the same. You must take into account the fact that red car is at the origin later than the blue car. If the time t is zero when the blue car is at the origin and t_0 is the time when the red car is at the origin, then the equations for the red car contain $t - t_0$ instead of just t.

$\left[\text{Ans: } x = t^2 \text{ and } x = 8(t - 2) + 1.5(t - 2)^2 \text{ with } x \text{ in meters and } t \text{ in seconds}\right]$

Hints for Problems

7 (a) The particle is momentarily stopped when its velocity is zero. Its velocity as a function of time is the derivative of its coordinate with respect to time.
(b) Substitute the value for the time of stopping into the function given for the coordinate.
(c) and (d) Put x equal to zero in the given function and solve for t. The equation is quadratic in t and so has two solutions, one positive and one negative.
(f) Suppose the function is $x = 4.0 - 6.0(t - a)^2$, where a is a constant. This is the same as the original function except it is shifted in time by a. If a is positive it is shifted in the positive t direction and if a is negative it is shifted in the negative t direction. The second function is identical to $x = 4.0 - a^2 + 12at - 6.0t^2$ as you can easily see by squaring the quantity in parentheses.
(g) Set the derivative of the new function equal to zero and solve for t, then substitute back into the function to find the coordinate at which the particle stops.

$\left[\text{Ans: (a) } 0; \text{ (b) } 4.0\,\text{m}; \text{ (c) } -0.82\,\text{s}; \text{ (d) } 0.82\,\text{s}; \text{ (f) } +20t; \text{ (g) increase}\right]$

13 (a) Use $v_{\text{avg}} = \Delta x / \Delta t$. To calculate the particle's coordinate at the beginning of the interval substitute $t = 2.00\,\text{s}$ into the equation for x and to calculate its coordinate at the end of the interval substitute $t = 3.00\,\text{s}$. Δx, of course, is the difference and Δt is $1.00\,\text{s}$
(b), (c), and (d) Differentiate the expression for the coordinate with respect to time and evaluate the result for $t = 2.00\,\text{s}$. Do the same for $t = 3.00\,\text{s}$ and for $t = 2.5\,\text{s}$.
(e) You must find the time when the particle is midway between the two positions. Find the midway point x_m by taking the average of the two coordinates found in part (a), then solve $x_m = 9.75 + 1.50t^3$ for t. Finally, substitute this value for t into the expression for the velocity as a function of time (the derivative of the coordinate).

$\left[\text{Ans: (a) } 28.5\,\text{cm/s}; \text{ (b) } 18.0\,\text{cm/s}; \text{ (c) } 40.5\,\text{cm/s}; \text{ (d) } 28.1\,\text{cm/s}; \text{ (e) } 30.3\,\text{cm/s}\right]$

17 (a) The position of the particle at any time is given by the function $x = 12t^2 - 2t^3$.
(b) The velocity at any time is given by the derivative of the coordinate with respect to time.
(c) The acceleration at any time is given by the derivative of the velocity with respect to time.
(d) and (e) When the maximum positive coordinate is reached the velocity is zero.
(f) and (g) When the maximum positive velocity is reached the acceleration is zero.
(h) Find the time for which the velocity of the particle is zero and substitute into the expression for the acceleration as a function of time.

(i) Use $v_{avg} = \Delta x / \Delta t$. Find the coordinate at the beginning and end of the interval and subtract the former from the latter to get Δx.

[Ans: (a) 54 m; (b) 18 m/s; (c) $-12\,\text{m/s}^2$; (d) 64 m; (e) 4.0 s; (f) 24 m/s; (g) 2.0 s; (h) $-24\,\text{m/s}^2$; (i) 18 m/s]

21 Use $v = v_0 + at$, where v is the electron's velocity at any time t, v_0 is its velocity at time $t = 0$, and a is its acceleration. Let $t = 0$ at the instant the electron's velocity is $+9.6$ m/s and evaluate the expression for (a) $t = -2.5$ s and for $t = +2.5$ s.

[Ans: (a) $+1.6$ m/s; (b) $+18$ m/s]

31 Because the acceleration is constant, the particle's coordinate at any time t is given by $x = x_0 + v_0 t + \frac{1}{2}at^2$, where x_0 is its coordinate at $t = 0$, v_0 is its velocity then, and a is its acceleration. According to the graph $x_0 = 0$. The slope of the curve at $t = 0$ appears to be zero so you may take v_0 to be zero. Thus $x = \frac{1}{2}at^2$. The graph also shows that $x = 6.0$ m at $t = 2.0$ s. Put this information into the equation and solve for a.

[Ans: (a) $4.0\,\text{m/s}^2$; (b) $+x$]

35 Put the origin of an x axis at the position of train A when it starts slowing and suppose the train has velocity v_{A0} ($= +40$ m/s from the graph) at that time. Its velocity as a function of time is given by $v_{A0} + a_A t$, where a_A is its acceleration. This is the slope of the upper line on the graph and is negative. Solve for the time when train A stops and use $x_A = v_{A0}t + \frac{1}{2}a_A t^2$ to find its position when it stops.

The velocity of train B is given by $v_B = v_{B0} + a_B t$, where v_{B0} is its velocity at $t = 0$ (-30 m/s from the graph), and a_B is its acceleration. This is the slope of the lower line on the graph and is positive. Solve for the time when train B stops and use $x_B = x_{B0} + v_{B0}t + \frac{1}{2}a_B t^2$ to find its position when it stops. Here $x_{B0} = 200$ m. The separation of the trains when both have stopped is the difference of the coordinates you have found.

[Ans: 40 m]

37 (a) Position the x axis of a coordinate system so the origin is at the front end of the passenger train when it starts slowing and suppose the train is moving in the positive x direction. The coordinate of the front end is then given by $x_P = v_{P0} + \frac{1}{2}at^2$, where v_{P0} is its velocity at $t = 0$ and a is its acceleration. Its velocity is given by $v_P = v_{P0} + at$. The coordinate of the back end of the locomotive is given by $x_L = D + v_L t$, where v_L is its velocity.

A collision is just avoided if, at the time the speed of the passenger train equals the speed of the locomotive, the coordinate of the front end of the passenger is just slightly less than the coordinate of the back end of the locomotive. Set the expression for v_P equal to the expression for v_L and the expression for x_P equal to the expression for x_L. Solve these equations simultaneously for a. (You can also solve for the time when these conditions are met but you are not asked for this.) Be sure to convert the given speeds to meters per second. The magnitude of the acceleration must be slightly greater than the magnitude of the value you calculated.

[Ans: (a) $0.994\,\text{m/s}^2$]

45 The coordinate of apple 2 is given by $y_2 = h + v_0(t - t_0) - \frac{1}{2}g(t - t_0)^2$, where v_{20} is the velocity with which it is thrown, h is the height of the bridge above the roadway, and t_0

(= 1 s) is the time it is thrown. Here the origin is placed at the roadway and y is taken to be positive in the upward direction. You need to know the value of h.

The coordinate of apple 1 is given by $y_1 = h - \frac{1}{2}gt^2$. Set y_1 equal to zero and solve for h using $t = 2$ s. Use this value and $t = 2.25$ s in the equation for y_2 and solve for v_{20}.

[Ans: 9.6 m/s]

<u>53</u> Put the origin at the nozzle and take the downward direction to be positive. Then the coordinate of a drop is given by $y = \frac{1}{2}g(t - t_0)^2$, where t_0 is the time the drop started. Suppose $t_0 = 0$ for the first drop, $t_0 = \Delta t$ for the second, $t_0 = 2\Delta t$ for the third, and $t_0 = 3\Delta t$ for the fourth. Each drop hits the floor when $y = h$, the height of the nozzle above the floor. For the first drop $h = \frac{1}{2}g(3\Delta t)^2$. Solve for Δt. The coordinate of the second drop at this time is $y_2 = \frac{1}{2}g(2\Delta t)^2$ and the coordinate of the third drop is $y_3 = \frac{1}{2}g(\Delta t)^2$.

[Ans: (a) 89 cm; (b) 22 cm]

<u>57</u> Divide the falling of the ball into two segments: from the top of the building to the top of the window and from the top of the window to the sidewalk. You need to find the lengths of each of these segments.

You need to know the velocity of the ball as it passes the top of the window going down. Take the origin of a coordinate system to be at the top of the window and suppose the downward direction is positive. Suppose further that the velocity of the ball is v_0 when falls past that point. If h_w is the top-to-bottom dimension of the window, then $h_w = v_0 t + \frac{1}{2}gt^2$, where t is the time to pass the window. Solve for v_0.

To find the length of the first segment solve for the distance the ball must fall to achieve a velocity of v_0. To find the length of the second segment solve for the distance the ball falls in 1.125 s, starting with a downward velocity of v_0. The time here is the time for the ball to pass the window plus the time for it to fall from the bottom of the window to the sidewalk.

[Ans: 20.4 m]

<u>61</u> The displacement over a given time interval is the integral of the velocity over that interval and this is the area under the curve of the velocity versus time. The curve of Fig. 2–30 can be divided into segments so that the region under each segment is either a rectangle, a right triangle, or a right triangle on top of a rectangle. The velocity is positive throughout so each segment contributes a positive amount to the integral. Recall that the area of a right triangle is half the product of the two perpendicular sides and the area of a rectangle is the product of two perpendicular sides.

[Ans: 100 m]

<u>67</u> Let v_0 be the initial velocity and t the time to reach the highest point. The velocity at the highest point is zero. Solve $0 = v_0 - gt$ for v_0, then use $y = v_0 t - \frac{1}{2}gt^2$ to compute the coordinate y of the highest point, relative to the top of the building. Lastly, use the same equation, but with $t = 6.00$ s to compute the coordinate of the ground relative to the top of the building. The magnitude of the coordinate is the height of the building.

[Ans: (a) 15.7 m/s; (b) 12.5 m; (c) 82.3 m]

<u>73</u> Solve $v^2 - v_0^2 = 2ax$ for the acceleration a. Here v is the final velocity (0) and V_0 is the initial velocity (200 km/h). You need to convert this to meters per second. Solve $x = v_0 T_b + \frac{1}{2}aT_b^2$

for the time of braking T_b. The distance traveled during the reaction time T_r is $v_0 T_r$.

[Ans: (a) $9.08\,\text{m/s}^2$; (b) $0.926g$; (c) 6.12 s; (d) $15.3 T_r$; (e) braking; (f) 5.56 m]

79 The average velocity is the change in the coordinate divided by the time for the change. Use the expression for x to compute the change in the coordinate. The average acceleration is the change in the velocity divided by the time for the change. Differentiate the expression for x to obtain an expression for the velocity, then evaluate it for the two times. The instantaneous acceleration is the derivative of the velocity with respect to time.

[Ans: (a) $14\,\text{m/s}$; (b) $18\,\text{m/s}$; (c) $6.0\,\text{m/s}$; (d) $12\,\text{m/s}^2$; (e) $24\,\text{m/s}$; (f) $24\,\text{m/s}^2$]

89 Use $v_B^2 - v_A^2 = -2g\,\Delta y$, where v_A is the velocity of the stone when it is at A, v_B is its velocity when it is at B, and Δy is the height of B above A. Put v_A equal to v and $v_B = v/2$, then solve for v. $-v_B^2 = -2gh$ to compute the maximum height h that the stone rises above point B.

[Ans: (a) $8.85\,\text{m/s}$; (b) 1.00 m]

105 Consider the motion to consist of two parts: the free-fall portion and the deceleration portion. The position and velocity at the end of the free-fall portion are the initial condition for the deceleration portion. During free fall the acceleration of the parachutist is downward and during the deceleration portion it is upward.

[Ans: (a) 17 s; (b) 290 m]

109 An expression for the velocity as a function of time is found by integrating the acceleration with respect to time and an expression for the coordinate is found by integrating the velocity, again with respect to time. Values of the initial coordinate and velocity are used to find the constants of integration. The maximum velocity occurs when the acceleration is zero.

[Ans: (a) $18\,\text{m/s}$; (b) 83 m]

113 You can calculate the speed of each rate in meters per second. Simply multiple the length of a step by the number of steps per minute. The result, which is in inches per minute, should be converted to meters per second. Find the total distance traveled and the total time taken for each sequence of rates, then divide the total distance by the total time to find the average velocity. For part (a) use $x = vt$ to find the distance traveled at each rate, then sum the distances to find the total distance. For part (b) use $x = vt$ to find the time for each rate, then sum the times to find the total time.

[Ans: (a) 32 m; (b) $1.6\,\text{m/s}$ (c) 24.5 s; (d) $1.3\,\text{m/s}$]

117 For parts (a) and (b) Use $y = h - v_0 t - \frac{1}{2}gt^2$, with $y = 0$ to find an expression for the time to reach the ground and $v = -v_0 - gt$ to find an expression for the velocity of the ball just before it hits the ground. Here v_0 is a speed, not a velocity. For parts (c) and (d) use $y = h + v_0 t - \frac{1}{2}gt^2$ and $v = v_0 - gt$.

[Ans: (a) $\sqrt{v_0^2 + 2gh}$; (b) $[\sqrt{v_0^2 + 2gh} - v_0]/g$; (c) same as (a); (d) $t = [\sqrt{v_0^2 + 2gh} + v_0]/g$, greater]

Quiz

Some questions might have more than one correct answer.

1. The magnitude of a particle's displacement during an interval
    A. might be greater than the distance traveled during that interval
    B. might be less than the distance traveled during that interval
    C. might be equal to the distance traveled during that interval
    D. must be equal to the distance traveled during that interval
    E. might be the negative of the distance traveled during that interval

2. A graph of a particle's velocity versus time can be used to find the position of a particle at the end of an interval if
    A. the coordinate at the beginning of the interval is known
    B. the velocity at the beginning of the interval is known
    C. the coordinate at the end of the intervals known
    D. nothing else is known
    E. the acceleration is known for every time in the interval

3. If the velocity of a particle moving on the x axis is positive,
    A. it is traveling from left to right
    B. it is traveling from right to left
    C. its coordinate is positive
    D. its coordinate is negative
    E. it is traveling in the positive x direction

4. If the acceleration of a particle moving on the x axis is negative
    A. its speed is decreasing
    B. its speed is increasing
    C. its speed might be increasing or decreasing
    D. its speed is not changing
    E. its speed must be a maximum

5. Which of the following quantities can be found from a graph of the velocity versus time for a particle moving along a straight line? (No other information is known.)
    A. its displacement during any interval
    B. its average velocity for any interval
    C. its average acceleration for any interval
    D. its acceleration at any instant
    E. its position at any instant

6. During any sufficiently long interval in which a particle moves along a straight line with constant (nonzero) acceleration

    A. its velocity cannot be zero
    B. its velocity can be zero, but at only one instant of time
    C. its velocity can be zero at several instants of time
    D. its speed must eventually increase
    E. it might stop and remain stopped

7. Two particles start at point P and end at another point Q but the average velocity of particle B is much greater than the average velocity of particle A. We must conclude that

    A. particle B traveled a greater distance than particle A
    B. at each instant the speed of particle B was greater than the speed of particle A
    C. the speed of particle B was greater than the speed of particle A for at least part of the trip
    D. the acceleration of particle B was greater than the acceleration of particle A
    E. particle B took a shorter time to make the trip than particle A

8. The equations for motion with constant acceleration ($x = x_0 + v_0 t + \frac{1}{2}at^2$ and $v = v_0 + at$) can be used to find

    A. the velocity v and coordinate x if all other quantities are given
    B. the velocity v, coordinate x, and time t if all other quantities are given
    C. the acceleration a and the coordinate x if all other quantities are given
    D. the acceleration a, the coordinate x, and the initial velocity v_0 if all other quantities are given
    E. the initial position x_0 and the initial velocity v_0 if all other quantities are given

9. Which of the following statements are true?

    A. heavier objects have a greater free-fall acceleration than lighter objects
    B. if a particle's velocity is zero at an instant then its acceleration must be zero at that instant
    C. the faster a particle moves the greater its acceleration must be
    D. the average velocity for any interval must equal the instantaneous velocity at some instant in the interval
    E. none of the above

10. Which of the following statements are true for a particle that is moving with constant acceleration?

    A. it cannot have a velocity of zero at any instant
    B. its direction of motion cannot reverse
    C. for any interval the distance it travels must equal the magnitude of its displacement
    D. its speed is greater at the end of end interval that at the beginning
    E. none of the above

11. The coordinates x of five particles are given below as functions of the time t. Which particles are moving with constant velocity?

A. $x = (5.0\,\text{m/s})t - (7.0\,\text{m/s}^2)t^2$
B. $x = (5.0\,\text{m/s})t - (7.0\,\text{m/s}^3)t^3$
C. $x = (5.0\,\text{m/s})t + (7.0\,\text{m/s}^2)t^2$
D. $x = (3.0\,\text{m/s}^2)t^2$
E. $x = (5.0\,\text{m/s})t - (7.0\,\text{m/s}^{-2})/t^2$

Answers: (1) B, C; (2) A, C; (3) E; (4) C; (5) C, D; (6) B, D; (7) E; (8) A, C; (9) D; (10) E; (11) A, C, D

Chapter 3
VECTORS

You will deal with vector quantities throughout the course. In this chapter, you will learn about their properties and how they are manipulated mathematically. A solid understanding of this material will pay handsome dividends later.

Important Concepts

- ☐ vector
- ☐ scalar
- ☐ component of a vector
- ☐ unit vector
- ☐ vector addition
- ☐ negative of a vector

- ☐ vector subtraction
- ☐ multiplication of a vector by a scalar
- ☐ scalar product of two vectors
- ☐ vector product of two vectors

Overview

3–2   Vectors and Scalars

☐ A **vector** quantity has a direction as well as a magnitude and obeys the rules of vector addition (discussed later in this chapter). In contrast, a **scalar** quantity has only a magnitude and obeys the rules of ordinary arithmetic. Displacement, velocity, acceleration, and force are vector quantities; mass, speed, charge, and temperature are scalar quantities.

☐ A vector is represented graphically by an arrow in the direction of the vector, with length proportional to the magnitude of the vector (according to some scale). As an algebraic symbol, a vector is written with an arrow over the symbol ($\vec{a}$). The magnitude of $\vec{a}$ is written a, in italics (without an arrow) or as $|\vec{a}|$. Be sure you follow this convention. It helps you distinguish vectors from scalars and components of vectors. It helps you communicate properly with your instructors and exam graders. Do *not* write $a + b$ when you mean $\vec{a} + \vec{b}$, for example. They have entirely different meanings!

☐ Displacement vectors are used as examples of vectors in this chapter. A displacement vector is a vector from the position of a particle at the beginning of a time interval to its position at the end of the interval. Note that a displacement vector tells us nothing about the path of the object, only the relationship between the initial and final positions.

3–3 Adding Vectors Geometrically

☐ If $\vec{d_1}$ is the displacement vector from point A to point B and $\vec{d_2}$ is the displacement vector from point B to point C, then the sum (written $\vec{d_1} + \vec{d_2}$) is the displacement vector from A to C, as shown on the diagram to the right.

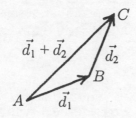

☐ To add two vectors, place the tail of the second vector at the head of the first, then draw the vector from the tail of the first to the head of the second. The order in which you draw the vectors is not important: $\vec{a} + \vec{b} = \vec{b} + \vec{a}$. It is important that the tail of one be at the head of the other and that the resultant vector be from the "free" tail to the "free" head, as in the diagram.

☐ Except in special circumstances, the magnitude of the resultant vector is *not* the sum of the magnitudes of the vectors entering the sum and the direction of the resultant vector is *not* in the direction of any of the vectors entering the sum.

☐ Remember you can reposition a vector as long as you do not change its magnitude and direction. Thus, if two vectors you wish to add graphically do not happen to be placed with the tail of one at the head of the other, simply move one into the proper position.

☐ The idea of the negative of a vector is used to define vector subtraction. The negative of a vector is a vector that is parallel to the original vector but in the opposite direction.

☐ Vector subtraction is defined by $\vec{a} - \vec{b} = \vec{a} + (-\vec{b})$. That is, you add $\vec{a}$ and $-\vec{b}$.

3–4 Components of Vectors

☐ To find the x **component** of a vector, draw lines from the head and tail to the x axis, both perpendicular to the axis. The x component of the vector is the projection of the vector on the axis and is indicated by the separation of the points where the lines meet the axis. Similarly, to find the y component draw lines from the head and tail to the y axis. The upper diagram on the right shows the components of a vector.

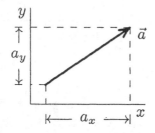

☐ The components of a vector are not vectors themselves but they can be either positive or negative. The vector in the upper diagram on the right has positive x and y components. The vector in the lower diagram has a positive x component and a negative y component.

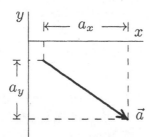

☐ Given the magnitude a of a vector in the xy plane and the angle θ it makes with the x axis, you can find the components using

$$a_x = a \cos\theta \qquad \text{and} \qquad a_y = a \sin\theta$$

For these expressions to be valid, θ must be measured counterclockwise from the positive x direction. If θ is between 0 and 90°, both the x and y components are positive; if θ is between 90° and 180°, the x component is negative and the y component is positive; if θ is between 180° and 270°, both the x and y components are negative; if θ is between 270° and 360°, the x component is positive and the y component is negative.

☐ You must also be able to find the magnitude and orientation of a vector when you are given its components. Suppose a vector $\vec{a}$ lies in the xy plane and its components a_x and a_y are given. In terms of the components, the magnitude of $\vec{a}$ is given by

$$a = \sqrt{a_x^2 + a_y^2}$$

and the angle $\vec{a}$ makes with the positive x direction is given by

$$\tan \theta = \frac{a_y}{a_x}.$$

The magnitude is always positive. There are two possible solutions to the equation for θ, the one given by your calculator and that angle plus $180°$. You must look at the orientation of the vector to see which makes physical sense. For example, if $a_y/a_x = 0.50$, then $\theta = 26.6°$ or $206.6°$. In the first case, both components are positive, while in the second, both are negative.

3–5 Unit Vectors

☐ The **unit vectors** $\hat{\imath}$, $\hat{\jmath}$, and $\hat{k}$ are used when a vector is written in terms of its components. These vectors have magnitude 1 and are in the positive x, y, and z directions respectively.

☐ $a_x \hat{\imath}$ is a vector parallel to the x axis with x component a_x, $a_y \hat{\jmath}$ is a vector parallel to the y axis with y component a_y, and $a_z \hat{k}$ is a vector parallel to the z axis with z component a_z. The vector $\vec{a}$ is given by $\vec{a} = a_x \hat{\imath} + a_y \hat{\jmath} + a_z \hat{k}$, where the rules of vector addition apply.

☐ Units are associated with the components a_x, a_y, and a_z of a vector but *not* with the unit vectors $\hat{\imath}$, $\hat{\jmath}$, and $\hat{k}$. Thus, the same unit vectors can be used to write any vector, no matter what its units.

3–6 Adding Vectors by Components

☐ Suppose $\vec{c}$ is the sum of two vectors $\vec{a}$ and $\vec{b}$ (i.e. $\vec{c} = \vec{a} + \vec{b}$). In terms of the components of $\vec{a}$ and $\vec{b}$: $c_x = a_x + b_x$, $c_y = a_y + b_y$, and $c_z = a_z + b_z$.

☐ Suppose $\vec{c}$ is the negative of $\vec{a}$ (i.e. $\vec{c} = -\vec{a}$). In terms of the components of $\vec{a}$: $c_x = -a_x$, $c_y = -a_y$, and $c_z = -a_z$.

☐ Suppose $\vec{c}$ is the difference of two vectors $\vec{a}$ and $\vec{b}$ (i.e. $\vec{c} = \vec{a} - \vec{b}$). In terms of the components of $\vec{a}$ and $\vec{b}$: $c_x = a_x - b_x$, $c_y = a_y - b_y$, and $c_z = a_z - b_z$.

3–7 Vectors and the Laws of Physics

☐ Many of the laws of physics are written in vector notation. For example, Newton's second law of motion tells us that a single force $\vec{F}$ acting on an object of mass m produces an acceleration $\vec{a}$ according to $\vec{F} = m\vec{a}$. Notice that both the right and left hand sides of the equation are vectors. The advantage to writing the law in vector form is that the equation is independent of the coordinate system used and no coordinate system need be specified when the equation is written.

3–8 Multiplying Vectors

☐ Vectors can be multiplied by scalars. If $\vec{a}$ is a vector and s a scalar. then, $s\vec{a}$ is another vector. If s is positive, its direction is the same as that of $\vec{a}$ and its magnitude is sa. If s is negative, the direction of $s\vec{a}$ is opposite that of $\vec{a}$ and its magnitude is $|s|a$.

☐ If $\vec{b} = s\vec{a}$, then in terms of components $b_x = sa_x$, $b_y = sa_y$, and $b_z = sa_z$.

☐ Division of a vector by a scalar is, of course, just multiplication by the reciprocal of the scalar.

☐ The **scalar product** (or dot product) of two vectors is defined in terms of the magnitudes of the two vectors and the angle between them when they are drawn with their tails at the same point: $\vec{a} \cdot \vec{b} = ab \cos \phi$. The geometry is shown on the diagram to the right.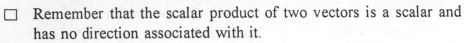

☐ Remember that the scalar product of two vectors is a scalar and has no direction associated with it.

☐ $\vec{a} \cdot \vec{b}$ is positive if ϕ is between 0 and 90°; it is negative if ϕ is between 90° and 180°; it is zero if $\phi = 90°$.

☐ In terms of components, $\vec{a} \cdot \vec{b} = a_x b_x + a_y b_y + a_z b_z$.

☐ The **vector product** (or cross product) of two vectors, written $\vec{a} \times \vec{b}$, is a vector. Its magnitude is given by $|\vec{a} \times \vec{b}| = ab \sin \phi$, where ϕ is the angle between $\vec{a}$ and $\vec{b}$ when they are drawn with their tails at the same point. ϕ is always in the range from 0 to 180° and the magnitude of the vector product is always positive.

☐ The direction of the vector product is always perpendicular to each of the factors. To find the direction of $\vec{a} \times \vec{b}$, draw the vectors with their tails at the same point and pretend there is a hinge at that point. Curl the fingers of your right hand so they rotate $\vec{a}$ into $\vec{b}$. Your thumb will then point in the direction of $\vec{a} \times \vec{b}$. Note that $\vec{b} \times \vec{a} = -\vec{a} \times \vec{b}$.

☐ The vector product of two vectors with given magnitudes is zero if the vectors are parallel or antiparallel ($\phi = 0$ or 180°); it has its maximum value if they are perpendicular to each other ($\phi = 90°$).

Hints for Questions

1 Notice that the x coordinate of the pit is the same as the x coordinate of the hole. Thus you cannot use $\vec{d_1} + \vec{d_2}$. One path the ball might take is along the y axis to a point opposite the hole and then parallel to the x axis to the hole. There is another path.

[Ans: Either the sequence $\vec{d_2}$, $\vec{d_1}$ or the sequence $\vec{d_2}$. $\vec{d_2}$, $\vec{d_3}$]

3 Draw the vectors $\vec{a} - \vec{b}$ and $\vec{b} - \vec{a}$.

[Ans: no]

5 (a) Substitute $\vec{a} + \vec{b} - \vec{c}$ for $\vec{d}$ in $\vec{a} - \vec{d}$ and see if the result is $\vec{c} - \vec{b}$.
(b) Substitute $\vec{a} + \vec{b} - \vec{c}$ for $\vec{d}$ in $-\vec{b} + \vec{d} + \vec{c}$ and see if the result is $\vec{a}$.
(c) Substitute $\vec{a} + \vec{b} - \vec{c}$ for $\vec{d}$ in $\vec{c} - \vec{d}$ and see if the result is $\vec{a} + \vec{b}$.

[Ans: (a) yes; (b) yes; (c) no]

7 Rotate the coordinate system so the z axis is upward and the y axis is to the right. If the x axis is then toward you the system is right handed; otherwise it is left handed.

[Ans: all but the system of (e)]

9 (a) The vectors are already drawn with their tails at the same point (or nearly so). Curl the fingers of your right hand so they tend to rotate $\vec{v}$ toward $\vec{B}$ through the smaller angle between them. Your thumb will then point in the direction of the vector product. Since q is positive $\vec{F}$ is in the same direction.

(b) Now q is negative, so in each situation $\vec{F}$ is in the direction opposite to that found in part (a).

[Ans: (a) positive x direction for (1), positive z direction for (2) and (3); (b) negative x direction for (1), negative z direction for (2) and (3)]

Hints for Problems

5 Use a coordinate system with its origin at the original position of the ship, its x axis positive to the east, and its y axis positive to the north. Let $\vec{d}_g$ ($= (120\,\text{km})\hat{\jmath}$) be the vector from the ship to its goal, $\vec{d}_w$ ($= (100\,\text{km})\hat{\imath}$) be the vector to the point where the wind blows the ship, and $\vec{d}$ be the vector describing the required displacement. The vector equation is $\vec{d}_g = \vec{d}_w + \vec{d}$. Solve for $\vec{d}$. The distance it must sail is the magnitude of $\vec{d}$ and you might describe the direction it must sail by calculating the angle $\vec{d}$ makes with north or with east.

[Ans: (a) 156 km; (b) 39.8° west of north]

9 Let $\vec{d}$ be the displacement vector from her initial position to her goal. She walks the minimum distance if she first walks east a distance that is equal to the component of $\vec{d}$ along an east-west axis and then walks north a distance that is equal to the component of $\vec{d}$ along a north-south axis.

[Ans: 4.74 km]

13 (a) and (b) The sum of the displacements involved in the four consecutive darts is equal to the overall displacement. Write the x and y components of this equation separately and solve for b_x and c_y.

(c) The magnitude of a vector is the square root of the sum of the squares of its components and the tangent of the angle it makes with the x axis is its y component divided by its x component. Draw a sketch to be sure you obtain the correct result when evaluate the inverse tangent.

[Ans: (a) −70.0 cm; (b) 80.0 cm; (c) 141 cm; (d) −172°]

21 Vectorially add the displacements for the three moves, then calculate the magnitude and direction of the result. The net displacement is five squares forward and two squares leftward. Each square represents a move of 1 m. You might place an x axis parallel to the forward direction and a y axis that runs left to right. Once you have the total displacement calculate its magnitude by taking the square root of the sum of the squares of its components. If you use the coordinate system suggested above, the tangent of the angle it makes with the forward direction is its y component divided by its x component. Sketch the displacement to be sure you get the correct value when you evaluate the inverse tangent.

$\left[\text{Ans: } 5.39\,\text{m at } 21.8° \text{ to the left of forward}\right]$

23 The sum of the displacements of beetle 1 must be the same as the sum of the displacements of beetle 2. First find the components of the two displacements of beetle 1, then add the vectors to find the total displacement. Find the components of the first displacement of beetle 2 and subtract this displacement from the total displacement of beetle 1. The result is the second displacement of beetle 2. Finally, take the square root of the sum of the squares to find the magnitude. The direction can be specified by finding the tangent of the angle between the displacement vector and one of the compass directions.

$\left[\text{Ans: (a) } 0.84\,\text{m; (b) } 79° \text{ south of west}\right]$

27 (a) Use $\vec{a} \cdot \vec{b} = ab\cos\phi$. Where the angle ϕ between them is the difference of the angles they make with the positive x axis.

(b) The magnitude of the vector product is $ab\sin\phi$. Use the right-hand rule to find the direction.

$\left[\text{Ans: (a) } -18.8; \text{ (b) } 26.9, \text{ in the positive } z \text{ direction}\right]$

29 Use $\vec{a} \times \vec{c} = \hat{\text{i}}(b_y c_z - b_z c_y) + \hat{\text{j}}(b_z c_x - b_x c_z) + \hat{\text{k}}(b_x c_y - b_y c_x)$ and a similar equation for $\vec{a} \times \vec{c}$. Use $\vec{a} \cdot \vec{b} = a_x b_x + a_y b_y + a_z b_z$ and a similar equation for $\vec{a} \cdot \vec{c}$. Also use $\vec{c} \cdot (\vec{b} \times \vec{c}) = a_x(\vec{b} \times \vec{c})_x + a_y(\vec{b} \times \vec{c})_y + a_z(\vec{b} \times \vec{c})_z$.

$\left[\text{Ans: (a) } -21; \text{ (b) } -9; \text{ (c) } 5\hat{\text{i}} - 11\hat{\text{j}} - 9\hat{\text{k}}\right]$

33 Use $\vec{a} \cdot \vec{b} = ab\cos\phi$, where ϕ is the angle between the vectors when they are drawn with their tails at the same point. Solve for $\cos\phi$, then take the inverse.

$\left[\text{Ans: } 70.5°\right]$

37 Notice that AB, AD, and BD form a right triangle. You are given the sides and asked for the hypotenuse. The vertical component of $\vec{\text{AB}}$ has the same magnitude as the vertical component of $\vec{\text{AD}}$.

$\left[\text{Ans: (a) } 27.8\,\text{m; (b) } 13.4\,\text{m}\right]$

39 The vertical component of the displacement of P equals a diameter of the wheel and the horizontal component is half the circumference of the wheel. The magnitude of the displacement is the square root of the sum of the squares of the components and the tangent of the angle it makes with the horizontal is the vertical component divided by the horizontal component.

$\left[\text{Ans: (a) } 168\,\text{cm; (b) } 32.5°\right]$

43 The scalar product of two vectors is the magnitude of either one of them multiplied by the component of the second vector along an axis that is parallel to the first. If the angle between the vectors is greater than 90° when they are drawn with their tails at the same point then the scalar product is negative. If it is less than 90° the scalar product is positive.

$\left[\text{Ans: (a) } 0; \text{ (b) } -16; \text{ (c) } -9\right.$

53 The x component of $\vec{r}$ is $d_{1x} - d_{2x} + d_{3x}$. Similar expression hold for the other components. The cosine of the angle between $\vec{r}$ and the z axis is given by r_z/r. The component of $\vec{d_1}$ along $\vec{d_2}$ is $\vec{d_1} \cdot \vec{d_2}/d_2$. The component of $\vec{d_1}$ that is perpendicular to $\vec{d_2}$ and in the plane of the two vectors is $|\vec{d_1} \times \vec{d_2}|/d_2$. These expressions follow from the definitions of the scalar and vector products.

[Ans: (a) $(9.0\,\text{m})\hat{i} + (6.0\,\text{m})\hat{j} - (7.0\,\text{m})\hat{k}$; (b) 123°; (c) $-3.2\,\text{m}$; (d) 8.2 m]

61 Carry out the specified operations in component form. Use $\vec{A} \cdot \vec{B} = A_x B_x + A_y B_y + A_z B_z$ and $\vec{A} \times \vec{B} = (A_y B_z - A_z B_y)\hat{i} + (A_z B_x - A_x B_z)\hat{j} + (A_x B_y - A_y B_z)\hat{k}$. In (a) and (b) you will need to add two vectors in order to obtain one of the vectors in the product. In (b) evaluate the vector product, then the scalar product.

[Ans: (a) $3.0\,\text{m}^2$; (b) $52\,\text{m}^3$; (c) $(11\,\text{m}^2)\hat{i} + (9.0\,\text{m}^2)\hat{j} + (3.0\,\text{m}^2)\hat{k}$]

69 If a vector is multiplied by a positive scalar its magnitude is multiplied by the scalar and its direction does not change. If it is multiplied by a negative scalar its magnitude is multiplied by the magnitude of the scalar and its direction reverses.

[Ans: (a) 10 m; (b) north; (c) 7.5 m; (d) south]

73 If the two vectors are perpendicular to each other then their scalar product is 0, if they are in the same direction then the scalar product is +1, and if they are in opposite directions then the scalar product is -1. If the vectors are in the same or opposite directions then the vector product is zero, if they are perpendicular to each other then the magnitude of the vector product is 1. Use the right-hand rule for vector products to determine the direction.

[Ans: (a) 0; (b) 0; (c) -1; (d) west; (e) up; (f) west]

Quiz

Some questions might have more than one correct answer.

1. The magnitude of the sum of two vectors must be

    A. greater than the magnitude of either of the vectors
    B. less than the magnitude of either of the vectors
    C. equal to the sum of the magnitudes of the vectors
    D. equal to the square root of the sum of the squares of the magnitudes of the vectors
    E. none of the above

2. If three vectors form a triangle, with the head of each vector at the tail of another,

    A. the magnitude of the sum of the three vectors is the square root of the sum of the squares of the three magnitudes
    B. the magnitude of the sum of the three vectors is the sum of the three magnitudes
    C. any one of the vectors is the sum of the other two
    D. the magnitude of any one of three vectors is the sum of the magnitudes of the other two
    E. the magnitude of the sum of the three vectors is zero

3. Which of the following is NOT a vector quantity?

   A. force
   B. velocity
   C. temperature
   D. acceleration
   E. displacement

4. Which of the following statements are true?

   A. unit vectors do not have units
   B. components of all vectors do not have units
   C. the sum of any two of the unit vectors $\hat{\imath}$, $\hat{\jmath}$, and $\hat{k}$ is parallel to a coordinate axis
   D. the difference of any two of the unit vectors $\hat{\imath}$, $\hat{\jmath}$, and $\hat{k}$ is zero
   E. any vector and its negative are in opposite directions

5. The y component of a vector in the xy plane is negative only if

   A. the angle between the vector and the positive x axis, measured counterclockwise from that axis, is less than 90°
   B. the angle between the vector and the positive x axis, measured counterclockwise from that axis, is greater than 90° and less than 180°
   C. the angle between the vector and the positive y axis, measured counterclockwise from that axis, is less than 90°
   D. the angle between the vector and the positive y axis, measured counterclockwise from that axis, is greater than 90°
   E. the x component of the vector is zero

6. Two vectors in the xy plane have the same x components but different y components. All components are positive. As a result,

   A. their sum is parallel to the x axis
   B. their sum is parallel to the y axis
   C. their difference is parallel to the x axis
   D. their difference is parallel to the y axis
   E. they make the same angle with the positive x axis

7. If the sum of two vectors is 0 then

   A. the vectors are in the same direction
   B. the vectors are in opposite directions
   C. the vectors have the same magnitude
   D. the magnitude of one of the vectors is twice the magnitude of the other
   E. the vectors are perpendicular to each other

8. Two vectors are equal if

   A. their x components are equal, their y components are equal, and their z components are equal
   B. their scalar product is equal to the square of the magnitude of either of them
   C. their vector product is zero
   D. their magnitudes are equal and they make the same angle with the positive x direction
   E. their vector product is equal to the square of the magnitude of either of them

9. Vectors $\vec{A}$ and $\vec{B}$ are perpendicular to each other. As a result the magnitude of their sum is

   A. $\sqrt{A^2 + B^2}$
   B. $A + B$
   C. $A - B$
   D. $\sqrt{A + B}$
   E. A/B

10. The product of a scalar and a vector

    A. has a magnitude that must be larger than the magnitude of the vector
    B. must be in the same direction as the vector
    C. is not defined
    D. has a magnitude that is equal to the product of the magnitude of the scalar and the magnitude of the original vector
    E. has components that are the products of the scalar and the components of the original vector

11. The scalar product of two vectors is negative if

    A. the angle between the vectors, when their tails are at the same point, is less than $90°$
    B. the angle between the vectors, when their tails are at the same point, is greater than $90°$
    C. the angle between the vectors, when their tails are at the same point, is exactly $90°$
    D. one of the vectors is the negative of the other
    E. both of the vectors are the negative of another vector

12. The vector product $(a\hat{i}) \times (-b\hat{j})$ is

    A. in the positive z direction
    B. in the negative z direction
    C. in the positive x direction
    D. in the negative y direction
    E. zero

13. If $\vec{A} \cdot \vec{B} = |\vec{A} \times \vec{B}|$ then

    A. the angle between $\vec{A}$ and $\vec{B}$ is 0
    B. the angle between $\vec{A}$ and $\vec{B}$ is 90°
    C. the angle between $\vec{A}$ and $\vec{B}$ is 45°
    D. $\vec{A}$ and $\vec{B}$ have the same magnitude
    E. the magnitude of $\vec{A}$ is the reciprocal of the magnitude of $\vec{B}$

Answers: (1) E; (2) E; (3) C; (4) A, E; (5) D; (6)D; (7) B, C; (8) A, D; (9) A; (10) D, E; (11) B, D; (12) B; (13) C

Chapter 4
MOTION IN TWO AND THREE DIMENSIONS

The ideas of position, velocity, and acceleration that were introduced earlier in connection with one-dimensional motion are now extended. You should pay close attention to the definitions and relationships discussed in this chapter. Three applications are discussed: projectile motion, circular motion, and relative motion.

Important Concepts

☐ position vector

☐ displacement vector

☐ average velocity

☐ (instantaneous) velocity

☐ speed

☐ average acceleration

☐ (instantaneous) acceleration

☐ projectile motion

☐ uniform circular motion

☐ relative motion

Overview

4–2 Position and Displacement

☐ The fundamental concept used to describe the motion of a particle moving in two or three dimensions is its **position vector**. The tail of this vector is always at the origin and at any instant the head is at the particle. The Cartesian components of the position vector are the coordinates of the particle. As the particle moves its position vector changes and so is a function of time.

☐ A **displacement vector** is used to describe a change in a position vector. If the particle has position vector $\vec{r}_1$ at time t_1 and position vector $\vec{r}_2$ at a later time t_2, then the displacement vector for this interval is $\Delta \vec{r} = \vec{r}_2 - \vec{r}_1$. If the particle has coordinates x_1, y_1, z_1 at time t_1 and coordinates x_2, y_2, z_2 at time t_2, then the components of the displacement vector are given by $(\Delta \vec{r})_x = \Delta x = x_2 - x_1$, $(\Delta \vec{r})_y = \Delta y = y_2 - y_1$, and $(\Delta \vec{r})_z = \Delta z = z_2 - z_1$. When using these equations, pay close attention to the order of the terms: a displacement vector is a position vector at a *later* time minus a position vector at an *earlier* time.

4–3 Average Velocity and Instantaneous Velocity

☐ In terms of the displacement vector $\Delta \vec{r}$, the **average velocity** of the particle in the interval from t_1 to t_2 is

$$\vec{v}_{\text{avg}} = \frac{\Delta \vec{r}}{\Delta t},$$

where $\Delta t = t_2 - t_1$. The average velocity has components given by

$$v_{\text{avg } x} = \frac{\Delta x}{\Delta t}, \qquad v_{\text{avg } y} = \frac{\Delta y}{\Delta t}, \qquad \text{and} \qquad v_{\text{avg } z} = \frac{\Delta z}{\Delta t}.$$

☐ To use the definition to calculate the average velocity over the time interval from t to t_2, you must know the position vector for the beginning and end of the interval.

☐ The **instantaneous velocity** $\vec{v}$ at any time t is the limit of the average velocity over a time interval that includes t, as the duration of the interval becomes vanishingly small. In terms of the position vector, it is given by the derivative

$$\vec{v} = \frac{d\vec{r}}{dt}.$$

In terms of the particle coordinates, its components are

$$v_x = \frac{dx}{dt}, \qquad v_y = \frac{dy}{dt}, \qquad \text{and} \qquad v_z = \frac{dz}{dt}.$$

The term "velocity" means the same as "instantaneous velocity".

☐ You should be aware that the instantaneous velocity, unlike the average velocity, is associated with a single instant of time. At any other instant, no matter how close, the instantaneous velocity might be different.

☐ To use the definition to calculate the instantaneous velocity, you must know the position vector as a function of time. This is identical to knowing the coordinates as functions of time. You should remember that the instantaneous velocity vector at any time is tangent to the path at the position of the particle at that time. If you are asked for the direction the particle is traveling at a certain time, you automatically calculate the components of its velocity for that time.

☐ **Speed** is the magnitude of the instantaneous velocity and, if the velocity components are given, can be calculated using

$$v = \sqrt{v_x^2 + v_y^2 + v_z^2}.$$

4–4 Average Acceleration and Instantaneous Acceleration

☐ In terms of the velocity $\vec{v_1}$ at time t_1 and the velocity $\vec{v_2}$ at a later time t_2, the **average acceleration** over the interval from t_1 to t_2 is given by

$$\vec{a}_{\text{avg}} = \frac{\vec{v_2} - \vec{v_1}}{t_2 - t_1} = \frac{\Delta \vec{v}}{\Delta t}.$$

In terms of velocity components, the components of the average acceleration are

$$a_{\text{avg } x} = \frac{\Delta v_x}{\Delta t}, \qquad a_{\text{avg } y} = \frac{\Delta v_y}{\Delta t}, \qquad \text{and} \qquad a_{\text{avg } z} = \frac{\Delta v_z}{\Delta t}.$$

Note that $\vec{v_1}$ and $\vec{v_2}$ are instantaneous, not average, velocities.

☐ To use the definition to calculate the average acceleration over the interval from t_1 to t_2, you must know the velocity at the beginning and end of the interval. This may mean you must differentiate the position vector with respect to time.

☐ The **instantaneous acceleration** $\vec{a}$ at any time t is the limit of the average acceleration over an interval that includes t, as the duration of the interval becomes vanishingly small. In terms of the velocity vector, it is given by

$$\vec{a} = \frac{d\vec{v}}{dt}.$$

In terms of the velocity components, its components are

$$a_x = \frac{dv_x}{dt}, \qquad a_y = \frac{dv_y}{dt}, \qquad \text{and} \qquad a_z = \frac{dv_z}{dt}.$$

☐ To use the definition to calculate the acceleration, you must know the velocity vector as a function of time. The terms "instantaneous acceleration" and "acceleration" mean the same thing.

☐ A non-zero velocity indicates that the position vector of the particle is changing with time. A non-zero acceleration indicates that the velocity vector of the particle is changing with time. Remember that these changes may be changes in magnitude, direction, or both.

4–5 Projectile Motion

☐ If air resistance is negligible, a projectile simultaneously undergoes free-fall vertical motion and constant-velocity horizontal motion. The acceleration of the projectile has magnitude g (the free-fall acceleration) and is downward, toward the Earth. Note that the vertical component is constant. Once the projectile is launched its acceleration does not change until it hits the ground or a target.

4–6 Projectile Motion Analyzed

☐ If $\vec{r}_0$ is the initial position (at time $t = 0$) and $\vec{v}_0$ is the initial velocity, then the coordinates of the projectile at time t are given by $x = x_0 + v_{0x}t$ and $y = y_0 + v_{0y}t - \frac{1}{2}gt^2$. The velocity components are given by $v_x = v_{0x}$ and $v_y = v_{0y} - gt$. Here the positive y direction is upward. Given the initial position and velocity of the projectile the these equations can be used to predict its position and velocity throughout its flight (until it hits something). In many cases they must be solved simultaneously.

☐ Note that the equations for x and v_x describe motion with constant velocity and that the equations for y and v_y describe free-fall motion (with constant acceleration g).

☐ If v_0 is the initial speed and θ_0 is the angle between the initial velocity vector and the horizontal, then $v_{0x} = v_0 \cos\theta_0$ and $v_{0y} = v_0 \sin\theta_0$. θ_0 (and v_{0y}) are negative if the projectile is launched downward.

☐ There are two special conditions you should remember. At the highest point of its trajectory the velocity of a projectile is horizontal and $v_y = 0$. To find the time when the projectile is at its highest point, solve $0 = v_0 \sin\theta_0 - gt$ for t. When a projectile returns to the original launch height, $y = y_0$. To find the time when the projectile returns to the launch height, solve $0 = (v_0 \sin\theta_0)t - \frac{1}{2}gt^2$ for t. The magnitude of the displacement from the launch point to the point at launch height is called the horizontal range of the projectile.

4–7 Uniform Circular Motion

☐ A particle in **uniform circular motion** moves around a circle with constant speed. Remember that the velocity vector is always tangent to the path and therefore continually changes direction. This means the acceleration is *not* zero.

☐ If the radius of the circle is r and the speed is v, the acceleration of the particle has magnitude

$$a = \frac{v^2}{r}$$

and the acceleration vector always points from the particle toward the center of the circle. The direction of the acceleration continually changes as the particle moves around the circle.

☐ The term "centripetal acceleration" is applied to this acceleration to indicate its direction. You should not forget it is the rate of change of velocity, as are all accelerations.

☐ The period T of the motion is the time for the particle to go around exactly once. Since the speed is constant $T = 2\pi r/v$.

4–8 Relative Motion in One Dimension

☐ The position or velocity of a particle is measured using two coordinate systems (or reference frames) that are moving relative to each other. Given the position and velocity of the particle in one frame and the relative position and velocity of the frames, you should be able to calculate the position and velocity in the other frame.

☐ Let x_{PA} be the coordinate of a particle P as measured in frame A (see the diagram on the right), let x_{PB} be the coordinate as measured in frame B, and let x_{BA} be the coordinate of the origin of frame B as measured in frame A. Then, $x_{PA} = x_{PB} + x_{BA}$.

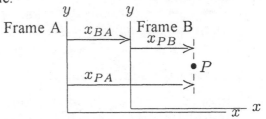

☐ Differentiate the coordinate equation with respect to time to obtain the relationship between the velocity of the particle as measured in one frame and the velocity as measured in the other. The result is $v_{PA} = v_{PB} + v_{BA}$. Here v_{PA} is the velocity as measured in A, v_{PB} is the velocity as measured in B, and v_{BA} is the velocity of frame B as measured in frame A.

☐ Take care with the subscripts. The first symbol in a subscript names an object (the particle or the origin of frame B) and the second names the coordinate frame used to measure the position or velocity of the object. You should say all the words as you read the symbols. That is, when you see x_{BA} you should say "the position vector of the origin of frame B relative to the origin of frame A." You will then get acquainted with the notation fast and won't get it mixed up later.

4–9   Relative Motion in Two Dimensions

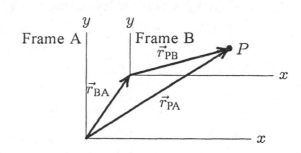

☐ According to the diagram on the right, at any instant of time the position of the particle P relative to the origin of coordinate frame A is given by $\vec{r}_{PA} = \vec{r}_{PB} + \vec{r}_{BA}$ in terms of its position $\vec{r}_{PB}$ relative to the origin of frame B and the position vector $\vec{r}_{BA}$ of the origin of frame B relative to the origin of frame A.

☐ The above expression can be differentiated with respect to time to obtain $\vec{v}_{PA} = \vec{v}_{PB} + \vec{v}_{BA}$ for the velocity of the particle in frame A in terms of the particle velocity $\vec{v}_{PB}$ in frame B and the velocity $\vec{v}_{BA}$ of frame B as measured in frame A. This expression is valid even if the two frames are accelerating relative to each other.

☐ Airplanes flying in moving air or ships sailing in moving water are often used as examples of relative motion. The airplane or ship is the particle, one coordinate frame moves with the air or water, and the other coordinate frame is fixed to the earth. The heading of the airplane or ship is in the direction of its velocity as measured relative to the air or water, *not* relative to the ground.

Hints for Questions

1 The position vector of the particle is given by $\vec{r} = x\hat{i} + y\hat{j} + z\hat{k}$, where x, y, and z are its coordinates. The diplacement is the final position vector minus the initial position vector.

[Ans: (a) $(7\,\text{m})\hat{i} + (1\,\text{m})\hat{j} - (2\,\text{m})\hat{k}$; (b) $(5\,\text{m})\hat{i} - (3\,\text{m})\hat{j} + (1\,\text{m})\hat{k}$; (c) $(-2\,\text{m})\hat{i}$]

3 The ball is still going up if the y component of its velocity is positive and is going down if that component is negative.

[Ans: yes; it is coming down]

5 (a) The launch speed is the magnitude of the launch velocity: the square root of the sum of the squares of the launch velocity components.
(b) The time of flight is given by $2v_{0y}/g$, where v_{0y} is the vertical component of the launch velocity. Notice that 1 and 2 are shot upward and that 3 and 4 are shot downward.

[Ans: (a) all tie; (b) 1 and 2 tie, then 3 and 4 tie

7 Since the bundle is dropped its initial velocity is the same as the velocity of the airplane. Recall that the horizontal motion of a projectile is a motion with constant velocity. Also recall that the time of flight depends only on the vertical component of velocity.

[Ans: (a) 0; (b) 350 km/h; (c) 350 km/h; (d) the same]

9 Recall that the height of the highest point and the time of flight depend only on the vertical component of the launch velocity. The range depends on the product of the horizontal and vertical components of the launch velocity. The initial speed is the square root of the sum of the squares of the launch velocity components.

[Ans: (a) all tie; (b) all tie; (c) 3, 2, 1; (d) 3, 2, 1]

11 As the train rounds a curve the magnitude of its acceleration is proportional to the reciprocal of the curve radius.

[Ans: 2, then 1 and 4 tie, then 3]

13 Remember that the acceleration of an object is the rate of change of its velocity and that the velocity might change in magnitude or direction or both. An object is accelerating when it rounds a curve even if its speed is not changing.

[Ans: (a) yes (just round a curve); (b) no (the direction of the velocity must be changing); (c) yes (if going with constant speed)

Hints for Problems

7 The average velocity during an interval is given by $\vec{v}_{avg} = (\vec{r}_f - \vec{r}_i)/\Delta t$, where $\vec{r}_i$ is the position at the beginning of the interval, $\vec{r}_f$ is the position at the end, and Δt is the duration of the interval.

[Ans: $(-0.70\,\text{m/s})\,\hat{\imath} + (1.4\,\text{m/s})\,\hat{\jmath} - (0.40\,\text{m/s})\,\hat{k}$]

19 (a) Take the time to be zero when the ball is at a height of 9.1 m. Put the origin of a coordinate system at this point and take the y axis to positive in the upward direction. Then $\vec{r}_0 = 0$ and $\vec{v}_0 = (7.6\,\text{m/s})\,\hat{\imath} + (6.1\,\text{m/s})\,\hat{\jmath}$. When the ball reaches its maximum height the vertical component of its velocity is zero. Solve $v_y = v_{0y} - gt$ for the time t when it reaches its maximum height and $y = v_{0y}t = \frac{1}{2}gt^2$ for the maximum height above the origin. Add 9.1 m to get the maximum height above the ground.
(b) The ball is at ground level when $y = 0$. Solve $y = v_{0y}t - \frac{1}{2}gt^2$, with $y = 0$, for the times when the ball is at ground level. The negative solution is the firing time and the positive solution is the time the ball returns to the ground. Evaluate $x = v_{0x}t$ for the coordinates of the firing and landing points. The total horizontal distance traveled is the distance between these points.
(c) and (d) Evaluate $v_y = v_{0y} - gt$ for the y component of the velocity just before landing. v_x, of course, is just v_{0x}.

[Ans: (a) 11 m; (b) 23 m; (c) 17 m/s; (d) 63°]

23 Put the origin of a coordinate system at the point where the decoy is released and take the time to zero when it is released. The initial velocity of the decoy is the same as the velocity of the plane.
(a) Solve $x = v_{0x}t$ for the flight time of the decoy.
(b) Evaluate $y = v_{0y}t - \frac{1}{2}gt^2$ for the y coordinate of the landing point. Its magnitude is the height of the plane above the ground when the decoy is released.

[Ans: (a) 10.0 s; (b) 897 m]

29 Take the x axis to be horizontal and the y axis to be vertical. Let v_{0x} and v_{0y} be the components of the initial velocity. At maximum height the speed of the projectile is v_x since it is then traveling horizontally and that component of the velocity is constant. Thus $\sqrt{v_{0x}^2 + v_{0y}^2} = 5v_{0x}$. Calculate v_{0y} and then the inverse tangent of v_{0y}/v_{0x}.

[Ans: 78.5°]

33 Following the hint given in the problem statement, suppose the ball is launched from the ground at an angle of 60°, a distance d from the edge of the building. Put the origin of a coordinate system there and take the time of launch to be zero. If the y axis is positive in the upward direction and the x axis is positive to the right, the coordinates of the ball are given by $x = v_{0x}t$ and $y = v_{0y}t - \frac{1}{2}gt^2$, where t is the time. The components of the initial velocity are given by $v_{0x} = v_0 \cos 60°$ and $v_{0y} = v_0 \sin 60°$, where v_0 is the speed.

(a) Solve $d = v_{0x}t$ for the time of flight and evaluate $h = v_{0y}t - \frac{1}{2}t^2$ for the height of the building.

(b) and (c) Evaluate $v_x = v_{0x}$ and $v_y = v_{0y} - gt$, with t equal to the time of flight. The speed with which the ball is thrown in the original problem is the square root of the sum of the squares of these velocity components. The angle to the horizontal with which the ball was thrown in the original problem s given by $\tan \theta_0 = v_y/v_x$.

(d) If v_y is positive the throwing angle is above the horizontal; if it is negative the throwing angle is below the horizontal.

[Ans: (a) 32.2 m; (b) 21.9 m/s; (c) 40.4°; (d) below the horizontal.]

35 Place the origin of a coordinate system at the launch point and take the time to be zero at launch. Take the y axis to vertical and positive in the upward direction. Take the x axis to be horizontal and positive to the right. Then the coordinates of the ball are given by $x = v_{0x}t$ and $y = v_{0y}t - \frac{1}{2}gt^2$, where t is the time. Eliminate t to obtain the equation for the trajectory of the ball:

$$y = \frac{v_{0x}v_{0y}}{x} - \frac{1}{2}g\frac{v_{0x}^2}{x^2}.$$

The equation of the ramp is $y = (d_2/d_1)x$.

Solve these equations simultaneously for x and y. If x is less than or equal to d_1 the ball lands on the ramp and the coordinates you calculated are the coordinates of the landing point. If x is greater than d_1 the ball lands on the plateau and you must rework the problem. Put y equal to d_2 and solve $y = v_{0y}t - \frac{1}{2}gt^2$ for the time of flight. Then evaluate $x = v_{0x}t$ for the other coordinate of the landing point.

In either case the magnitude of the displacement from the launch point to the landing point is $\sqrt{x^2 + y^2}$ and the tangent of the angle that this displacement makes with the horizontal is y/x.

[Ans: (a) it lands on the ramp; (b) 5.82 m; (c) 31.0°]

43 (a) The ball travels a horizontal distance of 50.0 m in 4.00 s. Calculate the horizontal component of its velocity. This is constant throughout the motion. Before it reaches the top of wall it travels horizontally for 1.00 s. Calculate the horizontal distance it travels from when it is hit to when it reaches the top of the wall. Since the motion is symmetric, the horizontal distance it travels from when it passes the top of the wall on the way down to when it is caught is the same. The total horizontal distance traveled is the sum of these three distances.

(b) and (c) The ball is in flight for 6.00 s. Use $y = y_0 + v_{0y}t - \frac{1}{2}gt^2$ to compute the y component of its initial velocity. Its initial speed is the square root of the sum of the squares of the velocity components and the tangent of the launch angle is the y component divided by the x component.

(d) Use $y = v_{0y}t - \frac{1}{2}gt^2$ to compute the height of the ball 1.00 s after it is hit.

[Ans: (a) 75.0 m; (b) 31.9 m/s; (c) 66.9°; (d) 25.5 m]

45 The period is the time to complete exactly one turn. The centripetal acceleration is given by v^2/r, where v is the speed of the woman and r is radius of her circular path. The speed is related to the period T by $v = 2\pi r/T$. The acceleration vector points from the woman toward the center of her path.

[Ans: (a) 12 s; (b) 4.1 m/s^2; (c) down; (d) 4.1 m/s^2; (e) up]

51 Find the angle between the two acceleration vectors and use this to determine the fraction of a period that has elapsed and finally the period itself. A diagram showing the position of the particle at the two times comes in handy. Use $a = 4\pi^2 r/T^2$, where r is the radius of the path, a is the magnitude of the acceleration, and T is the period. Solve for r.

[Ans: 2.92 m]

57 Let v_r be the man's running speed and v_w be the speed of the moving walk. When running in the direction of travel of the walk the speed of the man relative to the building is $v_r + v_w$ and when running in the opposite direction his speed relative to the building is $v_r - v_w$. The distance run in the first case is given by $t_1(v_r + v_w)$ and in the second by $t_2(v_r - v_w)$, where $t_1 = 2.5$ s and $t_2 = 10.0$ s. The two distances are the same. Solve for v_r/v_w.

[Ans: 5/3]

67 Set up a coordinate system with the x axis along the initial direction of travel and the y axis perpendicular to that direction. Calculate the components of the displacements in each of the intervals. The length of a displacement is the product of the speed and the duration of the interval. Sum the displacements to obtain the total displacement. The distance from the starting point is the magnitude of the total displacement and the tangent of the angle with the initial direction of travel is the y component of the total displacement divided by the x component.

[Ans: (a) 2.7 km; (b) 76° clockwise]

71 The average velocity in an interval is the displacement during the interval divided by the duration of the interval. Recall that the displacement depends on the end points and not on the path between.

[Ans: (a) 0.83 cm/s; (b) 0; (c) 0.11 m/s; (d) −63°]

73 The speed of the particle is $2\pi r/T$, where r is the radius of the circle and T is the period of the motion. The radius is the magnitude of $\vec{r}$. The velocity is tangent to the circle.

[Ans: (a) 0.83 cm/s; (b) 0; (c) 0.11 m/s; (d) −63°]

83 Place the origin at the launch point and take the positive y direction to be upward. Then $x = v_{0x}t$ and $y = v_{0y}t - \frac{1}{2}gt^2$. The component of the initial velocity are $v_{0x} = v_0 \cos\theta_0$ and $v_{0y} = v_0 \sin\theta_0$, where v_0 is the initial speed. The distance from the launch point to the burrito at any time is given by $r = \sqrt{x^2 + y^2}$.

[Ans: (c) 2.10 s; (d) 25.7 m; (e) 25.7 m; (f) 0; (g) 1.71 s; (h) 13.5 m; (i) 4.76 m; (j) 12.6 m]

89 The radial acceleration a and speed v are related by $a = v^2/r$, where r is the radius of the circular path. Solve for v.

[Ans: (a) 6.7 × 10^6 m/s; (b) 1.4 × 10^{-7} s]

91 Place the origin at the firing point and take the positive y direction to be upward. Then the coordinates of the ball at time t are $x = v_{0x}t$ and $y = v_{0y}t - \frac{1}{2}gt^2$, where v_{0x} and v_{0y} are the components of its initial velocity. The components are given by $v_x = v_0 \cos\theta_0$ and $v_{0y} = v_0 \sin\theta_0$, where v_0 is the initial speed and θ_0 is the firing angle. The ball lands when $y = -h$, where h is the height of the cannon above sea level. Use the equation for x to eliminate t from the equation for y, replace y with $-h$, and solve for x.

$\left[\text{Ans: } 3 \times 10^1 \,\text{m}\,\right]$

101 Let $\vec{v}_{PG}$ be the velocity of the plane relative to the ground, $\vec{v}_{PA}$ be the velocity of the plane relative to the air, and $\vec{v}_{AG}$ be the velocity of the air relative to the ground. Then $\vec{v}_{PG} = \vec{v}_{PA} + \vec{v}_{AG}$. Since $\vec{v}_{PG}$ is to the east and $\vec{v}_{AG}$ is to the south, the three vectors form a right triangle with sides of length v_{PG} and v_{AG} and a hypotenuse of length v_{PA}. Thus $v_{PA}^2 = v_{PG}^2 + v_{AG}^2$. Solve for v_{PG}.

$\left[\text{Ans: } 67 \,\text{km/h}\,\right]$

105 Use $v_x = v_{0x}$ and $v_y = v_{0y} - gt$ to compute the velocity components at time t. Here v_x $(= v_0 \cos\theta_0)$ and v_{0y} $(= v_0 \sin\theta_0)$ are the initial velocity components, v_0 is the initial speed and θ_0 is the launch angle. The magnitude of the velocity is the square root of the sum of the squares of the components and the angle θ that the velocity makes with the horizontal is determined from $\tan\theta = v_y/v_x$.

$\left[\text{Ans: (a) } 16\,\text{m/s; (b) } 23°; \text{(c) above; (d) } 27\,\text{m/s; (e) } 57°; \text{(f) below}\,\right]$

111 Since the particle moves on a straight line the acceleration is along that line. Thus the ratio of the acceleration components must be the same as the ratio of the displacement components. The distance traveled along the line during the interval Δt is $v_0\,\Delta t + \frac{1}{2}a(\Delta t)^2$, where v_0 is the speed at the beginning of the interval, and the speed at the end of the interval is $v_0 + a\,\Delta t$. The speed at the end of the first interval is the initial speed for the second interval. You should be able to eliminate Δt_1 and write an expression for the distance traveled during the second interval in terms of the distance traveled during the first.

$\left[\text{Ans: (a) } 1.5; \text{(b) } (36\,\text{m}, 54\,\text{m})\,\right]$

117 The initial speed of the ball relative to the ground is $30\,\text{m/s}$ and its initial height above the ground is $30\,\text{m}$. Use the free-fall equations to calculate the maximum height of the ball above the ground. The position of the elevator floor above the ground is $y_{E0} + v_E t$, where y_{E0} $(= 28\,\text{m})$ is the initial height of the elevator floor and v_E is the speed of the elevator. You want the time when the coordinates of the elevator floor and the ball are the same.

$\left[\text{Ans: (a) } 76\,\text{m; (b) } 4.2\,\text{s}\,\right]$

123 To avoid a collision the rear of the car must be on the far side of the train when the train gets to the crossing center. That is, the car must go a distance D equal to the sum of its initial distance from the crossing center, its length, and half the width of the train. Calculate the time for the train to get to the crossing center and the distance d actually traveled by the car in this time. Compare D and d. Compare the time for the car to go the distance D with the actual time.

$\left[\text{Ans: (a) yes; (b) } 0.16\,\text{s}\,\right]$

129 The magnitude v of the speed and the magnitude a of the acceleration are related by $a = v^2/r$, where r is the radius of the circular path. Solve for r. The acceleration points from the car toward the center of the circle.

[Ans: (a) 48 m, west of center; (b) 48 m, west of center]

133 Place the origin at the point of ejection, with the positive y direction upward and the positive x direction forward. The the initial velocity of the package relative to the ground is $\vec{v_0} = \vec{v}_{PH} + \vec{v}_{HG}$, where $\vec{v}_{PH}$ is the velocity of the package relative to the helicopter and $\vec{v}_{HG}$ is the velocity of the helicopter relative to the ground. At time t the coordinates of the package relative to the ground are given by $x = v_{0x}t$ and $y = -\frac{1}{2}gt^2$. Use the second equation to calculate the time when $y = -h$, where h is the altitude of the helicopter, then use the first equation to calculate x. The angle θ between the velocity vector and the horizontal is given by the ratio v_y/v_x of the velocity components just before impact of the package. Use $v_x = v_0$ and $v_y = -gt$ to compute the velocity components.

[Ans: (a) 5.8 m/s; (b) 17 m; (c) 67°]

Quiz

Some questions might have more than one correct answer.

1. The velocity vector for a particle

    A. is perpendicular to the path of the particle when the speed of the particle is changing
    B. is tangent to the path of the particle only when the speed of the particle is not changing
    C. is always perpendicular to the path of the particle
    D. is always tangent to the path of the particle
    E. is never perpendicular to the path of the particle

2. The speed of a particle is increasing whenever

    A. its velocity and acceleration are perpendicular to each other
    B. the scalar product of its velocity and acceleration is positive
    C. the scalar product of its velocity and acceleration is negative
    D. the scalar product of its velocity and acceleration is zero
    E. its velocity and acceleration are in the same direction

3. The direction of motion of a particle is changing without a change in speed when

    A. its velocity and acceleration are perpendicular to each other
    B. the scalar product of its velocity and acceleration is positive
    C. the scalar product of its velocity and acceleration is negative
    D. the scalar product of its velocity and acceleration is zero
    E. its velocity and acceleration are in the same direction

4. The acceleration vector of a particle is

    A. is always perpendicular to the path of the particle
    B. is always tangent to the path of the particle
    C. is never perpendicular to the path of the particle
    D. is never tangent to the path of the particle
    E. is tangent to the path of the particle only when the direction of motion of the particle is not changing

5. Which of the following situations are not possible?

    A. the speed of a particle is changing but its acceleration is zero
    B. the direction of motion of a particle is changing but its acceleration is zero
    C. the speed of a particle is not changing and its acceleration is not zero
    D. the direction of motion of a particle is not changing and its acceleration is not zero
    E. none of the above (all are possible)

6. At the highest point of its trajectory the acceleration of a projectile is

    A. zero
    B. g, upward
    C. g, downward
    D. g, tangent to the trajectory
    E. tangent to the trajectory with a magnitude other than g

7. the height above the launch point of the highest point on the trajectory of a projectile depends on

    A. the horizontal component of its launch velocity
    B. the vertical component of its launch velocity
    C. its launch speed for a given launch angle
    D. its launch angle for a given launch speed
    E. the local magnitude of the free-fall acceleration

8. A projectile is launched upward from the edge of a cliff and lands in a valley below the launch point. Its speed is the greatest

    A. just after launch
    B. when it is at the highest point on its trajectory
    C. when it reaches the launch height on the way down
    D. just before it lands
    E. at some other time

9. A projectile is fired over level ground with a launch angle of 45°. At the instant of firing a pellet is fired straight upward. If the pellet is to land at the same instant as the projectile its initial speed must be the same as

   A. the initial speed of the projectile
   B. the horizontal component of the initial velocity of the projectile
   C. the vertical component of the initial velocity of the projectile
   D. about half the initial speed of the projectile
   E. about twice the initial speed of the projectile

10. A particle is moving with a nonzero acceleration. It is viewed from two reference frames which are moving with constant velocity relative to each other and whose origins never coincide. Which of the following statements are true?

    A. the position vector of the particle is the same in the two frames
    B. the velocity vector of the particle is the same in the two frames
    C. the acceleration vector of the particle is the same in the two frames
    D. at some instant the velocity vector of the particle might be zero in one of the frames
    E. at some instant the acceleration vector of the particle might be zero in one of the frames

11. A particle is in uniform circular motion. Which of the following quantities are constant?

    A. the velocity of the particle
    B. the acceleration of the particle
    C. the speed of the particle
    D. the magnitude of the particle's acceleration
    E. none of these

12. The following four particles are in uniform circular motion. Their speeds and the radii of their paths are:

    I V, R
    II. $2V, R$
    III $V, 2R$
    IV $V, 3R$

    Rank them in order of their centripetal accelerations, least to greatest.

    A. I, II, III, IV
    B. IV, III, II, I
    C. II, I, IV, III
    D. IV, III, I, II
    E. III, II, I, IV

Answers: (1) D, E; (2) B; (3) A, D; (4) E; (5) A, B; (6) C; (7) B, C, D, E; (8) D; (9) C; (10) C; (11) C, D; (12) D

Chapter 5
FORCE AND MOTION —— I

In this, the most fundamental chapter in the mechanics section of the text, you start to study how objects influence the motion of each other. The fundamental problem of mechanics is to find the acceleration of an object, given the object and its environment. The problem is split into two parts, connected by the idea of a force: the environment of an object exerts forces on the object and the net force on it causes it to accelerate. In this chapter, you concentrate on the relationship between the net force on an object and its acceleration.

Important Concepts

- ☐ Newton's first law
- ☐ inertial reference frame
- ☐ force
- ☐ mass
- ☐ inertia
- ☐ Newton's second law

- ☐ force of gravity (weight)
- ☐ tension (in a string)
- ☐ normal force
- ☐ frictional force
- ☐ Newton's third law
- ☐ free-body diagram

Overview

5–3 Newton's First Law

- ☐ If the net force on a particle is zero, then the acceleration of the particle, as measured relative to an inertial frame, is also zero. Such a frame might be attached to a particle on which zero total force acts. Clearly the acceleration of the particle, as measured in that frame, is zero.
- ☐ Newton's laws as stated in this text are valid only for inertial frames.
- ☐ For phenomena that take place over short times we may take a reference frame attached to Earth to be an inertial frame, but strictly it is not.

5–4 Force

- ☐ A **force** is a push or pull exerted by one object on another. It is measured, in principle, by applying the push or pull (alone) to the standard (1 kg) mass and measuring the acceleration of the standard mass. If SI units are used, the magnitudes of these quantities are numerically equal. They are vectors in the same direction. That forces obey the laws of vector addition can be checked by simultaneously applying two forces in different directions and verifying that the result is the same as when the vector sum of the forces is applied as a single force. All measurements must be made using an inertial reference frame.
- ☐ The SI unit of force is the newton and is abbreviated N. In terms of the SI base units, $1\,\text{N} = 1\,\text{kg} \cdot \text{m/s}^2$.

5–5 Mass

☐ The **mass** of an object is measured, in principle, by comparing the accelerations of the object and the standard mass when the same force is applied each to them. In particular, the mass of the object is given by $m = (1\,\text{kg})(a_0/a)$, where a is the magnitude of the acceleration of the object and a_0 is the magnitude of the acceleration of the mass standard. The accelerations must be measured using an inertial frame.

☐ An object with a small mass acquires a greater acceleration than an object with a large mass when the same force is applied to each of them. Mass is said to measure **inertia** or resistance to changes in motion.

☐ Mass is a scalar and is always positive. The mass of two objects in combination is the sum of the individual masses.

5–6 Newton's Second Law

☐ This is the central law of classical mechanics. It gives the relationship between the net force $\vec{F}_{\text{net}}$ on an object and the acceleration $\vec{a}$ of the object:

$$\vec{F}_{\text{net}} = m\vec{a},$$

where m is the mass of the object.

☐ The Newton's second law equation is a vector equation. It is equivalent to the three component equations

$$F_{\text{net }x} = ma_x, \qquad F_{\text{net }y} = ma_y, \qquad \text{and} \qquad F_{\text{net }z} = ma_z.$$

☐ In these equations, $\vec{F}_{\text{net}}$ is the *vector* sum of all the individual forces. This means that in any given situation you must identify all the forces acting on the object and then sum them *vectorially*.

☐ Note that $\vec{F}_{\text{net}} = 0$ implies $\vec{a} = 0$. If the net force vanishes, then the object does not accelerate; its velocity as observed in an inertial reference frame is constant in both magnitude and direction. The resultant force may vanish because there are no forces on the object or because the forces on it sum to zero.

5–7 Some Particular Forces

☐ Every object attracts every other object with a **gravitational force**. In this part of the course you consider only the mutual gravitational attraction of Earth and some object of interest.

☐ The gravitational force Earth on an object is equal to the **weight** of the object and its magnitude is given by $W = mg$, where m is the mass of the object and g is the magnitude of the acceleration due to gravity at the position of the object. Near the surface of Earth the direction of the weight is toward the center of Earth.

☐ Be sure you understand that mass and weight are quite different concepts. Mass is a property of an object and does not change as the object is moved from place to place or even into outer space. It is a scalar. Weight, on the other hand, is a force. It varies as the object moves from place to place and vanishes when the object is far from all other objects, as in outer space. This is because $\vec{g}$, not the mass, varies from place to place.

□ Remember that the weight of an object is $m\vec{g}$ regardless of its acceleration. Weight is a force and, if appropriate, is included in the sum of all forces on the object. This sum equals $m\vec{a}$ and if other forces act, then $\vec{a}$ is different from $\vec{g}$.

□ The SI unit of weight is the newton.

□ A surface may exert a force on an object in contact with it. If the surface is frictionless, that force must be perpendicular to the surface. It is called a **normal force**. Unless adhesive glues the object to the surface a normal force can only push on the object. It must be directed away from the surface toward the interior of the object.

□ If the surface is at rest or moving with constant velocity, the normal force adjusts until the component of the object's acceleration perpendicular to the surface vanishes. We often use this condition to solve for the normal force. Set the sum of the normal components of the forces on the object equal to zero. Since the normal force is one of these the resulting equation can be solved for it in terms of the normal components of the other forces.

□ A surface may also exert a **frictional force** on an object in contact with it. This force is parallel to the surface.

□ If a string with negligible mass connects two objects, it pulls on each with a force of the same magnitude, called the **tension** in the string. You may think of the string as simply transmitting a force from one object to the other; the situation is exactly the same if the objects are in contact and exert forces on each other. Strings pull, not push, along their lengths, so a string serves to define the direction of the force.

5–8 Newton's Third Law

□ Newton's third law tells us something about forces. If the force of object A on object B is $\vec{F}_{BA}$, then according to the third law, the force of object B on object A is given by $\vec{F}_{AB} = -\vec{F}_{BA}$. Compared to the force of A on B, the force of B on A is the same in magnitude and opposite in direction. You should also be aware that these two forces are of the same type. That is, if the force of A on B is gravitational then the force of B on A is also gravitational.

□ The third law is useful in solving problems involving more than one object. If two objects exert forces on each other, we use the same symbol to represent their magnitudes and we remember the forces are in opposite directions when we write the second-law equations. In addition, we remember that the forces act on different objects. When we write Newton's second law for object A, one of the forces we include is the force of B on A, but emphatically *NOT* the force of A on B.

5–9 Applying Newton's Laws

□ A definite procedure has been devised to solve Newton's second law problems. It ensures that you consider only one object at a time, reminds you to include all forces acting on the object you are considering, and guides you in writing Newton's second law in an appropriate form. Follow it closely.

1. Identify the object to be considered. It is usually the object on which the given forces act or about which a question is posed.

2. Represent the object by a dot on a diagram. Do not include the environment of the object since this is replaced by the forces it exerts on the object.

3. On the diagram draw arrows to represent the forces exerted by the environment on the object. Try to draw them in roughly the correct directions. The tail of each arrow should be at the dot. Label each arrow with an algebraic symbol to represent the magnitude of the force, even if a numerical value is given in the problem statement.

The hard part is getting all the forces. If appropriate, don't forget to include the weight of the object, the normal and frictional forces of a surface on the object, and the forces of any strings or rods attached to the object. Carefully go over the sample problems in the text to see how to handle these forces.

Some students erroneously include forces that are not acting on the object. For each force you include, you should be able to point to something in the environment that is exerting the force. This simple procedure should prevent you from erroneously including a normal force, for example, when the object you are considering is not in contact with a surface.

4. Draw a coordinate system on the diagram. In principle, the placement and orientation of the coordinate system do not matter as far as obtaining the correct answer is concerned, but some choices reduce the work involved. If you can guess the direction of the acceleration, place one of the axes along that direction. The acceleration of an object sliding on a surface such as a table top or inclined plane, for example, is parallel to the surface. Once the coordinate system is drawn, label the angle each force makes with a coordinate axis. This will be helpful in writing down the components of the forces later.

The diagram, with all forces shown but without the coordinate system, is called a **free-body diagram** (or a force diagram). We add the coordinate system to help us carry out the next step in the solution of the problem.

5. Write the Newton's second law equation in component form: $F_{\text{net } x} = ma_x$, $F_{\text{net } y} = ma_y$, and, if necessary, $F_{\text{net } z} = ma_z$. The left sides of these equations should contain the appropriate components of the forces you drew on your diagram. You should be able to write the equations by inspection of your diagram. Use algebraic symbols to write them, not numbers; many problems give or ask for force magnitudes so you should usually write each force component as the product of a magnitude and the sine or cosine of an appropriate angle.

6. If more than one object is important, as when two objects are connected by a string, you might carry out the steps above separately for each object. Then, you consider additional conditions. When two objects are connected by a string, for example, the magnitudes of their accelerations might be the same. Be aware of these situations as you study the sample problems of the text.

7. Identify which quantities are known and which are unknown; solve for the unknowns.

Hints for Questions

1 For the block to remain stationary or move with constant velocity the vector sum of the forces must be zero. You can change the magnitudes of the forces but not their orientations.

[Ans: (a) 2 and 4; (b) 2 and 4]

3 To keep the lunch box sliding with constant velocity the vector sum of all forces on it must be zero. Since the gravitational force of Earth and the normal force of the floor are vertical the horizontal component of $\vec{F}_1$ must be equal to the magnitude of $\vec{F}_2$.

[Ans: increase]

5 (a) and (b) To see when the acceleration has an x component, add the x components of the forces. If they sum to zero, $a_x = 0$. Similarly, to see when the acceleration has a y component add the y components of the forces.

(c) If both the x and y components of the total force are positive the acceleration is in the first quadrant; if the x component is negative and the y component is positive the acceleration is in the second quadrant; if both components are negative the acceleration is in the third quadrant; if the x component is positive and the y component is negative the acceleration is in the fourth quadrant.

[Ans: (a) 2, 3, 4; (b) 1, 3, 4; (c) 1: in the positive y direction, 2: in the positive x direction, 3: in the fourth quadrant, 4: in the third quadrant]

7 The magnitude of the acceleration is proportional to the magnitude of the net force. Add the forces vectorially and compare their magnitudes.

[Ans: a, then b, c, and d tie]

9 Since the block is at rest, the vector sum of $\vec{F}$, the gravitational force of Earth, and the normal force of the floor must be zero. If $\vec{F}$ is downward $F + mg - F_N = 0$, where m is the mass of the block and F_N is magnitude of the normal force. If $\vec{F}$ is upward $F + F_N - mg = 0$. In each case tell what happens to F_N is F is increased.

[Ans: (a) increases from mg; (b) decreases from mg to zero]

11 (a), (b), and (c) $\vec{F}$ accelerates all three blocks. $\vec{F}_{21}$ accelerates blocks 2 and 3. $\vec{F}_{32}$ accelerates only block 3.

(d) The blocks move together.

(e) The force is proportional to the mass it accelerates and to the acceleration.

[Ans: (a) 17 kg; (b) 12 kg; (c) 10 kg; (d) all tie; (e) $\vec{F}$, $\vec{F}_{21}$, $\vec{F}_{32}$]

Hints for Problems

3 Use Newton's second law: $\vec{F}_{net} = m\vec{a}$, where $\vec{F}_{net}$ is the vector sum of all the forces on the block, m is the mass of the block, and $\vec{a}$ is its acceleration. Add the forces vectorially and divide by the mass.

[Ans: (a) 0; (b) $(4.0 \, \text{m/s}^2)\hat{\jmath}$; (c) $(3.0 \, \text{m/s}^2)\hat{\imath}$]

15 Use Newton's second law in component form. The x component of the acceleration is the slope of the left-hand graph and the y component is the slope of the right-hand graph. Multiply each of these components by the mass of the package to obtain the components of the force. The magnitude is the square root of the sum of the squares of the components and the tangent of the angle that the force makes with the positive x axis is the y component divided by the x component.

[Ans: (a) 11.7 N; (b) $-59.0°$]

21 Use Newton's second law. The forces on the firefighter are the downward gravitational force of Earth and the upward force of the pole. Their vector sum must equal the product of the firefighter's mass and his acceleration, which is downward. Solve for the force of the pole on the firefighter. Don't forget to calculate his mass from the given value of his weight. According to Newton's third law the force of the firefighter on the pole is equal in magnitude and opposite in direction to the force of the pole on the firefighter.

[Ans: (a) 494 N; (b) up; (c) 494 N; (d) down]

25 Draw a free-body diagram for Tarzan and put in the axes. Remember that the vine pulls, not pushes, on him. The x component of the force of the vine on Tarzan is given by $T\sin\theta$ and the y component is given by $T\cos\theta$, where T is the tension in the vine and θ is the angle the vine makes with the vertical. The net force is the vector sum of the tension force and the gravitational force of Earth on Tarzan (his weight). According to Newton's second law his acceleration is the net force divided by his mass.

[Ans: (a) $(285\,\text{N})\,\hat{\imath}+(705\,\text{N})\,\hat{\jmath}$; (b) $(285\,\text{N})\,\hat{\imath}-(115\,\text{N})\,\hat{\jmath}$; (c) 307 N; (d) $-22.0°$; (e) $3.67\,\text{m/s}^2$; (f) $-22.0°$]

35 Draw a free-body diagram for the elevator cab. The forces on it are the tension force of the cable (upward) and gravitational force of Earth (downward). Solve the Newton's second law equation for the tension. In (a) the acceleration is upward and in (b) it is downward.

[Ans: (a) 31.3 kN; (b) 24.3 kN]

37 Draw a free-body diagram for the bundle. The forces on it are the upward tension force of the cable and the downward gravitational force of Earth. Take the tension to have its maximum value and solve the Newton's second law equation for the acceleration. Use $v^2 = 2ah$ to find the speed v of the bundle when it hits the ground. Here h is the starting height of the bundle above the ground.

[Ans: (a) $1.4\,\text{m/s}^2$; (b) 4.1 m/s]

45 (a) The tension force T_3 accelerates the three blocks, considered to be a single object. Use Newton's second law in the form $T_3 = (m_1 + m_2 + m_3)a$ to find the acceleration a.
(b) The tension force T_1 accelerates only block 1. Use Newton's second law in the form $T_1 = m_1 a$ to find T_1.
(c) The tension force T_2 accelerates blocks 1 and 2, considered to be a single object. Use Newton's second law in the form $T_2 = (m_1 + m_2)a$ to find T_2.

[Ans: (a) $0.970\,\text{m/s}^2$; (b) 11.6 N; (c) 19.1 N]

47 Draw a free-body diagram for each of the blocks. The forces on the left-hand block are the force of gravity $m_1 g$, down, and the tension force of the cord T, up. The forces on the right-hand block are the force of gravity $m_2 g$ and the tension force T, up. Let a_1 be the acceleration of block 1 and a_2 be the acceleration of block 2, then write a Newton's second law equation for each block. Note that the tension force on block 1 has the same magnitude as the tension force on block 2 and that the accelerations are actually vertical components.

The magnitudes of the accelerations are the same since the blocks are connected by the cord. Their signs, however, depend on the coordinate system used. If, for example, you take the upward direction to be positive for both blocks then $a_1 = -a_2$. Substitute $a_1 = a$ and $a_2 = -a$ into the second law equations and solve them simultaneously for a and the cord tension.

$\Big[$Ans: (a) $3.6 \, \text{m/s}^2$; (b) $17 \, \text{N}\Big]$

55 The acceleration of the block at any instant is given by $a_x = F_x/m$, where m is the mass of the block. At time t the velocity of the block is given by

$$v = v_0 + \int_0^{11 \, s} a \, dt = v_0 + \frac{1}{m} \int_0^{11 \, a} F_x \, dt \, ,$$

The value of the integral is just the area under the curve. Don't forget you must subtract the magnitude of the area after $t = 6 \, \text{s}$ from the area before $t = 6 \, \text{s}$. If the result is positive the block is moving in the positive x direction; if it is negative the block is moving in the negative x direction.

$\Big[$Ans: (a) $8.0 \, \text{m/s}$; (b) positive x direction$\Big]$

61 If $F_2 + F_3$ is less than F_1, the least force (and so the least acceleration) occurs if the two forces are both in the negative x direction. In part (c) the sum of the magnitudes is greater than F_1 but the forces have the same magnitude, so they can be oriented so their y components sum to zero and the sum of their x components cancel F_1.

$\Big[$Ans: (a) 0; (b) $0.83 \, \text{m/s}^2$; (c) $0\Big]$

69 To answer (a) calculate what the tension should be to hold the performer stationary and see if it is greater or less than $425 \, \text{N}$. To answer (b) take the tension force to have a magnitude of $425 \, \text{N}$ and use Newton's second law to compute the acceleration of the performer.

$\Big[$Ans: (a) rope breaks; (b) $1.6 \, \text{m/s}^2\Big]$

75 Draw a free-body diagram for each of the boxes. Take the positive x direction to be to the right for box 1 and up the incline for box 2. Write the x component of Newton's second law for each box. The tension force of the cord pulls to the right on box 1 and pulls down the incline on box 2. Use the same symbol for the magnitude of the tension force on each box. Solve the two second law equations simultaneously for the magnitude of the tension force.

$\Big[$Ans: $4.6 \, \text{N}\Big]$

81 The mass is W/g, where W is the weight. The mass is the same no matter where the particle is but the weight depends on the local value of g.

$\Big[$Ans: (a) $11 \, \text{N}$; (b) $2.2 \, \text{kg}$; (c) 0; (d) $2.2 \, \text{kg}\Big]$

85 Consider the person and parachute to be a single object. The forces on it are the downward force of gravity (which is the total weight of the person and parachute) and the upward force of the air. Solve Newton's second law for the force of the air. The forces on the parachute are the downward force of gravity (which is the weight of the parachute alone), the downward force of the person, and the upward force of the air. Solve Newton's second law for the force of the person.

$\Big[$Ans: (a) $620 \, \text{N}$; (b) $580 \, \text{N}\Big]$

87 Since the force is in the positive y direction, the x component of the velocity is constant and the x coordinate at the end of time t is $x = v_{0x}t$. The y component of the acceleration is

$a_y = F/m$, where F is the force of the wind and m is the mass of the armadillo. At the end of time t the y component of the velocity is $v_y = a_y t$ and the y coordinate is $y = \frac{1}{2}a_y t^2$.

[Ans: (a) $(5.0\,\mathrm{m/s})\hat{\imath} + (4.3\,\mathrm{m/s})\hat{\jmath}$; (b) $(15\,\mathrm{m})\hat{\imath} + (6.4\,\mathrm{m})\hat{\jmath}$]

93 Use $v = v_0 + at$ to compute the acceleration a. Here v is the velocity after time t and v_0 is the initial velocity. Use Newton's second law in the scalar form $F = ma$ to compute the force.

97 Use $v^2 = v_0^2 + 2a\,\Delta x$, where v_0 ($= 0$) is the initial speed, v is the speed after the electron travels a distance Δx, and a is its acceleration. Solve for a. The force on the electron is $F = ma$, where m is the electron's mass, and the weight of the electron is mg.

[Ans: (a) $1.1 \times 10^{-15}\,\mathrm{N}$; (b) $8.9 \times 10^{-30}\,\mathrm{N}$]

Quiz

Some questions might have more than one correct answer.

1. Which of the following statements are true?

   A. an inertial frame cannot be accelerating relative to us
   B. a reference frame that is rotating relative to any other frame cannot be an inertial frame
   C. if a particle is not accelerating relative to a noninertial frame then the net force on the particle is not zero
   D. two inertial frames might be accelerating relative to each other
   E. the acceleration of a particle is different for different inertial frames

2. Object A has twice the mass of object B. For A to have the same acceleration as B the net force on it must be

   A. more than twice as great as the net force on B
   B. twice as great as the net force on B
   C. the same as the net force on B
   D. half as great as the net force on B
   E. less than half as great as the net force on B

3. The masses of four particles and the magnitudes of the net forces on them are:

   I. $m = 2\,\mathrm{kg}$, $F_{\mathrm{net}} = 6\,\mathrm{N}$
   II. $m = 3\,\mathrm{kg}$, $F_{\mathrm{net}} = 4\,\mathrm{N}$
   III. $m = 6\,\mathrm{kg}$, $F_{\mathrm{net}} = 3\,\mathrm{N}$
   IV. $m = 10\,\mathrm{kg}$, $F_{\mathrm{net}} = 2\,\mathrm{N}$

   Rank the magnitudes of their accelerations, least to greatest.

   A. I, III, IV, II
   B. I and II tied, III, IV
   C. IV, III, II, I
   D. III, II, IV, I
   E. I, II, III, IV

4. A block is sliding down an incline and you want to calculate its acceleration. Which of the following forces should you include in its free-body diagram?

   A. the gravitational force of Earth on the block
   B. the gravitational force of Earth on the incline
   C. the normal force of the incline on the block
   D. the normal force of the block on the incline
   E. the frictional force of the incline on the block

5. You are throwing a ball into the air and you want to calculate its acceleration while it is in your hand. Of the following forces, which should you include in the free-body diagram for the ball?

   A. the gravitational force of Earth on the ball
   B. the gravitational force of Earth on your hand
   C. the force of your hand on the ball
   D. the force of the ball on your hand
   E. the frictional force of the air on your hand

6. A man uses a rope to pull a child on a sled across a snowy field. To calculate the acceleration of man, child, and sled, taken together as an object which of the following forces should you include?

   A. the normal forces of the snow on the sled and man
   B. the frictional forces of the snow on the sled and man
   C. the force of the man on the rope
   D. the gravitational forces of Earth on the sled, man, and child
   E. the normal force of the snow on the child

7. A block is sliding with constant acceleration down an incline. Which of the following statements are true about the forces on the block?

   A. the gravitational and normal forces sum to zero
   B. the gravitational and frictional forces sum to zero
   C. the frictional and normal forces sum to zero
   D. the normal force is zero
   E. none of the above statements are true

8. Three forces of equal magnitude are applied to a particle. One is in the positive x direction, one is in the positive y direction, and one is in the negative y direction. The acceleration of the particle is

   A. zero
   B. in the positive y direction
   C. in the negative y direction
   D. in the positive x direction
   E. not parallel to either the x or y axis

9. Particle A, with a small mass, caroms off particle B, with a larger mass. Only the forces of the particles on each other are significant. During the time the blocks are in contact
   A. the acceleration of A is greater than the acceleration of B
   B. the acceleration of A is less than the acceleration of
   C. the acceleration of A is the same as the acceleration of B
   D. the force of particle A on B is less than the force of B on particle A
   E. the force of particle A on B is greater than the force of B on particle A

10. A block is sliding down an incline. Pairs of forces that are equal in magnitude and opposite in direction by virtue of Newton's third law are
    A. the normal force of the incline on the block and the gravitational force of Earth on the block
    B. the normal force of the incline on the block and the normal force of the block on the incline
    C. the gravitation force of Earth on the block and the normal force of the incline on the block
    D. the gravitational force of Earth on the block and the gravitational force of the block on Earth
    E. the frictional force of the incline on the block and the frictional force of the block on the incline

11. You hold a book motionless in your hand. Pairs of forces that are equal in magnitude and opposite in direction by virtue of Newton's third law are
    A. the gravitational force of Earth on the book and the gravitational force of the book on Earth
    B. the gravitational force of Earth on the book and the force of your hand on the book
    C. the force of your hand on the book and the force of the book on your hand
    D. the gravitational force of your hand on Earth and the force of the book on your hand
    E. the gravitation force of Earth on your hand and the force of your hand on the book

Answers: (1) B, C; (2) B; (3) C; (4) A, C, D; (5) A, C, E; (6) A, B, D; (7) E; (8) D; (9) A; (10) B, D, E; (11) A, C, E

Chapter 6
FORCE AND MOTION —— II

This chapter contains a great many applications of Newton's laws, with special emphasis on frictional and centripetal forces. Here's where your understanding of the fundamentals begins to pay off!

Important Concepts

- ☐ static friction
- ☐ coefficient of static friction
- ☐ kinetic friction
- ☐ coefficient of kinetic friction

- ☐ drag force
- ☐ drag coefficient
- ☐ terminal speed
- ☐ centripetal force

Overview

6–2 Friction

☐ A surface may exert a **frictional force** on an object that is in contact with it. Friction is unavoidable when the object is sliding on the surface, although lubricants and air films may make it small. Even when the object and surface are stationary with respect to each other, they exert frictional forces if other forces present would otherwise cause them to slide.

☐ A frictional force arises because microscopic welds form between protrusions from the two surfaces in contact. The macroscopic force of friction is actually the vector sum of a great many microscopic forces.

☐ When the two objects in contact are not moving relative to each other, the frictional force is labeled **static friction**. When the surfaces are moving relative to each other, the force is labeled **kinetic friction**.

☐ All frictional forces, whether static or kinetic, are parallel to the surfaces in contact.

6–3 Properties of Friction

☐ If an object is at rest on a motionless surface, the static frictional force is determined mathematically, via Newton's second law, by the condition that the component of acceleration parallel to the surface is zero. This condition is analogous to the condition used to determine the normal force. The difference is that the normal force of one object on another is perpendicular to the surface of contact while the frictional force is parallel to it.

☐ The magnitude f_s of the force of static friction exerted by one surface on another must be less than a certain value, determined by the nature of the surfaces and by the magnitude of the normal force one surface exerts on the other. In particular, $f_s \leq \mu_s F_N$, where μ_s is the

coefficient of static friction and F_N is the magnitude of the normal force. If the force of friction required to hold the surfaces at rest with respect to each other is greater than the maximum allowed, then the surfaces slide over each other and the frictional force is kinetic rather than static.

☐ The magnitude of the force of kinetic friction is given by $f = \mu_k F_N$, where μ_k is the **coefficient of kinetic friction**. If the surface on which an object rests is motionless, the force of kinetic friction is opposite the velocity of the object.

☐ The normal force that appears in the expressions for the force of kinetic friction and the maximum force of static friction must be computed for each situation using Newton's second law. As you know by now the magnitude of the normal force depends on the directions and magnitudes of other forces acting.

☐ To solve problems involving frictional forces proceed as before: draw the free-body diagram and write down Newton's second law in component form, just as for any other problem. Use an algebraic symbol, f say, for the frictional force. You must now decide if the frictional force is static or kinetic. If static friction is involved, f is probably an unknown but the acceleration is known or is related to other known quantities in the problem. If the object is at rest on a stationary surface, its acceleration is zero. If it is at rest relative to an accelerating surface, its acceleration is the same as that of the surface. Kinetic friction is involved if one surface is sliding on the other. Then, the magnitude of the frictional force is given by $\mu_k F_N$.

☐ If you do not know that the object is at rest relative to the surface, assume it is and use Newton's second law, with the acceleration of the object equal to the acceleration of the surface, to calculate both the force of static friction f_{rest} that will hold it at rest and the normal force F_N. Compare f_{rest} with $\mu_s F_N$. If $f_{\text{rest}} < \mu_s F_N$, the object remains at rest relative to the surface and the force of friction has the value you computed. That is, $f = f_{\text{rest}}$. If $f_{\text{rest}} > \mu_s F_N$, then the object moves relative to the surface. Go back to the second law equation and set $f = \mu_k F_N$, then solve for the acceleration.

6–4 The Drag Force and Terminal Speed

☐ When an object moves in a fluid, such as air, the fluid exerts a **drag force** on it. When the relative speed of the object and fluid is so great that the fluid flow around the object is turbulent, the force of the fluid on the object is proportional to the square of the relative speed. In fact, the magnitude of the drag force is given by: $D = \frac{1}{2}C\rho A v^2$, where v is the relative speed of the object, A is the effective cross-sectional area of the object, ρ is the mass density of the fluid, and C is a drag coefficient. The direction of the drag force is opposite the direction of the relative velocity of the object.

☐ When an object falls in a fluid, it approaches a constant speed, called the **terminal speed**. You can use Newton's second law to find a value for this speed. The force of gravity is mg, down, and the drag force is $\frac{1}{2}C\rho A v^2$, up. At terminal speed these sum to zero. Thus,

$$v_t = \sqrt{\frac{2mg}{C\rho A}}.$$

The larger the combination $C\rho A$ the smaller the terminal speed and the shorter the time taken to reach that speed from rest.

☐ When the object is dropped from rest its acceleration is g at first but as it picks up speed the drag force increases, thereby reducing the acceleration. The object continues to gain speed but at a lesser rate. At terminal speed its acceleration vanishes. From then on the acceleration remains zero, the speed does not change, and the drag force remains constant.

6–5 Uniform Circular Motion

☐ An object in uniform circular motion has a non-zero acceleration because the direction of its velocity changes with time. A force must be applied to the object in order to produce its acceleration. If m is the mass of the object, then the applied force must have magnitude $F = mv^2/r$, where r is the radius of the orbit and v is the speed of the object. The force is directed toward the center of the circular orbit and, because of its direction, is called a **centripetal force**.

☐ Acquire the habit of pointing out to yourself the object in the environment that exerts the force. It might be a string force, a gravitational force, or a frictional force, for example.

☐ Uniform circular motion problems are solved in much the same way as any other second law problem. Draw a free-body diagram. Place the coordinate system so one of the axes is in the direction of the acceleration, pointing from the object toward the center of its path. For most problems, you will want to substitute v^2/r for the magnitude of the acceleration. Here v is the speed of the object and r is the radius of its orbit.

Hints for Questions

<u>1</u> (a) Only two horizontal forces act: the frictional force and the applied force. Since the block remains stationary you know that the vector sum of all forces is zero.
(b) The maximum magnitude of the frictional force is the product of the coefficient of static friction and the magnitude of the normal force. Because the normal force is vertical it does not depend on the applied force.
[Ans: (a) $\vec{F}_1$, $\vec{F}_2$, $\vec{F}_3$; (b) all tie]

<u>3</u> (a) and (b) Draw a free-body diagram for the crate. The normal force must be perpendicular to the wall and the frictional force must be parallel to the wall.
(c) and (d) The frictional and normal forces adjust so that the net force on the crate is zero.
(d) The maximum magnitude of the friction force is equal to the product of the coefficient of static friction and the magnitude of the normal force.
[Ans: (a) upward; (b) horizontally away from the wall; (c) no change; (d) increases; (e) increases]

<u>5</u> (a) F_x is the horizontal component of $\vec{F}$ and is given by $F_x = F\cos\theta$. F does not change but θ increases.
(b) The frictional force adjusts so the horizontal component of the net force is zero. Thus $f_s = F_x$.
(c) The normal force adjusts so that the vertical component of the net force is zero.

(d) The maximum magnitude of the frictional force is equal tot he product of the coefficient of static friction and the magnitude of the normal force.

(e) Again use $f_s = F_x$ with $F_x = F \cos\theta$. Now F rather than θ increases.

[Ans: (a) decreases; (b) decreases; (c) decreases; (d) decreases; (e) decreases]

7 Draw a free-body diagram for the block. The forces on it are the force of gravity, the normal force of the ramp surface, and the frictional force of the ramp surface. Look at the component of the net force along the ramp. When F is small the frictional force adjusts so this component of the net force is zero. When F becomes slightly larger than the value that causes the magnitude of the frictional force to have its maximum possible value, the block begins to slide.

[Ans: At first $\vec{f}_s$ is directed up the ramp and its magnitude increases from $mg\sin\theta$ until it reaches $f_{s\,max}$. Thereafter the force is one of kinetic friction directed up the ramp and has magnitude f_k, a constant value less than $f_{s\,max}$.]

9 If the coefficient of static friction between the block and slab is large the block accelerates with the slab and you can consider the two to be a single object. If the coefficient of static friction is zero the block remains stationary and the slab accelerates from under it.

[Ans: (a) $5\,\mathrm{m/s^2}$ to $10\,\mathrm{m/s^2}$; (b) 0 to $5\,\mathrm{m/s^2}$]

11 (a) and (b) The magnitude of the centripetal acceleration depends only on the speed of the person and the radius of his path. The magnitude of the centripetal force is, of course, the product of the person's mass and the magnitude of the centripetal acceleration.

(c) The centripetal force is the vector sum of the force of gravity and the normal force. Draw a free-body diagram for the person at each of the positions.

[Ans: (a) all tie; (b) all tie; (c) 2, 3, 1]

Hints for Problems

1 The pan is an inclined plane that makes the (unknown) angle θ with the horizontal. Draw a free-body diagram for an egg. The force of gravity, with magnitude mg, is vertically downward, the frictional force is up the plane, and the normal force is perpendicular to the plane. Here m is the mass of the egg. Take the x axis to along the plane and the y axis to be perpendicular to the plane. Write the Newton's second law equation in component form. Suppose the egg is on the verge of sliding and write $f_s = \mu_s F_N$ for the magnitude of the frictional force, where μ_s is the coefficient of static friction and F_N is the magnitude of the normal force. Use one of the equations to eliminate F_N from the other and solve for $\tan\theta$, then θ itself.

[Ans: $2°$]

7 Draw a free-body diagram for the block. The forces on it are the applied force, the force of gravity, the normal force of the floor, and the frictional force of the floor. Write the Newton's second law equation in component form, taking the x axis to be horizontal and the y axis to be vertical. Solve one of the equations for the magnitude F_N of the normal force, then use $f_k = \mu_k F_N$ to calculate the magnitude of the frictional force. Here μ_k is the coefficient of

kinetic friction between the block and the floor. Once f_k is known you can solve the other second law equation for the acceleration.

$\left[\text{Ans: (a) } 11\,\text{N; (b) } 0.14\,\text{m/s}^2\,\right]$

9 You need to know the acceleration a of the book while it is being pushed. Use $v^2 = 2ax$, where v is the final speed of the book and x is the distance it travels while being pushed. Divide by the mass of the book to find the net force on it. Subtract the applied force to find the force of friction. Also, use Newton's second law to find the magnitude of the normal force of the floor on the book. Divide the magnitude of the frictional force by the magnitude of the normal force to find the coefficient of kinetic friction.

$\left[\text{Ans: } 0.58\,\right]$

17 Draw a free-body diagram for the box of sand. The forces on it are the force of gravity, the frictional force of the floor, the normal force of the floor, and the tension force of the cable, at some angle θ above the horizontal. Write the Newton's second law equation in component form, with the x axis horizontal and the y axis vertical. Take the acceleration to be zero and the crate to be on the verge of sliding, so the magnitude of the frictional force is equal to the product of the normal force magnitude and the coefficient of static friction. Use two of the equations to eliminate the normal and frictional forces from the third and solve for the mass. If the mass is to be a maximum its derivative with respect to the angle θ should be zero. Use this condition to find the angle, then use the value of the angle to calculate the mass. Multiply by g to obtain the weight.

$\left[\text{Ans: (a) } 19°; \text{ (b) } 3.3\,\text{kN}\,\right]$

19 In each case you must decide if the block moves or not. If it moves the frictional force is kinetic in nature; if it does not move the frictional force is static in nature. Assume the block does not move and find the frictional force that is needed to hold it stationary. Draw a free-body diagram for the block. The forces on it are the gravitational force, the normal force of the plane, the frictional force of the plane, and the applied force $\vec{P}$. Write the Newton's second law equation in component form, with the acceleration equal to zero. Calculate the frictional and normal forces and compare the magnitude of the frictional force with the product of the coefficient of static friction and the magnitude of the normal force. If it is less the block does not move and the frictional force you computed is the actual frictional force. If the frictional force is greater the block does move and the magnitude of the frictional force is the product of the coefficient of kinetic friction and the magnitude of the normal force. Go back and resolve the second aw equations, which should now include the kinetic frictional force.

$\left[\text{Ans: (a) } (17\,\text{N})\,\hat{\text{i}}; \text{ (b) } (20\,\text{N})\,\hat{\text{i}}; \text{ (c) } (15\,\text{N})\,\hat{\text{i}}\,\right]$

23 Let T_L be the tension in the left-hand cord. Newton's second law for the left-hand block gives $T_L = M(g + a)$, where a is the acceleration of the blocks. Let T_R be the tension in the right-hand cord. Newton's second law for the right-hand block gives $T_R = 2M(g - 1)$. Now draw a free-body diagram for the middle block. A force of magnitude T_L pulls to the left and a force of magnitude T_R pulls to the right. In addition, the force of gravity pulls downward, the normal force of the table pushes upward, and the frictional force of the table pulls to the left on the block. Write the Newton's second law equation in component form and set the magnitude of the frictional force equal to the product of the coefficient of kinetic friction and the magnitude of the normal force. Solve for the coefficient.

[Ans: 0.37]

29 If the smaller block does not slide the frictional force of the larger block on it must have magnitude mg and this must be less than the product of the coefficient of static friction between the blocks and the magnitude of the normal force of the blocks on each other. Write the horizontal component of the Newton's second law equation for the smaller block. The horizontal forces on it are the applied force and the normal force of the larger block. Do the same for the larger block. The only horizontal force on it is the normal force of the smaller block. According to Newton's third law the two normal forces have the same magnitude. These equations can be solved for the magnitude of the normal force and then the value of the maximum static frictional force can be computed.

[Ans: $4.9 \times 10^2\,\text{N}$]

37 To round the curve without sliding the frictional force must be equal to mv^2/r, where m is the mass of the bicycle and rider, v is their speed, and r is the radius of the curve, and it must be less than $\mu_s F_N$, where μ_s is the coefficient of kinetic friction between the tires and the road and F_N is the normal force of the road on the tires.

[Ans: 21 m]

39 If the cat goes around without sliding the frictional force on it must be equal to mv^2/r, where m is the mass of the cat, v is its speed, and r is the distance from the center of the merry-go-round to the cat, and it must be less than $\mu_s F_N$, where μ_s is the coefficient of static friction between the cat and the merry-go-round floor and F_N is the normal force of the floor on the cat. The cat goes a distance of $2\pi r$ every 6.0 s, so you can easily compute its speed.

[Ans: 0.60]

43 (a) The period of the motion is the time to go around exactly once. That is, he travels a distance of $2\pi r$, where r is the radius of the path, in a time interval equal to one period. Since you know his speed you can calculate the period.

(b) At the highest point the net force on him is downward and has a magnitude of $mg - F_N$, where m is the mass of the addict and F_N is the magnitude of the normal force. According to Newton's second law this must be equal to mv^2/r, where v is the speed of the addict. Solve for F_N.

(c) At the lowest point the net force is upward and has a magnitude of $F_N - mg$, but again this must equal mv^2/r. Solve for F_N.

[Ans: (a) 10 s; (b) $4.9 \times 10^2\,\text{N}$; (c) $1.1 \times 10^3\,\text{N}$]

57 Draw a free-body diagram for the block. The forces on it are the gravitational force of Earth, the frictional force of the plane, and the normal force of the plane. When the block is sliding down the plane the frictional force is up the plane and it is sliding up the plane the frictional force is down the plane. Take the x axis to be parallel to the plane with the positive direction down the plane and the y axis to be perpendicular to the plane. Write Newton's second law in component form. Substitute $\mu_k F_N$, where μ_k is the coefficient of kinetic friction and F_N

is the normal force, for the magnitude of the frictional force in the x component of the second law. Solve the y component of the second law equation for F_N and substitute for it in the x component equation. solve for the acceleration of the block. The sign of the acceleration tells you its direction.

$\left[\text{Ans: (a) } 7.5\,\text{m/s}^2; \text{(b) down; (c) } 9.5\,\text{m/s}^2; \text{(d) down }\right]$

59 Draw a free-body diagram for the mop head. The forces on it are the force $\vec{F}$ of the man, the frictional force of the floor, the normal force of the floor, and the gravitational force of Earth. Take the x axis to be horizontal and the y axis to be vertical and write Newton's second law for the mop head in component form. In (a) the magnitude of the frictional force is $\mu_k F_N$, where μ_k is the coefficient of kinetic friction and F_N is the normal force. Solve the y component of the second law for the normal force. After substituting into the x component equation and setting the acceleration equal to zero, solve for F.
(b) Assume the mop head is stationary and use f for the magnitude of the frictional force (not a coefficient of friction times the normal force). Solve the x component of the second law for f. If f is less than $\mu_s F_N$ the mop does not move.

$\left[\text{Ans: (a) } \mu_s mg/(\sin\theta - \mu_s \cos\theta); \text{(b) } \theta_0 = \tan^{-1}\mu_s \right]$

65 Draw a free-body diagram for the box. The forces on it are the gravitational force of Earth, the normal force of the belt, and the frictional force of the belt. Take the x axis to be parallel to the belt and the y axis to be perpendicular to it. Write Newton's second law in component form. Solve the x component equation for the frictional force. The sign tells you the direction.

$\left[\text{Ans: (a) } 3.0\,\text{N; (b) } 3.0\,\text{N; (c) } 1.6\,\text{N; (d) } 4.4\,\text{N; (e) } 1.0\,\text{N; (f) e }\right]$

75 Draw a free-body diagram for the child. The forces on him are the gravitational force of Earth, the normal force of the ground and the frictional force of the ground. Take the x axis to be parallel to the ground and the y axis to be perpendicular to it. Write Newton's second law in component form. Solve the y component equation for an expression that gives the normal force, then multiply by the coefficient of kinetic friction μ_k to obtain the frictional force and substitute the resulting expression into the x component equation. Solve for μ_s.

$\left[\text{Ans: } 0.76 \right]$

79 Draw a free-body diagram for each of the blocks. The forces on block 1 are the gravitational force of earth and the tension force of the cord. The forces on block 2 are the gravitational force of Earth, the tension forces of the two cords, the normal force of the table, and the frictional force of the table. The forces on block 3 are the gravitational force of Earth and the tension force of the cord. The tension is different in the two cords, so different symbols must be used. Take the positive direction to be upward for block 1, to the right for block 2, and downward for block 3. Write a Newton's second law equation for each block. With the choice of positive directions just given the same symbol can be used for each of the three accelerations. You will not need the vertical component of the second law for block 2. Use the equations for blocks 1 and 3 to eliminate the tension force from the equation for block 2, then solve for the frictional force.

$\left[\text{Ans: } 4.6\,\text{N} \right]$

83 (a) and (b) The skier starts from rest. Take the acceleration to be constant, so the distance traveled in time t is given by $\Delta x = \frac{1}{2}at^2$, where a is the acceleration. Solve for a.

(c) and (d) Draw a free-body diagram for the skier. The forces are the gravitational force of gravity, the normal force of the slope, and the frictional force of the snow. Take the x axis to be parallel to the slope and the y axis to be perpendicular to it. Write Newton's second law in component form. Solve the x component equation to obtain an expression for the magnitude f of the frictional force and the y component equation to obtain an expression for the magnitude F_N of the normal force. The coefficient of kinetic friction is given by $\mu_k = f/F_N$. The expressions for the forces contain the mass of the skier but this will cancel from the ratio.

$\Big[$Ans: (a) $0.11\,\mathrm{m/s^2}$; (b) $0.23\,\mathrm{m/s^2}$; (c) 0.041; (d) 0.029% $\Big]$

93 (a) – (f) Draw a free-body diagram for the chair and write Newton's second law in component form, with the x axis horizontal and the y axis vertical. If F is the magnitude of the applied force then the horizontal component is $F_h = F\cos\theta$. Use the vertical component of second law to find the magnitude of the normal force.

(g), (h), and (i) Assume the chair is at rest, so its acceleration is zero. Use the horizontal component of the second law to find the magnitude f of the frictional force of the floor on the chair. If f is less than $\mu_s F_N$, where μ_s is the coefficient of static friction and F_N is the magnitude of the normal force, then the chair does not move. Otherwise it does.

$\Big[$Ans: (a) $100\,\mathrm{N}$; (b) $245\,\mathrm{N}$; (c) $86.6\,\mathrm{N}$; (d) $195\,\mathrm{N}$; (e) $50.0\,\mathrm{N}$; (f) $158\,\mathrm{N}$; (g) at rest; (h) slides; (i) at rest $\Big]$

101 The seat exerts an upward normal force that is equal in magnitude to the weight of the passenger and a horizontal frictional force that gets the passenger around the curve. This force has magnitude v^2/R. The two forces are perpendicular to each other, so the magnitude of the net force is the square root of the sum of the squares of their magnitudes.

$\Big[$Ans: $874\,\mathrm{N}$ $\Big]$

107 Assume the speed of the stone is essentially constant over a revolution. The forces on the stone are the gravitational force of Earth and the tension force of the string. These add vectorially to produce a net force that is radially inward. Note that the forces are in the same direction when the stone is at the top of its circular path and are in opposite directions when the stone is at the bottom. According to Newton's second law the magnitude of the net force is equal to v^2/r, where v is the speed of the stone and r is the radius of the path.

$\Big[$Ans: (a) bottom of circle; (b) $9.5\,\mathrm{m/s}$ $\Big]$

Quiz

Some questions might have more than one correct answer.

1. If N is the magnitude of the normal force and μ_s is the coefficient of static friction for two surfaces in contact, the magnitude of the static frictional force is given by $\mu_s N$

   A. always
   B. whenever sliding does not occur
   C. whenever sliding does occur
   D. whenever the surfaces are on the verge of sliding
   E. never

2. A block rests on an incline that makes the angle θ with the horizontal. As θ is slowly increased, the magnitudes of the normal force and the static frictional force of the incline on the block

   A. both increase
   B. both decrease
   C. increase and decrease, respectively
   D. decrease and increase, respectively
   E. both remain the same

3. If you are trying to get a heavy crate to start sliding on a floor you can do so with the smallest force if you

   A. push horizontally
   B. push horizontally and slightly upward
   C. push horizontally and slightly downward
   D. push vertically
   E. pull horizontally

4. You hold a picture against a wall by pushing horizontally on it. The frictional force is

   A. vertically upward
   B. vertically downward
   C. horizontally to your right
   D. horizontally to your left
   E. at some angle between the horizontal and vertical

5. A block is placed on a uniform incline and the angle between the incline and the horizontal is increased until the block begins to slide. The block then

   A. slides a short distance and stops
   B. slides down the incline with constant speed
   C. slides down the incline with decreasing speed
   D. slides down the incline with increasing speed

6. When a car rounds a curve in a level road the centripetal force is provided by

    A. the normal force of the road on the tires
    B. the gravitational force of Earth on the car
    C. the force of the engine on the drive shaft
    D. the force of the brakes on the wheels
    E. the frictional force of the road on the tires

7. A girl rides on a Ferris wheel that revolves at a uniform rate. The force of the seat on the girl is least when she is

    A. at the top
    B. halfway down and descending
    C. at the bottom
    D. halfway up and ascending
    E. between halfway up and the top

8. As the Moon orbits Earth the centripetal force is

    A. the normal force of Earth on the Moon
    B. the gravitational force of the Sun on the Moon
    C. a tension force
    D. the gravitational force of Earth on the Moon
    E. the gravitational force of the Moon on Earth

9. A car of mass m, traveling with speed v, rounds a level curve of radius r without sliding. The coefficient of static friction between the tires and the road is μ_s and the coefficient of kinetic friction is μ_k. The magnitude of the total frictional force on the tires must be

    A. at least $\mu_s mg$
    B. equal to $\mu_s mg$
    C. equal to $\mu_s v^2/r$
    D. equal to mv^2/r
    E. equal to $\mu_k mg$

Answers: (1)D; (2) D; (3) B; (4) A; (5) D; (6) E; (7) A; (8) D; (9) D

Chapter 7
KINETIC ENERGY AND WORK

The central concept of this chapter is the idea of work. You should understand its definition, you should learn how to calculate the work done by forces in various situations, and you should learn how work is related to the change in the kinetic energy of a particle (the work-kinetic energy theorem).

Important Concepts

☐ kinetic energy

☐ work-kinetic energy theorem

☐ work

☐ work done by gravity

☐ force of an ideal spring

☐ spring constant

☐ work done by an ideal spring

☐ power

Overview

7–3   Kinetic Energy

☐ The **kinetic energy** of a particle with mass m and speed v is defined by $K = \frac{1}{2}mv^2$. For a particle moving in the xy plane, $K = \frac{1}{2}m(v_x^2 + v_y^2)$, where v_x and v_y are its velocity components. The two terms of the expression are *not* components of a vector. Kinetic energy is a scalar and so does not have a direction associated with it.

☐ The SI unit of kinetic energy is the joule (abbreviated J). In terms of SI base units, $1\,\mathrm{J} = 1\,\mathrm{kg \cdot m^2/s^2}$. A unit of energy called an electron volt is often used when dealing with atomic phenomena. In terms of the SI unit, $1\,\mathrm{eV}$ is $1.60 \times 10^{-19}\,\mathrm{J}$.

7–4   Work

☐ **Work** is energy that is being transferred to or from an object by a force on the object. It might be positive, indicating that the energy of the object increased, or negative, indicating that the energy of the object is decreasing.

7–5   Work and Kinetic Energy

☐ The **work-kinetic energy theorem** is: during any interval the *net* work done by *all* forces on a particle equals the change in the particle's kinetic energy. Let W be the net work done on a particle during some interval. If K_i is the initial kinetic energy and K_f is the final kinetic energy for that interval, then $W = K_f - K_i$. Be sure you perform the subtraction in the correct order.

- [ ] If the net work is negative, then the kinetic energy and speed of the particle both decrease; if it is positive, then the kinetic energy and speed both increase; if it is zero, then the kinetic energy and speed do not change.

- [ ] The work-kinetic energy theorem is a direct result of Newton's second law. The derivation is given in Section 7–8 of the text.

- [ ] The work-kinetic energy theorem is valid only for a single particle or an object that can be treated as a particle. If the object becomes distorted or its orientation changes, then the theorem cannot be applied directly to the object as a whole.

- [ ] If a constant force $\vec{F}$ acts on a particle as it moves through a displacement $\vec{d}$, the force does work $W = \vec{F} \cdot \vec{d} = Fd \cos \phi$, where ϕ is the angle between the force and the displacement when they are drawn with their tails at the same point. If the particle has displacement Δx, along the x axis, and the force has a constant x component F_x, then the expression for the work can be written $W = F_x \Delta x$.

- [ ] Work is a scalar. It does not have a direction associated with it. If several forces act on the particle, the net work done by all the forces is the algebraic sum of the individual works.

- [ ] Work is positive if the angle ϕ is less than 90°; it is negative if ϕ is greater than 90°. A force that is perpendicular to the displacement ($\phi = 0$) does no work.

- [ ] The SI unit of work is the same as that for energy, the joule.

7–6 Work Done by the Gravitational Force

- [ ] As an object of mass m moves from height y_i to height y_f near the Earth's surface, gravity does work $W = -mg(y_f - y_i)$. This expression is valid if $y_f > y_i$ (the object rises during the interval) or if $y_f < y_i$ (the object is falling during the interval). On the way down the sign of the work done by gravity is positive while on the way up it is negative.

- [ ] The work done by gravity is the same no matter what path is taken between the initial and final points. In fact, all that counts is the initial and final altitudes. The two positions need not have the same horizontal coordinate.

- [ ] The expression for the work done by gravity is valid even if the object experiences air resistance. If air resistance is present, this expression does not, of course, give the net work done by all forces.

7–7 Work Done by a Spring Force

- [ ] The force exerted by an ideal spring on an object is a variable force. Its direction depends on whether the spring is extended or compressed and its magnitude depends on the amount of extension or compression.

- [ ] Assume the spring is along the x axis with one end fixed and the other attached to a block, as shown to the right. When the block is at $x = 0$, the spring is neither extended or compressed. This is the equilibrium configuration. When the block is at any coordinate x, the force exerted by the spring is given by $F = -kx$. Here k is the **spring constant**, a property of the spring: the stiffer the spring, the greater the spring constant.

- [ ] When the spring is extended, the force it exerts is negative; when the spring is compressed the force is positive. In either case, the force tends to push the block toward the equilibrium point ($x = 0$). The force is called a *restoring force*.

- [ ] The spring constant of an ideal spring can be measured by applying a force F to the block and measuring the elongation x of the spring with the block at rest. In terms of F, x, and k, the net force on the block is given by $F - kx$ and since the block is at rest this must be zero. Thus, $k = F/x$. The SI unit of a spring constant is a newton per meter.

- [ ] As the block moves from an initial coordinate x_i to a final coordinate x_f the work done by the spring is

$$W = \int_{x_i}^{x_f} -kx\, dx = -\tfrac{1}{2}k\left(x_f^2 - x_i^2\right).$$

- [ ] The spring does positive work whenever the block moves toward the equilibrium point from either side and does negative work whenever the block moves away from the equilibrium point.

7–8 Work Done by a General Variable Force

- [ ] If a particle is subjected to a variable force $\vec{F}$, the work done by the force on the particle as it moves from $\vec{r}_i$ to $\vec{r}_f$ is given by the integral

$$W = \int_{\vec{r}_i}^{\vec{r}_f} \vec{F} \cdot d\vec{r}.$$

Here $d\vec{r}$ is an infinitesimal displacement along the path of the particle. This expression is the general definition of work. To evaluate the integral, the force must be known as a function of the particle position.

- [ ] If the particle is moving on the x axis subject, the work done by the force is

$$W = \int_{x_i}^{x_f} F_x\, dx.$$

This can be interpreted as the area under the graph of F_x as a function of the particle coordinate.

7–9 Power

- [ ] The **power** associated with a force is the rate with which it does work. Let the function $W(t)$ represent the work done by a force from time 0 to time t. Then, the instantaneous power delivered by the force is given by

$$P = \frac{dW}{dt}.$$

- [ ] The SI unit of power is the watt (abbreviated W): $1\,\text{W} = 1\,\text{J/s}$.

- [ ] If a particle is moving with velocity $\vec{v}$, then the power delivered to it by the force $\vec{F}$ is given by $P = \vec{F} \cdot \vec{v}$.

Hints for Questions

1 The kinetic energy of a particle is proportional to the square of the magnitude of its velocity. This means it is proportional to the sum of the squares of the velocity components.

[Ans: all tie]

3 According to the work-kinetic energy theorem the work done on the puck is equal to the change in its kinetic energy. This means it is proportional to the change in the square of the speed. The directions of the initial and final velocities are not relevant.

[Ans: c, b, a]

5 The work done by the gravitational force is mgh, where h is the vertical distance traversed by the pig.

[Ans: all tie]

7 (a) The greater the magnitude of the force, the greater the magnitude of the acceleration and the greater the slope of the velocity versus time graph. Note that two of the forces are in the negative x direction and produce negative accelerations while two are in the positive x direction and produce positive accelerations.

(b) The greater the magnitude of the acceleration, the greater the final kinetic energy. Note that kinetic energy cannot be negative.

[Ans: (a) A, $\vec{F}_2$; B, $\vec{F}_1$; C, $\vec{F}_3$; D, $\vec{F}_4$; (b) E, A and D; F, B and C; G and H are meaningless since kinetic energy cannot be negative]

9 If the glob is dropped its kinetic energy increases in proportion to the square of the time. If it is thrown downward it has an initial kinetic energy and its kinetic energy increases. If it is thrown upward its kinetic energy decreases from its initial value to zero and then increases. The kinetic energy cannot increase and then decrease, it cannot decrease over the whole trip, and it cannot be negative.

[Ans: e through h]

Hints for Problems

7 The force is constant so the work it does is given by $W = \vec{F} \cdot \vec{d}$, where $\vec{F}$ is the force and $\vec{d}$ is the displacement of the coin. Use $\vec{F} \cdot \vec{d} = F_x \, \Delta x + F_y \, \Delta y$ to evaluate the scalar product. Here $F_x = F \cos \theta$ and $F_y = F \sin \theta$, where θ is the angle the force makes with the positive x direction.

[Ans: 6.8 J]

9 Since the force is constant and in the same direction as the displacement, the work it does is given by $W = F \, \Delta x$, where F is the magnitude of the force and Δx is the displacement. Use Newton's second law to obtain $W = ma \, \Delta x$, where a is the acceleration of the body. Since the acceleration is constant and the body starts at the origin with zero velocity, $x = \frac{1}{2}at^2$. Solve for a and then calculate the work.

[Ans: 0.96 J]

11 The magnitude of the force is given by Newton's second law: $F = ma$, where m is the mass of the luge and rider and a is the magnitude of their acceleration. Since the force is constant and directed oppositely to the displacement, the work it does is $W = -Fd$, where d is the distance traveled while stopping. According to the work-kinetic energy theorem this must be the change in kinetic energy of the luge and rider and since the luge stops, $-\frac{1}{2}mv^2 = -Fd$, where v is the initial speed of the luge. Solve for d. Use $W = -\frac{1}{2}mv^2$ or $W = -Fd$ to calculate the work done by the force.

$\left[\text{Ans: (a) } 1.7 \times 10^2\,\text{N; (b) } 3.4 \times 10^2\,\text{m; (c) } -5.8 \times 10^4\,\text{J; (d) } 3.4 \times 10^2\,\text{N; (e) } 1.7 \times 10^2\,\text{N;}\right.$
$\left.\text{(f) } -5.8 \times 10^4\,\text{J}\,\right]$

15 Since the force is constant, the work it does is given by $\vec{F} \cdot \vec{d}$, where $\vec{F}$ is the force and $\vec{d}$ is the displacement. According to the definition of the scalar product this is $W = Fd \cos\phi$, where ϕ is the angle between the force and the displacement. According to the work-kinetic energy theorem the work is equal to the change in the kinetic energy. Use this theorem to solve for ϕ.

$\left[\text{Ans: (a) } 62.3°; \text{ (b) } 118°\,\right]$

23 The work done by the cable is given by $W = Td$, where T is the tension force of the cable and d is the distance the elevator cab travels (d_1 in part (a) and d_2 in part (b)). According to Newton's second law the acceleration of the cheese (and also of the elevator) is $a = F_N/m_c$, where m_c is the mass of the cheese, and the tension force of the cable is $T = m_e a$, where m_e is the mass of the elevator cab. (Strictly, it should be the mass of the cab and cheese together, but the mass of the cheese is so small it may be neglected here.) (a) Put d equal to d_1 ($= 2.40\,\text{m}$) and $F_N = 3.00\,\text{N}$, then solve for W. (b) Put $d = d_2$ ($= 10.5\,\text{m}$) and $W = 92.61 \times 10^3\,\text{J}$, then solve for F_N.

$\left[\text{Ans: (a) } 25.9\,\text{kJ; (b) } 2.45\,\text{N}\,\right]$

25 Since the block begins and ends its trip at rest, the sum of the work done by the spring force and the work done by the applied force is zero. Thus the work done by the spring is the negative of the work done by the applied force. Now the work done by the spring force is given by $W_s = -\frac{1}{2}k(x_f^2 - x_i^2)$, where k is the spring constant, x_i is the initial coordinate of the block, and x_f is its final coordinate. You can set this equal to the negative of the work of the applied force and solve for x_f if you know the value of the spring constant.

Let F be the magnitude of the applied force required to hold the block stationary at coordinate x. Then $F = k|x|$, a relationship that can be solved for k.

$\left[\text{Ans: } x = -4.9\,\text{cm or } x = +4.9\,\text{cm}\,\right]$

35 Use the work-kinetic energy theorem: the work done by the force is equal to the change in the kinetic energy of the object. Find the speed of the object at the beginning and end of the interval by differentiating its coordinate with respect to time.

$\left[\text{Ans: } 5.3 \times 10^2\,\text{J}\,\right]$

37 Since the acceleration is not constant we know that the force is variable. The work it does is given by the integral

$$W = \int_0^{x_f} F\,dx = m \int_0^{x_f} a\,dx\,.$$

Here x_f is the final coordinate of the particle, F is the x component of the force, m is the mass of the particle, and a is the x component of its acceleration. Newton's second law was used to obtain the second form. Evaluate the integral by computing the area under the curve.

Be careful about signs. Portions of the curve for which the acceleration is positive contribute positive amounts to the integral and portions for which the acceleration is negative contribute negative amounts.

After you have found the work done by the force use the work-kinetic energy theorem to find the final speed. Since the particle starts from rest $W = \frac{1}{2}mv^2$, where v is the final speed.

[Ans: (a) 42 J; (b) 30 J; (c) 12 J; (d) 6.5 m/s, positive x direction; (e) 5.5 m/s, positive x direction; (f) 3.5 m/s, positive x direction]

39 The x component of the tension force varies as the cart moves because the angle that the cord makes with the x axis changes. Let θ be that angle. Then the work done the tension force is given by

$$W = \int_{x_1}^{x_2} T_x \, dx = -T \int_{x_1}^{x_2} \cos\theta \, dx \,,$$

where T_x is the x component of the tension force. A little geometry shows that when the cart is at x, $\cos\theta = x/\sqrt{h^2 + x^2}$. Thus

$$W = -T \int_{x_1}^{x_2} \frac{x}{\sqrt{h^2 + x^2}} \, dx \,.$$

Evaluate the integral. (The integrand is the derivative of $\sqrt{h^2 + x^2}$.)

[Ans: +41.7 J]

47 (a) The work done by the force is given by $W = \vec{F} \cdot \Delta\vec{d}$, where $\Delta\vec{d} = \vec{d}_f - \vec{d}_i$. Use the component equation for the value of the scalar product: $\vec{F} \cdot \Delta\vec{d} = F_x(\Delta\vec{d})_x + F_y(\Delta\vec{d})_y + F_z(\Delta\vec{d})_z$.

(b) The average power is the work done by the machine's force divided by the time.

[Ans: (a) 1.0×10^2 J; (b) 8.4 W]

51 Draw a free-body diagram for the block. The forces on the block are the worker's force, the gravitational force of earth, and the normal force of the incline. Take the x axis to be parallel to the incline, with the positive direction down the incline. Write the x component of Newton's second law, with zero acceleration, and solve for the worker's force. You will need to calculate the angle between the incline and the horizontal using the dimensions of the incline. The work done by the worker's force is Fd, where F is the magnitude of the force and d is the distance the block moves. The work done by the gravitational force is $-mg\,\Delta y$, where Δy is the vertical distance moved by the block. The normal force is perpendicular to the displacement of the block.

[Ans: (a) 2.7×10^2 N; (b) -4.0×10^2 J; (c) 4.0×10^2 J; (d) 0; (e) 0]

53 The work done by the force is $W = F_a d \cos\phi$, where d is the distance the bead moves. If $\phi = 0$, $W = F_a d$ and this value can be read from the graph.

[Ans: (a) 11 J; (b) −21 J]

61 According to the work-kinetic energy theorem the kinetic energy of the block is equal to the sum of the work done by the spring and the work done by the applied force $\vec{F}$. This is $K = Fx - \frac{1}{2}kx^2$, where k is the spring constant. The values of F and k can be found by selecting two points on the graph and writing the equation for K twice, once for each point, then solving the equations simultaneously. The point $x = 1.0\,$m, $K = 4.0\,$J and $x = 2.0\,$m, $k = 0$ work well.

$\big[$Ans: (a) $8.0\,$N; (b) $8.0\,$N/m $\big]$

67 The kinetic energy of the box does not change, so you know that the net work done on it is zero. The net work is the sum of the work done by gravity and the work done by the belt and the work done by gravity is $-mgd_v$ when the box is going upward, 0 when the box is moving horizontally, and $+mgd_v$ when the box is going downward. Here m is the mass of the box and d_v is the vertical distance through which the box moves.

$\big[$Ans: (a) $1.7\,$W; (b) 0; (c) $-1.7\,$W $\big]$

69 According to the work-kinetic energy theorem the increase in the child's kinetic energy is the sum of the work done by gravity and the work done by the mother. The work done by the mother is $-Fd$, where F is the magnitude of the mother's force and d is the distance the child slides. If the child is not restrained, the increase in her kinetic energy is equal to the work done by gravity, which is the same as when she is restrained.

$\big[$Ans: (a) $2.1 \times 10^2\,$J; (b) $2.1 \times 10^2\,$J $\big]$

75 The work done as the box moves from $x = 0$ to some coordinate x is given by

$$W = \int_0^x F_{ax}\, dx' \,.$$

The case starts from rest so the work done on it is zero when it comes to rest again. At the position for which the work is a maximum its derivative with respect to x is zero.

$\big[$Ans: (b) $x = 3.00\,$m; (c) $13.5\,$J; (d) $x = 4.50\,$m; (e) $x = 4.50\,$m $\big]$

Quiz

Some questions might have more than one correct answer.

1. The work done by any force on an object is
   A. proportional to the displacement of the object
   B. proportional to the distance traveled by the object
   C. positive
   D. zero
   E. none of the above

2. Four objects have the following masses (m) and speeds (v):

      I. $m = 2\,\text{kg}$, $v = 2\,\text{m/s}$
    II. $m = 4\,\text{kg}$, $v = 2\,\text{m/s}$
   III. $m = 2\,\text{kg}$, $v = 4\,\text{m/s}$
   IV. $m = 2\,\text{kg}$, $v = 1\,\text{m/s}$

Order them according to their kinetic energies, least to greatest.

A. I, II, III, IV
B. IV, II, I, III
C. IV, III, I, II
D. IV, I, II, III
E. III, II, I, IV

3. Which of the following forces do work as a boy pushes a crate up an incline with constant velocity?

A. the force of the boy on the crate
B. the gravitational force of Earth on the crate
C. the normal force of the incline on the crate
D. the frictional force of the incline on the crate
E. none of the above

4. Which of the following forces do work as a girl carries a suitcase across a room, keeping the suitcase at a constant height above the level floor?

A. the gravitational force of Earth on the suitcase
B. the upward component of the force of the girl on the suitcase
C. the frictional force of the air on the suitcase
D. the horizontal component of the force of the girl on the suitcase
E. none of the above

5. A car of mass m travels a distance d up a hill, ending a vertical distance h above the starting point. The work done by the gravitational force of Earth on the car during the trip is

A. mgd
B. $-mgd$
C. mgh
D. $-mgh$
E. $mg(d + h)$

6. A block of mass m is attached to one end of a horizontal ideal spring with spring constant k. The other end of the spring is held fixed as a student uses a horizontal constant force F to pull the block horizontally a distance d from its equilibrium position. The work done by the force of the student is

   A. $\frac{1}{2}kd^2$
   B. $-\frac{1}{2}kd^2$
   C. Fd
   D. $-Fd$
   E. $-\frac{1}{2}kd^2 + Fd$

7. For which of the following situations is negative net work done on a block?

   A. the block is whirled by means of a string around a circle with constant speed
   B. the block slides a distance d up an incline with decreasing speed
   C. the block slides a distance d down an incline with increasing speed
   D. the block is on an airplane that is decelerating as it comes to a stop at a terminal
   E. the block is carried from rest to the top of a hill, where it is again at rest

8. A block of mass m on a level table top is attached to one end of a horizontal spring with spring constant k. The other end of the spring is fixed. At one instant the spring is stretched by d and the block has speed v. Later the block has zero velocity. What net work was done on the block to bring it to rest?

   A. kd^2
   B. $\frac{1}{2}kd^2$
   C. mgd
   D. mgv
   E. $\frac{1}{2}mv^2$

9. A block of mass m starts from rest and slides a distance d down a frictionless incline that makes the angle θ with the horizontal. At the end of the slide the kinetic energy of the block is

   A. $mgd \sin\theta$
   B. $mgd \cos\theta$
   C. $mgd \tan\theta$
   D. $mgd/\sin\theta$
   E. $mgd/\cos\theta$

10. A man pushes a crate with mass m a distance d up a frictionless loading ramp that makes the angle θ with the horizontal. If the crate has a constant speed the work done by the man is given by

    A. $mgd \sin\theta$
    B. $mgd \cos\theta$
    C. mgd
    D. $-mgd$
    E. $-mgd \cos\theta$

11. A block of mass m on a level table top is attached to one end of a horizontal spring with spring constant k. The other end of the spring is fixed. At one instant the block is displaced a distance d from its equilibrium point, is moving toward the equilibrium point, and has speed v. At that instant the rate with which the spring is doing work on the block is given by

A. $\frac{1}{2}kdv$

B. kdv

C. $-\frac{1}{2}kdv$

D. $-kdv$

E. kv

Answers: (1) E; (2) D; (3) A, B, D; (4) C, D; (5) D; (6) C; (7) B, D; (8) E; (9) A; (10) A; (11) B

Chapter 8
POTENTIAL ENERGY
AND CONSERVATION OF ENERGY

The closely related concepts of conservative force and potential energy are central to this chapter. Pay close attention to their definitions. If *all* forces exerted by objects in a system on each other are conservative and no net work is done on the objects by outside agents, then the mechanical energy of the system (the sum of the kinetic and potential energies) does not change. When non-conservative forces act, another energy, called the thermal energy of the system, must be included in the sum for the principle of energy conservation to hold. When external agents do work on the system, its energy is not conserved.

Important Concepts

- ☐ potential energy
- ☐ gravitational potential energy
- ☐ elastic (spring) potential energy
- ☐ mechanical energy
- ☐ conservative force
- ☐ nonconservative force
- ☐ potential energy curve

- ☐ turning point
- ☐ equilibrium points
- ☐ stable equilibrium
- ☐ unstable equilibrium
- ☐ neutral equilibrium
- ☐ thermal energy
- ☐ conservation of energy

Overview

8–2 Work and Potential Energy

☐ **Potential energy** is an energy associated with the configuration of two or more interacting objects; that is, it depends primarily on the positions of the objects. For example, the potential energy associated with the gravitational interaction between Earth and another object depends on the separation of Earth and the object. The greater the separation the greater the gravitational potential energy.

☐ Potential energy may be thought of as stored in the system. For example, the gravitational potential energy of a ball and Earth is stored in the Earth-ball system.

☐ Potential energy is associated with conservative forces only. For these forces the net work they do as the system configuration changes and then returns to the original configuration is zero, no matter what the system does in between. Gravitational and spring forces are conservative. Frictional forces are not.

☐ The change ΔU in the potential energy of a system when its configuration changes is the negative of the work done by internal forces: $\Delta U = -W$.

8–3 Path Independence of Conservative Forces

☐ As a system changes from some initial configuration to some final configuration the work done by a conservative force is the same regardless of the intermediate configurations. The work done by gravity, for example, depends only on the initial and final separations of Earth and the object, not on the paths they may take.

☐ This means there is a potential energy value associated with each configuration of the system and we can compute the work done by the conservative force by taking the difference in the values for the beginning and ending configurations.

☐ The work done by a frictional force, on the other hand, depends on the path, so we cannot associate potential energy values with the configurations.

8–4 Determining Potential Energy Values

☐ We use $\Delta U = -W$, where W is the work done by conservative forces, to compute changes in the potential energy.

☐ Only *changes* in the potential energy are physically meaningful. Usually the potential energy is chosen to have some value for a particular reference configuration of the system. Then, the work done by forces of the system as the system goes from the reference configuration to another configuration is computed. If U_{ref} represents the potential energy in the reference configuration and U represents the potential energy in the other configuration, then $U = U_{ref} - W$, where W is the work. U_{ref} is often taken to be zero.

☐ Potential energy is a scalar. If two or more conservative forces are present, the total potential energy is simply the algebraic sum of the individual potential energies.

☐ The SI unit for potential energy is the joule.

☐ The gravitational potential energy of a system consisting of Earth and an object of mass m, near its surface, is given by

$$U(y) = 0 - \int_0^y (-mg)\, dy = mgy ,$$

where the y axis is taken to be positive in the upward direction. U was chosen to be zero at $y = 0$. Gravitational potential energy is often ascribed to the object alone, but it is actually a property of the Earth-object system.

☐ Gravitational potential energy depends only on the altitude of the object above the surface of Earth, even if the object moves horizontally as well as vertically.

☐ If the altitude of an object above Earth is increased, the work done by gravity is negative and the potential energy change is positive. If the altitude is decreased, the work done by gravity is positive and the potential energy change is negative.

☐ Remember that the force of an ideal spring on an object attached to one end is given by $F = -kx$, where k is the spring constant and x is the coordinate of the object, measured from an origin at the point where the spring is neither elongated nor compressed. This force always pulls the object toward the equilibrium point ($x = 0$).

☐ The potential energy of a system consisting of an object attached to an ideal spring with spring constant k is given by

$$U(x) = 0 - \int_0^x (-kx)\,dx = \tfrac{1}{2}kx^2,$$

where U was chosen to be zero at $x = 0$. The greater the elongation or compression of the spring the greater the potential energy.

☐ An external agent can increase the spring potential energy by compressing or extending the spring. When a spring is elongated, the work done by the spring is negative and the change in the spring potential energy is positive. When the spring is compressed, the work done by the spring is again negative and the change in the spring potential energy is again positive. This stored potential energy is converted to kinetic energy when the spring is released.

8–5   Conservation of Mechanical Energy

☐ The **mechanical energy** of a system is the sum of its potential and kinetic energies: $E = K + U$. The mechanical energy of a system is **conserved** (remains constant) if a potential energy is associated with each of the forces between objects of the system and no external agents do work on the system. This follows directly from the work-kinetic energy theorem and the definition of potential energy.

☐ If a ball is thrown upward, the force of gravity slows it down; its kinetic energy decreases but the gravitational potential energy of the Earth-ball system increases. Similarly, as the ball falls its kinetic energy increases and the gravitational potential energy of the Earth-ball system decreases. If the gravitational force is the only force acting on the ball, the sum of the potential and kinetic energies remains constant during the flight of the ball; the mechanical energy of the Earth-ball system is conserved. $\tfrac{1}{2}mv^2 + mgy = $ constant can be used to solve for one of the quantities that appear in it.

☐ If a block on a horizontal frictionless surface is attached to an ideal spring and the block is pulled aside and released, it oscillates back and forth, repeatedly coming back to its original position. As the block moves toward the equilibrium point ($x = 0$) it speeds up and the spring becomes less extended or compressed; as it moves away from the equilibrium point it slows down and the spring becomes more extended or compressed. In the first case, the block gains kinetic energy and the spring loses potential energy; in the second case, the block loses kinetic energy and the spring gains potential energy. The sum of the potential energy stored in the spring and the kinetic energy of the block remains constant throughout; the mechanical energy of the spring-block system is conserved. $\tfrac{1}{2}mv^2 + \tfrac{1}{2}kx^2 = $ constant can be used to solve for one of the quantities that appear in it.

☐ If frictional forces act, the mechanical energy of a system decreases. Mechanical energy is said to be **dissipated**.

8–6   Reading a Potential Energy Curve

☐ If the potential energy function is known, the force can be computed by evaluating its derivatives with respect to coordinates. If $U(x)$ is the potential energy for an object moving on the

x axis and subjected to a conservative force F, then

$$F(x) = -\frac{dU(x)}{dx}.$$

On a graph of the potential energy as a function of x the force is the negative of the slope of the line that is tangent to the curve at the coordinate of the particle. A positive slope indicates a force in the negative x direction and a negative slope indicates a force in the positive x direction.

☐ When dealing with potential energy curves, assume that the agent exerting the force is stationary, so only the object whose motion is being considered has kinetic energy.

☐ A potential energy curve is shown to the right for an object that moves along the x axis with constant mechanical energy E, represented by a horizontal dotted line. The potential and kinetic energies when the object is at x_1 are indicated by arrows. The object cannot move to the left of x_{min} or to the right of x_{max} because the potential energy is greater than the mechanical energy in those regions. The kinetic energy and the velocity are zero at x_{min} and x_{max}.

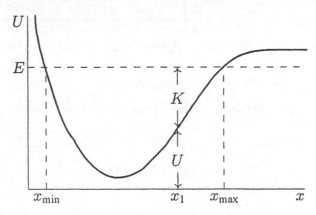

When the object is at x_{min}, the force acting on it is positive (the slope of the curve is negative) and the object is pushed to the right. When it is at x_{max}, the force acting on it is negative and the object is pushed to the left. x_{min} and x_{max} are called the **turning points** of the motion.

☐ Suppose the object starts at x_1 and travels in the negative x direction. Its speed increases until it gets to the coordinate corresponding to the minimum in the potential energy curve; then it slows. When it reaches x_{min}, its velocity is zero but a positive force acts on it and it starts moving in the positive x direction. Its speed increases, then decreases until it gets to x_{max}, where its velocity is zero. It then starts moving back again and the motion is repeated. If the energy is decreased, it oscillates over a narrower range of coordinates.

☐ **Equilibrium points** are places where the force vanishes. The slope of the potential energy curve is zero at these points and the potential energy is either a maximum or a minimum. When an object is released from rest near a point where the curve has a minimum (a point of **stable equilibrium**), its subsequent motion is toward the equilibrium point. When an object is released from rest near a point where the curve has a maximum (a point of **unstable equilibrium**), its subsequent motion is away from the equilibrium point. When an object is released from rest in a region where the curve is horizontal (a region of **neutral equilibrium**), it remains at rest.

8–7 Work Done on a System by an External Force

☐ If external agents do work on the objects of a system, the mechanical energy of the system is not conserved. If frictional forces are not involved the energy equation becomes

$$\Delta K + \Delta U = W,$$

where W is the total work done *on* objects of the system by outside agents. An external force doing work on objects in a system transfers energy to or from the system. For example, the work done by a person carrying a crate up some stairs increases the mechanical energy of the crate-Earth system. If the speed of the crate does not change, the increase in energy appears as potential energy; the crate is further away from Earth than previously. If the crate's speed increases, both the potential and kinetic energies increase.

☐ If a nonconservative force, such as friction, acts on objects of a system, the thermal energy E_{th} of the objects must be included in the total energy of the system and the energy equation becomes $\Delta E_{mec} + \Delta E_{th} = W$, where W is again the work done by external forces, ΔE_{mec} is the change in the mechanical energy of the system, and ΔE_{th} is the change in the thermal energy of the system.

☐ If a constant frictional force of magnitude f acts between two objects in the system $\Delta E_{th} = fd$, where d is the magnitude of the relative displacement of the objects. The thermal energy is shared by the two objects.

8–8 Conservation of Energy

☐ A system may have an internal energy E_{int} above and beyond its thermal energy. The kinetic energy of the pistons in an automobile engine is an example.

☐ When the thermal and other internal energies are included, the energy of a system of objects becomes

$$E = K + U + E_{th} + E_{int}.$$

Here K is the total kinetic energy of the objects, U is the total potential energy of their interactions with each other, E_{int} is the total thermal energy, and E_{int} is the total of the other internal energies. The energy equation is $\Delta E = W$, where W is the net work done by external forces. If $W = 0$ then the energy of the system is conserved.

☐ Whenever mechanical energy is not conserved, physicists have been able to restore the principle of energy conservation by defining other forms of energy and including them in the energy balance. The total is conserved if no external agents do work on the system. The energy may change form and may be transferred from one object to another in the system but the total remains constant.

Hints for Questions

<u>1</u> (a) The change in mechanical energy is the same for every path along which only a conservative force acts. For each of the indirect paths sum the changes in the mechanical energy for the various segments to find the change for the whole path. The result is the same for three of the indirect paths. This is the change in mechanical energy for the direct path.

(b) For the one path for which the change in mechanical energy is different the total change is the sum of the change due to the conservative force and the change due to nonconservative force. The change due to the conservative force is the same as for the other paths. Since you know the total you can calculate the change due to the nonconservative force.

[Ans: (a) 12 J; (b) −2 J]

<u>3</u> (a) Since mechanical energy is conserved for the system consisting of the block and Earth, the block comes to rest whenever its height above the ground is 3.0 m. Then the potential energy of the system is the same as its initial value and since the kinetic energy is initially zero it must also be zero when (and if) the cart reaches such a point.

(b) You can tell if the force on the cart is zero or not and the direction of the force (if the force is not zero) by looking at the slope of the ramp at the point where the cart comes to rest. This will tell you if the cart starts to move again and the direction of its motion. If it moves you should be able to tell where it stops again and what happens then.

(c) Since all hills have the same radius the centripetal acceleration is proportional to the square of the cart's speed when it gets to the top of the hill. That is, it is proportional to the cart's kinetic energy there. Since mechanical energy is conserved the cart has the greatest kinetic energy and centripetal acceleration at the top of the hill for which the potential energy is least.

(d) According to Newton's second law the difference of the gravitational and normal forces is proportional to the centripetal acceleration. The normal force is least where the centripetal acceleration is the greatest.

[Ans: (a) 4; (b) returns to its starting point and repeats the trip; (c) 1; (d) 1]

<u>5</u> (a) The force is proportional to the slope of the potential energy curve at the position of the particle.

(b) and (c) A particle is trapped in any region for which the potential energy at the boundaries is equal to the mechanical energy.

(d) and (e) The kinetic energy and speed are greatest if the particle is at a point for which the potential energy is least and are the least if the particle is at a point for which the potential energy is greatest.

[Ans: (a) AB, CD, then BC and DE tie (zero force); (b) 5 J; (c) 5 J; (d) 6 J; (e) FG; (f) DE]

<u>7</u> The change in the total energy of the system is equal to the work done by your force. It is also equal to the sum of the changes in the kinetic, potential, and thermal energies of the system.

[Ans: +30 J]

<u>9</u> In regions AB and BC the mechanical energy of the system consisting of the block, the track on which it slides, and Earth, is conserved. The kinetic energy increases in regions for which the potential energy decreases and decreases in regions for which the potential energy increases. In CD the potential energy does not change but the kinetic energy decreases as friction slows the block.

[Ans: (a) increasing; (b) decreasing; (c) decreasing; (d) constant in AB and BC, decreasing in CD]

Hints for Problems

<u>5</u> The work done by the gravitational force is given by $W = -mg\,\Delta y$, where m is the mass of the car and Δy is the change in its distance from the ground. The work is positive if the car is lower at the end of the interval than at the beginning and negative if it is higher. The

potential energy of the car-Earth system is given by $U = mgy$, where y is the height of the car above ground level. Notice that the potential energy is proportional to the mass of the cart.

[Ans: (a) 0; (b) 170 kJ; (c) 340 kJ; (d) 170 kJ; (e) 340 kJ; (f) increases]

__11__ Use conservation of mechanical energy. When the car is a distance y above ground level and is traveling with speed v the mechanical energy is given by $\frac{1}{2}mv^2 + mgy$, where m is the mass of the car. Write this expression for the initial values and for the values when the car is at another point (A, B, C, or the point where it stops on the last hill). Equate the two expressions to each other and solve for the unknown (either the speed or the height above the ground). Notice that the mass of the car cancels from the conservation of energy equation.

[Ans: (a) 17.0 m/s; (b) 26.5 m/s; (c) 33.4 m/s; (d) 56.7 m; (e) all the same]

__19__ (a) Use $U = -\int F\,dx$ and select the constant of integration so that $U = 27\,\mathrm{J}$ for $x = 0$.
(b) The force is zero at the value of x for which the potential energy is maximum.
(c) and (d) Set the expression for U equal to zero and solve for x.

[Ans: (a) $U = 27 + 12x - 3x^2$; (b) 39 J; (c) -1.6 m; (d) 5.6 m]

__21__ (a) and (b) Two forces act on the block: the force of gravity and the normal force of the track. The force of gravity has magnitude mg, where m is the mass of the block, and is downward. At Q the normal force is horizontal and to the left. It provides the centripetal acceleration and must have magnitude mv^2/R, where v is the speed of the block at that point. Use conservation of mechanical energy to find the speed. When the block is at P its kinetic energy is zero and you may take the potential energy to be mgh. When the block is at Q its kinetic energy is $\frac{1}{2}mv^2$ and the potential energy is mgR. Thus $mgh = \frac{1}{2}mv^2 + mgR$. solve for v^2.
(c) At the top of the loop the normal force F_N (if any) is downward, so $mg + F_N = mv^2/R$. Since the normal force cannot be upward, the block loses contact with the track if it going so slowly that mv^2/R is less than mg. Use conservation of mechanical energy to find the value of h so that $mv^2/R = mg$.

[Ans: (a) 2.5 N; (b) 0.31 N; (c) 30 cm]

__27__ (a) Use conservation of mechanical energy. Take the potential energy to be zero when the stone is at the lowest point. When the string makes an angle of $60°$ with the vertical the stone is $L(1 - \cos 60°)$ above the lowest point and the potential energy is $mgL(1 - \cos 60°)$. The kinetic energy at the lowest point is $\frac{1}{2}mv_0^2$, where v_0 is its speed there, and the kinetic energy when the string makes an angle of $60°$ with the vertical is $\frac{1}{2}mv^2$, where v is its speed there. Thus $\frac{1}{2}mv_0^2 = \frac{1}{2}mv^2 + mgL(1 - \cos 60°)$. Solve for v.
(b) The speed and kinetic energy of the stone are zero when the stone is at the highest point it reaches. Let θ be the angle made by the string with the vertical when the stone is at the highest point and solve $\frac{1}{2}mv_0^2 = mgL(1 - \cos \theta)$ for θ.
(c) Since the potential energy is zero when the stone is at the lowest point, the mechanical energy is just the kinetic energy when the stone is there.

[Ans: (a) 5.0 m/s; (b) 79°; (c) 64 J]

__35__ Take the gravitational potential energy to be zero when the whole cord is stuck to the ceiling. Now compute the potential energy when the cord is hanging by one end. Consider an infinitesimal segment of cord with mass dm a distance y from the ceiling. The potential energy associated with this segment is $dU = -gy\,dm$. If the length of the segment is dy, then $dm = (dy/L)M$, where L is the length of the cord and M is its mass. The total potential energy is the sum over segments:

$$U = -\int_0^L gy\frac{M}{L}\,dy\,.$$

Evaluate the integral.

$\left[\text{Ans: } -18\,\text{mJ}\,\right]$

39 (a) The mechanical energy is the sum of the kinetic and potential energies.

(c) and (d) The least and greatest values of x the particle can reach are those for which $U = E_{\text{mec}}$, where E_{mec} is the mechanical energy you found in part (a).

(e) and (f) The particle has the greatest kinetic energy when it is at the point where the potential energy is the least.

(g) The force is the negative of the derivative of the potential energy with respect to x.

(h) Solve $F(x) = 0$ for x.

$\left[\text{Ans: (a) } -3.7\,\text{J; (c) } 1.3\,\text{M; (d) } 9.1\,\text{m; (e) } 2.2\,\text{J; (f) } 4.0\,\text{m; (g) } (4 - x)e^{-x/4}; \text{ (h) } 4.0\,\text{m}\,\right]$

49 Use $W = \Delta E_{\text{mec}} + +\Delta E_{\text{th}}$, where W is the work done by the applied force $\vec{F}$, ΔE_{mec} is the change in mechanical energy, and ΔE_{th} is the change in the thermal energy of the block-incline system. The work done by $\vec{F}$ is Fd, where d is the distance the block slides between A and B, the change in thermal energy is fd, and the change in mechanical energy is $\Delta K + \Delta U$, where K is the kinetic energy of the block and ΔU is the change in the gravitational potential energy. This last change is just the negative of the work done by the gravitational force.

$\left[\text{Ans: } 75\,\text{J}\,\right]$

57 (a) The work done by the spring force is given by $W_s = -\frac{1}{2}kd^2$, where d is the spring compression and k is the spring constant.

(b) The change in the thermal energy is given by $\Delta E_{\text{th}} = fd$, where f is the magnitude of the frictional force. This is the product of the magnitude of the normal force of the floor on the block and the coefficient of kinetic friction. Use Newton's second law to obtain the magnitude of the normal force.

(c) Use the energy equation $W = \Delta E_{\text{mec}} + \Delta E_{\text{th}}$, where W is the work done by external forces and ΔE_{mec} is the change in the mechanical energy. Take the system to be composed of the block and the spring. Then no external forces do work and the mechanical energy is the sum of the kinetic energy of the block and the potential energy stored in the spring. Solve for the initial kinetic energy and then the initial speed.

$\left[\text{Ans: (a) } -0.90\,\text{J; (b) } 0.46\,\text{J; (c) } 1.0\,\text{m/s}\,\right]$

65 To find the length of the slide in part (a) draw a diagram showing the circular arc of the slide and its center. Draw radii from the center to the bottom and top of the slide. Draw the horizontal line a distance h from the ground and show that $h = R(1 - \cos\theta)$, where R is the radius and θ is the angle subtended by the slide. Solve for θ in radians. The length of the slide is then $R\theta$. Use $\Delta K + \Delta U + \Delta E_{\text{th}} = 0$, where K is the kinetic energy, U is the potential energy, and E_{th} is the thermal energy. Replace ΔE_{th} with $f\ell$, where f is the average frictional force on the child and ℓ is the length of the slide. Solve for f. In part (b) the angle subtended by the slide is given by $\sin\theta = h/R$.

[Ans: (a) 10 m; (b) 49 N; (c) 4.1 m; (d) 1.2×10^2 N]

75 Let F_P be the force of the road on the tires. This is the force that accelerates the car. According to Newton's second law $F_P - F = ma$, so $F_P = F + ma$. The power required is $P = F_P v = (F + ma)v$.

[Ans: 69 hp]

79 Use conservation of mechanical energy Take the potential energy to be zero when the can is on the ground. The initial kinetic energy is $\frac{1}{2}mv_0^2$, where m is the mass of the can and v_0 is its initial speed. The initial potential energy is mgh, where h is the height above the ground from which the can is thrown. The potential energy is zero at the end of the can's fall and is mgh' when it is h' above the ground. Use $y = y_0 + v_0 t - \frac{1}{2}gt^2$ with $y = 0$ to find the time when the can reaches the ground, then use the same equation to find the can's position 2.00 s earlier.

[Ans: (a) 109 J; (b) 60.3 J; (c) 68.2 J; (d) 41.0 J]

87 Use $K = \frac{1}{2}mv^2$, where m is the mass of the balloon and v is its speed, to compute the kinetic energy. During the full ascent the gravitational force does just the right amount of work to reduce the kinetic energy to zero; the initial kinetic energy is converted to potential energy. The potential energy for the balloon at any height h is the value for the balloon at the reference level plus the change in potential energy as the balloon moves from the reference level to h. Use $U = mgh$, to compute the height h above a reference level.

[Ans: (a) 6.75 J; (b) -6.75 J; (c) 6.75 J; (d) 6.75 J; (e) -6.75 J; (f) 0.4459 m]

91 Use conservation of mechanical energy. The potential energy increases by $\Delta U = mgh$, where m is the mass of the clown and h is the distance of the landing point above the launch point. The change in the kinetic energy is $\Delta K = -\Delta U$.

[Ans: 5.4 kJ]

97 Calculate the change in the mechanical energy during the ascent. Since the kinetic energy is zero when the banana is at the highest point, the change in the kinetic energy is equal to the negative of the initial kinetic energy. The change in the potential energy is $\Delta U = +mgh$, where m is the mass of the banana and h is the height of the highest point above the throwing point.

[Ans: 80 mJ]

103 The power required to drive the ship with speed v is $P = Fv$, where F is the driving force.

[Ans: 5.5×10^6 N]

109 The change in potential energy is $\Delta U = -mgh$, where m is the mass of the performer and h is the height of the starting position above ground level. If friction is negligible then mechanical energy is conserved and the change in the kinetic energy is $\Delta K = -\Delta U$. If friction is present the change in the kinetic energy is $\Delta K = -\Delta U - \Delta E_{th}$, where E_{th} is the thermal energy of the performer and pole. Use $\Delta E_{th} = -fh$, where f is the frictional force, to calculate the change in the thermal energy.

[Ans: (a) 2.35×10^3 J; (b) 352 J]

121 The volume of water that falls on the continental United States in a year is the product of the area and the yearly rainfall depth. Multiply by $1000\,\mathrm{kg/m^3}$ to find the mass of this water. One-third of the water flows to the oceans and loses potential energy mgh, where m is the mass of the water and h is average distance through which it falls. Divide by the time interval $(1\,\mathrm{y})$ to obtain the average power.

$\left[\text{Ans: } 3.1 \times 10^{11}\,\mathrm{W}\right]$

123 The change in the gravitational potential energy is $\Delta U_g = mgy$, the change in the elastic potential energy is $\Delta U_e = \frac{1}{2}ky^2$, and since mechanical energy is conserved, the change in the kinetic energy is $\Delta K = -\Delta U_g - \Delta U_e$.

$\left[\text{Ans: (a) } 0.75\,\mathrm{J}; \text{ (b) } -1.0\,\mathrm{J}; \text{ (c) } 0.25\,\mathrm{J}; \text{ (d) } 1.0\,\mathrm{J}; \text{ (e) } -2.0\,\mathrm{J}; \text{ (f) } 1.0\,\mathrm{J}; \text{ (g) } 0.75\,\mathrm{J}; \text{ (h) } -3.0\,\mathrm{J};\right.$
$\left.\text{(i) } 2.3\,\mathrm{J}; \text{ (j) } 0; \text{ (k) } -4.0\,\mathrm{J}; \text{ (l) } 4.0\,\mathrm{J}\right]$

135 The potential energy function is the negative of the integral of the force: $U = \int F(x)\,dx$. Use the condition that $U(x) \longrightarrow 0$ as $x \longrightarrow 0$ to evaluate the constant of integration. The work done by the outside agent is equal to the change in the potential energy.

$\left[\text{Ans: (a) } U(x) = -Gm_1m_2/x; \text{ (b) } Gm_1m_2d/x_1(x_1 + d)\right]$

Quiz

Some questions might have more than one correct answer.

1. Which of the following statements are true?

   A. a potential energy is defined for all forces

   B. a potential energy is defined for all conservative forces

   C. a potential energy is defined for all conservative forces that are internal to the system of interest

   D. a potential energy cannot be defined for a system if external frictional forces are present

   E. a potential energy cannot be defined for a system if internal frictional forces are present

2. A potential energy is defined for which of the following forces?

   A. gravitational forces

   B. spring forces

   C. frictional forces

   D. drag forces

   E. forces that depend only on the coordinates of the particles

3. The mechanical energy of a system is conserved if

   A. the net external work is zero

   B. all the internal forces are conservative

   C. all the external forces are conservative

   D. all the internal forces are conservative and the net external work is zero

   E. all forces, both internal and external, are conservative

4. A block is hung from the ceiling by means of a spring, then is pulled downward and released. Air resistance is negligible. The potential energy (gravitational and elastic) of the block-spring-Earth system is greatest

    A. only when the block is at the highest point
    B. only when the block is at the lowest point
    C. when the block is either at the highest or lowest point
    D. when the block is at its equilibrium point
    E. when the block is at some other point

5. As a block slides down an incline the incline exerts a frictional force on it. For which system is energy conserved?

    A. the system consisting of the block and incline
    B. the system consisting of the block and Earth
    C. the system consisting of the incline and Earth
    D. the system consisting of the block alone
    E. the system consisting of the block, incline, and earth

6. A block is hung from the ceiling by means of a spring, then is pulled downward and released. It oscillates, subject to air resistance. For which system is energy conserved?

    A. the system consisting of the block alone
    B. the system consisting of the block and spring
    C. the system consisting of the block, the spring, and Earth
    D. the system consisting of the block, spring, Earth, and the air
    E. the system consisting of the block, Earth, and the air

7. A pendulum hangs from the ceiling and swings back and forth. Which of the following statements is true if air resistance can be neglected?

    A. the Earth-pendulum system has its greatest potential energy when it is at its highest point
    B. the Earth-pendulum system has its greatest kinetic energy when it is at its highest point
    C. the Earth-pendulum system has its greatest potential energy when it is at its lowest point
    D. the Earth-pendulum system has its greatest kinetic energy when it is at its lowest point
    E. both the kinetic and potential energies of the Earth-pendulum system remain constant throughout the swing

8. If both the potential and kinetic energies of a system decrease it is possible that:

    A. external forces are doing negative work on the system
    B. external forces are doing positive work on the system
    C. some internal forces are not conservative
    D. all external forces are conservative
    E. all internal forces are conservative

9. A system consists of several objects. It starts with certain values of its kinetic and potential energies. Sometime later it returns to its starting configuration but with less kinetic energy. If no external forces act on the objects of the system, we must conclude that

   A. the objects exert nonconservative forces on each other and the thermal energy of the system has increased
   B. the objects exert nonconservative forces on each other and the thermal energy of the system has decreased
   C. the objects exert nonconservative forces on each other and the thermal energy of the system has not changed
   D. the objects exert conservative forces on each other and the thermal energy of the system has increased
   E. the objects exert conservative forces on each other and the thermal energy of the system has decreased

10. A crate slides across a level floor. The frictional force of the floor on the crate

    A. increases both the kinetic and thermal energies of the crate
    B. increases the kinetic energy of the crate and decreases its thermal energy
    C. decreases the kinetic energy of the crate and increases its thermal energy
    D. decreases both the kinetic and thermal energies of the crate
    E. does not change the thermal energy of the crate

11. A crate is given a push across a level floor and it comes to rest in distance d because the floor exerts a frictional force of magnitude f. Which of the following statements are true?

    A. the thermal energy of the crate increases by fd
    B. the thermal energy of the crate decreases by fd
    C. the total thermal energy of the crate and floor increases by fd
    D. the total thermal energy of the crate and floor decreases by fd
    E. the total thermal energy of the crate and floor does not change

Answers: (1) C; (2) A, B; (3) D; (4) A; (5) E; (6) D; (7) A, D; (8) A, C; (9) A; (10) C; (11) C

Chapter 9
CENTER OF MASS AND LINEAR MOMENTUM

This chapter is about systems of more than one particle. The center of mass is important for describing the motion of the system as a whole, and its velocity is closely related to the total momentum of the system. External forces acting on the system accelerate the center of mass and when the net external force vanishes the velocity of the center of mass is constant. The total momentum of the system is then conserved. The principle of momentum conservation is applied to collisions between objects. Pay special attention to the role played by the impulses the objects exert on each other. Also take careful notice of when energy is conserved in a collision and when it is not.

Important Concepts

- [ ] center of mass
- [ ] Newton's second law
    for the center of mass
- [ ] linear momentum
- [ ] impulse
- [ ] conservation of
    linear momentum

- [ ] collision
- [ ] inelastic collision
- [ ] completely inelastic
    collision
- [ ] elastic collision

Overview

9–2   The Center of Mass

- [ ] The motion of an object might be quite complicated: the object might rotate and deform as it moves. Nevertheless, the motion of a special point, called the **center of mass** of the object, is much less complicated than the motion of other parts of the object. It moves like a point particle that is subjected only to the external forces applied to the object. No matter how a thrown object twists and turns, for example, its center of mass follows the parabolic trajectory of an ideal projectile.

- [ ] The coordinates of the center of mass of a collection of particles are given by

$$x_{\text{com}} = \frac{1}{M} \sum_{i=1}^{N} m_i x_i, \qquad y_{\text{com}} = \frac{1}{M} \sum_{i=1}^{N} m_i y_i, \qquad \text{and} \qquad z_{\text{com}} = \frac{1}{M} \sum_{i=1}^{N} m_i z_i,$$

where M is the total mass of the system, m_i is the mass of particle i, and x_i, y_i, z_i are its coordinates. In vector notation, the position vector of the center of mass is given by

$$\vec{r}_{\text{com}} = \frac{1}{M} \sum_{i=1}^{N} m_i \vec{r}_i \,,$$

where $\vec{r}_i$ is the position vector of particle i. The center of mass is not necessarily at the position of any particle in the system.

☐ For a *continuous* distribution of mass, the coordinates of the center of mass are computed by dividing the object into a large number of infinitesimal regions, each with mass dm. Each region is treated like a particle and the contributions are summed. The results are written as the integrals

$$x_{\text{com}} = \frac{1}{M} \int x \, dm \,, \qquad y_{\text{com}} = \frac{1}{M} \int y \, dm \,, \qquad \text{and} \qquad z_{\text{com}} = \frac{1}{M} \int z \, dm \,.$$

☐ If the mass is distributed symmetrically about some point or line, then the center of mass is at that point or on that line. For example, the center of mass of a uniform spherical shell is at its center, the center of mass of a uniform sphere is at its center, the center of mass of a uniform cylinder is at the midpoint of its central axis, the center of mass of a uniform rectangular plate is at its center, and the center of mass of a uniform triangular plate is at the intersection of the lines from the vertices to the midpoints of the opposite sides and halfway through its thickness.

☐ Sometimes an object with a complicated shape can be divided into parts such that the center of mass of each part can be found easily. Then, each part is replaced by a particle with mass equal to the mass of the part, positioned at the center of mass of the part. The center of mass of the whole object is at the position of the center of mass of these particles. This idea can also be used to find the center of mass of an object with a hole.

9–3 Newton's Second Law for a System of Particles

☐ As the particles of a system move, the coordinates of the center of mass might change. Consider a system of discrete particles and differentiate the expression for the center of mass position vector to find the velocity of the center of mass in terms of the individual particle velocities:

$$\vec{v}_{\text{com}} = \frac{1}{M} \sum_{i=1}^{N} m_i \vec{v}_i \,,$$

where $\vec{v}_i$ is the velocity of particle i.

☐ As the particles of a system accelerate the center of mass might accelerate. Differentiate the expression for the velocity of the center of mass to find the acceleration of the center of mass in terms of the accelerations of the individual particles:

$$\vec{a}_{\text{com}} = \frac{1}{M} \sum_{i=1}^{N} m_i \vec{a}_i \,,$$

where $\vec{a}_i$ is the acceleration of particle i.

☐ Since each particle of a system obeys Newton's second law, $\vec{F}_{net} = M\vec{a}_{com}$, where $\vec{F}_{net}$ is the vector sum of all forces on all particles of the system. Some of the forces might be due to other particles in the system and some might be due to the environment of the system but, because Newton's third law is valid, $\vec{F}_{net}$ reduces to the vector sum over all *external* forces acting on particles of the system, i.e. those due to the environment of the system. If two particles of the system exert forces on each other, the two forces are equal in magnitude and opposite in direction. Both occur in the sum over all forces and so cancel each other. Thus, the center of mass obeys a Newton's second law equation:

$$\vec{F}_{net} = M\vec{a}_{com}.$$

The mass that appears in this law is the total mass of the system (the sum of the masses of the individual particles) and the force that appears is the vector sum of all *external* forces acting on all particles of the system.

☐ If the total external force acting on a system is zero, then the acceleration of the center of mass is zero and the velocity of the center of mass is constant. If the center of mass of the system is initially at rest and the total external force acting on the system is zero, then the velocity of the center of mass is always zero and the center of mass remains at its initial position.

☐ These statements are true no matter what forces the particles of the system exert on each other. The particles might, for example, be fragments that are blown apart in an explosion, they might be objects that collide violently, or they might be objects that interact with each other from afar via gravitational or electrical forces.

9–4   Linear Momentum

☐ The momentum of a particle of mass m, moving with a velocity $\vec{v}$ is given by $\vec{p} = m\vec{v}$. Momentum is a vector. Its SI units are kg · m/s.

☐ In terms of momentum, Newton's second law for a particle is $\vec{F}_{net} = d\vec{p}/dt$, where $\vec{F}_{net}$ is the net force acting on the particle.

9–5   The Linear Momentum of a System of Particles

☐ The total momentum $\vec{P}$ of a system of particles is the vector sum of the individual momenta. Since $\vec{P} = m_1\vec{v}_1 + m_2\vec{v}_2 + \ldots$, the total momentum is given by $\vec{P} = M\vec{v}_{com}$, where M is the total mass of the system. That is, the total momentum of the system is identical to the momentum of a single particle with mass equal to the total mass of the system, moving with a velocity equal to the velocity of the center of mass.

☐ Newton's second law for a system of particles can be written in terms of the total momentum: $\vec{F}_{net} = d\vec{P}/dt$.

9–6   Collision and Impulse

☐ In a collision, two objects exert relatively strong forces on each other for a relatively short time. A collision takes place during a well-defined time interval: the objects do not interact with each other before or after.

☐ During a collision each object exerts a force on the other. What is important is not the force alone or its duration alone but a combination called the **impulse** of the force. If one body acts on the other with a force $\vec{F}(t)$ for a time interval from t_i to t_f, the impulse of the force is given by

$$\vec{J} = \int_{t_i}^{t_f} \vec{F}(t)\, dt \,.$$

Don't confuse impulse and work. Impulse is an integral of force with respect to time and work is an integral of force with respect to coordinate. Impulse is a vector, work is a scalar. The SI unit of impulse is $kg \cdot m/s$; the SI unit of work is $kg \cdot m^2/s^2$.

☐ In terms of the average force $\vec{F}_{avg}$ that acts during a collision, the impulse is given by $\vec{J} = \vec{F}_{avg} \Delta t$, where Δt is the duration of the collision. This expression is often used to calculate the average force, which is useful as an estimate of the strength of the interaction.

☐ Because Newton's second law is valid, the total impulse acting on an object gives the change in the momentum of the object: $\vec{J} = \vec{p}_f - \vec{p}_i$, where $\vec{p}_i$ is the initial momentum and $\vec{p}_f$ is the final momentum. Because Newton's third law is valid the impulse of one object on the other is the negative of the impulse of the second object on the first. If external impulses can be neglected, the total momentum of the two objects is conserved during a collision. This means the velocity of the center of mass of the colliding bodies is constant.

☐ For most collisions, the impulse of either colliding object on the other is usually much greater than any external impulse and we may neglect impulses exerted by the environment of the colliding bodies. We may, for example, neglect the effects of gravity and air resistance during the time a baseball is in contact with a bat.

☐ Consider a stream of objects that collide with another object. Examples are bullets hitting a target or the molecules of a gas hitting the walls of a container. Suppose each of the "bullets" has mass m and each suffers the same change in velocity $\Delta \vec{v}$. If n "bullets" hit in time Δt, then during that interval the change in momentum of the "target" is given by $\Delta \vec{P} = -nm\,\Delta \vec{v}$ and the average force exerted by the "bullets" on the "target" is given by $\vec{F}_{avg} = \Delta \vec{P}/\Delta t = -(n/\Delta t)m\,\Delta \vec{v}$.

9–7 Conservation of Linear Momentum

☐ The principle of momentum conservation is: If the total external force acting on a system of particles vanishes, then the total momentum of the system is constant. Total momentum is said to be conserved. Since momentum is a vector this principle is equivalent to the three equations: $P_x =$ constant if $F_{net\ x} = 0$, $P_y =$ constant if $F_{net\ y} = 0$, and $P_z =$ constant if $F_{net\ z} = 0$. One component of momentum might be conserved even if the others are not.

☐ If the momentum of a system is conserved, then the acceleration of the center of mass of the system is zero and the velocity of the center of mass is constant. If the momentum of the system is zero, then its center of mass does not move.

9–8 Momentum and Kinetic Energy in Collisions

☐ The total linear momentum of objects involved in a collision is always conserved.

☐ The total kinetic energy of the colliding objects may or may not be conserved in a collision. If it is, the collision is said to be **elastic**. If it is not, the collision is said to be **inelastic**.

9–9 Inelastic Collisions in One Dimension

☐ If the total kinetic energy of the colliding objects is not conserved, the collision is said to be **inelastic**. The kinetic energy might increase, as when an explosion takes place, or it might decrease, as when it is converted to internal energy. An explosion in which an object splits into two or more parts as a result of internal forces can be handled in exactly the same manner as a collision: total momentum is conserved but total kinetic energy increases.

☐ After a **completely inelastic collision** the two bodies stick together and move off with the same velocity. During such a collision the loss in total kinetic energy is as large as conservation of momentum allows. That is, the bodies cannot lose a larger fraction of their original kinetic energy and still retain all the original total momentum.

☐ Consider a completely inelastic collision between two objects moving along the x axis. Before the collision object 1, with mass m_1, moves with velocity v_1 and object 2, with mass m_2, moves with velocity v_2. After the collision both objects move along the x axis with velocity V. Since momentum is conserved during the collision $m_1 v_1 + m_2 v_2 = (m_1 + m_2)V$, so $V = (m_1 v_1 + m_2 v_2)/(m_1 + m_2)$. Notice that the final speed of the combination *must* be less than the initial speed of object 1.

☐ Kinetic energy is lost during a completely inelastic collision, transformed to internal energy, radiation, etc. Suppose object 2 is initially at rest. The initial total kinetic energy is $K_i = \frac{1}{2}m_1 v_1^2$ and the final total kinetic energy is $K_f = \frac{1}{2}(m_1 + m_2)V^2$. The fractional energy loss is

$$\frac{K_i - K_f}{K_i} = \frac{m_1 v_1^2 - (m_1 + m_2)V^2}{m_1 v_1^2} = \frac{m_2}{m_1 + m_2},$$

where the expression for V in terms of v_1 was substituted.

9–10 Elastic Collisions in One Dimension

☐ Suppose object 1, with mass m_1 moves with velocity v_{1i} along the x axis and collides elastically with object 2, which has mass m_2 and is initially at rest. After the collision object 1 moves along the x axis with velocity v_{1f} and object 2 moves along the x axis with velocity v_{2f}. The equation that expresses conservation of momentum during the collision is

$$m_1 v_{1i} = m_1 v_{1f} + m_2 v_{2f},$$

and the equation that expresses conservation of kinetic energy during the collision is

$$\tfrac{1}{2}m_1 v_{1i}^2 = \tfrac{1}{2}m_1 v_{1f}^2 + \tfrac{1}{2}m_2 v_{2f}^2.$$

The solution to these two equations is

$$v_{1f} = \frac{m_1 - m_2}{m_1 + m_1} v_{1i} \quad \text{and} \quad v_{2f} = \frac{2m_1}{m_1 + m_2} v_{1i}.$$

□ Some special cases:

  1. If the two masses are the same, then $v_{1f} = 0$ and $v_{2f} = v_{1i}$. The objects have interchanged their velocities.

  2. If the target object, object 2, is very massive compared to the incident object, object 1, then $v_{1f} \approx -v_{1i}$ and $v_{2f} \approx (2m_1/m_2)v_{1i}$. The incident object bounces off the target with only a slight loss of speed and the target moves slowly away from the collision.

  3. If object 1 is very massive compared to object 2, then $v_{1f} \approx v_{1i}$ and $v_{2f} \approx 2v_{1i}$. The incident object continues with only a slight loss in speed and the target moves away with a speed that is twice the original speed of the incident object.

□ The velocity of the center of mass of the two-object system is not changed by the collision. If the target object is initially at rest, the velocity of the center of mass is given by $v_{\text{com}} = m_1 v_{1i}/(m_1 + m_2)$.

□ If both the target and incident object are initially moving, then

$$v_{1f} = \frac{m_1 - m_2}{m_1 + m_2} v_{1i} + \frac{2m_2}{m_1 + m_2} v_{2i} \quad \text{and} \quad v_{2f} = \frac{2m_1}{m_1 + m_2} v_{1i} + \frac{m_2 - m_1}{m_1 + m_2} v_{2i}.$$

9–11 Collisions in Two Dimensions

□ A two-dimensional elastic collision is diagramed on the right. Object 2 is initially at rest and object 1 impinges on it with speed v_{1i} along the x axis. Object 1 leaves the collision with speed v_{1f} along a line that is below the x axis and makes the angle θ_1 with that axis. Ob-

Before Collision   After Collision

ject 2 leaves the collision with speed v_{2f} along a line that is above the x axis and makes the angle θ_2 with that axis. Take the y axis to be upward in the diagram. Conservation of total momentum $\vec{P}$ leads to the equations $m_1 v_{1i} = m_1 v_{1f} \cos\theta_1 + m_2 v_{2f} \cos\theta_2$ (conservation of P_x) and $0 = -m_1 v_{1f} \sin\theta_1 + m_2 v_{2f} \sin\theta_2$ (conservation of P_y). If the collision is elastic, then $\frac{1}{2}m_1 v_{1i}^2 = \frac{1}{2}m_1 v_{1f}^2 + \frac{1}{2}m_2 v_{2f}^2$. These equations can be solved for three unknowns.

□ Note that the masses and the initial velocity of object 1 alone do not determine the outcome of a two-dimensional elastic collision. The magnitude and direction of the impulses acting on the object during the collision also play important roles. Often these are not known, so the outcome cannot be predicted. Experimenters usually measure θ_1 or θ_2, then use the equations to calculate other quantities.

□ If a two-dimensional collision is completely inelastic, the objects stick together after the collision and kinetic energy is not conserved. Suppose that, before the collision object 1 has velocity $\vec{v}_{1i}$ and object 2 has velocity $\vec{v}_{2i}$. After the collision both objects have velocity $\vec{v}_f$. Conservation of total linear momentum leads to $m_1 \vec{v}_{1i} + m_2 \vec{v}_{2i} = (m_1 + m_2)\vec{v}_f$. The two corresponding component equations can be solved for two unknowns. Unlike an elastic collision, the masses and initial conditions completely determine the outcome.

9–12 Systems with Varying Mass: A Rocket

☐ The conservation of momentum principle can be used to find an equation for the acceleration of a rocket if the system includes both the rocket and the fuel it expels. Any external forces, such as gravity, are assumed to be negligible.

☐ Suppose that at time t a rocket with mass M (including fuel) is moving with velocity v. It will eject fuel of mass $-\Delta M$ (a positive quantity) in time Δt. Originally the fuel is traveling with the rocket at velocity v but after it is ejected it has velocity U and the rocket has velocity $v + \Delta v$. Carefully note that v, $v + \Delta v$, and U are all measured relative to an inertial frame and that ΔM is taken to be negative, so the mass of the rocket after ejection is $M + \Delta M$. Also note that the rocket exerts a force on the fuel to eject it and the fuel exerts a force on the rocket of the same magnitude but in the opposite direction.

☐ The rocket and the ejected fuel form a constant mass system on which no external forces act, so the total momentum Mv before the fuel is ejected is equal to the total momentum $-dM\,U + (M + dM)(v + dv)$ after the fuel is ejected. Divide by dt and take the limit as dt becomes vanishingly small. Products of two infinitesimals vanish and the result is $(U - v)\,dM/dt = M\,dv/dt$.

☐ This expression is usually written in terms of the fuel speed *relative to the rocket*. Assume the direction of fuel ejection is opposite the direction of travel of the rocket and let u be the relative speed. Then, $u = v - U$ and $-u\,dM/dt = M\,dv/dt$. The rate at which the rocket loses mass is $R = -dM/dt$ and dv/dt is the acceleration a of the rocket, so the rocket equation becomes $Ru = Ma$. This equation is used to find the acceleration of the rocket in terms of the rate of fuel consumption. The quantity Ru is called the **thrust** of the rocket engine.

☐ Integrate $dv = -u\,dM/M$ from some initial velocity v_i and mass M_i to some final velocity v_f and mass M_f to obtain

$$v_f - v_i = u \ln \frac{M_i}{M_f}.$$

This expression gives the change in the rocket's velocity in terms of the initial and final masses. $M_i - M_f$ is the mass of the expended fuel.

Hints for Questions

1 (a) The two particles must be moving in opposite directions.

(b) The two particles must be moving in opposite directions, be the same distance from the origin, and be on opposite sides of the origin.

(c) The vector sum of the two velocities cannot be zero and the midpoint between them must pass through the origin as they move.

[Ans: (a) ac, cd, bc; (b) bc; (c) bd, ad]

3 Since the particles have equal masses the speed of the center of mass is the magnitude of the vector sum of the velocities divided by the number of particles in the group.

[Ans: d, c, a, b (zero)]

5 The impulse is the integral of the force with respect to time. In each case it is therefore the area under the curve shown on the graph.

[Ans: all tie]

7 The sum of the momenta after the explosion must equal the momentum before the explosion.

[Ans: a, c, e, f]

9 (a) and (b) Look at the slopes of the lines on the graph.

(c) At each instant of time the two objects must be on opposite sides of their center of mass.

[Ans: (a) positive; (b) positive; (c) 2 and 3]

11 If the center of mass is at A then block 1 is more massive than block 2; if it is at B then the two blocks have the same mass; if it is at C then block 2 is more massive than block 1. Recall from your study of one-dimensional elastic collisions that the incident object continues to move forward if it more massive than the target object, bounces back if it is less massive, and stops if it has the same mass.

[Ans: (a) forward; (b) stationary; (c) backward]

Hints for Problems

5 Use $\vec{r}_{com} = (1/M) \sum m_i \vec{r}_i$ to compute the center of mass of the molecule. Here m_i is the mass of particle i, $\vec{r}_i$ is its location, and M is the total mass.

(a) Draw the base triangle, with the given x axis and origin. Find the x coordinates of each of the hydrogen atoms. Note that the line from a hydrogen atom through the triangle center is perpendicular to the opposite side and makes an angle of $30°$ with each of the adjacent sides. Use the right triangle that is formed by the x axis, the line joining the two hydrogen atoms on the left, and the line from one of these atoms through the triangle center. You should be able to show that for each of the two left-hand hydrogen atoms $x = -d \cos 30° = -d/2$. The x coordinate of the nitrogen atom is 0.

(b) The y coordinate of each hydrogen atom is 0. To find the y coordinate of the nitrogen atom use the right triangle that is formed by the x axis, the y axis, and the line joining the nitrogen atom and the right-hand hydrogen atom. You should be able to show that $y = \sqrt{L^2 - d^2}$.

[Ans: (a) 0; (b) 3.13×10^{-11} m]

7 Replace each of the box sides and bottom with a particle that has a mass equal to the mass of the side or bottom and is located at the center of the side or bottom. Then use $\vec{r}_{com} = (1/M) \sum m_i \vec{r}_i$ to compute the center of mass of the system. Here m_i is the mass of a side or bottom i, $\vec{r}_i$ is the location of its center, and M is the total mass. The sides and bottom have equal masses.

[Ans: (a) 20 cm; (b) 20 cm; (c) 16 cm]

11 Suppose the vehicles move along an x axis with its origin at the traffic signal. The x coordinate of the center of mass of the two-vehicle system is given by $x_{om} = (m_a x_a +$

$m_t x_t)/(m_a + m_t)$, where m_a is the mass of the automobile, x_a is its coordinate at time $t = 3.0\,\text{s}$, m_t is the mass of the truck, and x_t is its coordinate at the same time. The coordinate of the automobile is given by $x_a = \frac{1}{2}at^2$ and the coordinate of the truck is given by $x_t = v_t t$, where a is the acceleration of the automobile and v_t is the velocity of the truck. (b) The velocity of the center of mass of the two-vehicle system is given by $v_{om} = (m_a v_a + m_t v_t)/(m_a + m_t)$, where v_a is the velocity of the automobile at $t = 3.0\,\text{s}$. It is given by $v_a = at$.

$\Big[$Ans: (a) 22 m; (b) 9.3 m/s $\Big]$

13 (a) Use $\vec{a}_{\text{com}} = (m_1 \vec{a}_1 + m_2 \vec{a}_2)/(m_1 + m_2)$, where $\vec{a}_1$ is the acceleration of block 1 and $\vec{a}_2$ is the acceleration of block 2. First use Newton's second law to find the magnitude of the acceleration of the blocks. If T is the tension force of the cord, block 1 obeys $T = m_1 a$ and block 2 obeys $m_2 g - T = m_2 a$. Eliminate T and solve for a. Then $\vec{a}_1 = a\hat{i}$ and $\vec{a}_2 = -a\hat{j}$.
(b) Since the acceleration is constant and the system is released from rest the velocity of the center of mass as a function of time t is $\vec{v}_{\text{com}} = \vec{a}_{\text{com}}t$.
(d) The center of mass follows a straight line. If θ is the angle between the path and the x axis, then $\tan\theta = a_{\text{com }y}/a_{\text{com }y}$.

$\Big[$Ans: (a) $(2.35\,\text{m/s}^2)\hat{i} - (1.57\,\text{m/s}^2)\hat{j}$; (b) $\Big[(2.35\,\text{m/s}^2)\hat{i} - (1.57\,\text{m/s}^2)\hat{j}\Big]\,t$; (d) straight, at a downward angle of $34°$ $\Big]$

21 The initial angle is given by $\tan\theta_0 = p_{0y}/p_{0x}$, where p_{0x} is the horizontal component of the momentum and p_{0y} is the vertical component of the moment, both at the launch time $t = 0$. At the highest point on its trajectory the vertical component of the momentum vanishes and since the horizontal component is constant $p_{0x} = p_h$, where p_h is the magnitude of the momentum when the ball is at its highest point. The initial vertical component can be found using $p_{0y} = \sqrt{p_0^2 - p_{0x}^2}$, where p_0 is the magnitude of the momentum at $t = 0$.

$\Big[$Ans: $48°$ $\Big]$

31 (a) The impulse is given by the integral $J = \int F\,dt$ over the duration of a single impact. This is just the area enclosed by one of the triangles on the graph. The area, of course, is half the product of the base ($= 10\,\text{ms}$) and the altitude ($= 200\,\text{N}$).
(b) The average force is given by $F_{\text{avg}} = J/\Delta t$, where Δt is the duration of the impact.
(c) Use $F_{\text{avg}} = \Delta p/\Delta t$, where Δp is the change in the momentum of the snowball stream in time Δt, a time interval that includes many impacts. Since the snowballs stick to the wall, the momentum of each snowball changes by mv, where v is its speed. If N snowballs hit per unit time, the change in the total momentum is $\Delta p = mvN\,\Delta t$.

$\Big[$Ans: (a) $1.00\,\text{N}\cdot\text{s}$; (b) $100\,\text{N}$; (c) $20\,\text{N}$ $\Big]$

39 The momentum of the system consisting of the mess kit (before the explosion) and its parts (after the explosion) is conserved. Thus $M\vec{V} = m\vec{v}_1 + m\vec{v}_2$, where M is the mass of the kit and $\vec{V}$ is its velocity before the explosion, m is the mass of each of the parts and $\vec{v}_1$ and $\vec{v}_2$ are the velocities of the parts after the explosion. Take the x axis to run from west to east and the y axis to run from south to north, then solve the conservation of momentum equation for $\vec{V}$. The original speed of the kit is the square root of the sum of the squares of the components of $\vec{V}$.

[Ans: 3.5 m/s]

49 The momentum of the bullet-bock 1 system is conserved in the encounter of the bullet with that block. Let m be the mass of the bullet and v_0 be its initial velocity. Let v_{int} be the velocity of the bullet after it emerges from block 1 but before it strikes block 2. Let M_1 be the mass of block 1 and let V_1 be its velocity after the bullet emerges from it. Then $mv_0 = mv_{int} + M_1V_1$.

Momentum is also conserved in the encounter of the bullet with block 2. Let M_2 be the mass of that block and let V_2 be its velocity after the bullet becomes embedded in it. Then $mv_{int} = (m + M_2)V_2$. Solve the second equation for v_{int} and then the first for v_0.

[Ans: (a) 721 m/s; (b) 937 m/s]

53 The total momentum of the two-block system is conserved. Let m_1 be the mass of block 1 and v_1 be its initial velocity. Let m_2 be the mass of block 2 and v_2 be its initial velocity. Let V be the velocity of the blocks when the spring has maximum compression. Then $m_1v_1 + m_2v_2 = (m_1 + m_2)V$. Solve for V and then compute the change in the total kinetic energy of the system. The kinetic energy lost by the blocks is stored as elastic potential energy in the compressed spring and is equal to $\frac{1}{2}kx^2$, where k is the spring constant and x is the compression of the spring. Solve for x.

[Ans: 25 cm]

65 The total momentum of the two-particle system is conserved. The x component of the conservation of momentum equation is $m_1v_{1i} = m_1v_{1f}\cos\theta_1 + m_2v_{2f}\cos\theta_2$ and the y component is $0 = -m_1v_{1f}\sin\theta_1 + m_2v_{2f}\sin\theta_2$. Solve the second equation for v_{1f} and the first for v_{1i}. Notice that only the ratio of the masses enters, so you do not need to convert the units.

[Ans: 4.15×10^5 m/s; (b) 4.84×10^5 m/s]

77 The distance block R travels in time t is given by v_Rt, where v_R is its speed relative to the floor. The momentum of the two-block system is conserved, so $m_L\vec{v}_L + m_R\vec{v}_R = 0$, where $\vec{v}_L$ is the velocity of block L relative to the floor. In part (a) this is $\vec{v}_L = -(1.20 \text{ m/s})\hat{\imath}$, where the positive x direction is taken to be to the right. In part (b) $\vec{v}_L = \vec{v}_{rel} + \vec{v}_R$, where $\vec{v}_{rel}$ is the velocity of block L relative to block R. Solve for v_R.

[Ans: (a) 1.92 m; (b) 0.640 m]

83 Momentum is conserved, so $m_1\vec{v}_1 + m_2\vec{v}_2 + m_3\vec{v}_3 = 0$, where $\vec{v}_1$ is the velocity of part1, $\vec{v}_2$ is the velocity of part 2, and $\vec{v}_3$ is the velocity of part 3. Solve the x component of this equation for the x component of the velocity of part 3 and the y component for the y component of its velocity. The speed is the square root of the sum of the squares of the components and the angle between the velocity and the positive x axis is the y component divided by the x component.

[Ans: (a) 11.4 m/s; (b) 95.1°]

87 The momentum of the object is given by $\vec{p} = m\vec{v}$, where m is its mass and $\vec{v}$ $(= d\vec{r}/dt)$ is its velocity. According to Newton's second law the net force on it is given by $\vec{F}_{net} = d\vec{p}/dt$.

[Ans: (a) $(-4.0 \times 10^4$ kg · m/s; (b) west; (c) 0]

97 The total momentum of the cat and sleds is conserved during each takeoff and landing. Just after the cat leave the left sled $m_s\vec{v}_L + m_c\vec{v}_c = 0$, where m_s is the mass of the sled, $\vec{v}_L$ is the velocity of the sled, and $\vec{v}_c$ is the velocity of the cat. For the landing on the right sled $m_c \vec{v}_c = (m_c + m_s)\vec{v}_R$, where $\vec{v}_R$ is the velocity of the sled after the cat lands. Now analyze the jump back to the left sled in the same way.

[Ans: (a) 0.841 m/s; (b) 0.975 m/s]

109 The momentum of the ball-gun system is conserved. That is, $m_b \vec{v}_i = (m_b + m_g)\vec{v}_f$, where m_b is the mass of the ball, m_g is the mass of the gun, $\vec{v}_i$ is the velocity of the ball before it enters the gun, and $\vec{v}_g$ is the velocity of the gun (and ball) after the spring has reached maximum compression. Conservation of energy tells us that the energy stored in the spring is the difference between the original kinetic energy of the ball and the final kinetic energy of the ball and gun together.

[Ans: (a) 4.4 m/s; (b) 0.80]

113 Use Eqs. 9–67 and 9–68 for the velocities of the blocks after they collide. The final velocity of block 2 becomes its initial velocity for its collision with the wall. Take the mass of the wall to be infinite. You now have expressions for the final velocities of the two blocks in terms of $2\vec{i}$ and m_2. Set the two expressions equal to each other and solve for m_2.

[Ans: 2.2 kg]

119 The impulse is equal to the change in momentum. That is, it is the momentum $\vec{p}_i$ before hitting the wall minus the momentum $\vec{p}_f$ after hitting the wall. Since the momentum simply reverses direction $\vec{p}_f = -\vec{p}_i$ and $\Delta \vec{p} = 2\vec{p}_f$. The average force on the ball is the impulse divided by the time of contact of the ball with the wall.

[Ans: (a) 2.18 kg · m/s; (b) 575 N]

123 The linear momentum of the rocket (and its pieces) is conserved. Let M be the mass of the rocket, m_1 be the mass of the first piece, m_2 be the mass of the second piece, $\vec{V}$ be the velocity of the rocket and $\vec{p}_2$ be the momentum of the second piece. Then $M\vec{V} = m_1 \vec{v}_1 + \vec{p}_2$. Solve for $\vec{p}_2$. The kinetic energy of the second piece is $K_2 = p_2^2 / 2m_2$. The kinetic energy produced by the explosion is the final total kinetic energy $(K_1 + K_2)$ of the two pieces minus the initial kinetic energy of the rocket.

[Ans: (a) $(24.0 \text{ kg} \cdot \text{m/s})\hat{i} - (180 \text{ kg} \cdot \text{m/s})\hat{j} + (30.0 \text{ kg} \cdot \text{m/s})\hat{k}$; (b) 4.23 kJ; (c) 4.30 kJ]

127 The horizontal component of the total momentum of the flatcar and wrestler is conserved. Let M be the mass of the flatcar, m be the mass of the wrestler, $\vec{v}_i$ be the initial velocity of the wrestler, $\vec{v}_f$ be the final velocity of the wrestler, and $\vec{V}_f$ be the final velocity of the flatcar. All velocities are relative to the ground. Momentum conservation yields $m\vec{v}_i = m\vec{v}_f + M\vec{V}_f$. You need only the horizontal component of this equation and you want to solve it for the horizontal component of $\vec{V}_f$. For part (a) $\vec{v}_f = \vec{V}_a$. For parts (b) and (c) $\vec{v}_f = \vec{V}_f + \vec{v}_{\text{rel}}$, where $\vec{v}_{\text{rel}}$ is the velocity of the wrestler relative to the flatcar. The relative velocity is in different directions in parts (b) and (c).

[Ans: (a) 0.54 m/s; (b) 0; (c) 1.1 m/s]

137 The total momentum of the two cars is conserved. Let m_1 be the mass of the heavier car and $\vec{v}_{1i}$ be its velocity before the collision. Let m_2 be the mass of the lighter car and $\vec{v}_{2i}$ be its velocity before the collision. To solve part (a) let $\vec{V}$ be the common velocity of the cars after the collision. Then $m_1 \vec{v}_{1i} + m_2 \vec{v}_{2i} = (m_1 + m_2)\vec{V}$. For parts(c) and (d) use Eqs. 9–67 and 9–68 to calculate the final velocities of the cars.

[Ans: (a) 0; (b) 2.25 kJ; (b) 2.25 kJ; (c) 1.61 m/s; (d) 1.00 m/s]

Quiz

Some questions might have more than one correct answer.

1. The center of mass of any object is

    A. inside the object
    B. on the surface of the object
    C. the average position of particles in the object
    D. at the most massive particle in the object
    E. none of the above

2. If we consider any imaginary plane through the center of mass of any object

    A. there is just as much volume of the object on one side of the plane as on the other side
    B. there is just as much surface area of the object on one side of the plane as on the other side
    C. there are just as many particles on one side of the plane as on the other side
    D. there is just as much mass on one side of the plane as on the other side
    E. none of the above are true

3. The center of mass of an object obeys an equation like the Newton's second law equation, where the mass in the equation is

    A. the average mass of a particle in the object
    B. the mass of the most massive particle in the object
    C. the mass of the least massive particle in the object
    D. the average of the least and most massive particles in the object
    E. the total mass of the object

4. The center of mass of an object obeys an equation like the Newton's second law equation, where the force in the equation is

    A. the average force of particles in the object on other particles in the object
    B. the net force of particles in the object on other particles in the object
    C. the average force of particles outside the object on particles in the object
    D. the net force of particles outside the object on particles in the object
    E. none of the above

5. The product of the total mass of an object and the velocity of its center of mass is

    A. the total linear momentum of the object
    B. the total kinetic energy of the object
    C. the average linear momentum of particles in the object
    D. the average kinetic energy of particles in the object
    E. the acceleration of the object

6. Object A has twice the momentum of object B. Stopping object A requires

    A. twice the force needed to stop object B
    B. half the force needed to stop object B
    C. twice the impulse needed to stop object B
    D. half the impulse needed to stop object B
    E. more than twice the force need to stop object B

7. In all two-body collisions for which external forces are negligible

    A. the total kinetic energy of the colliding objects is conserved
    B. the total linear momentum of the colliding objects is conserved
    C. the velocity of the center of mass of the two colliding objects does not change
    D. the accelerations of the two objects are equal in magnitude and opposite in direction
    E. the forces of the colliding objects on each other are equal in magnitude and opposite in direction

8. Collisions are labeled elastic or inelastic according to

    A. the total momentum lost during the collision
    B. the loss in the speed of the center of mass of the colliding objects
    C. the total mass lost during the collision
    D. the total kinetic energy lost during the collision
    E. the kinetic energy of the incident object lost during the collision

9. In a completely inelastic collision

    A. the two colliding objects move with the same velocity after the collision
    B. all kinetic energy is lost
    C. all linear momentum is lost
    D. as much kinetic energy is lost as is consistent with the conservation of linear momentum
    E. as much linear momentum is lost as is consistent with the conservation of energy

10. During a collision involving two objects with unequal mass, the magnitude of the change in the momentum of the more massive object is

    A. greater than the magnitude of the change in the momentum of the less massive object
    B. less than the magnitude of the change in the momentum of the less massive object
    C. equal to the magnitude of the change in the momentum of the less massive object
    D. zero
    E. determined by the impulse on it during the collision

11. During every two-body collision

    A. the center of mass of the two-body system does not move
    B. the velocity of the center of mass of the two-body system does not change
    C. the acceleration of the center of mass of the two-body system is in the direction of the total momentum
    D. the acceleration of the center of mass of the two-body system is zero
    E. none of the above

12. A rocket accelerates because
    A. its mass is reduced as the exhaust gases leave
    B. the exhaust gases push against the atmosphere
    C. the exhaust gases push against the rocket
    D. the exhaust gases expand after they leave the rocket
    E. the total momentum of the rocket and its exhaust gases is not conserved

Answers: (1) E; (2) D; (3)E; (4) D; (5) A; (6) C; (7) B, C, E; (8) D; (9) A, D; (10) C, E; (11) B, D; (12) C

Chapter 10
ROTATION

You now begin the study of rotational motion. The pattern is similar to that used for the study of linear motion: first learn how to describe the motion, then learn what changes it. Take special care to understand the definitions of the various angular quantities and the roles they play. Concentrate on Newton's second law for rotation and on the work-kinetic energy theorem for rotation.

Important Concepts

- ☐ angular position
- ☐ angular displacement
- ☐ average angular velocity
- ☐ (instantaneous) angular velocity
- ☐ average angular acceleration
- ☐ (instantaneous) angular acceleration
- ☐ rotational kinetic energy

- ☐ rotational inertia
- ☐ torque
- ☐ Newton's second law for rotation
- ☐ work done by a torque
- ☐ work-kinetic energy theorem for rotation

Overview

10–2   The Rotational Variables

- ☐ If a rigid body undergoes pure rotation about a fixed axis, then the path followed by each point on the body is a circle, centered on the axis of rotation.

- ☐ The **angular position** of the body is described by giving the angle between a **reference line**, fixed to the body, and a non-rotating coordinate axis. If the body rotates around the z axis, the reference line rotates in the xy plane and the x axis might be chosen as the fixed axis. Usually the angle θ that gives the angular position is measured counterclockwise from the fixed coordinate axis.

- ☐ If the body rotates from θ_1 to θ_2, then its **angular displacement** is $\Delta\theta = \theta_2 - \theta_1$. If θ_2 is greater than θ_1, then the angular displacement is positive. According to the convention stated above, the body rotated in the counterclockwise direction. If θ_2 is less than θ_1, then the angular displacement is negative and the body rotated in the clockwise direction.

- ☐ If the body rotates through more than one revolution, θ continues to increase beyond 2π rad. An angle of 2π rad is not equivalent to 0.

□ If the body undergoes an angular displacement $\Delta\theta$ in time Δt, its **average angular velocity** during the interval is

$$\omega_{\text{avg}} = \frac{\Delta\theta}{\Delta t} .$$

Its **instantaneous angular velocity** at any time is given by the derivative

$$\omega = \frac{d\theta}{dt} .$$

According to the convention given above, a positive value for ω indicates rotation in the counterclockwise direction and a negative value for ω indicates rotation in the clockwise direction. Instantaneous angular velocity is usually called simply angular velocity.

□ If the angular velocity of the body changes with time, then the body has a non-vanishing **angular acceleration**. If the angular velocity changes from ω_1 to ω_2 in the time interval Δt, then the average angular acceleration in the interval is

$$\alpha_{\text{avg}} = \frac{\Delta\omega}{\Delta t} ,$$

where $\Delta\omega = \omega_2 - \omega_1$. The **instantaneous angular acceleration** at any time t is given by the derivative

$$\alpha = \frac{d\omega}{dt} .$$

Instantaneous angular acceleration is usually called simply angular acceleration.

□ Common units of angular velocity are deg/s, rad/s, rev/s, and rev/min. Corresponding units of angular acceleration are deg/s^2, rad/s^2, rev/s^2, and rev/min^2.

□ A positive angular acceleration does not necessarily mean the rotational speed is increasing and a negative angular acceleration does not necessarily mean the rotational speed is decreasing. The rotational speed decreases if ω and α have opposite signs and increases if they have the same sign, no matter what the signs are.

□ The values of $\Delta\theta$, ω, and α are the same for every point in a rigid body. The angular positions of different points may be different, of course, but when one point rotates through any angle, all points rotate through the same angle. All points rotate through the same angle in the same time and their angular velocities change at the same rate.

10–3   Are Angular Quantities Vectors?

□ Angular velocity and angular acceleration, but not angular position or displacement, can be considered to be vectors along the axis of rotation. To determine the direction of $\vec{\omega}$, use the right hand rule: curl the fingers of your right hand around the axis in the direction of rotation. Your thumb then points in the direction of $\vec{\omega}$.

□ If the body rotates faster as time goes on, then $\vec{\alpha}$ and $\vec{\omega}$ are in the same direction. If it rotates more slowly, then $\vec{\alpha}$ and $\vec{\omega}$ are in opposite directions.

□ Although directions can be assigned arbitrarily to angular displacements, they do not add as vectors. The result of two successive angular displacements around different axes that are not parallel, for example, depends on the order in which the angular displacements are carried out. Thus, the rules of vector addition are not obeyed.

10-4 Rotation with Constant Angular Acceleration

☐ If a body is rotating around a fixed axis with constant angular acceleration α, then as functions of the time t its angular velocity is given by

$$\omega(t) = \omega_0 + \alpha t$$

and its angular position is given by

$$\theta(t) = \theta_0 + \omega_0 t + \tfrac{1}{2}\alpha t^2 \ .$$

Here ω_0 is the angular velocity at $t = 0$ and θ_0 is the angular position at $t = 0$. The reference line can usually be placed so $\theta_0 = 0$. These two equations can be solved simultaneously to find values for two of the symbols that appear in them. Another useful equation can be obtained by using one of the equations to eliminate t from the other:

$$\omega^2 - \omega_0^2 = 2\alpha\theta \ ,$$

where θ_0 was taken to be 0.

☐ Any consistent set of units can be used. Thus, θ in degrees, ω in degrees/s, and α in degrees/s^2 as well as θ in radians, ω in radians/s, and α in radians/s^2 or θ in revolutions, ω in revolutions/s, and α in revolutions/s^2 can be used. Do not mix units, however.

10-5 Relating the Linear and Angular Variables

☐ Each point in a rotating body has a translational velocity and acceleration and these are related to the angular variables. Consider a point in the body a perpendicular distance r from the axis of rotation. If the body turns through the angle θ (in radians), the point moves a distance $s = \theta r$ along its circular path. The speed of the point is $v = ds/dt = (d\theta/dt)r = \omega r$. The units of ω MUST be rad/s. The tangential component of the acceleration of the point is $a_t = dv/dt = (d\omega/dt)r = \alpha r$, where the units of α MUST be rad/s^2.

☐ Because the point is moving in a circular path its acceleration also has a radial component: $a_r = v^2/r$. In terms of the angular velocity, $a_r = \omega^2 r$. In terms of the components a_r and a_t, the magnitude of the total acceleration is given by $a = \sqrt{a_r^2 + a_t^2}$.

☐ Notice that s, v, a_t, and a_r are proportional to r. Compared to a point on the rim of a rotating wheel, for example, a point halfway out travels half the distance, has half the speed, has half the tangential acceleration, and has half the radial acceleration.

10-6 Kinetic Energy of Rotation

☐ Suppose a rigid body, rotating about a fixed axis, is made up of N particles. The total kinetic energy is given by $K = \sum \tfrac{1}{2}m_i v_i^2$, where m_i is the mass of particle i, v_i is its speed, and the sum is over all particles in the body. Substitute $v_i = \omega r_i$, where r_i is the distance of particle i from the axis, and obtain $K = \tfrac{1}{2}\left(\sum m_i r_i^2\right)\omega^2$. ω MUST be in rad/s.

☐ The sum $\sum m_i r_i^2$ is called the **rotational inertia** of the body and is denoted by I. Thus, the kinetic energy of a rotating body is $K = \tfrac{1}{2}I\omega^2$.

10–7 Calculating the Rotational Inertia

☐ Rotational inertia is a property of a body that depends on the distribution of mass in the body and on the axis of rotation. A particle far from the axis of rotation contributes more to the rotational inertia than a particle of the same mass closer to the axis.

☐ The sum $\sum m_i r_i^2$ may be difficult to evaluate for a body with more than a few particles. Some bodies, however, can be approximated by a continuous distribution of mass and techniques of the integral calculus can be used to find the rotational inertia. The body is divided into a large number of small regions, each with mass dm. The contribution of each region to the rotational inertia is computed as if the region were a particle; then, the results are summed: $I = \int r^2\, dm$. Table 10–2 of the text gives the rotational inertias of several bodies.

☐ The defining equation for the rotational inertia is a sum over all particles in the body. If we like, we can consider the body to be composed of two or more parts and calculate the rotational inertia of each part about the same axis, then add the results to obtain the rotational inertia of the complete body.

☐ The defining equation for rotational inertia is used to prove the parallel-axis theorem. Consider two identical bodies, one rotating about an axis through the center of mass and the other rotating about an axis that is parallel to the first axis but is a distance h from the center of mass. The rotational inertia I for the second body is related to the rotational inertia I_{com} for the first by $I = I_{\text{com}} + Mh^2$, where M is the total mass of the body.

☐ The parallel-axis theorem tells us that of all the places we can position the axis of rotation, the one that leads to the smallest rotational inertia is the one through the center of mass and that the rotational inertia increases as the axis moves away from the center of mass (remaining parallel to its original orientation, of course).

10–8 Torque

☐ A **torque** is associated with any force that is applied to an object and is not along the line from the axis of rotation to the point of application. If the force is perpendicular to the rotation axis, then the torque is given by $\tau = rF_t$, where r is the distance from the axis of rotation to the point of application and F_t is the component of the force perpendicular to the line 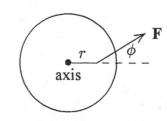 from the axis to the application point. In the diagram on the right, the torque associated with **F** is $\tau = rF \sin\phi$. Notice that the magnitude of a torque depends on the distance from the axis to the point of application. A force produces a greater torque if it is applied far from the axis than if it applied closer.

☐ For rotation about a fixed axis, a torque is taken to be positive if it tends to turn the object counterclockwise and negative if it tends to turn the body clockwise. This convention is consistent with the one introduced earlier for the signs of the angular velocity and acceleration.

10–9 Newton's Second Law for Rotation

☐ This law relates the net torque τ_{net} acting on a rigid body to the angular acceleration α of the body:

$$\tau_{\text{net}} = I\alpha,$$

where I is the rotational inertia of the body about the axis of rotation. It is as important to the study of rotational motion about a fixed axis as $F_{net} = ma$ is to the study of one-dimensional translational motion. It can be derived directly from Newton's second law.

☐ Be careful that you use the net torque in this equation. You must identify and sum all the torques that are acting and you must be careful about signs. You must also be careful to use radian measure for the angular acceleration α.

10–10 Work and Rotational Kinetic Energy

☐ Consider a particle traveling counterclockwise in a circular orbit, subjected to a force with tangential component F_t. As it moves through an infinitesimal arc length ds the magnitude of the work done by the force is d$W = F_t$ ds. Since d$s = r$ dθ, where dθ is the infinitesimal angular displacement, and $\tau = F_t r$, the expression for the work becomes d$W = \tau$ dθ. As the particle travels from θ_1 to θ_2 the work done by the torque is given by the integral

$$W = \int_{\theta_1}^{\theta_2} \tau \, d\theta \, .$$

☐ The work done by a torque may be positive or negative. If the torque and angular displacement are in the same direction, both clockwise or both counterclockwise, then the work is positive; if they are in opposite directions, one clockwise and one counterclockwise, then the work is negative. If more than one torque acts, sum the works done by the individual torques to find the total work.

☐ In terms of the torque τ and angular velocity ω, the power supplied by the torque is given by $P = dW/dt = \tau\omega$.

☐ A work-kinetic energy theorem is valid for rotational motion: the total work done by all torques acting on a body equals the change in its rotational kinetic energy.

Hints for Questions

<u>1</u> (a), (b), and (c) The angular velocity is the derivative with respect to time of the angular position. This is the slope of the curve shown on the graph.
(d) If the slope is algebraically increasing with time the acceleration is positive; if it is decreasing the acceleration is negative.
[Ans: (a) positive; (b) zero; (c) negative; (d) negative]

<u>3</u> (a) and (b) The direction of rotation reverses at a point where the slope of the graph is zero. The angular speed is zero at such a point.
(c) Determine if the slope is increasing or decreasing with time.
[Ans: (a) 1: counterclockwise (positive); 2: counterclockwise (positive); 3: at $\theta = 0$; (b) 1: before; 2: at $t = 0$; 3: after; (c) 1: positive; 2: negative; 3: positive]

<u>5</u> If the bar is not to turn the net torque must remain zero as the direction of $\vec{F}_2$ changes. This means that the magnitude of $\vec{F}_1$ must be equal to the perpendicular component of $\vec{F}_2$.
[Ans: larger]

7 The magnitude of the torque is the product of the magnitude of the force and its moment arm, the perpendicular distance from P to the line of the force. Since all the forces have the same magnitude you should rank the torques according to their moment arms.

$\left[\text{Ans: } \vec{F}_5, \vec{F}_4, \vec{F}_2, \vec{F}_1, \vec{F}_3 \text{ (zero)}\right]$

9 (a) In each case the magnitude of the torque is the product of the magnitude of the force and the distance from the center of the disk to the point of application of the force.

(b) The rotational inertia is given by $\sum m_i r_i^2$, where m_i is the mass of particle i in the disk and r_i is its distance from the disk center. The sum is over all particles in the disk. Look to see how the mass is distributed in the disks. Is more mass near the rotation axis for one disk than for the other, for example?

(c) The magnitude of the angular acceleration is the magnitude of the torque divided by the rotational inertia.

$\left[\text{Ans: (a) 1 and 2 tie, then 3; (b) 1 and 3 tie, then 2; (c) 2, 1, 3}\right]$

Hints for Problems

11 Use the equation for constant-angular acceleration rotation: $\theta = \omega_0 t + \frac{1}{2}\alpha t^2$, where θ is the angular position at time t (measured from the angular position at $t = 0$), ω_0 is the angular velocity at $t = 0$, and α is the angular acceleration. Solve for α.

(b) the average angular velocity is the angular displacement divided by the time interval.

(c) Evaluate the constant-angular acceleration equation $\omega = \omega_0 + \alpha t$.

(d) Use $\theta = \omega_0 t + \frac{1}{2}\alpha t^2$. You might take the time to be zero at the end of the first 5 seconds and take ω_0 to be the answer to part (c).

$\left[\text{Ans: (a) } 2.0\,\text{rad/s}^2; \text{ (b) } 5.0\,\text{rad/s; (c) } 10\,\text{rad/s; (d) } 75\,\text{rad}\right]$

15 (a) Use the equations for rotation with constant angular acceleration: $\theta = \omega_0 t + \frac{1}{2}\alpha t^2$ and $\omega = \omega_0 + \alpha t$, where θ is the angular position at time t (measured from the angular position at $t = 0$), ω is the angular velocity at time t, ω_0 is the angular velocity at $t = 0$, and α is the angular acceleration. Use the second equation to eliminate α from the first and obtain $\theta = \frac{1}{2}(\omega + \omega_0)t$. Solve for t. When you evaluate the result, convert 40 rev to radians.

(b) Solve $\omega = \omega_0 + \alpha t$ for α.

(c) Solve $\theta = \omega_0 t + \frac{1}{2}\alpha t^2$ for t.

$\left[\text{Ans: (a) } 3.4 \times 10^2\,\text{s; (b) } -4.5 \times 10^{-3}\,\text{rad/s}^2; \text{ (c) } 98\,\text{s}\right]$

17 (a) The reference line is at the maximum turning angle when the angular velocity is zero. Use $\omega = \omega_0 + \alpha t$, where ω is the angular velocity at time t, ω_0 is the angular velocity at $t = 0$, and α is the angular acceleration. Solve for t and use the result to evaluate $\theta = \omega_0 t + \frac{1}{2}\alpha t^2$, where θ is the angular position of the reference line at time t.

(b) and (c) Solve $\frac{1}{2}\theta_{max} = \omega_0 t + \frac{1}{2}\alpha t^2$ for t. There are two solutions. Pick the one that makes physical sense for this problem.

(d) and (e) Solve $-10.5\,\text{rad} = \omega_0 t + \frac{1}{2}\alpha t^2$ for t. There are two solutions. Again pick the one that makes physical sense for this problem.

$\left[\text{Ans: (a) } 44\,\text{rad; (b) } 5.5\,\text{s; (c) } 32\,\text{s; (d) } -2.1\,\text{s; (e) } 40\,\text{s}\right]$

23 (a) The angular velocity ω and the speed v are related by $v = \omega r$, where r is the radius of the path. To obtain the result in radians per second substitute the speed in kilometers per second and the radius in kilometers.

(b) The radial component of the acceleration is given by $a_r = \omega^2 r$. The angular velocity must be in radians per second.

(c) The tangential component of the acceleration is given by $a_t = \alpha r$. The angular acceleration must be in radians per second squared.

$\left[\text{Ans: (a) } 2.50 \times 10^{-3}\,\text{rad/s; (b) } 20.2\,\text{m/s}^2; \text{ (c) } 0\right]$

25 The tangential component of the acceleration is given by $a_t = \alpha r$ and the radial component is given by $a_r = \omega^2 r$, where α is the angular acceleration, ω is the angular velocity, and r is the distance from the rotation axis to the point. Use $\omega = d\theta/dt$ to find ω and $\alpha = d\omega/dt$ to calculate α.

$\left[\text{Ans: (a) } 6.4\,\text{cm/s}^2; \text{ (b) } 2.6\,\text{cm/s}^2\right]$

31 (a) The tangential component of the acceleration is zero since the angular acceleration of the turntable is zero. The radial component is given by $a_r = \omega^2 r$, where ω is the angular speed in radians per second. Use $1\,\text{rev} = 2\pi\,\text{rad}$ and $1\,\text{min} = 60\,\text{s}$ to convert $33\frac{1}{3}\,\text{rev/min}$ to radians per second.

(b) The magnitude of the frictional force of the turntable on the seed is $f = ma_r$, where m is the mass of the seed. If the seed does not slip this is less than or equal to $\mu_s F_N$, where μ_s is the coefficient of static friction between the turntable and the seed and F_N is the normal force of the turntable on the seed. Show that $F_N = mg$, then calculate μ_s so that $f = \mu_s F_N$.

(c) The maximum angular speed in the interval occurs for $t = 0.25\,\text{s}$ and is given by $\omega = \alpha t$, where α is the angular acceleration. Use this value to compute f ($= m\omega^2 r$), then use $f = \mu_s F_N$ to compute μ_s.

$\left[\text{Ans: (a) } 73\,\text{cm/s}^2; \text{ (b) } 0.075; \text{ (c) } 0.11\right]$

39 (a) The rotational inertia of a rod of length d and mass M, rotating about an axis through its center and perpendicular to it, is $\frac{1}{12}Md^2$ (see Table 10–2). Use the parallel-axis theorem to find the rotational inertias of the rods in this problem. For one rod the rotation axis is $d/2$ from its center and for the other it is $3d/2$ from its center. The rotational inertia of either of the particles is the product of its mass and the square of the its distance from the rotation axis. Sum the rotational inertias to obtain the total rotational inertia. (b) The rotational kinetic energy is given by $K = \frac{1}{2}I\omega^2$, where I is the total rotational inertia.

$\left[\text{Ans: (a) } 0.023\,\text{kg}\cdot\text{m}^2; \text{ (b) } 11\,\text{mJ}\right]$

51 Newton's second law for rotation is $\tau_{\text{net}} = I\alpha$, where I is the rotational inertia of the disk, α is its angular acceleration, and τ_{net} is the net torque on it. Use $\omega = \alpha t$ to find the angular acceleration. Here ω is the final angular velocity. Use $I = \frac{1}{2}MR^2$ to find the rotational inertia. Here M is the mass of the disk and R is its radius. The net torque is $\tau_{\text{net}} = F_2 R - F_1 R$, where counterclockwise rotation was taken to be positive. Substitute these expressions (or values) in the second law equation and solve for F_2.

$\left[\text{Ans: } 0.140\,\text{N}\right]$

53 Use $\tau_{\text{net}} = I\alpha$, where τ_{net} is the net torque on the cylinder, I is its rotational inertia, and α is its angular acceleration. The net torque is the sum of the individual torques. The magnitude of the torque associated with $\vec{F}_1$ is F_1R; the magnitude of the torque associated with $\vec{F}_2$ is F_2R; the magnitude of the torque associated with $\vec{F}_3$ is F_3r; the magnitude of the torque associated with $\vec{F}_4$ is 0. If the torque, acting alone, tends to produce an angular acceleration in the counterclockwise direction it enters the sum as a positive value; if it tends to produce an angular acceleration in the clockwise direction it enters the sum as a negative value. The rotational inertia is $I = \frac{1}{2}MR^2$, where M is the mass of the cylinder. Solve for α. If the result is positive the angular acceleration is in the counterclockwise direction; if it is negative the angular acceleration is in the clockwise direction.

$\big[$Ans: (a) $9.7\,\text{rad/s}^2$; (b) counterclockwise $\big]$

59 Take the system to consist of the pulley, the falling block, and Earth. Use conservation of mechanical energy. Initially the pulley and block are at rest, so the kinetic energy of the system is zero. Take the (gravitational) potential energy to be zero. When the block has fallen a distance h, the block has speed v and the pulley has angular speed ω. The kinetic energy is then $\frac{1}{2}mv^2 + \frac{1}{2}I\omega^2$, where I is the rotational inertia of the pulley, which you may take to be $\frac{1}{2}MR^2$. The potential energy is $-mgh$. Now the speed of the block is the same as the speed of the cord, which is also the speed of a point on the rim of the pulley. Thus $v = \omega R$. Use this relationship to replace ω in the expression for the kinetic energy, then solve the energy equation for v. Notice that R cancels from the algebraic solution.

$\big[$Ans: (a) $1.4\,\text{m/s}$; (b) $1.4\,\text{m/s}$ $\big]$

61 (a) The rotational kinetic energy is $K = \frac{1}{2}I\omega^2$, where I is the rotational inertia of the rod and ω is its angular speed in radians per second at the lowest point. The rotational inertia of a rod with mass M and length L, rotating on an axis through its center and perpendicular to it, is $\frac{1}{12}ML^2$. See Table 10–2. Use the parallel-axis theorem to show that if the rod is pivoted at one end its rotational inertia is $I = \frac{1}{12}ML^2 + \frac{1}{4}ML^2 = \frac{1}{3}ML^2$.
(b) If the center of mass rises a distance h, the potential energy increases by Mgh and the kinetic energy decreases to zero. The conservation of energy equation becomes $\frac{1}{2}I\omega^2 = Mgh$. Solve for h.

$\big[$Ans: (a) $0.63\,\text{J}$; (b) $0.15\,\text{m}$ $\big]$

65 Use conservation of mechanical energy. The initial kinetic energy is zero. Take the potential energy to be zero when the hoop and rod are level with the rotation axis. Then the initial potential energy of the rod is $mgL/2$ and the initial potential energy of the hoop is $mg(L+R)$. (Remember that the potential energy is proportional to the height of the center of mass above the reference level.) The final kinetic energy is $\frac{1}{2}(I_r + I_h)\omega^2$, where ω is the angular speed of the assembly as it passes the lowest point. Here I_r is the rotational inertial of the rod and I_h is the rotational inertia of the hoop. The final potential energy of the rod is $-mgL/2$ and the final potential energy of the hoop is $-mg(L + R)$.
You must now find the rotational inertias of the hoop and rod. Table 10–2 gives $\frac{1}{12}mL^2$ for the rotational inertia of a rod of mass m and length L about a rotation axis through its center and perpendicular to it. The rod of this problem rotates about an end. Use the parallel axis theorem to show that $I_r = \frac{1}{3}mL^2$. Table 10–2 gives $\frac{1}{2}mR^2$ for the rotational inertia of a hoop of mass m and radius R about a diameter. The hoop of this problem rotates about an axis that is $R + L$ from its center, so $I_h = \frac{1}{2}mR^2 + m(R + L)^2$.

Now you should be able to solve the energy equation for ω.

$\left[\text{Ans: } 9.82\,\text{rad/s}\,\right]$

75 Your angular speed is given by $\omega = v/r$, where v is your speed and r is the radius of your path. The linear speed of the cheetah is $v_c = \omega r_c$, where r_c is the radius of the cheetah's path. For this equation to be valid ω must be in radians per second.

$\left[\text{Ans: (a) } 0.32\,\text{rad/s; (b) } 1.0 \times 10^2\,\text{km/h}\,\right]$

79 The angular acceleration is $\Delta\omega/\Delta t$, where $\Delta\omega$ is the change in the angular velocity during the time interval Δt. Use Newton's second law for rotation to find the retarding torque. All of the initial rotational kinetic energy is converted to thermal energy. Use $\Delta\theta = \omega_0 t + \frac{1}{2}\alpha t^2$ to find the angular displacement in time t. Here ω_0 is the initial angular velocity and α is the angular acceleration. Because the rod is slowing you know that ω_0 and α have opposite signs.

$\left[\text{Ans: (a) } -7.66\,\text{rad/s}^2\text{; (b) } -11.7\,\text{N}\cdot\text{m; (c) } 4.59 \times 10^4\,\text{J; (d) } 624\,\text{rev; (e) } 4.59 \times 10^4\,\text{J}\,\right]$

81 The linear speed v of a point that is a distance r from the wheel center is related to the angular speed ω of the wheel by $v = \omega r$. Use the angular speed of pulley A to compute the linear speed of a point on belt 1, then use that speed to compute the angular speed of pulley B. The angular speed of pulley B' is the same as the angular speed of pulley B. Use that angular speed to compute the linear speed of a point on belt 2 and use that linear speed to compute the angular speed of pulley C.

$\left[\text{Ans: (a) } 1.5 \times 10^2\,\text{cm/s; (b) } 15\,\text{rad/s; (c) } 15\,\text{rad/s; (d) } 75\,\text{cm/s; (e) } 3.0\,\text{rad/s}\,\right]$

87 Use $\omega = \omega_0 + \alpha t$ to compute the angular acceleration required to bring a sphere from angular speed ω_0 to angular speed ω in time t. According the Newton's second law for rotation the torque is $\tau = I\alpha$, where I is the rotational inertia. A tangential force with magnitude F, applied at the equator, produces a torque Fr, where r is the radius of the sphere. See Table 10–2 for the rotational inertia of a sphere that rotates about a diameter.

$\left[\text{Ans: (a) } 0.689\,\text{N}\cdot\text{m; (b) } 3.05\,\text{N; (c) } 9.84\,\text{N}\cdot\text{m; (d) } 11.5\,\text{N}\,\right]$

93 The rotational inertia is mr^2, where m is the mass of the ball and r is the distance from the ball to the rotation axis. If the angular speed of the system is constant, you know that the net torque is zero. Thus the frictional and applied torques have the same magnitude.

$\left[\text{Ans: (a) } 0.791\,\text{kg}\cdot\text{m}^2\text{; (b) } 1.79 \times 10^{-2}\,\text{N}\cdot\text{m}\,\right]$

99 Draw a diagram and use a little geometry to find the distance of each particle from the rotation axis. The rotational inertia is the sum $\sum m_i r_i^2$, where m is the mass of particle i and r_i is the distance from the particle to the rotation axis.

$\left[\text{Ans: (a) } 0.17\,\text{kg}\cdot\text{m}^2\text{; (b) } 0.22\,\text{kg}\cdot\text{m}^2\text{; (c) } 0.10\,\text{kg}\cdot\text{m}^2\,\right]$

105 Solve $\theta_2 = \theta_1 + \omega_1 t + \frac{1}{2}\alpha t^2$ and $\omega_2 = \omega_1 + \alpha t$ for ω_1 and α. Here ω_1 is the angular velocity when the angular position is θ_1 and ω_2 is the angular velocity when the angular position is θ_2. Solve $0 = \omega_1 + \alpha t$ for the time t when the disk is at rest, then evaluate $\theta = \theta_1 + \omega_1 t + \frac{1}{2}\alpha t^2$ for the angular position then.

$\left[\text{Ans: (a) } 5.00\,\text{rad/s; (b) } 1.67\,\text{rad/s}^2\text{; (c) } 2.50\,\text{rad}\,\right]$

109 The angular velocity is given by the integral $\int \alpha\,dt$ and the angular position is given by $\int \theta\,dt$. To evaluate the constants of integration use the condition that $\omega = \omega_0$ when $t = 0$ and

take θ to be θ_0 when $t = 0$.

[Ans: (a) $\omega_0 + at^4 - bt^3$; (b) $\theta_0 + \omega_0 t + at^5/5 - bt^4/4$]

<u>117</u> The speed of the car at time t is $v = a_t t$, where a_t (= $0.500\,\text{m/s}^2$) is the magnitude of its tangential acceleration. The radial acceleration is $a_r = v/r^2$, where r is the radius of the track. Since these two acceleration components are perpendicular to each other the magnitude of the net acceleration is the square root of the sum of their squares. The tangential acceleration is $a_t = a\cos\phi$ and the radial acceleration is $a\sin\phi$, where a is the magnitude of the net acceleration and ϕ is the angle between the net acceleration and the tangent to the path. The velocity is tangent to the path but you must decide if the angle is less than $90°$ or greater than $90°$.

[Ans: (a) $1.94\,\text{m/s}^2$; (b) $75.1°$]

Quiz

Some questions might have more than one correct answer.

1. Here are expressions for the angular positions θ as functions of time t for four rigid bodies that are rotating on fixed axes. θ is in radians and t is in seconds.

   I. $\theta = 6 - 3t^2 + 4t^3$
   II. $\theta = 7 - 3t + 5t^2$
   III. $\theta = 25 + 2t + 4t^2$
   IV. $\theta = 18 - 4t - 3t^2$

   Rank them according to the magnitudes of their angular velocities at $t = 0$, least to greatest.

   A. I, II, III, IV
   B. II, III, I, IV
   C. IV, I, III, II
   D. III, I, IV, II
   E. I, III, II, IV

2. Here are expressions for the angular position θ as a function of time t for five rigid bodies that are rotating about a fixed axis. θ is in radians and t is in seconds. Which expressions describe a body that has constant angular acceleration?

   A. $\theta = 5t - 6t^2$
   B. $\theta = -4t^2$
   C. $\theta = 5t - 6t^3$
   D. $\theta = 5t + 6t^2$
   E. $\theta = 3t + 6t^2 - 2t^3$

3. A rigid body that is rotating with constant angular acceleration about a fixed axis
    A. can never have an angular velocity of zero
    B. cannot reverse its direction of rotation
    C. cannot reverse its direction of rotation twice
    D. cannot have an angular velocity and an angular acceleration with opposite signs
    E. cannot go around more than once

4. Wheel I is a uniform thin cylinder. Four other wheels are identical except that wheel II has a hole drilled through its center, wheel III has an identical hole drilled near the rim, wheel IV has a small amount of material added near its center, and wheel V has a small amount of material added near its rim. Rank the wheels according to their rotational inertias for rotation about a fixed axes through their centers, least to greatest.
    A. I, II, III, IV, V
    B. V, IV, III, II, I
    C. V, IV, I, II, III
    D. II, III, I, IV, V
    E. III, II, I, IV, V

5. The rotational inertia of a disk of mass M and radius R rotating about an axis that is through its center and perpendicular to its faces is $\frac{1}{2}MR^2$. The disk is suspended from the ceiling by means of a string of length L attached to its rim. The rotational inertia of the disk about the point of attachment at the ceiling is
    A. $\frac{1}{2}MR^2$
    B. $\frac{3}{2}MR^2$
    C. $\frac{1}{2}M(R^2 + L^2)$
    D. $\frac{1}{2}MR^2 + M(R+L)^2$
    E. $\frac{1}{2}M(R+L)^2$

6. In three different experiments a force is applied to an object that is free to rotate on a fixed axis. In each case the force is applied the same distance from the rotation axis but has a different magnitude and a different orientation in different experiments. Values of the magnitude F of the force and the angle ϕ it makes with the radial line from the rotation axis to the point of application are
    I. $F = 30\,\text{N}$, $\phi = 0$
    II. $F = 20\,\text{N}$, $\phi = 45°$
    III. $F = 10\,\text{N}$, $\phi = 90°$
    Rank the torques according to their magnitudes, least to greatest.
    A. I, II, III
    B. I, III, II
    C. II, I, III
    D. III, I, II
    E. II, III, I

7. Starting from rest, a constant torque is applied to a wheel that is free to rotate on a fixed axis. At the end of time t it has rotated though the angle θ. If the torque is doubled the wheel would rotate through the same angle in time

  A. $2t$
  B. $t/2$
  C. t
  D. $\sqrt{2}t$
  E. $t/\sqrt{2}$

8. The rotational inertia of a uniform cylinder that is free to rotate about the cylinder axis is given by $\frac{1}{2}MR^2$, where M is its mass and R is its radius. When a force with magnitude F is applied tangentially to the rim of the cylinder, its angular acceleration is α. When the same force is applied tangentially to a cylinder made of the same material and with the same length but with a radius that is twice as great, its angular acceleration is

  A. 2α
  B. $\alpha/2$
  C. 4α
  D. $\alpha/4$
  E. $\alpha/8$

9. Four cylinders that are free to rotate about their cylinder axes have the following rotational inertias I and angular velocities ω:

  I. $I = I_0$, $\omega = \omega_0$
  I. $I = 2I_0$, $\omega = \omega_0/2$
  I. $I = I_0/2$, $\omega = 2\omega_0$
  I. $I = I_0/3$, $\omega = 3\omega_0$

Rank them according to their rotational kinetic energies, least to greatest.

  A. I, II, III, IV
  B. I, II and III tied, IV
  C. II, I, III, IV
  D. IV, II and III tied, I
  E. II, III, IV, I

10. Constant torque τ is applied to a object that starts from rest and is free to rotate on a fixed axis. At the end of time t the rotational kinetic energy of the object is K. If the process is repeated but with the torque doubled, the kinetic energy at the end of time t is

  A. $K/2$
  B. K
  C. $2K$
  D. $4K$
  E. $8K$

Answers: (1) E; (2) A, B, D; (3) C, D; (4) E; (5) D; (6) B; (7) E; (8) E; (9) C; (10) D

Chapter 11
ROLLING, TORQUE, AND ANGULAR MOMENTUM

This chapter starts with a discussion of objects that roll; that is, they simultaneously translate and rotate. Learn the relationship between the velocity of the center of mass and the angular velocity when an object rolls without sliding. Next, rotational motion is discussed using the concept of angular momentum. Learn the definition for a single particle and learn how to calculate the total angular momentum of a collection of particles. Newton's second law for rotation describes the change in angular momentum brought about by the net external torque on a body. Angular momentum is at the heart of one of the great conservation laws of mechanics. Pay attention to the conditions for which this law is valid.

Important Concepts

☐ rolling

☐ (vector) torque

☐ angular momentum

☐ Newton's second law
   in angular form

☐ conservation of
   angular momentum

Overview

11–2 Rolling as Translation and Rotation Combined

☐ A **rolling** object, such as a wheel, rotates as its center of mass moves along a line. The center of mass obeys Newton's second law: $\vec{F}_{net} = M\vec{a}_{com}$, where $\vec{F}_{net}$ is the net force on the object, M is the mass of the object, and $\vec{a}_{com}$ is the acceleration of the center of mass. Rotation about an axis through the center of mass is governed by Newton's second law for rotation: $\tau_{net} = I\alpha$, where τ_{net} is the net torque on the object, I is the rotational inertia of the object, and α is its angular acceleration.

☐ The two motions are related by the forces that act on the wheel. The net force accelerates the center of mass and the net torque (derived from the forces) produces an angular acceleration. If, in addition to the mass and rotational inertia of the body, you know all the forces acting and their points of application you can calculate the acceleration of the center of mass and the angular acceleration about the center of mass. Don't forget the force of friction that may act at the point of contact between the rolling object and the surface on which it rolls.

☐ The velocity of the point on a rolling body in contact with the surface is given by the vector sum of the velocity due to the motion of the center of mass and the velocity due to rotation: $v = v_{com} - \omega R$, where the forward direction was taken to be positive. If the wheel slides, then $v \neq 0$ and v_{com} is different from ωR. If it does not slide, then $v = 0$ and $v_{com} = \omega R$.

Furthermore, if the wheel accelerates without sliding, the magnitude of the acceleration of the center of mass a_{com} is related to the magnitude of the angular acceleration around the center of mass by $a_{com} = \alpha R$.

☐ An object that is rolling without sliding on a surface may be considered to be in pure rotation about an axis through the point of contact with the surface.

11–3   The Kinetic Energy of Rolling

☐ A rolling wheel, sliding or not, has both translational and rotational kinetic energy. If the center of mass has velocity v_{com} and the wheel is rotating with angular velocity ω about an axis through the center of mass, then the total kinetic energy is given by $K = \frac{1}{2}Mv_{com}^2 + \frac{1}{2}I\omega^2$. If the wheel is not sliding, then $v_{com} = \omega R$ can be used to write both parts of the kinetic energy in terms of v_{com} or in terms of ω.

11–4   The Forces of Rolling

☐ If a wheel is accelerating as it rolls on a horizontal surface. a frictional force acts the point of contact between the wheel and the surface. The force is one of kinetic friction if the wheel is not sliding. Then the magnitude of the force is $f = \mu_k F_N$, where μ_k is the coefficient of kinetic friction between the surface and the wheel and F_N is the magnitude of the normal force of the surface on the wheel. If the wheel is not sliding the force is one of static friction and has whatever magnitude is required to make the $a_{com} = \alpha r$, provided that it is less than $\mu_s F_N$, where μ_s is the coefficient of static friction.

☐ If the wheel is going faster and not sliding the direction of the frictional force is opposite that of the acceleration and if the wheel is going slower and not sliding the direction of the frictional force is the same as that of the center of mass acceleration.

11–5   The Yo-Yo

☐ A yo-yo is another example of combined rotation and translation. The net force acting (the vector sum of the force of gravity and the force of the string), when entered into Newton's second law for linear motion, gives the acceleration of the center of mass. The net torque acting (due only to the string), when entered into Newton's second law for rotation, gives the angular acceleration about the center of mass. The second law equations, augmented by the condition that the string does not slide ($a = R_0\alpha$, where R_0 is the radius of the axle), can be solved for the acceleration of the center of mass and the angular acceleration about the center of mass.

11–6   Torque Revisited

☐ When a force $\vec{F}$ is applied to an object at a point with position vector $\vec{r}$, relative to some origin, then the **torque** $\vec{\tau}$ associated with the force is given by the vector product $\vec{\tau} = \vec{r} \times \vec{F}$. The magnitude of the torque is given by $\tau = rF\sin\phi$, where ϕ is the angle between $\vec{F}$ and $\vec{r}$ when they are drawn with their tails at the same point. A right-hand rule gives the direction of the torque: curl the fingers of the right hand so they rotate $\vec{r}$ into $\vec{F}$ about their tails, through the smallest angle between them. The thumb then points in the direction of the torque.

- ☐ The torque acting on an object is different for different choices of the origin.

- ☐ For an object rotating around a fixed axis, only the component of the torque along the axis is important. For a force that lies in a plane perpendicular to the axis, this component is given by $RF_\perp$, where R is the distance from the axis of rotation to the point of application of the force and $F_\perp$ is the tangential component of the force and is perpendicular to the axis of rotation.

11–7 Angular Momentum

- ☐ If a particle has momentum $\vec{p}$ and position vector $\vec{r}$ relative to some origin, then its **angular momentum** is defined by the vector product $\vec{\ell} = \vec{r} \times \vec{p}$. Notice that the angular momentum depends on the choice of origin. In particular, if the origin is picked so $\vec{r}$ and $\vec{p}$ are parallel at some instant of time, then $\vec{\ell} = 0$ at that instant.

- ☐ If a particle of mass m is traveling around a circle of radius R with speed v and the origin is placed at the center of the circle, then the magnitude of the angular momentum is $\ell = mRv$. Curl the fingers of the right hand around the axis in the direction the particle is traveling. Then, the thumb points along the axis in the direction of the angular momentum.

- ☐ For rotation about a fixed axis, the component of the angular momentum along the axis of rotation is the only component of interest. If the orbit of the particle is parallel to the xy plane and the center is on the z axis but is not at the origin, then the angular momentum is not along the z axis. Nevertheless, the z component is $\ell_z = mRv$.

- ☐ A particle moving along a straight line that does not pass through the origin has a non-zero angular momentum. Its magnitude is given by $\ell = mvd$, where m is the mass of the particle, v is its speed, and d is the distance from the origin to the line of motion.

11–8 Newton's Second Law in Angular Form

- ☐ If a net torque acts on a particle, then its angular momentum changes with time. Recall that, in terms of the momentum of a particle, Newton's second law can be written $\vec{F} = d\vec{p}/dt$, where $\vec{F}$ is the net force on the particle. Take the vector product of this equation with $\vec{r}$ to obtain $\vec{r} \times \vec{F} = \vec{r} \times \vec{p}/dt$. Note that $\vec{r} \times \vec{F} = \vec{\tau}$, the net torque acting on the particle. With a small amount of mathematical manipulation you can show that $\vec{r} \times (d\vec{p}/dt) = d\vec{\ell}/dt$. Thus, in terms of torque and angular momentum, Newton's second law becomes $\vec{\tau} = d\vec{\ell}/dt$.

11–9 The Angular Momentum of a System of Particles

- ☐ The total angular momentum $\vec{L}$ of a system of particles is the vector sum of the individual angular momenta of all the particles in the system: $\vec{L} = \sum \vec{\ell}_i$, where $\vec{\ell}_i$ is the angular momentum of particle i.

- ☐ For a system of particles, Newton's second law for rotation becomes

$$\vec{\tau}_{\text{net}} = \frac{d\vec{L}}{dt},$$

where $\vec{\tau}_{\text{net}}$ is the net external torque on particles of the system. This is the fundamental equation for a system of particles. Notice that only an external torque can change the total angular momentum of a system. Torques exerted by particles of the system on each other might change the individual angular momenta but not the total.

11–10   The Angular Momentum of a Rigid Body Rotating About a Fixed Axis

☐ For an object rotating with angular velocity ω about the z axis, the z component of the angular momentum, in terms of ω, the mass m_i, and distance $r_{\perp i}$ of each particle from the axis, is the sum $L_z = \sum m_i r_{\perp i} v_i = \sum m_i r_{\perp i}^2 \omega$. In terms of the rotational inertia I of the object, it is $L_z = I\omega$.

☐ Since $L_z = I\omega$ for a body rotating about the z axis, Newton's second law for fixed axis rotation becomes $\tau_{\text{net } z} = d(I\omega)/dt$. In some cases, the body is rigid and I does not change with time. Then, Newton's second law for fixed axis rotation becomes $\tau_{\text{net } z} = I\alpha$, where α is the angular acceleration of the body. In this chapter, you will encounter some situations for which I does change.

11–11   Conservation of Angular Momentum

☐ If the net external torque acting on a body is zero, then the change over any time interval of its angular momentum is zero. Angular momentum is then said to be conserved. This is a vector law. One component of the angular momentum may be conserved while other components are not. When attempting to use the conservation principle, you must first check to be sure the net external torque or one of its components vanishes.

☐ If the z component of the net external torque acting on a body rotating about the z axis vanishes, then $I\omega$ is a constant. Consider an ice skater, initially rotating on the points of her skates with her arms extended. By dropping her arms to her sides she decreases her rotational inertia and her angular velocity must increase for her angular momentum to remain the same.

☐ If the rotational inertia of the body changes, write $I_i\omega_i$ for the angular momentum in terms of the initial rotational inertia and angular velocity and $I_f\omega_f$ for the angular momentum in terms of the final rotational inertia and angular velocity. If no external torques act, these are equal. Solve $I_i\omega_i = I_f\omega_f$ for an unknown.

☐ If two objects exert torques on each other and no external torques act, write the total angular momentum in terms of the initial angular velocities ($I_1\omega_{1i} + I_2\omega_{2f}$), then write it again in terms of the final angular velocities ($I_1\omega_{1f} + I_2\omega_{2f}$). Equate the two expressions and solve for the unknown. For the spacecraft and flywheel example in the text, the total angular momentum is zero, so $I_{\text{sc}}\omega_{\text{sc}} + I_{\text{fw}}\omega_{\text{fw}} = 0$.

☐ In some problems, one object is initially moving along a straight line, as when a child runs and jumps on a merry-go-round. Don't forget to include the angular momentum of this object.

11–12   Precession of a Gyroscope

☐ An example of a gyroscope is a wheel that is free to turn at one end of a horizontal axle. The axle is supported only at the other end and the support ends in a pivot that allows the axle to turn. If the wheel is given a large angular momentum the effect of the torque of gravity is to cause the axle to rotate (precess) in the horizontal plane.

☐ The angular speed of precession is given by $\Omega = Mgr/I\omega$, where M is the mass of the wheel, r is the distance from the pivot point to the center of mass of wheel, I is the rotational inertia of the wheel, and ω is the angular speed of its rotation about the axle.

Hints for Questions

1 The magnitude of the torque is given by $\tau = rF \sin\phi$, where ϕ is the angle between the directions of $\vec{r}$ and $\vec{F}$ when their tails are at the same point.

[Ans: (a) 0 or 180°; (b) 90°]

3 (a) The torque associated with force $\vec{F}$ is $\vec{\tau} = \vec{r} \times \vec{F}$, where $\vec{r}$ is the vector from the origin to the point of application of the force. Test each force using the right-hand rule for vector products: move $\vec{r}$ parallel to itself so its tail is at the tail of $\vec{F}$, then curl the fingers of your right hand so they tend to rotate $\vec{r}$ toward $\vec{F}$ through the smallest angle between them. Your thumb then points in the direction of the vector product.
(b) The magnitude of a torque is given by $\tau = rF \sin\phi$, where ϕ is the angle between $\vec{r}$ and $\vec{F}$ when they are drawn with their tails at the same point.

[Ans: (a) 5 and 6; (b) 1 and 4 tie, then the rest tie]

5 The magnitude of the angular momentum is given by $\ell = rv \sin\phi$, where r is the distance from one of the points a through e, v is the speed of the particle, and ϕ is the angle between the vector from the point to the particle and the particle's velocity when they are drawn with their tails at the same point. It might help to draw the vectors from the points to the particle.

[Ans: b, then c and d tie, then a and e tic (zero)]

7 According to Newton's second law for rotation the torque τ and the angular momentum ℓ are related by $\tau = d\ell/dt$. Differentiate the expressions for ℓ with respect to time.

[Ans: (a) 3; (b) 1; (c) 2; (d) 4]

9 Assume the total angular momentum of the gum and slab is conserved. The angular momentum of the gum before it sticks is mvd, where d is the perpendicular distance from the path of the gum to the origin. The angular momentum of the slab and gum after the gum sticks is $(I + mr^2)\omega$, where r is the distance from the sticking point to the origin and ω is the angular speed of the slab and gum.

[Ans: (a) 4, 6, 7, 1, then 2, 3, and 5 tie (zero)]

Hints for Problems

7 (a) Draw a force diagram for the cylinder. The force of gravity has magnitude mg and pulls downward effectively at the center of the cylinder. Here m is the mass of the cylinder. A frictional force acts at the point of contact of the cylinder with the roof and the roof pushes upward with a normal force that is perpendicular to the roof and acts at the point of contact. Take the x axis to be parallel to the roof and write Newton's second law for translation of the center of mass in component form. Also write Newton's second law for rotation. The only torque on the cylinder is the torque of friction. Finally write the relationship between the acceleration a of the center of mass and the angular acceleration α about the center of mass: $a = \alpha R$, where R is the radius of the cylinder. You will also need the equation $I = \frac{1}{2}mR^2$ for the rotational inertia of the cylinder. Use two of these equations to eliminate the frictional force and the acceleration from these equations, then solve for the angular acceleration.

(b) The problem is now a projectile motion problem, with the initial velocity equal to the final velocity of part (a). That is, $v_0 = \omega R$, where ω is the angular velocity of the cylinder as it leaves the roof.

$\left[\text{Ans: (a) } 63 \text{ rad/s; (b) } 4.0 \text{ m}\right]$

9 Use conservation of energy. Draw a horizontal line on the graph at $E = 75 \text{ K}$. If it intersects the potential energy curve the point of intersection gives the turning point. If it does not, read the value of the potential energy for $x = 0$ or $x = 13 \text{ m}$, then use $E = K + U$ to calculate the kinetic energy at those points. The kinetic energy is $\frac{1}{2}mv^2 + \frac{1}{2}I\omega^2$, where m is the mass of the ball, I is its rotational inertia, v is its speed, and ω is its angular speed. The speed and angular speed are related by $v = \omega R$, where R is the radius. Use this relationship to eliminate ω from the energy equation. Also use $I = \frac{2}{5}mR^2$ (see Table 10–2). Solve for v

$\left[\text{Ans: (a) } 2.0 \text{ m; (b) } 7.3 \text{ m/s}\right]$

13 At the loop bottom the vector sum of the gravitational and normal forces must equal the product of the mass of the ball and the centripetal acceleration. That is, $F_N - Mg = Mv^2/r$, where v is the speed of the ball and r is the radius of the loop. Use conservation of energy to find an expression for v^2. Take the potential energy to be zero at the bottom of the loop. Then the initial potential energy is Mgh. The initial kinetic energy is zero and the final kinetic energy is $\frac{1}{2}Mv^2 + \frac{1}{2}I\omega^2$, where ω is the angular speed of the ball. Since the ball does not slide the speed and angular speed are related by $v = \omega R$. Use this relationship to eliminate ω from the energy equation, then solve for v^2. Substitute this expression and $F_N = 2.00Mg$ into $F_N - Mg = Mv^2/r$ and solve for I/MR^2.

$\left[\text{Ans: } 0.50\right]$

25 The angular momentum of a particle is given by $\vec{\ell} = m\vec{r} \times \vec{v}$, where m is its mass, $\vec{r}$ is its displacement from the origin, and $\vec{v}$ is its velocity. The magnitude is $\ell = mrv \sin\phi$, where ϕ is the angle between $\vec{r}$ and $\vec{v}$. For each of the particles in this problem $\vec{r}$ and $\vec{v}$ are perpendicular to each other, so $\ell = mrv$. To find the direction use the right hand rule: Move $\vec{r}$ so its tail is at the tail of $\vec{v}$, then curl the fingers of your right hand so they tend to rotate $\vec{r}$ toward $\vec{v}$. Your thumb points in the direction of the vector product. Find the angular momentum of each particle and vectorially sum the angular momenta.

$\left[\text{Ans: (a) } 9.8 \text{ kg} \cdot \text{m}^2/\text{s; (b) positive } z \text{ direction}\right]$

29 (a) Use Newton's second law to find the acceleration: $\vec{a} = \vec{F}/m$, where m is the mass of the object.

(b) The angular momentum is $\vec{\ell} = m\vec{d} \times \vec{v}$. Use the component form of the vector product:

$$\vec{d} \times \vec{v} = (yv_z - zv_y)\hat{i} + (zv_x - xv_z)\hat{j} + (xv_y - yv_x)\hat{k}.$$

(c) The torque is $\vec{\tau} = m\vec{d} \times \vec{F}$. Use the component form of the vector product:

$$\vec{d} \times \vec{F} = (yF_z - zF_y)\hat{i} + (zF_x - xF_z)\hat{j} + (xF_y - yF_x)\hat{k}.$$

(d) Equate the two forms of the scalar product, $\vec{v} \cdot \vec{F} = vF \cos\phi$ and $\vec{v} \cdot \vec{F} = v_x F_x + v_y F_y + v_z F_z$, then solve for ϕ, the angle between $\vec{v}$ and $\vec{F}$.

$\left[\text{Ans: (a) } (3.00 \text{ m/s}^2)\hat{i} - (4.00 \text{ m/s}^2)\hat{j} + (2.00 \text{ m/s}^2)\hat{k}; \text{ (b) } (42.0 \text{ kg} \cdot \text{m}^2/\text{s})\hat{i} + (24.0 \text{ kg} \cdot \text{m}^2/\text{s})\hat{j} + (60.0 \text{ kg} \cdot \text{m}^2/\text{s})\hat{k}; \text{ (c) } (-8.00 \text{ N} \cdot \text{m})\hat{i} - (26.0 \text{ N} \cdot \text{m})\hat{j} - (40.0 \text{ N} \cdot \text{m})\hat{k}; \text{ (d) } 127°\right]$

<u>39</u> (a) The rotational inertia is the sum of the rotational inertias of the hoop and the four rods. According to Table 10–2 the rotational inertia of the hoop about a diameter is given by $\frac{1}{2}mR^2$. Use the parallel axis theorem to find the rotational inertia about a tangent axis, a distance R from the parallel diameter. According to Table 10–2 the rotational inertia of one of the rods about an axis through it center and perpendicular to it is given by $\frac{1}{12}mR^2$. Use the parallel axis theorem to find the rotational inertia about a parallel axis through the end of the rod. This method gives the rotational inertia of the upper and lower rods. The right-hand rod has a rotational inertia of zero since it is along the rotation axis and the left-hand rod has a rotational inertia of mR^2 since all its mass is a distance R from the rotation axis.

(b) The magnitude of the angular momentum is $L = I\omega$, where I is the rotational inertia of the system and ω is its angular speed. Curl the fingers of your right hand around the rotation axis in the direction of rotation. Your thumb then points in the direction of the angular momentum.

[Ans: (a) $1.6\,\text{kg}\cdot\text{m}^2$; (b) $4.0\,\text{kg}\cdot\text{m}^2/\text{s}$, upward in the diagram]

<u>47</u> (a) The skaters must remain equidistance from the center of the circle and they are $3.0\,\text{m}$ apart.

(b) The total angular momentum of the skaters is conserved. While the skaters are skating along straight lines, each has an angular momentum of $mvd/2$ about the center of the circle. Here d is the separation of the skaters. After they start skating around the circle each has an angular momentum of $m(d/2)^2\omega$, where ω is their angular speed. Equate the two expression for the total angular momentum and solve for ω.

(c) The kinetic energy is given by $K = \frac{1}{2}I\omega^2$, where the rotational inertia of the two-skater system is $2m(d/2)^2$.

(d) The angular momentum of the system is conserved as the skaters pull along the pole. Thus $I_{\text{new}}\omega_{\text{new}} = I_{\text{old}}\omega_{\text{old}}$. Solve for ω_{new}.

(e) The kinetic energy is now $K = \frac{1}{2}I\omega_{\text{new}}^2$.

(f) The total energy of the system is the sum of the kinetic, potential, and internal energies and remains constant. The kinetic energy changed but the potential energy did not.

[Ans: (a) $1.5\,\text{m}$; (b) $0.93\,\text{rad/s}$; (c) $98\,\text{J}$; (d) $8.4\,\text{rad/s}$; (e) $8.8 \times 10^2\,\text{J}$; (f) internal energy of the skaters]

<u>51</u> The total angular momentum of the rod-bullet system is conserved during the collision. Before the collision the angular momentum about the rotation axis is $mv(d/2)\sin\theta$, where v is the speed of the bullet, m is its mass, and d is the length of the rod. After the collision it is $[I + m(d/2)^2]\omega$, where ω is the angular speed of the rod and I is its rotational inertia. According to Table 10–2 the rotational inertia of a rod about an axis through its center and perpendicular to it is $\frac{1}{12}Md^2$, where M is its mass. Equate the two expressions for the angular momentum and solve for v.

[Ans: $1.3 \times 10^3\,\text{m/s}$]

<u>55</u> (a) The angular momentum of the two-disk system is conserved as the small disk slides. Let I_1 be the rotational inertia of the large disk, I_2 be the rotational inertia of the small disk in its initial location, and I_2' be the rotational inertia of the small disk in its final location, all about the axis through the center of the large disk. If ω_i is the initial angular velocity of the

disks and ω_f is their final angular velocity, then conservation of angular momentum yields $(I_1 + I_2)\omega_i = (I_1 + I_2')\omega_f$. The center of the small disk moved a distance of $2r$. The parallel axis theorem tells us that $I_2' = I_2 + m(2r)^2$. According to Table 10–2, $I_1 = \frac{1}{2}(10m)(3r)^2$ and $I_2 = \frac{1}{2}mr^2$. You can now solve for ω_f.

(b) The old kinetic energy is $K_0 = \frac{1}{2}(I_1+I_2)\omega_i^2$ and the new kinetic energy is $K = \frac{1}{2}(I_1+I_2')\omega_f^2$.

[Ans: (a) 18 rad/s; (b) 0.92]

61 (a) Use Newton's second law for rotation in the form $\vec{\tau} = d\vec{L}/dt$, where $\vec{\tau}$ is the torque about the pivot point and $\vec{L}$ is the angular momentum about the same point. Show that the magnitude of the torque is $mgd \sin\theta$, where d is the distance from the pivot point to the center of mass. Now $dL = L_h\, d\phi$, where L_h is the horizontal component of the angular momentum and $d\phi$ is an infinitesimal precession angle. Use $L_h = L\sin\theta$ and $L = I\omega$, where I is the rotational inertia of the top and ω is its spin angular velocity. Thus $dL/dt = L\sin\theta\, d\phi/dt$. Set this equal to the torque and solve for the precession angular speed $d\phi/dt$.

(b) The tip of the angular momentum vector moves in the direction of the torque. Draw an overhead view that shows the directions of $\vec{L}$ and $\vec{\tau}$. To find the direction of $\vec{L}$ curl the fingers of your right hand around the rotation axis in the direction of the spin. Your thumb then points in the direction of $\vec{L}$. To find the direction of the torque use the right hand rule for vector products. Move the vector $\vec{r}$ from the pivot point to the center of mass so its tail is at the center of mass, then curl the fingers of your right hand so they tend to rotate $\vec{r}$ toward the direction of the gravitational force. Your thumb is in the direction of the torque.

[Ans: (a) 0.33 rev/s; (b) clockwise]

63 If $\vec{v}_1$ is the velocity of particle 1 before the collision and m_1 is its mass, $\vec{v}_2$ is the velocity of particle 2 before the collision and m_1 is its mass, and $\vec{V}$ is the velocity of both particles after the collision, then conservation of linear momentum tells us that $m_1\vec{v}_1+m_2\vec{v}_2 = (m_1+m_2)\vec{V}$. Solve for $\vec{V}$. The angular momentum of the stuck-together particles is given by the vector product $\vec{L} = \vec{r}\times\vec{V}$, where $\vec{r}$ is the position vector of the point where the collision occurred. Recall that $\hat{\imath}\times\hat{\imath} = 0$, $\hat{\imath}\times\hat{\jmath} = \hat{k}$, $\hat{\jmath}\times\hat{\imath} = -\hat{k}$, and $\hat{\jmath}\times\hat{\jmath} = 0$.

[Ans: $(5.55\,\text{kg}\cdot\text{m}^2/\text{s})\,\hat{k}$]

69 The magnitude of the angular momentum is given by mvs, where m is the mass of the particle and v is its speed. Use the kinematic equation $v_2 = 2g\Delta y$ to find the speed v a distance δy below the release point.

[Ans: $0.47\,\text{kg}\cdot\text{m}^2/\text{s}$]

73 Use Newton's second law for rotation $\tau = dL/dt$ to find the angular momentum L as a function of time. Solve for the time when $L = 0$.

[Ans: 12 s]

81 Use conservation of energy. The change in the kinetic energy is $\Delta K = -\frac{1}{2}mv^2 - \frac{1}{2}I\omega^2$, where I is the rotational inertia of the body and ω is its initial angular speed. Since it rolls without sliding $v = \omega R$. Use this to eliminate ω from the expression for ΔK. The change in the potential energy is mgh. Solve $\Delta K + \Delta U = 0$ for I in terms of m and R and compare the result with the expressions given in Table 10–2.

[Ans: (a) $mR^2/2$; (b) a solid cylinder]

<u>87</u> Let I be the rotational inertia of the wheel, turntable, and man together and I_w be the rotational inertia of the wheel alone. If ω_0 is the initial angular velocity of the wheel and ω is the angular velocity of the system when all parts are rotating together, then conservation of angular momentum tells you the $I_w\omega_0 = I\omega$. The rotational inertia of the wheel is given by $I_w = mR^2$, where m is its mass and R is its radius. If ω as the same sign as ω_0 the turntable spins in the same direction as the wheel. Otherwise it spins in the opposite direction.

[Ans: (a) 12.7 rad/s; (b) clockwise]

Quiz

Some questions might have more than one correct answer.

1. A wheel is on a level surface. The surface must exert a frictional force on the wheel at the point of contact
   A. always
   B. whenever the wheel is moving
   C. whenever the wheel is rolling without sliding
   D. whenever the wheel is rolling without sliding and its center of mass is speeding up or slowing down
   E. never

2. If there is a frictional force between a wheel and the surface on which it moves, that force is one of
   A. static friction if the wheel is rolling without sliding
   B. kinetic friction if the wheel is rolling without sliding
   C. static friction if the wheel is rolling and sliding
   D. kinetic friction if the wheel is rolling and sliding
   E. kinetic friction if the wheel is motionless

3. If a particle of mass m is moving with constant speed v around a circle of radius R, its angular momentum about the center of the circle
   A. has magnitude mRv and is perpendicular to the plane of the circle
   B. has magnitude mRv and is in the plane of the circle
   C. has magnitude mR^2v and is perpendicular to the circle
   D. has magnitude mR^2v and is in the plane of the circle
   E. has magnitude mRv and is not perpendicular or parallel to the plane of the circle

4. A child of mass m is running with constant speed v along a straight line that is tangent to the rim of a merry-go-round. The merry-go-round is turning about its center with constant angular velocity ω so that the point on the rim where the child will jump on is coming toward the child. The rotational inertia of the merry-go-round for rotation about its center is I. The total angular momentum of the system consisting of the child and the merry-go-round, about the center of the merry-go-round, is

    A. $I\omega$
    B. $I\omega + mv$
    C. $I\omega - mv$
    D. $I\omega + mRv$
    E. $I\omega - mRv$

5. A particle of mass m_1 is traveling clockwise with constant speed v around a circle of radius R_1, in the xy plane and centered at the origin. A second particle, with mass m_2, is traveling counterclockwise with the same speed around a concentric circle of smaller radius R_2. The magnitude of the total angular momentum about the origin of the two-particle system is

    A. zero
    B. $(m_1 R_1 + m_2 R_2)v$
    C. $|m_1 R_1 - m_2 R_2|v$
    D. $(m_1 + m_2)(R_1 + R_2)v$
    E. $(m_1 + m_2)|R_1 - R_2|v$

6. A particle of mass m_1 is traveling in the positive x direction with constant speed v along the line $y = d_1$, $z = 0$. A second particle, with mass m_2, is traveling in the negative x direction with the same speed along the line $y = d_2$, $z = 0$. Both d_1 and d_2 are positive. The magnitude of the total angular momentum about the origin of the two-particle system is

    A. zero
    B. $(m_1 d_1 + m_2 d_2)v$
    C. $|m_1 d_1 - m_2 d_2|v$
    D. $(m_1 + m_2)(d_1 + d_2)v$
    E. $(m_1 + m_2)\vec{d_1} - d_2|v|$

7. Which of the following statements are true?

    A. it is possible for the total angular momentum of a system to be conserved even if its total linear momentum is not
    B. it is possible for the total linear momentum of a system to be conserved even if its total angular momentum is not
    C. it is possible for the total energy of a system to be conserved even if its total angular momentum is not
    D. it is possible for the total angular momentum of a system to be conserved even if its total energy is not
    E. none of the above statements are true

8. The total angular momentum of a system of particles is conserved if

   A. the torques of all particles in the system on other particles in the system sum to zero
   B. the torques of all particles outside the system on particles in the system sum to zero
   C. the rotational inertia of the system is constant
   D. the angular velocity of the center of mass of the system is constant
   E. the angular velocities of all particles in the system are constant

9. An object is rotating on a fixed axis when its rotational inertia increases. There are no external torques on the object. This means that

   A. the angular momentum of the object must increase
   B. the angular momentum of the object must decrease
   C. the angular speed of the object must increase
   D. the angular speed of the object must decrease
   E. the object must stop rotating

10. The angular momentum about the origin is zero for a particle that

    A. moves with constant speed around a circle that is centered at the origin
    B. moves with constant speed along the line $x = 5.0\,\text{m}$, $y = 5.0\,\text{m}$
    C. moves along the line $x = y$, $z = 0$
    D. moves around a circle of radius R, centered at $x = R$, $y = 0$, $z = 0$
    E. moves along a parabolic trajectory (like a projectile) through the origin

11. A wheel lies in the xy plane with its center at the origin. It is free to rotate on a frictionless axle through its center and along the z axis. A force $\vec{F_1} = F\hat{\imath}$ is applied to the rim at $x = 0$, $y = R$. For which choices of a second force $\vec{F_2}$ is the angular acceleration of the disk zero?

    A. $\vec{F_2} = F\hat{\imath}$, applied at $x = 0$, $y = -R$
    B. $\vec{F_2} = F\hat{\imath}$, applied at $x = R$, $y = 0$
    C. $\vec{F_2} = F\hat{\imath}$, applied at $x = R$, $y = 0$
    D. $\vec{F_2} = -F\hat{\imath}$, applied at $x = 0$, $y = -R$
    E. $\vec{F_2} = -F\hat{\jmath}$, applied at $x = -R$, $y = 0$

12. Two wheels are free to rotate on separate axles. Starting with the wheels at rest torques are applied to them and they reach the same angular speed in the same time. Which of the following quantities must be the same for the wheels?

    A. their rotational inertias
    B. the magnitudes of the torques applied to them
    C. the magnitude of their angular momenta at any given time
    D. their masses
    E. their angular accelerations

13. Two disks are free to rotate on the same axle. Disk 1 has rotational inertia I_1 and angular speed ω_1. Disk 2 has rotational inertia I_2 and angular speed ω_2. The spin of disk 1 is clockwise and the spin of disk 2 is counterclockwise. Disk 2 slides along the axle until it makes contact with disk 1 and then, through frictional forces between them, the two disks come to the same angular speed. They are spinning clockwise only if

A. $omega_1$ is greater than ω_2
B. ω_1 is less than ω_2
C. I_1 is greater than I_2
D. I_1 is less than I_2
E. $\omega_1 I_1$ is greater than $\omega_2 I_2$
F. $\omega_1 I_1$ is less than $\omega_2 I_2$

Answers: (1) D; (2) A, D; (3) A; (4) E; (5) C; (6) C; (7) A, B, C, D; (8) B; (9) D; (10) C; (11) A, E; (12) E; (13) E

where U was taken to be zero at infinite separation. This energy is associated with the *pair* of masses, NOT with either mass alone.

☐ Potential energy is a scalar. The total gravitational potential energy of a system of particles is the sum of the potential energies of each *pair* of particles in the system. This is the work done by an external agent to assemble the particles from infinite separation, starting and ending with the particles at rest. The external agent must do negative work since the mass already in place attracts any new mass being brought in and the agent must pull back on it.

☐ If the particles of a system interact only via gravitational forces and no external forces act, then the total mechanical energy of the system is conserved. As the particles move, the sum of their kinetic energies and the total gravitational potential energy remains the same.

☐ For a two-particle system, the total mechanical energy consists of three terms, corresponding to the kinetic energy of each particle and the potential energy of their interaction. Suppose that at some instant particle 1 (with mass m_1) has speed v_1, particle 2 (with mass m_2) has speed v_2, and the particles are a distance r apart. Then, the mechanical energy of the system is given by

$$E = \tfrac{1}{2}m_1 v_1^2 + \tfrac{1}{2}m_2 v_2^2 - \frac{GMm}{r}.$$

At another time v_1, v_2, and r may have different values but the sum of the three terms is the same.

☐ The **escape speed** is the minimum initial speed that an object must be given at the surface of Earth (or other large mass) in order to get far away. The initial gravitational potential energy is $-GMm/R$, where M is the mass of Earth, R is its radius, and m is the mass of the object. The final potential energy is zero (the object is far away). The initial kinetic energy is $\tfrac{1}{2}mv^2$, where v is the escape speed, and the final kinetic energy is zero if v is to have its minimum value. Conservation of energy yields $\tfrac{1}{2}mv^2 - GMm/R = 0$, so

$$v = \sqrt{\frac{2GM}{R}}.$$

13–7 Planets and Satellites: Kepler's Laws

☐ Motions of the planets are controlled by gravity, due chiefly to the Sun. If we neglect the influence of other planets, the motion of a planet is described by Kepler's three laws:

1. **Law of orbits**: The orbit of a planet is an ellipse, with the Sun at one focus. This law is a direct consequence of the $1/r^2$ nature of the force law.

2. **Law of areas**: The line that joins the planet and the Sun sweeps out equal areas in equal times. This law is a direct consequence of the principle of angular momentum conservation.

3. **Law of periods**: The square of the period is proportional to the cube of the semimajor axis of the orbit.

☐ A planetary orbit can be described by giving its semimajor axis a, which is half the greatest distance across the orbit, and its eccentricity e, which is defined so that ea is half the distance between the foci. The point of closest approach to the Sun is called the **perihelion** and the distance of this point from the Sun is given by $R_p = a(1 - e)$. The point of maximum distance from the Sun is called the **aphelion** and the distance of this point from the Sun is given by $R_a = a(1 + e)$. For circular orbits, $e = 0$ and $R_a = R_p = a$. An eccentricity of nearly 1 corresponds to an ellipse that is much longer than it is wide.

☐ The law of areas can be used to relate a planet's speed at one point to its speed at another point. The simplest relationship holds for points that are the greatest and least distances from the Sun because at these points the velocity is perpendicular to the position vector from the Sun. In terms of the mass m of the planet, its distance r from the Sun, and its speed v, the magnitude of the angular momentum at one of these points is $L = mrv$, where the origin was placed at the Sun. If v_p is the speed at perihelion and v_a is the speed at aphelion, then conservation of angular momentum yields $R_a v_a = R_p v_p$.

☐ The acceleration of a planet in a circular orbit of radius r is given by $a = v^2/r = 4\pi^2 r/T^2$, where $2\pi r/T$ was substituted for v. Here T is the period. The gravitational force is $F = GMm/r^2$, where M is the mass of the Sun and m is the mass of the planet. Newton's second law yields $GMm/r^2 = 4\pi^2 mr/T^2$, so $T^2 = (4\pi^2/GM) r^3$. This equation is valid for elliptical orbits if r is replaced by the semimajor axis a.

☐ Asteroids and recurring comets in orbit around the Sun and satellites (including the Moon) in orbit around Earth or another planet also obey Kepler's laws. In reality each travels in an elliptical orbit around the center of mass.

13–8 Satellites: Orbits and Energy

☐ For a satellite in a circular orbit of radius r around a massive central body of mass M, the acceleration is v^2/r and Newton's second law yields $GMm/r^2 = mv^2/r$. Thus, the kinetic energy is $K = \frac{1}{2}mv^2 = GMm/2r$. If the potential energy is zero for infinite separation, then the potential energy for an orbit of radius r is $U = -GMm/r$ and the mechanical energy is

$$E = K + U = \frac{GMm}{2r} - \frac{GMm}{r} = -\frac{GMm}{2r}.$$

E is negative, indicating that the satellite is bound to the central body and does not have enough kinetic energy to escape.

☐ If the orbit is not circular, the mechanical energy is given by $E = -GMm/2a$, where a is the semimajor axis of the orbit. The kinetic energy is

$$K = E - U = \left(\frac{GMm}{r}\right) \left(\frac{1}{r} - \frac{1}{2a}\right).$$

As the satellite moves, its distance from the Sun varies; both the kinetic and potential energies vary but the sum remains constant.

Hints for Questions

1 Each of the fixed particles attracts the third particle with a force that is along the line joining the particles. The force has a magnitude that is proportional to the masses of the particles and to the reciprocal of the square of the distance between them. Note that the ratio of the magnitudes of the two forces due to the fixed particles does not depend on the mass of the third particle.

[Ans: (a) between, closer to the less massive particle; (b) no; (c) no]

3 Pair the particles in the rings so that the two particles of a pair exert forces of the same magnitude but opposite directions on the central particle. You should find only one particle left over.

[Ans: Gm^2/r^2, upward]

5 (a) The three forces on the central particle add vectorially. Notice that because the particles are symmetrically placed on the circle the vertical component of the net force is zero. Order the configurations according to the horizontal component of the net force.
(b) Assume the potential energy is zero if the particles are far from each other. It negative when they are close and the closer they are the more negative it is.

[Ans: (a) c, b, a; (b) a, b, c]

7 Draw vector arrows to indicate the forces on A from B and C. One arrow should be longer than the other. Now draw the arrow that represents the vector sum of these two forces. Particle D should be placed so its force on A is the negative of this vector sum.

[Ans: yes, in the second quadrant closer to the negative y axis than to the x axis]

9 The free-fall acceleration at a point is given by $g = a_g - \omega^2 R$, where a_g is the gravitational acceleration, ω is the angular speed of the planet, and R is the radius of the circular orbit of the point. The gravitational acceleration is the same for the three planets. R is the radius of the plane for points on the equator and is zero for points at a pole. The angular speed is proportional to the reciprocal of the period.

[Ans: b, d, and f all tie, then e, c, a]

11 The gravitational force is conservative. This means that the work it does depends only on the end points of the rocket's trip and not on the path taken. The change in the rocket-moon system is the negative of the work done by the gravitational force.

[Ans: (a) all tie; (b) all tie]

Hints for Problems

3 The magnitude of the gravitational force is $F = Gm(M - m)/r^2$, where r is the distance between the two parts. Set the derivative of this expression with respect to m equal to zero and solve for m/M.

[Ans: 1/2]

9 Symmetry tells us that the horizontal component of the net force on the central sphere is zero. A little geometry shows that the distance from a vertex to the center of the triangle is $\sqrt{3}L/4$, where L is the length of a triangle side. Furthermore, the angle between a side and the line from a vertex to the center is $30°$ and the sine of this angle is $1/2$. Thus vertical component of the force on the central sphere is

$$F_{\text{net, }y} = \frac{GMm_4}{(\sqrt{3}L/4)^2} - \frac{2Gmm_4(1/2)}{(\sqrt{3}L/4)^2}.$$

Set this expression equal to zero and solve for M. Notice that the algebraic result for M does not depend on the value of m_4.

[Ans: (a) m; (b) 0]

13 In vector notation the force of a particle of mass M at $\vec{r}$ on a particle of mass m at the origin is

$$\vec{F} = -\frac{GMm}{r^3}\vec{r}.$$

In terms of the components this is

$$F_x = \frac{GMm}{r^3}x, \qquad F_y = \frac{GMm}{r^3}y, \qquad \text{and} \qquad F_z = \frac{GMm}{r^3}z.$$

Sum the force and set each component of the net force equal to zero. Solve the resulting equations simultaneously for x, y, and z.

[Ans: (a) $-1.88d$; (b) $-3.90d$; (c) $0.489d$]

19 The gravitational force on a particle that is located at the equator has magnitude $F = GmM/R^2$, where M is the mass of the star and R is its radius. For the particle to remain on the surface this must equal the centripetal force $m\omega^2 R$ required to keep the particle on its circular path. Here ω is the angular speed of the star. Set these two expressions equal to each other and solve for M.

[Ans: 5×10^{24} kg]

27 At launch from the surface of a planet of radius R and mass M the gravitational potential energy is $-GmM/R$, where m is the mass of the rocket. This is relative to the potential energy when the rocket is far away. Suppose the kinetic energy at launch is K. To escape the total mechanical energy must be zero. Thus $K = GmM/R$. The ratio of the kinetic energy required to escape from the Moon to that required to escape from Earth is $K_M/K_E = M_M R_E/M_E R_M$, where M_M is the mass of the Moon, M_E is the mass of Earth, R_M is the radius of the Moon, and R_E is the radius of Earth. A similar expression holds for Jupiter. Values can be found in Appendix C.

[Ans: (a) 0.0451; (b) 28.5]

29 If $U(R_s)$ is the gravitational potential energy when the projectile is on the surface of the planet and K is its initial kinetic energy, then to escape the planet the mechanical energy must be at least zero. Set the mechanical energy equal to zero and solve for K.

[Ans: 5.0×10^9 J]

37 Consider first just two particles. If their masses are m_1 and m_2 and their separation is r then their gravitational potential energy is $-Gm_1m_2/r$. There are six pairs of particles in the system of the problem. If the particles are labeled 1, 2, 3, and 4 the pairs are 1 and 2, 1 and 3, 1 and 4, 2 and 3, 2 and 4, and 3 and 4. The total potential energy is the sum of the potential energies of these pairs.

[Ans: -4.82×10^{-13} J]

39 The period T and orbit radius r are related by the law of periods: $T^2 = (4\pi^2/GM)r^3$, where M is the mass of Mars. Convert the given period to seconds and solve for M.

[Ans: 6.5×10^{23} kg]

43 (a) If r is the radius of the orbit then the magnitude of the gravitational force on the satellite is given by GMm/r^2, where M is the mass of Earth and m is the mass of the satellite. The magnitude of the acceleration of the satellite is given by v^2/r, where v is its speed. Newton's second law yields $GMm/r^2 = mv^2/r$. Since the radius of Earth is 6.37×10^6 m the orbit radius is $r = 6.37 \times 10^6$ m $+ 160 \times 10^3$ m $= 6.53 \times 10^6$ m. Solve for v. (b) Since the circumference of the circular orbit is $2\pi r$, the period is $T = 2\pi r/v$.

[Ans: (a) 7.82 km/s; (b) 87.5 min]

47 (a) The semimajor axis is given by

$$a = \left(\frac{GMT^2}{4\pi^2} \right)^{1/3},$$

where M is the mass of the Sun (which can be found in Appendix C), and T is the period of the motion. Be sure to substitute the period in seconds.
(b) The eccentricity e of the orbit is related to the aphelion distance R_a by

$$e = 1 - \frac{R_a}{a}.$$

The mean orbital radius of Pluto can also be found in Appendix C.

[Ans: (a) 1.9×10^{13} m; (b) $3.6R_p$]

51 Let a be the radius of the orbit of either one of the stars. This is $R_E/2$, where R_E is the radius of Earth's orbit around the sun. The stars are separated by $2a$ so each is pulled toward the center of mass by a gravitational force of magnitude $F = GM^2/(2a)^2$. According to Newton's second law this must be equal to $M\omega^2 a$, where ω is the angular speed of the star. Since $\omega = 2\pi/T$, where T is the period of revolution, $GM^2/(2a)^2 = M(2\pi/T)^2 a$. Solve for T^2 and replace a with $R_E/2$. Then recognize that $4\pi^2 R^3/GM$ is the period of Earth's revolution and so is one year.

[Ans: 0.71 y]

59 The period T is given by the Kepler's law of periods: $T^2 = (4\pi^2/GM)r^3$. The speed is given by $v = 2\pi r/T$, the kinetic energy by $K = \frac{1}{2}Mv^2$, and the magnitude of the angular momentum by mrv.

[Ans: (a) $r^{3/2}$; (b) $1/r$; (c) $\sqrt{r}$; (d) $1/\sqrt{r}$]

63 Take the potential energy of the Earth-asteroid system to be $-GMm/r$, where M is the mass of Earth and m is the mass of the asteroid. Since the energy of the system is conserved, $K - GMm/r$ is a constant. Write the equation twice using different sets of values and solve the equations simultaneously for m. The speed of the asteroid is given by $v = \sqrt{2K/m}$.

[Ans: (a) 1.0×10^3 kg; (b) 1.5 km/s]

71 The gravitational acceleration at the surface is given by $a_g = GM/R^2$, where M is the mass of the star (1.99×10^{30} kg) and R is its radius. When the object falls from rest to the surface its kinetic energy changes by $\Delta K = \frac{1}{2}mv^2$, where m is its mass and v is its final speed, and the potential energy of the star-object system changes by $\Delta U = -GMm[(1/R) - (1/r)]$, where r is the distance from the star center to the starting point. You may take this to be $r = R + 1.0\,\text{m}$. To evaluate the quantity in brackets without loss of significance use $(1/R) - (1/r) = (r - R)/Rr$.

[Ans: (a) $1.3 \times 10^{12}\,\text{m/s}^2$; (b) $1.6 \times 10^6\,\text{m/s}$]

73 Vectorially add the forces of spheres A, C, and D on sphere B. The magnitude of any one of the other spheres on sphere B is given by $F = Gmm_B/r^2$, where m is the mass of the other sphere and r is the distance from sphere B to the other sphere. The x component of the force is $F\cos\theta$ and the y component is $F\sin\theta$, where θ is the angle between the x axis and the line from sphere B to the other sphere. Since sphere B is at the origin $\cos\theta = x/r$ and $\sin\theta = y/R$, where x and y are the coordinates of the other sphere.

[Ans: $(0.37\,\mu\text{N})\hat{\jmath}$]

83 Use conservation of energy. Take the initial potential energy to be $U_i = -GMm/r_i$, where M is the mass of Earth, m is the mass of the rocket, and r_i is the initial distance of the rocket from Earth's center (the radius of earth plus the initial altitude). The final potential energy is $U_f = -GMm/r_f$, where r_f is the final distance of the rocket from Earth's center. The initial kinetic energy is $K_i = \frac{1}{2}mv_i^2$, where v_i is the initial speed of the rocket. Denote the final kinetic energy by K_f. For part (a) solve for K_f. For part (b) put K_f equal to zero and solve for r_f, then subtract Earth's radius.

[Ans: (a) $38.3\,\text{MJ}$; (b) $1.03 \times 10^3\,\text{km}$]

87 Use Kepler's law of periods to calculate the orbit radius r. Use $v = 2\pi r/T$, where T is the period, to compute the speed.

[Ans: (a) $1.9 \times 10^{11}\,\text{m}$; (b) $4.6 \times 10^4\,\text{m/s}$]

93 If F is the magnitude of the gravitational force on a particle of mass m, then the gravitational acceleration is $a_g = F/m$. At an eclipse of the Moon, the forces of the Sun and the Moon are in the same direction while at an eclipse of the Moon they are in opposite directions. The percentage change is $(100\%)(a_{gS} - a_{gM})/a_{gS}$, where a_{gS} is the gravitational acceleration at an eclipse of the Sun and a_{gM} is the gravitational acceleration at an eclipse of the Moon.

[Ans: 1.1%]

101 The force of the sun on a planet in a circular orbit must be a centripetal force, so its magnitude is $F = mv^2/r$, where v is the speed of the planet and r is the radius of its orbit. Write an expression for F in terms of the period T by substituting $v = 2\pi r/T$, then substitute for T from Kepler's law of periods. You should find that the magnitude of the force is proportional to the reciprocal of r^2.

Quiz

Some questions might have more than one correct answer.

1. When two uniform spheres are a distance r apart the gravitational force of each on the other has magnitude F. When their separation is reduced to $r/2$ the magnitude of the gravitational force is
    A. $F/4$
    B. $F/2$
    C. F
    D. $2F$
    E. $4F$

2. The acceleration produced by the gravitational force on a small object does not depend on the mass of the object because
    A. the gravitational force is independent of the mass of the object and the object obeys Newton's second law
    B. the gravitational force is proportional to the mass of the object and the object obeys Newton's second law
    C. the gravitational force is proportional to the square of the mass of the object and the object obeys Newton's second law
    D. the gravitational force is inversely proportional to the mass of the object and the object obeys Newton's second law
    E. the gravitational force is inversely proportional to the square of the mass of the object and the object obeys Newton's second law

3. The gravitational force on a particle placed inside a uniform spherical shell of matter depends on
    A. the mass of the particle
    B. the mass of the shell
    C. the radius of the shell
    D. the position of the particle within the shell
    E. none of these

4. The gravitational force on a particle placed outside a uniform spherical shell of matter depends on
    A. the mass of the particle
    B. the mass of the shell
    C. the radius of the shell
    D. the distance of the particle from the shell center
    E. none of these

5. When two particles are far apart their gravitational potential energy is zero; when they are separated by a distance d their gravitational potential energy is U. When they are separated by a distance $2d$ their gravitational potential energy is

    A. $U/4$

    B. $U/2$

    C. U

    D. $2U$

    E. $4U$

6. The escape speed for an object shot from the surface of a planet depends on

    A. the radius of the planet

    B. the mass of the object

    C. the mass of the planet

    D. the firing angle

    E. the speed of the planet

7. That planetary orbits are ellipses follows directly from

    A. the conservation of energy principle

    B. the conservation of momentum principle

    C. the conservation of angular momentum principle

    D. the algebraic form of the gravitational force as a function of the Sun-planet distance

    E. none of the above

8. The planetary law of areas (the position vector of a planet, with its tail at the Sun, sweeps out equal areas in equal times) is a direct consequence of

    A. the conservation of energy principle

    B. the conservation of momentum principle

    C. the conservation of angular momentum principle

    D. the algebraic form of the gravitational force as a function of the Sun-planet distance

    E. none of the above

9. If G is the universal gravitation constant, M is the mass of Earth, and m is the mass of an artificial satellite, then the work required to take the satellite from a circular orbit with radius R_i to a circular orbit with radius R_f is

    A. $GMm[R_f - R_i]$

    B. $GMm[R_f - R_i]$

    C. $GMm\left[\dfrac{1}{R_i} - \dfrac{1}{R_f}\right]$

    D. $\dfrac{1}{2}GMm\left[\dfrac{1}{R_i} - \dfrac{1}{R_f}\right]$

    E. $\dfrac{1}{2}GMm[R_f - R_i]$

Answers: (1) E, (2) B; (3) E; (4) A, B, D; (5) B; (6) A, C; (7) D; (8) C; (9) D

Chapter 14
FLUIDS

Here you study gases and liquids, first at rest and then in motion. Learn well the definitions of pressure and density, then pay particular attention to the variation of pressure with depth in a fluid. Use the concepts to understand two of the most basic principles of fluid statics: Archimedes' and Pascal's principles. The continuity and Bernoulli equations are fundamental for understanding fluids in motion. They express the relationship between pressure, velocity, density, and height at points in a moving fluid. To understand these equations, you must first understand the ideas of streamlines and tubes of flow.

Important Concepts

- ☐ fluid
- ☐ density
- ☐ pressure (absolute and gauge
- ☐ Pascal's principle
- ☐ Archimedes' principle
- ☐ buoyant force
- ☐ steady, incompressible flow

- ☐ streamline
- ☐ tube of flow
- ☐ mass flow rate
- ☐ volume flow rate
- ☐ equation of continuity
- ☐ Bernoulli's equation

Overview

14–2 What is a Fluid?

☐ Both liquids and gases are classified as **fluids**. Both conform to the sides and bottom of any container in which they are placed. Solids, on the other hand, retain their shapes even when they are not in containers.

14–3 Density and Pressure

☐ If a small volume ΔV of fluid has mass Δm, then its **density** ρ at that place is given by $\rho = \Delta m / \Delta V$. The definition should include a limiting process in which the volume shrinks to a point. Thus, density is defined at each point in a fluid and may vary from point to point. If the density is uniform, then $\rho = M/V$, where M is the mass and V is the volume of the entire fluid. Density is a scalar.

☐ The SI unit of density is kg/m^3.

☐ The fluid in any region exerts an outward force on the fluid or container wall that bounds the region. The force on any small surface area ΔA is proportional to the area and is perpendicular to the surface. If ΔF is the magnitude of the force on the area ΔA, then the

pressure p exerted by the fluid at that place is defined by the scalar relationship $p = \Delta F/\Delta A$. Strictly, the pressure is the limit as the area ΔA tends toward zero. Thus, pressure is defined at each point and may vary from point to point.

☐ The SI unit of pressure (N/m^2) is called a pascal and is abbreviated Pa. Other units are: 1 atmosphere (atm) = 1.013×10^5 Pa and 1 mm of Hg = 1 torr = 133.3 Pa.

☐ If a fluid is compressible, its density depends on the pressure. Great pressure is required to change the densities of most liquids. Gases, on the other hand, are readily compressible.

14–4   Fluids at Rest

☐ Pressure varies with depth in a fluid subjected to gravitational forces. Consider an element of fluid at height y in a larger body of fluid. Suppose the upper and lower faces of the element each have area A and the element has thickness Δy. If the fluid has density ρ, the mass of the element is $\rho A \, \Delta y$ and the force of gravity on it is $\rho g A \, \Delta y$. The pressure at the upper face is $p(y + \Delta y)$ so the downward force of the fluid there is $p(y + \Delta y)A$. The pressure at the lower face is $p(y)$ so the upward force of the fluid there is $p(y)A$. Since the element is in equilibrium the net force must vanish: $p(y + \Delta y)A + \rho g A \, \Delta y - p(y)A = 0$. In the limit as Δy becomes infinitesimal, this equation yields

$$\frac{dp}{dy} = -\rho g \,.$$

The negative sign indicates that the pressure is less at points higher in the fluid than at lower points.

☐ If the fluid is incompressible, then the density is the same everywhere and the pressure difference between any two points in the fluid, at heights y_1 and y_2 respectively, is given by

$$p_2 - p_1 = -\rho g(y_2 - y_1) \,.$$

If p_0 is the pressure at the upper surface of an incompressible fluid, then

$$p = p_0 + \rho g h$$

is the pressure a distance h below the surface.

☐ The pressure is same at any points that are at the same height in a homogeneous fluid. If the fluid is inhomogeneous, as it is if it consists of layers of immiscible fluids with different densities, you must apply the expression for $p(y)$ to each layer separately.

☐ Although atmospheric pressure varies with height, the variation is negligible over distances on the order of meters. The upper surface in each arm of an open U-tube, for example, is at atmospheric pressure, even if the surfaces are at different heights.

14–5   Measuring Pressure

☐ A mercury barometer consists of a vertical tube with its open end inserted beneath the surface of a mercury pool with its surface exposed to the air. The region of the tube above the mercury column is evacuated and the pressure there is zero. The height h of the mercury column in the tube is measured from the top of the pool. If the atmospheric pressure is p_0, then $p_0 = \rho g h$, where ρ is the density of mercury. The height h is directly proportional to the atmospheric pressure.

☐ An open-tube manometer is essentially a U-tube that is open at both ends and is filled with an incompressible fluid. One end is inserted into the region where the pressure is to be measured while the other end is held vertically. The difference h in the heights of the fluid in the two arms is measured. If the pressure to be measured is p and atmospheric pressure is p_0, then $p - p_0 = \rho g h$, where ρ is the density of the fluid in the manometer.

☐ A mercury barometer measures **absolute pressure** while an open-tube manometer measures **gauge pressure** (the difference between absolute and atmospheric pressure.

14–6 Pascal's Principle

☐ **Pascal's principle** is: If the external pressure applied to the surface of a fluid changes, then the pressure everywhere in the fluid changes by the same amount.

☐ A hydraulic jack is essentially a tube that is narrow at one end and wide at the other, containing an incompressible fluid and fitted with pistons at both ends. When a small force F_i is applied to the piston at the narrow end, the fluid exerts a much larger force F_o on the piston at the wide end. The change in pressure at the narrow end is F_i/A_i, where A_i is the area of the tube there. The change in pressure at the wide end is F_o/A_o, where A_o is the area there. According to Pascal's principle, these changes must be the same, so $F_i/A_i = F_o/A_o$. Since A_o is much greater than A_i, F_o is much greater than F_i. Hydraulic jacks are used to lift heavy objects, such as automobiles, and to operate brake systems on cars.

☐ Suppose the distances moved by pistons of a hydraulic jack are d_i (at the input end) and d_o (at the output end). If the fluid is incompressible, then the volume of fluid does not change and $d_i A_i = d_o A_o$. Multiply the left side of this equation by F_i/A_i and the right side by the equal quantity F_o/A_o to obtain $F_i d_i = F_o d_o$. The work done by the input force is the same as the work done by the output force.

14–7 Archimedes' Principle

☐ **Archimedes' principle** is: A fluid exerts an upward buoyant force on any object that is partially or wholly immersed in it and the magnitude of the force is equal to the weight of the fluid displaced by the body.

☐ The buoyant force on an object is a direct result of the pressure exerted on it by the fluid and depends on the variation of pressure with depth. The pressure at the bottom of the object is greater than the pressure at the top and the net buoyant force is upward. If the submerged portion of an object is replaced by an equal volume of fluid, the fluid would be in equilibrium and the net force associated with the pressure of surrounding fluid would equal the weight of the fluid that replaced the object.

☐ To calculate the buoyant force acting on a given object, first find the submerged volume. If the object is completely surrounded by fluid, this is the volume of the object. If the object is floating on the surface, it is the volume that is beneath the surface. If the submerged volume is V_s, then the magnitude of the buoyant force is

$$F_b = \rho_f g V_s ,$$

where ρ_f is the density of the fluid.

14–8 Ideal Fluids in Motion

☐ Fluid flow can be categorized according to whether it is steady or nonsteady, compressible or incompressible, viscous or nonviscous, and rotational or irrotational. This chapter deals chiefly with ideal flow: steady, incompressible, nonviscous, and irrotational.

☐ The velocity and density of a fluid in **steady flow** do not depend on the time. If we follow all the particles that eventually get to any selected point, we find they all have the same velocity as they pass that point, regardless of their velocities when they are elsewhere. In addition, particles flow into and out of any volume in such a way that the mass in the volume at any time is the same as at any other time.

☐ If the flow is **incompressible**, the density does not depend on either time or position. It is the same everywhere in the fluid and retains the same value through time. If the flow is nonviscous, no part of the fluid exerts resistive forces on neighboring parts and the fluid does not exert a resistive force on an object in it. If the flow is irrotational, a small test body placed in it does not rotate about an axis through it.

14–9 The Equation of Continuity

☐ A **streamline** is the path traced out by a moving fluid particle. In steady flow, streamlines are fixed curves in space. Every fluid particle on the same streamline passes through the same sequence of points and at each point has the same velocity as other particles when they are at the point. At any point the fluid velocity is tangent to the streamline through the point.

☐ A **tube of flow** is a tube bounded by streamlines. No particles ever cross the boundaries of a tube of flow. Since streamlines do not cross the boundaries of a tube of flow, they crowd closer together in narrow portions of a tube than in wide portions.

☐ If the fluid passing through the cross section of a narrow tube of flow has density ρ and speed v, then the mass of fluid that passes the cross section per unit time is given by ρAv, where A is the cross-sectional area of the tube. The volume of fluid that passes the cross section per unit time is given by Av. The former quantity is called the **mass flow rate** and the later is called the **volume flow rate** and is denoted by R_V.

☐ In steady flow, the mass of fluid in any volume does not change, so the mass flow rate must be the same for every cross section of a tube of flow. Thus, $\rho Av = $ constant along a tube of flow. This is the **equation of continuity** in its most general form. For incompressible flow, density is the same everywhere along a tube of flow and the equation becomes $Av = $ constant along a tube of flow.

☐ The equation of continuity implies that fluid particles have greater speed in a narrow portion of a tube than in a wider portion. Because streamlines crowd closer together in narrow portions of tubes of flow than in wider portions we conclude that a high concentration of streamlines corresponds to a high fluid speed and a low concentration corresponds to a low fluid speed.

14–10 Bernoulli's Equation

☐ **Bernoulli's equation** for ideal flow tells us that the quantity $p + \frac{1}{2}\rho v^2 + \rho gy$ has the same value at every point along a streamline. Here p is the pressure, ρ is the fluid density, v is the fluid speed, and y is the height of the fluid above some (arbitrary) reference level. The value of $p + \frac{1}{2}\rho v^2 + \rho g y$ may be different for different streamlines.

☐ This equation is derived by applying the work-kinetic energy theorem to the fluid in a narrow tube of flow. The term p appears because the force exerted by neighboring portions of fluid do work, the term $\rho g y$ appears because gravity does work if the height of the tube varies, and the term $\frac{1}{2}\rho v^2$ appears because the kinetic energy of the fluid changes if work is done on it.

☐ If the tube does not change height, the gravitational term is not needed and Bernoulli's equation becomes $p + \frac{1}{2}\rho v^2 =$ constant along a streamline. This equation indicates that the pressure is high where the fluid speed is low and is low where the fluid speed is high. Combining this result with the continuity equation, we conclude that for steady, incompressible, horizontal flow, the pressure is low where a tube of flow is narrow and high where it is wide.

Hints for Questions

<u>1</u> All the floors and ceilings have the same area, so the force of the water on any of them is proportional to the pressure, which in turn is proportional to the depth of the floor or ceiling from the top of the tank.

[Ans: e, then b and d tie, then a and c tie]

<u>3</u> (a) The U-tube is open. This means for static equilibrium to occur the mass of fluid must be the same in each arm.

(b) Consider the volume of fluid in each arm above the lower dotted line. If the density of the red fluid is greater than that of the gray fluid there is less red fluid than gray fluid above the line. Similarly, if the density of the red fluid is less than that of the gray fluid there is more red fluid than gray fluid above the line. If the densities are the same there are equal amounts of the two fluids above the line.

[Ans: (a) 2; (b) less dense in 1, equally dense in 2, more dense in 3]

<u>5</u> Notice that the upper water surfaces are at the same height and that the ducks are not on the bottoms of the containers. In each of (b) and (c) the duck displaces a water with weight equal to the weight of the duck and this water has been removed from the container.

[Ans: all tie]

<u>7</u> The buoyant force is proportional to the density of the fluid and this in turn is proportional to the slope of the gauge pressure versus depth graph.

[Ans: a, b, c]

<u>9</u> According to the equation of continuity the water speed is inversely proportional to the square of the radius of the pipe. The kinetic energy is proportional to the square of the water speed.

[Ans: B, C, A]

Hints for Problems

<u>5</u> Let V_f be the volume of the fish with its air sacs collapsed and V_a be the volume of air used to expand the sacs. The mass of the fish alone is $m_f = \rho_f V_f$ and the mass of air is

$m_a = \rho_a V_a$. The density of the fish with its air sacs expanded is $(m_f + m_a)/(V_f + V_a)$, which is $(\rho_f V_f + \rho_a V_a)/(V_f + V_a)$ once substitutions are made for the masses. This density should equal the density of water ($\rho_w = 0.998 \times 10^3 \, \text{kg/m}^3$). You want to calculate $V_a/(V_f + V_a)$. Solve the density equation for V_a in terms of V_f and then substitute into $V_a/(V_f + V_a)$.

[Ans: 0.074]

9 The pressure p in a fluid at depth h below the surface is $p = p_0 + \rho g h$, where p_0 is the pressure at the surface and ρ is the density of the fluid.

[Ans: $1.90 \times 10^4 \, \text{Pa}$]

13 The pressure on the bottom of the seal is the same as the pressure in the left arm of the tube at a height d above the bottom. If the water in the left arm is a distance $d + h$ above the bottom, then that pressure is $p = p_0 + \rho g h$, where ρ is the density of water ($0.998 \times 10^3 \, \text{kg/m}^3$) and p_0 is atmospheric pressure. The pressure on the top of the seal is p_0 and the net force on the seal is $F = A(p - p_0)$, where A is the cross-sectional area of the tube. Solve $F = A\rho g h$ for h, then calculate $d + h$.

[Ans: 2.80 m]

17 Divide the wall into horizontal strips of length W (= 8.0 m) and width dy. The magnitude of the force on a strip is $pW\,dy$, where p is the pressure. The pressure at a depth y below the surface of the water is $p = p_0 + \rho g y$, where ρ is the density of water. If h is the depth of water in the aquarium then the force on the wall is

$$F = \int_0^h W(p_0 + \rho g y)\,dy \,.$$

[Ans: $4.69 \times 10^5 \, \text{N}$]

23 The force on the output piston has magnitude $F_o = kx$, where k is the spring constant and x is the distance the spring is compressed. The force on the input piston has magnitude $F_i = mg$, where m is the mass of sand in the container. The forces obey Pascal's principle: $F_o/A_0 = F_i/A_i$, where A_o is the cross-sectional area of the output piston and A_i is the cross-sectional area of the input piston. Replace F_o with kx and F_i with mg, then solve for m.

[Ans: 8.50 kg]

27 Suppose there are N logs, each with length L and radius R. If the density of the logs is ρ_L then the total weight of the logs is $N\pi R^2 L \rho_L g$ and the total weight of the raft and children is $N\pi R^2 L \rho_L g + 3W$, where W is the weight of a child. Suppose the raft floats with its top at the surface of the water. Then the weight of the water it displaces is $N\pi R^2 L \rho_w g$, where ρ_w is the density of water. According to Archimedes' principle this is the magnitude of the buoyant force of the water on the raft. Since the raft is in equilibrium the magnitude of the buoyant force must equal the weight of the raft and children. Solve for N and round up to the nearest integer.

[Ans: five]

37 (a) Let V be the volume of water displaced by the car and ρ_w (= $0.998 \times 10^3 \, \text{kg/m}^3$) be the density of water. Then the magnitude of the buoyant force on the car is $\rho_w g V$. Since the car is essentially in equilibrium this must equal the weight mg of the car, where m is its mass. solve $mg = \rho_w g V$ for V.

(b) Let V_w be the volume of water in the car. The total weight of the car and the water in it is $mg + \rho_w g V_w$. The magnitude of the buoyant force is now $\rho_w g V_{total}$, where V_{total} is the total volume of the car $(= 5.00\,\text{m}^3 + 0.75\,\text{m}^3 + 0.800\,\text{m}^3 = 6.55\,\text{m}^3)$. Solve $mg + \rho_w g V_w = \rho_w g V_{total}$ for V_w.

$\left[\text{Ans: } 4.75\,\text{m}^3\right]$

39 The weight of the weights that are removed is equal to the buoyant force of the water on the model and, according to Archimedes' principle this is the weight of a volume of water equal to the volume of the model. If W is the weight of the removed weights then $W = \rho g V$, where ρ is the density of water and V is the volume of the model.

(b) The volume of the model is $(1/20)^3$ times the actual volume.

(c) The mass is ρV.

$\left[\text{Ans: (a) } 637.8\,\text{cm}^3; \text{ (b) } 5.102\,\text{m}^3; \text{ (c) } 5.102 \times 10^3\,\text{kg}\right]$

47 (a) Use the equation of continuity: $v_b A_b = v_2 A_2$, where v_b is the water speed and A_b is the cross-sectional ares of the pipe at the basement and v_2 is the speed and A_2 is the cross-sectional area of the pipe at the second floor. If d is the diameter of the pipe its cross-sectional area is $\pi d^2 /4$. Solve for v_2.

(b) Use the Bernoulli equation: $p_b + \frac{1}{2}\rho v_b^2 = p_2 + \frac{1}{2}v_2^2 + h$, where p_b is the pressure at the basement, p_2 is the pressure at the second floor, h is the height of the second floor above the basement, and ρ is the density of water.

$\left[\text{Ans: (a) } 3.9\,\text{m/s}; \text{ (b) } 88\,\text{kPa}\right]$

51 The work is the product of the net force and the distance moved. The net force is the product of the pressure difference and the cross-sectional area of the pipe. Thus $W = AL\,\Delta p$, where A is the cross-sectional area of the pipe, L is the distance moved, and Δp is the pressure difference. Now AL is the volume of water, so $W = V\,\Delta p$.

$\left[\text{Ans: } 1.5 \times 10^5\,\text{J}\right]$

55 (a) Use Bernoulli's principle and the projectile motion equations to develop an expression for x as a function of h. The pressure inside the tank at the hole is $p_0 + \rho g h$, where p_0 is atmospheric pressure and ρ is the density of water. The pressure outside the tank at the hole is p_0. Let v_0 be the speed of the water as it leaves the hole. The speed of the water inside the tank is quite small and may be taken to be zero. Then the principle yields $p_0 + \rho g h = p_0 + \frac{1}{2}\rho v_0^2$. Algebraically solve this equation for v_0. If it takes time Δt for water to fall from the hole to the ground, a distance $H - h$, then $H - h = \frac{1}{2}g(\Delta t)^2$. Solve for Δt. The distance from the tank to the point on the ground where the water hits is $x = v_0\,\Delta t$. Evaluate the expression you developed for x.

(b) Solve the expression for h. There are two solutions. One is 10 cm. You want the other one.

(c) Set the derivative of x with respect to h equal to zero and solve for h.

$\left[\text{Ans: (a) } 35\,\text{cm}; \text{ (b) } 30\,\text{cm}; \text{ (c) } 20\,\text{cm}\right]$

61 Bernoulli's principle gives $p_A = p_B + \frac{1}{2}\rho_{air}v^2$. The pressure difference is $p_A - p_B = \rho g h$.

Use this to substitute for $p_A - p_B$ in the Bernoulli equation, then solve for v.

[Ans: (b) 63.3 m/s]

73 Use the Bernoulli equation. The pressure at the opening is atmospheric pressure. The pressure above the water in the keg is atmospheric pressure in part (a) and is 1.40 atm in part (b). Assume that the cross-sectional area of the can is much larger than the area of the opening, so the speed of the liquid as it leaves the opening is much greater than the speed of the liquid at the liquid surface in the can.

[Ans: (a) 3.1 m/s; (b) 9.5 m/s]

77 The volume flow rate, which is the product of the fluid speed and the cross-sectional area of the pipe, is uniform along the pipe. The cross-sectional area is πR^2, where R is the radius of the pipe. For any two points along the pipe $\pi R_1^2 v_1 = \pi R_2^2 v_2$.

[Ans: 1.00×10^{-2} m/s]

81 Assume the top of the can is level with the water surface. Then the buoyant force of the water on the can is $\rho_w g V$, where ρ is the density of water and V is the volume of the can. This must be balanced by the weight of the can and lead shot. You do not need to know the density of the lead shot.

[Ans: 1.07×10^3 g]

83 Use the apparent weight in water to find the volume of the object. The weight of the displaced water is 10 N. Use the apparent weight in the other liquid to find the density of that liquid. The weight of the displaced liquid is 6 N and its volume is the volume of the object.

[Ans: 6.0×10^2 kg/m^3]

87 The pressure at depth h is $p_0 + \rho g h$, where p_0 is atmospheric pressure (1.013×10^5 Pa) and ρ is the density of seawater.

[Ans: 60 MPa]

93 The weight of the water is $W = \rho g A h$, where ρ is the density of the water, A is the horizontal cross-sectional area of the submarine hull, and h is the depth of the submarine. The pressure is $p_0 + \rho g h$, where p_0 is atmospheric pressure.

[Ans: (a) 6.06×10^9 N; (b) 20 atm; (c) no]

Quiz

Some questions might have more than one correct answer.

1. A vessel has curved sides and is filled with water. At any point on a side the force of the water on the side is

    A. horizontally outward
    B. horizontally inward
    C. vertical
    D. perpendicular to the side and outward
    E. perpendicular to the side and inward

2. A dam is thicker at the bottom than at the top. This is because
   A. the water pressure is greater at the top than at the bottom
   B. the water pressure is greater at the bottom than at the top
   C. the water density is greater at the top than at the bottom
   D. the water density is greater at the bottom than at the top
   E. the bottom has to hold up the portion of the dam above

3. Before a gate on a canal lock is opened, small windows at the bottom of the gate are opened to allow the water level to become the same on the two sides. As a result the gate is easier to open because
   A. the pressure becomes the same on the two sides
   B. the density becomes the same on the two sides
   C. the water flow is less turbulent
   D. scum on the surface of the water does not pass through the lock
   E. boats in the lock cannot obstruct the gate

4. A piece of cork is embedded in an ice cube and the ice cube floats on water in vessel A. A piece of iron is embedded in another ice cube and the ice cube floats on water in vessel B. When both cubes have melted
   A. both water levels are the same as before
   B. both water levels are lower than before
   C. the water level in A is the same and the water level in B is lower than before
   D. the water level in A is higher and the water level in B is lower than before
   E. the water level in A is lower and the water level in B is higher than before

5. An object floats at the top surface of a fluid. When it is pushed down so it is completely submerged
   A. it displaces a greater volume of fluid than before
   B. it displaces a greater mass of fluid than before
   C. the net force of the fluid on it is greater than before
   D. the net force of the fluid on it is less than before
   E. the net force of the fluid on it is the same as before

6. Pascal's principle is valid for
   A. all fluids
   B. all incompressible fluids
   C. all incompressible fluids at rest
   D. all fluids in motion
   E. all incompressible fluids in motion

7. Water is pumped up a vertical pipe that is wider at the top than at the bottom. The pressure at the top

    A. might be greater than the pressure at the bottom
    B. might be less than the pressure at the bottom
    C. might be the same as the pressure at the bottom
    D. is definitely greater than the pressure at the bottom
    E. is definitely less than the pressure at the bottom

8. The volume flow rate of water at one end of pipe can be increased by

    A. making that end more narrow than the other end
    B. making that end less narrow than the other end
    C. increasing the pressure at the other end
    D. orienting the pipe so the water flows down hill
    E. none of the above

9. Water streams downward from an open facet. Compared the upper portion of the stream

    A. the pressure is greater in the lower portion and the cross-sectional area of the stream is less
    B. the pressure is less in the lower portion and the cross-sectional area of the stream is greater
    C. the pressure is the same in the lower portion and the cross-sectional area of the stream is less
    D. the pressure is the same in the lower portion and the cross-sectional area of the stream is greater
    E. the pressure is less in the lower portion and the cross-sectional area of the stream is greater

10. Water flows in a horizontal pipe that has narrow and wide sections.

    A. both the fluid speed and pressure are greater in a narrow section than in a wide section
    B. the fluid speed is greater and the pressure is less in a narrow section than in a wide section
    C. the fluid speed is less and the pressure is greater in a narrow section than in a wide section
    D. both the fluid speed and the pressure are less in a narrow section than in a wide section
    E. the fluid speed is greater in a narrow section than in a wide section but the pressure is the same

Answers: (1) D; (2) B; (3) A; (4) C; (5) A, B, C; (6) C; (7) A, B, C; (8) C, D; (9) C; (10) B

Chapter 15
OSCILLATIONS

This chapter is about objects whose motions are repetitive. Pay attention to the meanings of the terms used to describe simple harmonic motion: amplitude, period, frequency, angular frequency, and phase constant; pay attention to the transfer of energy from kinetic to potential and back again as the object moves; and also pay attention to the form of the force law that leads to this type of motion.

Important Concepts

- [ ] oscillation
- [ ] simple harmonic motion
- [ ] amplitude
- [ ] angular frequency
- [ ] frequency
- [ ] period
- [ ] phase

- [ ] phase constant
- [ ] angular oscillator
- [ ] simple pendulum
- [ ] physical pendulum
- [ ] damped oscillations
- [ ] forced oscillations
- [ ] resonance

Overview

15–2 Simple Harmonic Motion

- [ ] If an object moves along the x axis in **simple harmonic motion**, its co-ordinate x as a function of time t is given by

$$x(t) = x_m \cos(\omega t + \phi),$$

where x_m, ω, and ϕ are constants. A possible function $x(t)$ is graphed above .

- [ ] The object moves back and forth between $x = -x_m$ and $x = +x_m$. x_m is called the **amplitude** of the oscillation.

- [ ] The constant ω is called the **angular frequency** of the oscillation. Since ωt is measured in radians, the unit of ω is rad/s. The **frequency** f of the oscillation gives the number of times the object moves through a complete cycle per unit time. It is measured in hertz ($1\,\text{Hz} = 1\,\text{s}^{-1}$). The frequency and angular frequency are related by $\omega = 2\pi f$. The **period** T of the oscillation is the time for one complete cycle. It is related to the angular frequency and frequency by $T = 1/f$ and $T = 2\pi/\omega$.

☐ The combination $\omega t + \phi$ is called the **phase** of the oscillation and ϕ is called the **phase constant** or **phase angle**. A change in the phase constant simply moves the curve shown on the graph above left or right along the t axis.

☐ An expression for the velocity of the object as a function of time can be found by differentiating the expression for $x(t)$ with respect to time:

$$v(t) = \frac{dx(t)}{dt} = -\omega x_m \sin(\omega t + \phi).$$

The maximum speed of the object is given by $v_m = \omega x_m$. The object has maximum speed when its coordinate is $x = 0$. The velocity of the object is zero when its coordinate is $x = -x_m$ and also when its coordinate is $x = +x_m$.

☐ An expression for the acceleration of the object as a function of time can be found by differentiating $v(t)$ with respect to time:

$$a(t) = \frac{dv(t)}{dt} = -\omega^2 x_m \cos(\omega t + \phi) = -\omega^2 x(t).$$

The magnitude of the maximum acceleration is $\omega^2 x_m$. The object has maximum acceleration when its coordinate is $x = -x_m$ and also when its coordinate is $x = +x_m$. This is where the velocity vanishes. The acceleration is zero when the coordinate is $x = 0$. This is where the speed is a maximum.

☐ The amplitude x_m and phase constant ϕ are determined by the initial conditions (at $t = 0$). Since $x(t) = x_m \cos(\omega t + \phi)$ the initial coordinate is given by $x_0 = x_m \cos\phi$ and the initial velocity is given by $v_0 = -\omega x_m \sin\phi$.

x_0 is positive for $-\pi/2 \,\text{rad} < \phi < \pi/2 \,\text{rad}$.

x_0 is negative for $\pi/2 \,\text{rad} < \phi < 3\pi/2 \,\text{rad}$.

v_0 is positive for $\pi \,\text{rad} < \phi < 2\pi \,\text{rad}$.

v_0 is negative for $0 < \phi < \pi \,\text{rad}$.

☐ The equations $x_0 = x_m \cos\phi$ and $v_0 = -\omega x_m \sin\phi$ can be solved for x_m and ϕ. To obtain an expression for x_m, solve the first equation for $\cos\phi$ and the second for $\sin\phi$, then use $\cos^2\phi + \sin^2\phi = 1$. To obtain an expression for ϕ, divide the second equation by the first and solve for $\tan\phi$. The results are

$$x_m = \sqrt{x_0^2 + \frac{v_0^2}{\omega^2}} \qquad \text{and} \qquad \tan\phi = -\frac{v_0}{\omega x_0}.$$

☐ Be careful when you evaluate the expression for ϕ. There are always two angles that are the inverse tangent of any quantity, but your calculator only gives the one closest to 0. The other is 180° or π radians away. Always check to be sure $x_m \cos\phi$ gives the correct initial coordinate and $-\omega x_m \sin\phi$ gives the correct initial velocity. If they do not, add π rad to the value you used for ϕ.

15–3 The Force Law for Simple Harmonic Motion

☐ Newton's second law tells us that the force that must be applied to an object of mass m to produce simple harmonic motion is $F(t) = ma(t) = -m\omega^2 x(t)$. The force must be proportional to the displacement from equilibrium and the constant of proportionality must be negative.

☐ Consider a block on the end of spring, moving on a frictionless horizontal surface. If the origin is taken to be at the position of the block when the spring is neither extended nor compressed, the force of the spring on the block is given by $F = -kx$, where k is the spring constant. Since the force is proportional to the displacement and the constant of proportionality is negative we know the motion is simple harmonic. Furthermore, a comparison of $F = -kx$ with $F = -m\omega^2 x$ tells us that the angular frequency of the motion is $\omega = \sqrt{k/m}$. The period is $T = 2\pi/\omega = 2\pi\sqrt{m/k}$.

15–4 Energy in Simple Harmonic Motion

☐ For the block on the end of a spring, an expression for the potential energy as a function of time can be found by substituting $x(t) = x_m \cos(\omega t + \phi)$ into $U = \frac{1}{2}kx^2$:

$$U(t) = \tfrac{1}{2}kx_m^2 \cos^2(\omega t + \phi).$$

An expression for the kinetic energy as a function of time can be found by substituting $v = -\omega x_m \sin(\omega t + \phi)$ into $K = \frac{1}{2}mv^2$:

$$K(t) = \tfrac{1}{2}m\omega^2 x_m^2 \sin^2(\omega t + \phi).$$

If $\omega^2 = k/m$ is used, this can also be written

$$K(t) = \tfrac{1}{2}kx_m^2 \sin^2(\omega t + \phi).$$

☐ Both the potential and kinetic energies vary with time. The potential energy is a maximum when the coordinate is $x = -x_m$ and also when it is $x = +x_m$. Then, the speed is zero and the kinetic energy vanishes. The potential energy is a minimum when $x = 0$. Then, the speed is the greatest and the kinetic energy is a maximum. Notice that the maximum kinetic energy has exactly the same value as the maximum potential energy.

☐ Although both the potential and kinetic energies vary with time, the total mechanical energy $E = K + U$ is constant, as you can see by adding the expressions given above and using $\cos^2(\omega t + \phi) + \sin^2(\omega t + \phi) = 1$. All of the following expressions for the mechanical energy give the same value:

$$E = \tfrac{1}{2}mv^2(t) + \tfrac{1}{2}kx^2(t) = \tfrac{1}{2}kx_m^2 = \tfrac{1}{2}mv_m^2.$$

You will solve some problems by equating two of these expressions to each other and solving for one of the quantities in them.

15–5   An Angular Simple Harmonic Oscillator

□ An angular simple harmonic oscillator consists of an object suspended by a wire that exerts a torque when it is twisted. The torque is proportional to the angle of twist and the constant of proportionality is negative; it is a restoring torque. If the angular position θ of the object is measured from its position when the wire is not twisted, then $\tau = -\kappa\theta$, where κ is called the torsion constant of the wire. Newton's second law for rotation becomes $-\kappa\theta = I\alpha$, where I is the rotational inertia of the object. This equation is exactly like the equation for a mass on a spring, except that θ has replaced x, α has replaced a, I has replaced m, and κ has replaced k. The object rotates back and forth in simple harmonic motion with angular frequency $\omega = \sqrt{\kappa/I}$ and period $T = 2\pi/\omega = 2\pi\sqrt{I/\kappa}$.

15–6   Pendulums

□ A **simple pendulum** consists of a mass m suspended by a string. After the mass is pulled aside and released, it swings back and forth along the arc of a circle with radius equal to the length L of the string. When the string makes the angle θ with the vertical, the tangential component of the gravitational force acting on the mass is $F = -mg\sin\theta$, where the negative sign indicates that the force is pulling the mass toward the $\theta = 0$ position. Newton's second law becomes $-mg\sin\theta = ma$. In terms of the distance s along the arc of the mass from its equilibrium point (the bottom of the arc), $\theta = s/L$ if the angle is measured in radians. Thus, $-g\sin(s/L) = a$, where the mass has been canceled from both sides. The acceleration is not proportional to the displacement and the motion is not strictly simple harmonic. However, if s is much less than L, then $\sin(s/L)$ can be approximated by s/L itself and the equation becomes $-(g/L)s = a$. If s is always small, the motion is very nearly simple harmonic, the angular frequency is $\omega = \sqrt{g/L}$, and the period is $T = 2\pi/\omega = 2\pi\sqrt{L/g}$.

□ A **physical pendulum** consists of an object that is pivoted about some point other than its center of mass. If h is the distance from the pivot point to the center of mass, the torque acting on the object is $\tau = -mgh\sin\theta$, where θ is the angle between the vertical and the line that joins the pivot and center of mass. Newton's second law for rotation becomes $-mgh\sin\theta = I\alpha$, where I is the rotational inertia of the object. If θ is small, $\sin\theta$ can be approximated by θ itself, in radians. Then, $-mgh\theta = I\alpha$. The angular acceleration is proportional to the angular displacement and the constant of proportionality is negative, so the motion is simple harmonic. The angular frequency is $\omega = \sqrt{mgh/I}$ and the period is $T = 2\pi\sqrt{I/mgh}$.

15–7   Simple Harmonic Motion and Uniform Circular Motion

□ When a particle moves around a circle with constant speed, the projection of its position vector on the x axis performs simple harmonic motion. Suppose the angular speed of the particle is ω and the radius of the circle is R. Put the origin of the coordinate system at the center of the circle and measure angles counterclockwise from the x axis. If the initial angular position of the particle is ϕ, then the x component of its position vector is given by $x(t) = R\cos(\omega t + \phi)$. You can identify the angular speed ω of the particle with the angular frequency of the oscillation, the radius of the circle with the amplitude of the oscillation, and the initial angular position with the phase constant.

15–8 Damped Simple Harmonic Motion

☐ Many oscillating systems in nature are damped by a force that is proportional to the velocity. Consider a mass m on the end of spring with spring constant k and subject to the damping force $-bv$, where b is a damping constant and v is the velocity. Newton's second law becomes $-kx - bv = ma$. The solution is

$$x(t) = x_m e^{-bt/2m} \cos(\omega_d t + \phi),$$

where

$$\omega_d = \sqrt{\left(\frac{k}{m}\right)^2 - \frac{b^2}{4m^2}}.$$

The motion is oscillatory but the amplitude $(x_m e^{-bt/2m})$ decreases exponentially with time. See Fig. 15–16 of the text for a graph of $x(t)$.

☐ The total mechanical energy of a damped oscillator is not constant. As time goes on, the damping force dissipates energy.

15–9 Forced Oscillations and Resonance

☐ A mass on the end of spring can be driven to oscillate in simple harmonic motion by applying an external force of the form $F_m \cos \omega_d t$. In this section, the angular frequency of the force is denoted by ω_d and the natural angular frequency $(\sqrt{k/m})$ of the spring-mass system is denoted by ω.

☐ The angular frequency of a forced oscillation is the same as the angular frequency of the driving force and is NOT necessarily the natural angular frequency of the oscillator.

☐ The amplitude of the oscillation depends on the driving frequency, as well as on the driving amplitude F_m. The driving frequency for which the velocity amplitude is a maximum is the same as the natural frequency of the oscillator and is called the **resonance frequency**. It is nearly the same as the frequency for which the amplitude is a maximum. Fig. 15–17 of the text shows how the amplitude depends on the driving frequency.

☐ The amplitude of the motion does not decay with time even though a resistive force is present and energy is being dissipated. The mechanism that drives the oscillator and provides the external force does work on the system and supplies the energy required to keep it going at constant amplitude.

Hints for Questions

<u>1</u> In simple harmonic motion the acceleration is proportional to the displacement and the constant of proportionality is negative.

[Ans: c]

<u>3</u> The graph is negative at $x = 0$ and its slope is positive there. The function is $x = x_m \cos(\omega t + \phi)$ and its derivative is $dx/dt = -\omega x_m \sin(\omega t + \phi)$. Chose ϕ so that $\cos \phi$ and $\sin \phi$ are both negative.

[Ans: a and b]

5 A shift of 2π (360°) in the phase constant corresponds to a shift of one cycle on the graph. An increase in ϕ corresponds to a shift to the left and a decrease in ϕ corresponds to a shift to the right. In (a) the shift is one-half of a cycle to the right, in (b) it is one-quarter of a cycle to the right, and in (c) it is three-quarters of a cycle to the right or one-quarter of a cycle to the left.

$\left[\text{Ans: (a) } -\pi, -180°; \text{ (b) } -\pi/2, -90°; \text{ (c) } +\pi/2, +90°\right]$

7 At time $t = 0$ the acceleration is positive (the block is going left and slowing) and it is increasing (it reaches its maximum value when the block gets to $-x_m$). You want to choose ϕ so that $\cos\phi$ is positive and $\sin\phi$ is negative.

$\left[\text{Ans: (a) between B and C; (b) between } 3\pi/2 \text{ and } 2\pi \text{ rad}\right]$

9 The maximum kinetic energy of a harmonic oscillator is given by $K_{max} = \frac{1}{2}kx_m^2$, where k is the spring constant. The period is given by $T = 2\pi\sqrt{m/k}$, where m is the mass.

$\left[\text{Ans: (a) A, B, C; (b) C, B, A}\right]$

11 The period of a physical pendulum is given by $T = 2\pi\sqrt{I/mgh}$, where I is its rotational inertia about the pivot point, m is its mass, and h is the distance from the pivot point to the center of mass. Let d be the length of each rod. In (a) $I = md^2$ and $h = d$. In (b) $I = 2md^2$ and $h = 0$. In (c) $d = 5md^2$ and $h = d$.

$\left[\text{Ans: (a) b (infinite period, does not oscillate), c, a}\right]$

Hints for Problems

7 The points of zero velocity are the end points of the motion, so it takes half a period to travel between them and the distance between them is twice the amplitude. The frequency is the reciprocal of the period.

$\left[\text{Ans: (a) 0.50 s; (b) 2.0 Hz; (c) 18 cm}\right]$

9 The expression for the displacement has the form $x = x_m \cos(\omega t + \phi)$, where $x_m = 6.0$ m, $\omega = 3\pi$ rad/s, and $\phi = \pi/3$ rad. The velocity is given by $a = dx/dt = -\omega x_m \sin(\omega t + \phi)$ and the acceleration by $a = dv/dt = -\omega^2 x_m \cos(\omega t + \phi)$. The angular frequency ω and the frequency f are related by $\omega = 2\pi f$ and the period is the reciprocal of the frequency.

$\left[\text{Ans: (a) 3.0 m; (b) } -49 \text{ m/s; (c) } -2.7 \times 10^2 \text{ m/s}^2; \text{ (d) 20 rad; (e) 1.5 Hz; (f) 0.67 s}\right]$

17 (a) The displacement x and acceleration a at any instant of time are related by $x = -\omega^2 a$, where ω is the angular frequency. The frequency is $f = \omega/2\pi$.
(b) The angular frequency, mass m, and spring constant k are related by $\omega^2 = k/m$.
(c) If x_m is the amplitude then we may take the displacement to be $x = x_m \cos(\omega t)$. The velocity is $v = -\omega x_m \sin(\omega t)$. Use the trigonometric identity $\sin^2(\omega t) + \cos^2(\omega t) = 1$ to find an expression for x_m.

$\left[\text{Ans: (a) 5.58 Hz; (b) 0.325 kg; (c) 0.400 m}\right]$

19 Take the displacement to be $x = x_m \cos(\omega t + \phi)$. Then the velocity is $v = -\omega x_m \sin(\omega t + \phi)$. Here ω is the angular frequency and is given by $\omega = \sqrt{k/m}$, where k is the spring constant and m is the mass.

(a) Use the trigonometric identity $\sin^2(A)+\cos^2(A) = 1$, with $A = \omega t+\phi$, to find an expression for x_m.

(b) and (c) Use the trigonometric identity $\tan(A) = \sin(A)/\cos(A)$ to find an expression for $\tan(\omega t + \phi)$, then calculate ϕ itself. Make sure you obtain the solution for which $\cos(\omega t + \phi)$ is positive and $\sin(\omega t + \phi)$ is negative. Then substitute into $x = x_m \cos(\omega t + \phi)$ and $v = -\omega x_m \sin(\omega t + \phi)$.

[Ans: (a) 0.500 m; (b) -0.251 m; (c) 3.06 m/s]

25 Suppose the smaller block is on the verge of slipping when the acceleration of the blocks has its maximum value. This occurs when the displacement of the blocks is equal to the amplitude x_m of their oscillation and the value is $a_m = \omega^2 x_m$, where ω is the angular frequency of the oscillation. The magnitude of the force of friction is $f = ma_m = m\omega^2 x_m$. Since the block is on the verge of slipping $f = \mu_s N = \mu_s mg$, where $N\ (= mg)$ is the magnitude of the normal force of either block on the other. Solve for x_m. The angular frequency is given by $\omega = \sqrt{k/(m+M)}$.

[Ans: 22 cm]

27 (a) When the sum of the forces on it is zero, the block remains at rest. Draw a free-body diagram for the block. The forces on it are the force of the spring, the force of gravity, and the normal force of the incline. Consider components along an axis that is parallel to the incline. If the direction down the incline is taken to be the positive direction the component of the spring force $-kx$ and the component of the gravitational force is $W\sin\theta$, where k is the spring constant, x is the extension of the spring from its unstretched length, and W is the weight of the block. Set the sum equal to zero and solve for x. You must add x to the unstretched length of the spring to obtain the distance of the block from the top of the incline.

(b) Newton's second law gives $-kx + W\sin\theta = ma$, where m is the mass of the block and a is its acceleration. The equilibrium extension of the spring is $x_0 = (W/k)\sin\theta$. Let $x' = x - x_0$. This is the extension or compression of the spring from its unstretched length. Substitute $x = x' + x_0$ into the second law equation. You should find that the acceleration is proportional to x'. The constant of proportionality is the square of the angular frequency ω. The period is $2\pi/\omega$.

[Ans: (a) 0.525 m; (b) 0.686 s]

33 (a) Use $f = (1/2\pi)\sqrt{k/m}$, where k is the spring constant and m is the mass of the object.
(b) The potential energy is given by $U = \frac{1}{2}kx^2$.
(c) The kinetic energy is given by $K = \frac{1}{2}mv^2$, where v is the speed of the object.
(d) The mechanical energy, which is the sum of the kinetic and potential energies, is $\frac{1}{2}kx_m^2$, where x_m is the amplitude of the oscillation.

[Ans: (a) 2.25 Hz; (b) 1125 J; (c) 250 J; (d) 86.6 cm]

37 (a) If the block moves only a negligible distance while the bullet becomes embedded in it the linear momentum of the bullet-block system is conserved. If V is the velocity of the block immediately after the collision $mv = (M+m)V$. Solve for V.

(b) The mechanical energy of the bullet-block-spring system is $\frac{1}{2}(M+m)V^2$ and, since mechanical energy is conserved during the oscillation, this must be the same as $\frac{1}{2}kx_m^2$, where

k is the spring constant and x_m is the amplitude of the oscillation.

[Ans: (a) 1.1 m/s; (b) 3.3 cm]

41 The period of a physical pendulum with rotational inertia I and mass m, suspended from a point that is a distance h from its center of mass, is $T = 2\pi\sqrt{I/mgh}$. The distance between the point of suspension and the center of oscillation is the same as the length of a simple pendulum with the same period. The period of a simple pendulum of length L_0 is $T = 2\pi\sqrt{L_0/g}$. Set these two expressions for the period equal to each other and solve for L_0.

47 The period of a physical pendulum with rotational inertia I and mass M, suspended from a point that is a distance h from its center of mass, is $T = 2\pi\sqrt{I/Mgh}$. For this pendulum the center of mass is the distance $h = (m\ell/2)/2m = \ell/4$ from A. Here m is the mass of either one of the sticks and ℓ is its length. The rotational inertia is the sum of the contributions of the two sticks. The horizontal stick is pivoted at its center of mass, so according to Table 10–2 its rotational inertia is $(1/12)m\ell^2$. The vertical stick is pivoted at one end, so according to the parallel-axis theorem its rotational inertia is $(1/12)m\ell^2 + m(\ell/2)^2 = (1/3)m\ell^2$. The total mass is $M = 2m$.

[Ans: 1.83 s]

53 At time $t = 0$ the angle is $\theta_0 = \theta_m \cos(\phi)$ and its rate of change is $(d\theta/dt)_0 = -\omega\theta_m \sin(\phi)$, where ω (= 4.44 rad/s) is the angular frequency of oscillation. Solve the first equation for $\cos(\phi)$ and the second for $\sin(\phi)$. Use the results to find an expression for $\tan(\phi)$ in terms of θ_0, $(d\theta/dt)_0$, and ω. Then find θ itself. Be sure you obtain the correct result for ϕ. Both the sine and cosine should be positive. Now add the expressions for the square of the sine and the square of the cosine to find an expression for θ_m.

[Ans: (a) 0.845 rad; (b) 0.0602 rad]

61 Use $\omega = \sqrt{g/L}$, where L is the pendulum length, to compute the natural angular frequencies of the pendulums. The driving angular frequency is between 2.00 rad/s and 4.00 rad/s so those pendulums with natural angular frequencies in that range are in resonance with the driving force and show large amplitude oscillations.

[Ans: d and e]

63 Assume the springs of the car act as a single ideal spring with spring constant k. The oscillations of the car are in resonance with the driving force provided by the corrugations in the road and high points in the road is hit at intervals of duration d/v, where d is the distance between corrugations and v is the speed of the car. This must be the period of oscillation of the car. The period of a mass m on a spring with spring constant k is given by $2\pi\sqrt{m/k}$. Solve $2\pi\sqrt{m/k} = d/v$ for k. Use the total mass of the car and the four passengers for m and substitute for v in meters per second.

The compression of the spring adjusts so the force of the spring on the car matches the force of gravity in magnitude. This means that if mass Δm is removed from the car the length of the spring changes by $\Delta x = \Delta m\,g/k$.

[Ans: 5.0 cm]

67 The maximum kinetic energy is $K_{max} = \frac{1}{2}mv_m^2$, where m is the mass of the pendulum and v_m is its maximum speed. Since the small-angle approximation is valid here you can use

$v_m = L\omega\theta_m$ and $\omega = \sqrt{g/L}$ to write the expression for k_{max} in terms of the length L of the pendulum and its amplitude θ_m. Values for K_{max} and θ_m can read from the graph.

[Ans: 1.53 m]

73 The period of oscillation is given by $T = 2\pi\sqrt{I/\kappa}$, where I is the rotational inertia and κ is the torsion constant. Since the magnitude τ of the torque on the pendulum is related to the angle of rotation θ, you can find κ as the slope of the graph in Fig. 15–54(a). You can find the period from the graph in Fig. 15–54(b). The maximum angular speed is $\omega\theta_m$, where ω is the angular frequency, given in terms of the period T by $2\pi/T$, and θ_m is the amplitude. T and θ_m can be found from the graph.

[Ans: (a) 8.11×10^{-5} kg · m^2; (b) 3.14 rad/s]

83 The period of oscillation is given by $T = 2\pi\sqrt{I/mgh}$, where I is the rotational inertia of the disk, m is its mass, and h is the distance from the center of mass to the pivot point. In this case h is a radius of the disk. Look in Table 10–2 for the rotational inertia for rotation about an axis through the disk center and use the parallel-axis theorem to find the rotational inertia for rotation about a parallel axis that is a distance R away. To find the position of another pivot point for which the period is the same, solve $T = 2\pi\sqrt{(I_{com} + mr^2)/mgr}$ for r. Here I_{com} is the rotational inertia for rotation about an axis through the center of mass.

[Ans: (a) 0.873 s; (b) 6.3 cm]

89 The forces of gravity and the spring balance, both before and after the smaller body is attached. Use this to show that $mg = kx$, where m is the mass of the smaller body, x is the extra amount the spring stretches when that body is attached, and k is the spring constant. Solve for k. With the smaller body removed the period is $2\pi\sqrt{M/k}$, where M is the mass of the large block.

[Ans: (a) 147 N/m; (b) 0.733 s]

99 Use the formula $T = 2\pi\sqrt{I/\kappa}$ for the period of a torsion pendulum. Here I is the rotational inertia and κ is the torsion constant. Solve for I. The period is the time to oscillate through one cycle.

[Ans: 0.079 kg · m^2]

101 Use a vertical y axis with the origin at the position of the block when the spring is unstretched and take the upward direction to be positive. When the block has coordinate y the force on it is $F = -mg - ky$, where k is the spring constant. This can be written $F = -ky'$, where $y' = y + (m/k)g$. Thus $k = m\omega^2$, where ω is the angular frequency. The energy is $E = \frac{1}{2}ky^2 + mgy$, where the gravitational potential energy is taken to be zero when the block is at $y - 0$, its position when the spring is unstretched. Find the values of y for which $E = 2.00$ J. The kinetic energy is $K = E - \frac{1}{2}ky^2 - mgy$. Find the value of y for which this is a maximum, then substitute the value into the expression for K to find the maximum kinetic energy.

[Ans: (a) 0.45 s; (b) 0.10 m above and 0.20 m below; (c) 0.15 m; (d) 2.3 J]

Quiz

Some questions might have more than one correct answer.

1. If the frequency of oscillation is 10 Hz then the angular frequency and period are
    A. 3.1 rad/s and 10 ms
    B. 10 rad/s and 6.3 ms
    C. 6.3 rad/s and 20 ms
    D. 6.3 rad/s and 100 ms
    E. 3.1 rad/s and 200 ms

2. The speed of an object in one-dimensional simple harmonic motion is greatest when the object
    A. is at the equilibrium point
    B. is between the equilibrium point and either end of its path
    C. is at either end of its path
    D. at one end of its path but not the other
    E. has zero acceleration

3. The acceleration of an object in one-dimensional simple harmonic motion is greatest when the object is
    A. is at the equilibrium point
    B. is between the equilibrium point and either end of its path
    C. is at either end of its path
    D. has its greatest speed
    E. has zero speed

4. The acceleration of an object in one-dimensional simple harmonic motion is greatest when the velocity of the object is
    A. has its maximum magnitude and is toward the equilibrium point
    B. has its maximum magnitude and is away from the equilibrium point
    C. is zero
    D. has a magnitude that is half its maximum value
    E. has a magnitude that is one-fourth its maximum value

5. The oscillation frequency of an object on the end of a spring is determined by
    A. its initial speed
    B. its initial position
    C. the spring constant of the spring
    D. its mass
    E. the position of the equilibrium point

6. The amplitude of an object in one-dimensional simple harmonic motion is determined by
    A. its initial speed
    B. its initial position
    C. the spring constant of the spring
    D. its mass
    E. the position of the equilibrium point

7. A block is in simple harmonic motion. When it is at an end point of its path another block is placed rapidly and securely on top of it. As a result which of the following changes?
    A. the amplitude
    B. the frequency
    C. the spring constant
    D. the phase constant
    E. the mechanical energy

8. Two blocks are in simple harmonic motion on parallel lines in front of you. They have identical frequencies but block A lags block B by 90° in phase. When block B is at the right end of its path, block B is
    A. at its equilibrium point, moving to the left
    B. between its equilibrium point and the left end of its path
    C. at its equilibrium point, stationary
    D. at the left end of its path
    E. at the equilibrium point, moving to the right

9. When a block is hung from the ceiling via a spring and brought to rest, the spring is elongated from its unstretched length by 2.45 m. When the block is set into oscillation its angular frequency is
    A. 2.0 rad/s
    B. 4.0 rad/s
    C. 20 rad/s
    D. 40 rad/s
    E. 10 rad/s

10. The mechanical energy of an oscillating block-spring system is determined by
    A. the initial speed of the block
    B. the initial position of the block relative to its equilibrium position
    C. the spring constant of the spring
    D. the mass of the block
    E. the position of the block at equilibrium

11. A block is attached to a horizontal spring and oscillates in simple harmonic motion with another block on top of it. The upper block is most likely to skid off the lower block when

    A. the kinetic energy of the two-block system has its maximum value
    B. the kinetic energy of the two-block system is zero
    C. as the blocks pass through the equilibrium point
    D. the acceleration of the blocks has its maximum value
    E. when the acceleration of the blocks is zero

12. A simple pendulum of length L is drawn aside so the string makes the angle θ_0 with the vertical. It is then released. The speed of the pendulum bob as it passes the lowest point on its path is

    A. $\sqrt{gL(1 - \cos\theta_0)}$
    B. $\sqrt{gL\cos\theta_0)}$
    C. $\sqrt{2gL(1 - \cos\theta_0)}$
    D. $\sqrt{2gL\cos\theta_0)}$
    E. $2gL\cos\theta_0$

Answers: (1) D; (2) A, E; (3) C, E; (4) C; (5) C, D; (6) C, D; (7) B; (8) E; (9)A; (10) A, B, C, D; (11) B, D; (12) C

Chapter 16
WAVES —— I

In this chapter, you study wave motion, a mechanism by which a disturbance (or distortion) created at one place in a medium propagates to other places. The general ideas are specialized to a mechanical wave on a taut string. Learn what determines the speed of the wave. Also learn about sinusoidal waves, for which the string has the shape of a sine or cosine function. Two or more waves present at the same time and place give rise to what are called interference effects. Special combinations of traveling waves result in standing waves. Learn what these phenomena are and how to analyze them.

Important Concepts

- [ ] wave
- [ ] transverse wave
- [ ] longitudinal wave
- [ ] angular wave number
- [ ] wavelength
- [ ] angular frequency
- [ ] period
- [ ] wave speed
- [ ] power in a wave
- [ ] superposition of waves

- [ ] interference of waves
- [ ] fully constructive interference
- [ ] fully destructive interference
- [ ] intermediate interference
- [ ] phasor
- [ ] standing wave
- [ ] node of a standing wave
- [ ] antinode of a standing wave
- [ ] resonant frequency

Overview

16–2 Types of Waves

- [ ] Wave motion pervades nature. A mechanical wave arises from the motion of particles. Waves on a string and sound waves are examples. An electromagnetic wave arises from disturbances in an electromagnetic field. Radio waves and light are examples. Quantum mechanics deals with matter waves, which are associated with the probability of finding a particle in any region of space.

16–3 Transverse and Longitudinal Waves

- [ ] Waves are sometimes classified as transverse or longitudinal, according to the direction the medium moves relative to the direction the wave moves. In a **longitudinal** wave, the medium is displaced in a direction that is parallel to the direction the wave travels and in a **transverse** wave it is displaced in a direction that is perpendicular to the direction the wave travels. A wave on a string is transverse; sound waves in air are longitudinal. Many waves, water waves among them, are neither transverse nor longitudinal.

16–4   Wavelength and Frequency

☐ A wave on a string is described by giving the transverse string displacement $y(x, t)$ as a function of position and time, measured from the position of the string when no wave is present. To find the displacement of any point on the string at any time, substitute the value of the coordinate x of the point and the value of the time t into the function.

☐ A sinusoidal wave traveling in the positive x direction has the form $y(x, t) = y_m \sin(kx - \omega t)$, where y_m, k, and ω are constants. Notice the negative sign. The maximum displacement y_m is called the **amplitude** of the wave: every portion of the string moves back and forth between $y = -y_m$ and $y = +y_m$. The constant k is called the **angular wave number** and is related to the shape of the string. The constant ω is the **angular frequency**: every point on the string moves in simple harmonic motion with angular frequency ω. The **frequency** of its motion is given by $f = \omega/2\pi$ and the **period** of its motion is given by $T = 1/f = 2\pi/\omega$.

☐ The graph on the right shows a sinusoidal wave at time $t = 0$, with mathematical form $y(x, 0) = y_m \sin(kx)$. The graph indicates the amplitude y_m and the **wavelength** λ, the distance over which the pattern of the string 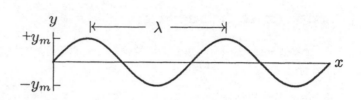 displacement repeats. The wavelength and angular wave number are related by $k = 2\pi/\lambda$.

16–5   The Speed of a Traveling Wave

☐ The speed v of a wave is related to the angular wave number k and angular frequency ω by $v = \omega/k$. If the wave is moving in the positive x direction and if the point at x_1 on the string has a given displacement y at time t_1, then at time t_2 the point at $x_2 = x_1 + v(t_2 - t_1)$ will have the same displacement, y. For these two points and times, the phase $kx - \omega t$ has the same value: $kx_1 - \omega t_1 = kx_2 - \omega t_2$, as you can prove by substituting $x_1 + v(t_2 - t_1)$ for x_2 and ω/k for v.

☐ When $2\pi/T$ is substituted for ω and $2\pi/\lambda$ is substituted for k, $v = \omega/k$ becomes $v = \lambda/T$. That is, the wave moves a distance equal to one wavelength in a time equal to one period.

☐ If the wave is traveling in the negative x direction, the string displacement has the form $y(x, t) = y_m \sin(kx + \omega t)$. Notice the positive sign.

☐ The wave speed is associated with the motion of a distortion in the string shape and is distinct from the velocity of the string itself. The transverse velocity $u(x, t)$ of the point on the string with coordinate x can be found by differentiating $y(x, t)$ with respect to time, treating x as a constant. If the displacement is given by $y_m \sin(kx - \omega t)$, then $u(x, t) = -\omega y_m \cos(kx - \omega t)$.

☐ Whether the wave is sinusoidal or not, the distortion of the string at $t = 0$ can be described by a function $y = f(x)$. If the wave is traveling with speed v in the positive x direction, then $y(x, t) = f(x - vt)$; if the wave is traveling in the negative x direction, then $y(x, t) = f(x + vt)$. All waves traveling along the x axis are functions of $x - vt$ or $x + vt$, never of x and t in any other combination.

16–6 Wave Speed on a Stretched String

☐ Newton's second law leads directly to an expression for the wave speed in terms of the tension τ in the string and the linear density μ of the string: $v = \sqrt{\tau/\mu}$. The tension in the string is usually determined by the external forces applied at its ends to hold it taut.

☐ The wave speed does NOT depend on the frequency or wavelength. If the frequency is increased (with the tension and linear mass density retaining their values), the wavelength must decrease so that $v = \lambda f$ has the same value.

16–7 Energy and Power of a Wave Traveling Along a String

☐ The mass in a segment of string of infinitesimal length dx is $dm = \mu\, dx$, where μ is the linear mass density. If u is the speed of the segment, then its kinetic energy is $dK = \frac{1}{2}\mu u^2\, dx$. This energy is transferred to a neighboring segment in time $dt = dx/v$, where v is the wave speed. Thus, the rate with which kinetic energy is carried by the string is $dK/dt = \frac{1}{2}\mu v u^2$. For a sinusoidal wave, with $y(x,t) = y_m \sin(kx - \omega t)$, $dK/dt = \frac{1}{2}\mu v \omega^2 y_m^2 \cos^2(kx - \omega t)$. The average rate over an integer number of periods is given by $\left(dK/dt\right)_{\text{avg}} = \frac{1}{4}\mu v \omega^2 y_m^2$, since the average value of $\cos^2(kx - \omega t)$ is $1/2$.

☐ The string has potential energy because it stretches as the wave passes by. The average rate with which potential energy is transported is exactly the same as the rate with which kinetic energy is transported so, for a sinusoidal wave, the average rate of energy transport is given by

$$P_{\text{avg}} = \left(\frac{dK}{dt}\right)_{\text{avg}} + \left(\frac{dU}{dt}\right)_{\text{avg}} = \frac{1}{2}\mu v \omega^2 y_m^2 .$$

It is proportional to the square of the amplitude and to the square of the angular frequency.

16–8 The wave Equation

☐ Newton's second law can be used to show that

$$\frac{\partial^2 y}{\partial x^2} = \frac{\mu}{\tau}\frac{\partial^2 y}{\partial t^2} .$$

Every wave form $y(x,t)$ must satisfy this equation, called the **wave equation**.

☐ The wave equation can be used to show that the wave speed is $v = \sqrt{\tau/\mu}$, no matter what the wave form.

☐ By studying the derivation of the wave equation you should learn that it is the tension in the string that causes a distortion to move along as a wave, without a change in shape.

16–9 The Principle of Superposition for Waves

☐ When two waves, one with displacement $y_1(x,t)$ and the other with displacement $y_2(x,t)$, are simultaneously on the same string, the displacement of the string is given by their sum: $y(x,t) = y_1(x,t) + y_2(x,t)$, provided the amplitudes are small. This is the **superposition principle**. Displacements, not transmitted powers, add.

16–10 Interference of Waves

☐ The trigonometric identity

$$\sin\alpha + \sin\beta = 2\sin\left[\tfrac{1}{2}(\alpha+\beta)\right]\cos\left[\tfrac{1}{2}(\alpha-\beta)\right],$$

valid for any angles α and β, is used to derive some results in this and the section on standing waves.

☐ The superposition of waves leads to **interference** phenomena. Suppose two sinusoidal waves $y_1(x,t) = y_m\sin(kx - \omega t + \phi)$ and $y_2(x,t) = y_m\sin(kx - \omega t)$ are on the same string. The waves are identical except that at every instant the second is shifted along the x axis from the first by an amount that depends on the value of the phase constant ϕ. The trigonometric identity given above can be used to show that the resultant string displacement is

$$y(x,t) = 2y_m\cos(\tfrac{1}{2}\phi)\sin(kx - \omega t + \tfrac{1}{2}\phi).$$

The composite wave is sinusoidal with the same frequency and wavelength as either of the constituent waves.

☐ The amplitude of the resultant wave is $2y_m\cos(\tfrac{1}{2}\phi)$, a function of the phase constant ϕ. The largest possible amplitude occurs if ϕ is 0 or a multiple of 2π rad and its value is then $2y_m$. Maximum amplitude occurs when ϕ is adjusted so the crests of one of the constituent waves fall exactly on the crests of the other. This condition is called **fully constructive interference**.

☐ The smallest possible amplitude occurs when ϕ is an odd multiple of π rad; its value is zero. Minimum amplitude occurs when ϕ is adjusted so the crests of one of the constituent waves fall exactly on the troughs of the other. This condition is called **fully destructive interference**. Other values of ϕ produce **intermediate interference**.

16–11 Phasors

☐ A phasor is a rotating arrow that is used to represent a traveling sinusoidal wave. The length of the arrow, in some scale, is taken to be the amplitude of a sinusoidal traveling wave and the angular velocity is taken to be the angular frequency of the wave. At any instant the angle made by the arrow with the horizontal axis is $kx - \omega t + \phi$. The projection of the arrow on an axis through its tail point behaves like the displacement in the wave.

☐ Phasors can be used to sum waves. Two phasors on the diagram to the right correspond to waves with amplitudes y_{1m} and y_{2m}, respectively. The waves have the same frequency and wavelength but differ in phase by ϕ. The phasor representing the resultant wave is labeled y_m. The law of cosines gives $y_m^2 = y_{1m}^2 + y_{2m}^2 + 2y_{1m}y_{2m}\cos\phi$.

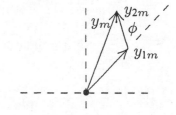

16–12 Standing Waves

☐ In a standing wave, each part of the string oscillates back and forth but the wave does not move. No energy is transmitted from place to place.

☐ A standing wave can be constructed as the superposition of two traveling waves with the same amplitude and frequency, but moving in opposite directions. Let $y_1(x, t) = y_m \sin(kx - \omega t)$ and $y_2(x, t) = y_m \sin(kx + \omega t)$ represent the two traveling waves. The trigonometric identity given above can be used to show that the sum is

$$y = y_1 + y_2 = 2y_m \sin(kx) \cos(\omega t).$$

Each point on the string vibrates in simple harmonic motion with an amplitude that varies with position along the string. In fact, the amplitude of the oscillation of the point with coordinate x is given by $2y_m \sin(kx)$.

☐ At certain points, called **nodes**, the amplitude is zero. Since their displacements are always zero, these points on the string do not vibrate at all. For the standing wave given above, nodes occur at positions for which kx is a multiple of π rad or, what is the same thing, x is a multiple of half a wavelength.

☐ At other points, called **antinodes**, the amplitude is a maximum. Its value is $2y_m$. For the standing wave given above, these points occur at positions for which kx is an odd multiple of $\pi/2$ rad or, what is the same thing, x is an odd multiple of $\lambda/4$.

☐ A standing wave can be generated by reflecting a sinusoidal wave from an end of the string. If such a wave is reflected from a *fixed* end, the reflected and incident waves at that end have opposite signs and cancel each other there. The fixed end is a node of the standing wave pattern. On the other hand, if the same wave is reflected from a *free* end, the incident and reflected waves have the same signs there and the free end of string is an antinode of the standing wave pattern.

16–13 Standing Waves and Resonance

☐ If both ends of a string are fixed there are nodes at the ends. This means that the traveling waves producing the pattern may have only certain wavelengths, determined by the condition that the length L of the string must be a multiple of $\lambda/2$, where λ is the wavelength. If v is the wave speed for traveling waves on the string, then the standing wave frequencies are $f = v/\lambda = (v/2L)n$, where n is a positive integer, called the **harmonic number**.

☐ Standing wave frequencies are called the **resonant frequencies** of the string. If the string is driven at one of its resonant frequencies by an applied sinusoidal force, the amplitude at the antinodes becomes large. The driving force is said to be in resonance with the string. Of course, the string can be driven at other frequencies but then the amplitude remains small.

Hints for Questions

<u>1</u> One-quarter of a wavelength is the distance from a peak to a neighboring zero. The wave speed depends only on the linear mass density of the string and the tension in the string. The angular frequency is $2\pi v/\lambda$, where v is the wave speed and λ is the wavelength.

[Ans: (a) 3, then 1 and 2 tie; (b) all tie; (c) 1 and 2 tie, then 3]

3 The waves have the form $y = y_m \sin(kx - \omega t)$. The wave speed is given by ω/k. You can read the values of k and ω from the expressions for the waves. Calculate the tension from the linear mass density of the string and the wave speed.

[Ans: (a) 1, 4, 2, 3; (b) 1, 4, 2, 3]

5 A shift of 5.4 wavelengths is the same as a shift of 0.4 wavelengths and a shift of 0.5 wavelengths produces completely destructive interference.

[Ans: intermediate, closer to fully destructive than to constructive]

7 If the phase difference is zero the amplitude of the resultant wave is the sum of the amplitudes of the interfering waves. If the phase difference is π rad the amplitude of the resultant wave is the magnitude of the difference in the amplitudes of the interfering waves.

[Ans: a and d tie, then b and c tie]

9 For the seventh harmonic there are seven loops between the ends of the string. They are separated by nodes and there is a node at each end. For the sixth harmonic there are six loops in the same length of string. The frequency is inversely proportional to the wavelength.

[Ans: (a) 8; (b) antinode; (c) longer; (d) lower]

11 What is the value of $\sin(kx + \phi)$ at $x = 0$? If it is zero there is a node there and if it is 1 there is an antinode.

[Ans: (a) node; (b) antinode]

Hints for Problems

3 Calculate the time when the displacement is each of the given values, then take the difference.

[Ans: 1.1 ms]

7 The amplitude y_m and the maximum string speed v_m are related by $v_m = \omega y_m$, where ω is the angular frequency of the wave. Since the string at $x = 0$ and $t = 0$ is not moving it is at an end point of its motion and the magnitude of its displacement must be the amplitude of the wave. The frequency is $f = \omega/2\pi$. The wavelength is $\lambda = v/f$, where v is the wave speed. The angular wave number is $k = 2\pi/\lambda$. Since the wave is moving in the positive x direction the sign in front of ω is negative. The string displacement at $x = 0$ and $t = 0$ is given by $y_m \sin(\phi)$ and the string velocity is given by $-\omega y_m \cos(\phi)$. The last equation means that ϕ is either $\pi/2$ rad or $3\pi/2$ rad. Choose the value of ϕ that makes the displacement positive.

[Ans: (a) 64 Hz; (b) 1.3 m; (c) 4.0 cm; (d) 5.0 m^{-1}; (e) 4.0×10^2 s^{-1}; (f) $\pi/2$ rad; (g) negative sign]

11 Determine the sign in front of ω from the direction of travel of the wave. At $t = 0$ the displacement at $x = 0$ is $y_m \sin(\phi)$. Solve this for ϕ and, of the possible solutions, choose the value that makes $\sin(-\omega t + \phi)$ a positive sine function. Then the displacement for $t = 0$ is $y_m \sin(kx + \phi)$. At $x = 0$ the wave is given by $y = y_m \sin(\pm\omega t + \phi)$, where you should use the appropriate sign in front of ω.

The amplitude is the maximum displacement of the string from $y = 0$. The angular wave number is $k = 2\pi/\lambda$, where λ is the wavelength. The angular frequency is $\omega = 2\pi/T$, where

T is period, which can be read from the graph. The wave speed is given by $v = \omega/k$. The transverse particle velocity is the derivative of y with respect to t.

[Ans: (a) negative sine function; (b) 4.0 cm; (c) 0.31 cm^{-1}; (d) 0.63 s^{-1}; (e) π rad; (f) negative sign; (g) 2.0 cm/s; (h) -2.5 cm/s]

15 (a) The wave speed is given by $v = \lambda/T = \omega/k$, where λ is the wavelength, T is the period, ω is the angular frequency $(2\pi/T)$, and k is the angular wave number $(2\pi/\lambda)$. The displacement has the form $y = y_m \sin(kx + \omega t)$, so you can read values for k and ω from the given function.
(b) Use $v = \sqrt{\tau/\mu}$, where τ is the tension in the string and μ is the linear mass density of the string to compute the tension in the string.

[Ans: (a) 15 m/s; (b) 0.036 N]

25 The rate with which kinetic energy passes a point on the cord is given by

$$\frac{dK}{dt} = \tfrac{1}{2}\mu v \omega^2 y_m^2 \cos^2(kx - \omega t),$$

where μ is the linear mass density of the cord, v is the wave speed, ω is the angular frequency, and y_m is the amplitude. A maximum on the graph has the value $\tfrac{1}{2}\mu v \omega^2 y_m^2$. Read this value from either graph. The linear mass density of the cord is given. The period T can be read from the second graph and $\omega = 2\pi/T$ can be used to compute ω. The wavelength λ can be read from the first graph and the wave speed can be computed using $v = \lambda/T$. You can now compute y_m.

[Ans: 3.2 mm]

27 The wave equation is

$$v^2 \frac{\partial^2 y}{\partial x^2} = \frac{\partial^2 y}{\partial t^2},$$

where v is the wave speed. Differentiate the given function twice with respect to x, treating t as a constant, and twice with respect to t, treating x as a constant. Substitute the results into the wave equation and solve for v.

[Ans: 0.20 m/s]

31 The amplitude y_m is given in the problem statement. The angular wave number is given by $k = 2\pi/\lambda$, where λ is the wavelength, which is the same for both waves and for the resultant. It can therefore be read from the graph. The angular frequency is given by $\omega = v/k$, where v is the wave speed, which you can easily calculate since you know that the wave travels 57.0 cm in 8.0 ms. The resultant amplitude, which can be read from the graph, is related to the phase constant ϕ_2 by $y_{\text{result, }m} = 2y_m \cos(\phi_2/2)$. Solve for ϕ_2. The sign in front of ω is determined by the direction of travel of the wave.

[Ans: (a) 9.0 mm; (b) 16 m^{-1}; (c) 1.1×10^3 s^{-1}; (d) 2.7 rad; (e) positive sign]

37 (a) and (b) Draw a phasor diagram. Place the phasor for wave 1 along the horizontal axis and the phasor for wave 2 with its tail at the head of the first. The second phasor should make an angle of 0.80π rad with the horizontal axis. Now calculate the magnitude of the vector sum. You should be able to show that the square of this magnitude is $y_{1m}^2 + y_{2m}^2 + 2y_{1m}y_{2m}\cos\phi$, where $\phi = 0.80\pi$ rad. The tangent of the phase constant of the resultant wave is the vertical component of the resultant phasor divided by the horizontal component.

(c) The third wave should be in phase with the resultant of the first two waves.

[Ans: (a) 3.29 mm; (b) 1.55 rad; (c) 1.55 rad]

43 (a) The wave speed is determined by the linear mass density of the string and the tension in it.

(b) The wavelength is twice the distance between nodes of the standing wave pattern. In this case one and one-half wavelengths fit into the distance between the string ends.

(c) The frequency f is given by $f = v/\lambda$, where v is the wave speed and λ is the wavelength.

[Ans: (a) 144 m/s; (b) 60.0 cm; (c) 241 Hz]

51 (a) and (b) Since $x = 0$ is an antinode the spatial factor in the wave form must be $\cos(kx)$, where k is the angular wave number. Since the string at $x = 0$ and $t = 0$ has zero displacement the time factor must be $\sin(\omega t)$, where ω is the angular frequency. Furthermore, the displacement at $x = 0$ is decreasing at $t = 0$. Thus the wave form is $y = -\cos(kx)\sin(\omega t)$. Since the distance between a node and an adjacent antinode if one-quarter of a wavelength, you can find the wavelength λ and use $k = 2\pi/\lambda$ to find k. Read the period T from the graph and use $\omega = 2\pi/T$ to find the angular frequency.

(c) and (d) The transverse velocity of the string is the derivative of y with respect to time, treating x as a constant.

[Ans: (a) +4.0 cm; (b) 0; (c) 0; (d) −0.13 m/s]

57 For $x = 0$ and $t = 0$ the equation for the wave gives $y = y_m \sin \phi$ and $dy/dt = -\omega y_m \cos \phi$. According to the graph $y_m = 6.0$ mm, $y = 2.0$ mm, and the slope at $t = 0$ is positive. You want to pick ϕ so that $\sin \phi = y/y_m$ and $\cos \phi$ is negative.

[Ans: 2.8 rad or −3.5 rad]

63 The displacement at $x = 0$ is given by $y = y_m \sin \phi$ and the transverse velocity is given by $u = -\omega y_m \cos \phi$, where y_m is the amplitude and ω is the angular frequency. Thus $\tan \phi = -\omega y/u$. Check your result to be sure the value of ϕ you obtain results in positive y and u.

[Ans: 1.2 rad]

71 Draw a phasor diagram for the two waves. The phasor for the second wave is $70°$ advanced from the phasor for the first wave. Sum the phasors vectorially to find the resultant amplitude and the angle the resultant phasors makes with the first phasor.

[Ans: (a) 6.7 mm; (b) 45°]

79 The wave has the form $y = y_m \sin(kx - \omega t)$, where y_m is the amplitude, k is the angular wave number, and ω is the angular frequency.. Read the values of y_m, k, and ω from the given equation. The frequency is $f = \omega/2\pi$, the wavelength is $\lambda = 2\pi/k$, the wave speed is λf (or ω/k) and the maximum transverse speed is $u = \omega y_m$.

[Ans: (a) 2.0 mm; (b) 95 Hz; (c) +30 m/s; (d) 31 cm; (e) 1.2 m/s]

83 Since the string oscillates as a standing wave with four loops, four half-wavelengths must fit into the string length. Use this to calculate the wavelength. Calculate the wave speed v as the product of the wavelength and frequency. Use $v = \sqrt{\tau/\mu}$, where τ is the tension and μ is the linear mass density, to find an expression for the tension. Use $\mu = m/L$, where m is the mass of the string and L is its length, to evaluate your expression for τ.

[Ans: 36 N]

91 Pick a peak on the wave form and identify it on each of your three graphs. Read the distance d the peak travels in 0.10 s, then divide d by this time to find the wave speed. Has the peak moved in the positive x direction or in the negative x direction?

[Ans: (c) 2.0 m/s; (d) $-x$]

Quiz

Some questions might have more than one correct answer.

1. The speed of a traveling sinusoidal wave on a taut string depends on
   A. the wavelength of the wave
   B. the amplitude of the wave
   C. the frequency of the wave
   D. the tension in the string
   E. the linear mass density of the string

2. If a traveling sinusoidal wave is on a taut string the maximum speed of any part of the string is proportional to
   A. the wave speed
   B. the amplitude of the wave
   C. the frequency of the wave
   D. the phase constant of the wave
   E. the tension in the string

3. In a time interval equal to a period a traveling sinusoidal wave moves a distance equal to
   A. the amplitude of the wave
   B. half the amplitude of the wave
   C. the wavelength of the wave
   D. twice the amplitude of the wave
   E. twice the wavelength of the wave

4. A traveling sinusoidal wave is created by shaking one end of a taut string so it moves in simple harmonic motion with a frequency f. If the frequency is doubled
   A. the wave speed is doubled
   B. the wave speed is halved
   C. the wavelength of the wave is doubled
   D. the wavelength of the wave is halved
   E. the wavelength of the wave remains the same

5. The rate with which energy is transported by a traveling sinusoidal wave is proportional to
   A. the frequency of the wave
   B. the square of the frequency of the wave
   C. the amplitude of the wave
   D. the square of the amplitude of the wave
   E. the wave speed

6. If two traveling sinusoidal waves are simultaneously on the same string
   A. their frequencies must be the same
   B. their wave speeds must be the same
   C. the displacement of the string at any point and time is the sum of the displacements due to the two waves at that point and time
   D. the resultant amplitude is the sum of the amplitudes of the two waves
   E. the phase constant is the sum of the phase constants of the two waves

7. Two traveling sinusoidal waves with the same frequency and moving in the same direction are simultaneously on a taut string. The resultant
   A. is a traveling sinusoidal wave
   B. has an amplitude equal to the magnitude of the difference of the two amplitudes if the waves are π rad out of phase with each other
   C. has an amplitude equal to the sum of the two amplitudes if the waves are π rad out of phase with each other
   D. has an amplitude equal to the sum of the two amplitudes if the waves are $\pi/2$ rad out of phase with each other
   E. has an amplitude equal to the magnitude of the difference of the two amplitudes if the waves are $\pi/2$ rad out of phase with each other

8. A standing wave on a string is created by
   A. two sinusoidal waves with the same frequency, traveling in opposite directions
   B. two sinusoidal waves with the same frequency, traveling in the same direction
   C. a traveling sinusoidal wave and its reflection from a string end
   D. two traveling sinusoidal waves with slightly different frequencies
   E. two traveling sinusoidal waves with slightly different wave speeds

9. A standing wave is on a string with both ends fixed. There are 4 nodes counting the end points and the frequency is f. The frequency of the fundamental standing wave mode is
   A. $f/4$
   B. $f/3$
   C. $f/2$
   D. $2f$
   E. $3f$

Answers: (1) D, E; (2) B, C; (3) C; (4) D; (5) B, D, E; (6) B, C; (7) A, B; (8) A, C; (9) E

Chapter 17
WAVES —— II

In this chapter, the ideas of wave motion introduced in the last chapter are applied to sound waves. Pay particular attention to the dependence of the wave speed on properties of the medium in which sound is propagating. Completely new concepts include beats, the Doppler effect, and shock waves. Be sure you understand what these phenomena are and how they originate.

Important Concepts

☐ sound wave
☐ displacement wave
☐ pressure wave
☐ sound intensity
☐ sound level

☐ standing sound waves
☐ beats
☐ beat angular frequency
☐ Doppler effect
☐ shock wave

Overview

17–2 Sound Waves

☐ **Sound waves** are propagating distortions of a medium. In fluids, they are longitudinal: particles of the fluid move back and forth along the line of motion of the wave. In solids, they may be longitudinal, transverse, or neither.

☐ Since particles at slightly different positions move different amounts, the medium becomes compressed or rarefied as a sound wave passes. A sound wave is the propagation of a local increase or decrease in density. Since a change in pressure is associated with a change in density a sound wave is also the propagation of a deviation in local pressure from the ambient value.

17–3 The Speed of Sound

☐ The volume of any fluid element changes when the pressure on the element changes. The bulk modulus, a property of the fluid, relates the two changes. If V is the original volume, Δp is the change in pressure, and ΔV is the change in volume, then the bulk modulus of the fluid is defined by $B = -V\,\Delta p/\Delta V$.

☐ The speed of sound in any fluid, including air, is determined by the density ρ and bulk modulus B of the fluid:

$$v = \sqrt{\frac{B}{\rho}}\,.$$

This expression is a direct result of the forces neighboring portions of the fluid exert on each other as the sound wave propagates and can be derived using Newton's second law.

17-4 Traveling Sound Waves

☐ Any of three related quantities can be used to describe a traveling sound wave: the particle displacement s, the deviation $\Delta\rho$ of the density from its ambient value, and the deviation Δp of the pressure from its ambient value.

☐ Suppose a sound wave is traveling along the x axis through a fluid and that the displacement of the fluid at coordinate x and time t is given by a known function $s(x, t)$. In the presence of the wave, the fractional change in the volume of any element of the fluid is given by $\Delta V/V = \partial s/\partial x$, the fractional change in the density is given by $\Delta\rho/\rho = -\Delta V/V = -\partial s/\partial x$, and the change in pressure is given by $\Delta p = -B\,\Delta V/V = -B\,\partial s/\partial x$. These expressions can be derived by considering the expansion or contraction of the fluid element in the presence of the wave.

☐ Suppose the displacement is given by $s(x, t) = s_m \cos(kx - \omega t)$, where k is the wave number and ω is the angular frequency. Then, the deviation of the pressure from its ambient value is given by

$$\Delta p(x, t) = -B\,\frac{\partial s}{\partial x} = Bks_m \sin(kx - \omega t),$$

which is sometimes written

$$\Delta p = \Delta p_m \sin(kx - \omega t),$$

where $\Delta p_m = Bks_m = v^2\rho ks_m = v\rho\omega s_m$, where $v^2 = B/\rho$ was used to obtain the second form and $k = \omega/v$ was used to obtain the third.

☐ The pressure wave is not in phase with the displacement wave. A fluid element in the neighborhood of a displacement maximum is neither compressed or elongated and the deviation of the pressure from its ambient value is zero there. The elongation or compression is greatest in the neighborhood of a displacement zero and the deviation of the pressure from its ambient value is greatest there.

17-5 Interference

☐ Sound waves obey a superposition principle. Two waves with the same frequency, traveling in the same direction in the same region interfere constructively if their phase difference ϕ has any of the values $2n\pi$ rad, where n is an integer. The interference is completely destructive if ϕ has any of the values $n\pi$ rad, where n is an odd integer.

☐ If the waves are generated by sources that are in phase but they travel to the detector along different paths, they may have different phases at the detector. If one wave travels a distance x to the detector and the other travels a distance $x + \Delta d$, then the phase difference at the detector is given by $\phi = k\,\Delta d$, where k is the angular wave number. Since $k = 2\pi/\lambda$, where λ is the wavelength, $\phi = 2\pi\,\Delta d/\lambda$. Fully constructive interference occurs if Δd is a multiple of λ; fully destructive interference occurs if Δd is an odd multiple of $\lambda/2$.

17-6 Intensity and Sound Level

☐ The **intensity** of a sound wave is the average rate of energy flow per unit cross-sectional area and, for the sinusoidal wave discussed above, is given by

$$I = \tfrac{1}{2}\rho v\omega^2 s_m^2.$$

This expression can be derived by considering the energy in an infinitesimal element of fluid. Energy is transported with the wave, at the wave speed. The SI units for intensity are W/m^2.

☐ For an isotropic point source that emits sound energy at the rate P_s the intensity a distance r away is given by $I = P_s/4\pi r^2$. Note that it decreases in proportion to $1/r^2$.

☐ The **sound level** associated with intensity I is defined by $\beta = (10\,dB)\log(I/I_0)$, where I_0 is the standard reference intensity ($10^{-12}\,W/m^2$), roughly at the threshold of human hearing. Sound level is measured in units of decibels, abbreviated dB.

☐ If $I = I_0$, the sound level is zero. If the intensity is increased by a factor of 10, the sound level increases by 10 dB.

17–7   Sources of Musical Sound

☐ Two sinusoidal traveling sound waves with the same frequency and amplitude but traveling in opposite directions combine to form a standing wave. At a point with coordinate x the particle displacement oscillates with an amplitude that is given by $2s_m \sin(kx)$, where s_m is the amplitude and k is the angular wave number of either of the traveling waves. At a displacement node $kx = 2n\pi$, where n is an integer, and the displacement is always zero. At a displacement antinode $kx = n\pi/2$, where n is an odd integer, and the displacement oscillates between $-2s_m$ and $+2s_m$.

☐ Standing sound waves are created in pipes by the superposition of a sinusoidal wave and its reflection from the end of the pipe. A displacement node exists at a closed end of a pipe. An open end is nearly a displacement antinode.

☐ If both ends of a pipe are open, the wavelengths associated with possible standing waves are such that the pipe length L is a multiple of $\lambda/2$. They are $\lambda = 2L/n$, where n is an integer, and the corresponding standing wave frequencies are

$$f = \frac{v}{\lambda} = \frac{nv}{2L},$$

where v is the speed of sound for the fluid that fills the pipe.

☐ If one end of a pipe is open and the other is closed, the standing wave wavelengths are such that the pipe length is an odd multiple of $\lambda/4$. They are $\lambda = 4L/n$, where n is an odd integer, and the corresponding standing wave frequencies are

$$f = \frac{v}{\lambda} = \frac{nv}{4L},$$

where v is the speed of sound for the fluid that fills the pipe.

☐ A string of a stringed instrument or the air in an organ pipe can vibrate with any superposition of its standing waves. Which combination is present depends on how the vibration is generated. Usually the lowest frequency dominates but higher frequency sound is mixed in. The admixture of higher frequencies gives any instrument the quality of sound peculiar to that instrument and, for example, allows us to distinguish a violin from a piano that are playing the same note.

☐ The lowest frequency is called the **fundamental frequency** or **first harmonic**, while higher frequencies are higher harmonics.

17–8 Beats

☐ **Beats** are created by the combination of two sound waves with nearly the same frequency. Let $s_1 = s_m \cos(\omega_1 t)$ represent the displacement at some point due to one of the waves and $s_2 = s_m \cos(\omega_2 t)$ represent the displacement at the same point due to the other wave. The resultant displacement is the sum of the two and, once the trigonometric identity given Chapter 16 is used, it can be written as

$$s(t) = 2s_m \cos(\omega' t) \cos(\omega t),$$

where $\omega = (\omega_1 + \omega_2)/2$ and $\omega' = (\omega_1 - \omega_2)/2$. There are two time dependent factors, both periodic. The angular frequency of one is the average of the two constituent angular frequencies. This is the greater of the two angular frequencies and if the two constituent frequencies are nearly the same, it is essentially equal to either of them.

☐ The angular frequency of the other time dependent factor is $\omega' = (\omega_1 - \omega_2)/2$. It depends on the difference in the two constituent frequencies. If ω_1 and ω_2 are nearly the same, this factor is slowly varying. We may think of it as a slow variation in the amplitude of the faster vibration. The effect can be produced, for example, by blowing a note on a horn at the angular frequency ω, but modulating it so it is periodically loud and soft.

☐ A **beat** is a maximum in the *intensity* and occurs each time $\cos(\omega' t)$ is $+1$ or -1. Thus, the **beat angular frequency** is $2\omega'$; that is,

$$\omega_{\text{beat}} = |\omega_1 - \omega_2|.$$

17–9   The Doppler Effect

☐ Suppose a sustained note with a well-defined frequency f is played by a stationary trumpeter. If you move rapidly *toward* the trumpeter, you will hear a note with a higher frequency. If you move rapidly *away* from the trumpeter, you will hear a note with a lower frequency. Similar effects occur if you are stationary and the trumpeter is moving: the note has a higher frequency if the trumpeter is moving toward you and a lower frequency if the trumpeter is moving away from you. These are examples of the **Doppler effect**.

☐ If a source emits a sound of frequency f and is moving with speed v_S, then the frequency detected by a detector moving with speed v_D is given by

$$f' = f \frac{v \pm v_D}{v \mp v_S}.$$

You can easily determine which signs to use in any particular situation by remembering that motion of the source toward the detector or the detector toward the source results in detecting a higher frequency while motion of the source away from the detector or the detector away from the source results in hearing a lower frequency than would be heard if both source and detector were stationary.

☐ The expression for f' can be derived by considering the number of wave crests that are intercepted by the detector per unit time.

☐ The velocities in the Doppler effect equation are measured relative to the medium in which the wave is propagating (the air, for example). What counts is not the motion of the source relative to the detector but the motions of both the source and detector relative to the medium of propagation. The Doppler effect equation given above is valid only if the motion is along the line joining the source and detector. For motion in other directions, v_D and v_S must be interpreted as components of the velocities along that line.

17–10   Supersonic Speeds, Shock Waves

☐ If a source of sound is moving through a medium faster than the speed of sound in the medium, a **shock wave** is produced. Then, a wavefront has the shape of a cone, with the source at its apex, as shown in Fig. 17–23 of the text. The half angle θ of the cone is given by $\sin\theta = v/v_S$, where v is the speed of sound and v_S is the speed of the source. A shock wave is not produced if $v_S < v$.

Hints for Questions

<u>1</u> Divide the distance from each source to P by the wavelength to find the number of wavelengths in each of the distances, then calculate the difference. If the difference is a multiple of a wavelength then fully constructive interference occurs; if it is an odd multiple of half a wavelength then completely destructive interference occurs.

[Ans: (a) 2.0 wavelengths; (b) 1.5 wavelengths; (c) fully constructive interference in (a), fully destructive interference in (b)]

<u>3</u> The average rate of energy transport is proportional to the square of the amplitude of the resultant wave. If the phase difference is zero or a multiple of the wavelength the resultant amplitude is large; in fact it is twice the amplitude of one of the waves. If the phase difference is an odd multiple of half a wavelength the resultant amplitude is zero. For other phase differences the resultant is intermediate between zero and twice the single-wave amplitude.

[Ans: (a) 0, 0.2 wavelength, .5 wavelength (zero); (b) $4P_{\text{avg, 1}}$]

<u>5</u> No matter if the pipe has one or two open ends the harmonic frequencies are all multiples of the fundamental frequency and every harmonic frequency is the average of the two adjacent harmonic frequencies on either side.

[Ans: 150 Hz and 450 Hz]

<u>7</u> The fundamental frequency for the string is given by $v/2L_s$, where L_s is the length of the string. For the pipes with one end closed the harmonic frequencies are given by $nv/4L_p$, where L_p is the length of the pipe and n is an odd integer. For the pipe with both ends open the harmonic frequencies are given by $nv/2L_p$, where n is an integer. Set the expressions for the frequencies equal to each other and solve for the length of the pipe, then see if any of the lengths match the given pipes.

[Ans: d, fundamental]

<u>9</u> The detected frequency depends on the component of the source velocity along the line that joins the source and detector, being greater for greater velocity components toward the detector and less for greater velocity components away from the detector.

[Ans: 1, 4, 3, 2]

Hints for Problems

<u>13</u> Use the relationship between the pressure amplitude Δp_m and the displacement amplitude s_m: $\Delta p_m = v\rho\omega s_m$, where v is the speed of sound, which is given, ρ is the density of air, which is also given, and ω is the angular frequency. The angular frequency is $2\pi/T$, where the period T can be read from the graph. Use $v = \omega/k$ to calculate k.

[Ans: (a) 6.1 nm; (b) 9.2 m^{-1}; (c) 3.1×10^3 s^{-1}; (d) 5.9 nm; (e) 9.8 m^{-1}; (f) 3.1×10^3 s^{-1}]

<u>15</u> The wave from S_1 to point A travels $(1.75\,\text{m})/(0.50\,\text{m}) = 3.5$ wavelengths further than the wave from S_2 to A. The waves are exactly out of phase there. The two waves travel the same distance to point B and so are in phase there. This means that on the arc between A and B there is a point for which one wave travels 1 wavelength further than the other, another point where it travels 2 wavelengths than the other, and still another point where it travels 3 wavelengths further than the other. At all these points the waves arrive in phase. Analyze the other quadrants of the circle in the same way and add up the number of points where they arrive in phase. There is a point where they arrive exactly out of phase between every pair of points where they arrive in phase.

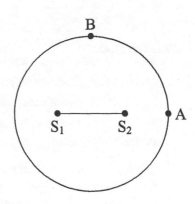

[Ans: (a) 14; (b) 14]

<u>27</u> The sound level in decibels is given by $\beta = (10\,\text{dB})\log(I/I_0)$, where I is the intensity and I_0 is the reference intensity (1×10^{-12} W/m^2). Solve for I. The intensity is related to the sound displacement amplitude s_m by $I = \frac{1}{2}\rho v\omega^2 s_m^2$, where ρ is the density of air (1.21 kg/m^3), ω is the angular frequency (2π times the frequency), and v is the speed of sound (343 m/s).

[Ans: (a) 10 μW/m^2; (b) 0.10 μW/m^2; (c) 70 nm; (d) 7.0 nm]

<u>29</u> (a) Let β_1 be the sound level at distance r_1 for either source and β_2 be the sound level at distance r_2 for the same source. Since the intensity is inversely proportional to the square of the distance $\beta_2 - \beta_1 = (10\,\text{dB})\log(r_1^2/r_2^2)$. Write this equation twice, once for each source, and subtract one from the other to show that the difference in sound levels for the sources does not depend on r_1 or r_2 (as long as the same distances are used for both sources). Read the difference in sound levels from the graph for any value of r.

(b) The intensity at any distance r is related to the power P of the source by $I = P/4\pi r^2$. Thus the difference in the sound levels of the sources at any distance is $\beta_A - \beta_B = (10\,\text{dB})\log(P_A/P_B)$.

[Ans: (a) 5 dB; (b) 3.2]

31 (a) The rate of energy transport is given by $\frac{1}{2}\rho v \omega^2 s_m^2 A$, where ρ is the density of air (1.21 kg/m^3), s_m is the displacement amplitude (12.0 nm), ω is the angular frequency, v is the speed of sound (343 m/s), and A is the cross-sectional area of the tube. Use $A = \pi R^2$, where R is the internal radius of the tube, to compute A.

(b) The two waves together carry twice as much energy as one wave alone.

(c), (d), and (e) Since the waves interfere the amplitude is now $2s_m \cos(\phi/2)$, where ϕ is the phase difference.

$\left[\text{Ans: (a) } 0.34 \text{ nW; (b) } 0.68 \text{ nW; (c) } 1.4 \text{ nW; (d) } 0.88 \text{ nW; (e) } 0 \right]$

33 If the tube is open at one end the harmonic frequencies are given by $nv/4L$, where n is an odd integer, v is the speed of sound, and L is the length of the tube. The ratio of a harmonic frequency to the next lowest harmonic frequency is $n/(n-2)$. Solve for n for each of the tubes and see if it is an odd integer.

If the tube is open at both ends the harmonic frequencies are given by $nv/2L$, where n is an integer. The ratio of a harmonic frequency to the next lowest harmonic frequency is $n/(n-1)$. Solve for n for each of the tubes and see if it is an integer.

$\left[\text{Ans: (a) open at both ends; (b) open at one end}\right]$

41 The top of the water is a displacement node and the top of the well is a displacement antinode. At the lowest resonant frequency exactly one-fourth of a wavelength fits into the depth of the well. If d is the depth and λ is the wavelength then $\lambda = 4d$. The frequency is $f = v/\lambda = v/4d$, where v is the speed of sound. The speed of sound is given by $v = \sqrt{B/\rho}$, where B is the bulk modulus and ρ is the density of air in the well.

$\left[\text{Ans: } 12.4 \text{ m}\right]$

53 Use the Doppler shift equation twice. First, the alarm is the source and is stationary and the intruder is the detector and is moving away from the source. Second, the intruder is the source and is moving away from the detector and the alarm is the detector and is stationary. The detected frequency of the first event is the source frequency of the second. The beat frequency is the difference of the original source and final detected frequencies.

$\left[\text{Ans: } 155 \text{ Hz}\right]$

55 Use the Doppler shift equation. In the reflector frame the speed of the source is 29.9 m/s + 65.8 m/s = 95.7 m/s, the speed of the detector (the reflector) is zero, and the speed of sound is 329 m/s + 65.8 m/s = 394.8 m/s. The source is moving toward the detector. In the source frame the speed of the source is zero, the speed of the detector is 95.7 m/s and the speed of sound is 329 m/s + 29.9 m/s = 358.9 m/s. The detector is moving toward the source. In each case the wavelength is the speed of sound divided by the frequency.

$\left[\text{Ans: (a) } 1.58 \text{ kHz; (b) } 0.208 \text{ m; (c) } 2.16 \text{ kHz; (d) } 0.152 \text{ m}\right]$

61 (a) Use the Doppler shift equation. One train is the source and is moving toward the detector. The other train is the detector and is moving toward the source. Both trains have a speed of 30.5 m/s

(b) and (c) You must use the speeds of the trains relative to the moving air. In part (b) the source has a speed of 30.5 m/s + 30.5 m/s = 61 m/s and the detector is stationary relative to the air. In part (c) the source is stationary and the detector has a speed of 61 m/s.

$\left[\text{Ans: (a) } 598 \text{ Hz; (b) } 608 \text{ Hz; (c) } 589 \text{ Hz}\right]$

65 Use the Doppler shift equation for the detected frequency: $f' = f(v \pm v_D)/(v \pm v_S$, where f is the emitted frequency, v is the speed of sound, v_D is the speed of the detector, and v_S is the speed of the source. Here $v_S = 0$. As the detector approaches the source the detected frequency is greater than the emitted frequency and you use the positive sign in the numerator. As the detector recedes from the source the detected frequency is less than the emitted frequency and you use the negative sign in the numerator. Develop an expression for $(f'_{app} - f'_{rec})/f$, set it equal to 0.500 and solve for v_D/v.

$\left[\text{Ans: } 0.250\,\right]$

79 The resonant frequencies for a pipe that is open at both ends are given by $nv/2L$, where v is the speed of sound, L is the length of the pipe, and n is a positive integer. The resonant frequencies for a pipe that is open at only one end are given by $f_n = nv/4L$, where n is an odd integer. Both sets of frequencies are proportional to $1/L$. Plot f_n versus $1/L$ for the given data. Use the slope of the graph to compute the speed of sound for the two possible cases and see which is within 25% of 1400 m/s.

$\left[\text{Ans: (a) } 5.5 \times 10^2\,\text{m/s; (b) } 1.1 \times 10^3\,\text{m/s; (c) } 1\,\right]$

85 Use the Doppler shift equation to compute the detected frequencies of the two sources. In (a) one source is moving toward the detector (the person) and one is moving away. The detector is stationary. In (b) the detector is moving toward one source and away from the other. Both sources are stationary. The beat frequency is the difference of the two detected frequencies.

$\left[\text{Ans: (a) } 7.70\,\text{Hz; (b) } 7.70\,\text{Hz}\,\right]$

99 The separation of adjacent wavefronts is a wavelength, which is the speed of sound divided by the frequency. When the source is moving use the Doppler shift equation to compute the detected frequency, then divide the speed of sound by the result. In (b) the source is moving toward the detection point, so the detected frequency is greater than the emitted frequency. In (c) the source is moving away from the detection point, so the detected frequency is less than the emitted frequency.

$\left[\text{Ans: (a) } 0.50\,\text{m; (b) } 0.34\,\text{m; (c) } 0.66\,\text{m}\,\right]$

109 The sound level is given by $\beta = (10\,\text{dB})\log(I/I_0)$, where I is the intensity and I_0 is the reference intensity. The difference in sound level for a barely audible intensity I_a and a minimally painful intensity I_p is $\beta_p - \beta_a = (10\,\text{dB}\log(I_p/I_a)$. The intensity is proportional to the reciprocal of the square of the distance from the source. Let r_a be the distance for which the sound is barely audible and r_p be the distance for which the sound first becomes painful. Then $\beta_p - \beta_a = (10\,\text{dB})\log(r_a^2/r_p^2)$. Solve for r_p.

$\left[\text{Ans: 1 cm}\,\right]$

Quiz

Some questions might have more than one correct answer.

1. The greatest pressure in a sound wave occurs

    A. where the density is the greatest
    B. where the displacement is the greatest
    C. where the density is the least
    D. where the displacement is the least
    E. at none of these places

2. The greatest density in a sound wave occurs where

    A. the pressure is the greatest
    B. the displacement is the greatest
    C. the pressure is the least
    D. the displacement is the least
    E. at none of these places

3. Sound from a string with both ends fixed and vibrating in a standing wave mode reaches a detector via two separate paths, one 11.25 m long and the other 11.75 m long. The interference at the detector is completely destructive. If the speed of sound is 340 m/s the frequency of the string might be

    A. 170 Hz
    B. 340 Hz
    C. 510 Hz
    D. 680 Hz
    E. 1020 Hz

4. An organ pipe is open at one end and closed at the other. The lowest frequency standing wave has a frequency of 60 Hz. Which of the following frequencies can it also produce?

    A. 120 Hz
    B. 180 Hz
    C. 240 Hz
    D. 300 Hz
    E. 3600 Hz

5. The sound level of sound wave A is 10 dB and the sound level of sound wave B is 20 dB. If I_A is the intensity of wave A and I_B is the intensity of wave B, then

    A. $I_B = 2I_A$
    B. $I_B = I_A/2$
    C. $I_B = 10I_A$
    C. $I_B = I_A/10$
    E. $I_B = 100I_A$

6. At a distance of 10 m from an isotropic point source of sound the sound level is 10 dB. At a distance of 20 m the sound level is

   A. 9.4 dB
   B. 10.6 dB
   C. 9.7 dB
   D. 10.3 dB
   E. 6.0 dB

7. Two strings have both their ends fixed and vibrate in their fundamental standing wave modes. The wave on string A has a frequency of 70 Hz and together their sound waves produce a 4-Hz beat. When the tension in string A is increased the beat frequency becomes 6 Hz. The frequency of the wave on string B is

   A. 64 Hz
   B. 66 Hz
   C. 70 Hz
   D. 74 Hz
   E. 76 Hz

8. A beat is produced if two sinusoidal sound waves simultaneously arrive at the same place and these waves

   A. have slightly different frequencies and are traveling in opposite directions
   B. have the same frequency and are traveling in opposite directions
   C. have slightly different frequencies and are traveling in the same direction
   D. have the same frequency and are traveling in the same direction
   E. have the same frequency and are traveling in perpendicular directions

9. The Doppler effect equation

$$f' = f\frac{v + v_D}{v + v_S},$$

where v is the speed of sound, v_D is the speed of the detector, v_S is the speed of the source, f is the frequency produced by the source, and f' is the frequency detected, is valid

   A. if the source is moving toward the detector and the detector is moving away from the source
   B. if the source is moving toward the detector and the detector is moving toward the source
   C. if the source is moving away from the detector and the detector is moving away from the source
   D. if the source is moving away from the detector and the detector is moving toward the source
   E. for none of these situations

Answers: (1) A, D; (2) A, D; (3) B; (4) B, D; (5) C; (6) A; (7) B; (8) C; (9) D

Chapter 18
TEMPERATURE, HEAT,
AND THE FIRST LAW OF THERMODYNAMICS

You now begin the study of thermodynamics. Be sure you understand how temperature is defined and measured. Pay particular attention to the zeroth law of thermodynamics, which codifies the characteristic of nature that allows temperature to be defined. You will also study the phenomenon of thermal expansion, the familiar change in the dimensions of an object that occurs when its temperature is changed. The first law of thermodynamics, derived from the conservation of energy principle, governs the exchange of energy between a system and its environment. Be sure to distinguish between the two mechanisms of energy transfer, work and heat. Learn how to calculate the energy absorbed or rejected as heat in terms of the heat capacity of the system and learn how to calculate the work done by a system, given the pressure as a function of volume.

Important Concepts

- ☐ thermal equilibrium
- ☐ temperature
- ☐ constant-volume
  gas thermometer
- ☐ Kelvin temperature scale
- ☐ Celsius temperature scale
- ☐ Fahrenheit temperature scale
- ☐ thermal expansion
- ☐ coefficient of linear expansion
- ☐ coefficient of volume expansion
- ☐ heat
- ☐ heat capacity
- ☐ specific heat, molar specific heat

- ☐ heat of fusion
- ☐ heat of vaporization
- ☐ thermodynamic state
- ☐ work done by a gas
- ☐ first law of thermodynamics
- ☐ adiabatic process
- ☐ cyclic process
- ☐ free expansion
- ☐ heat conduction
- ☐ heat convection
- ☐ heat radiation
- ☐ rate of heat transfer
- ☐ thermal conductivity

Overview

18–2   Temperature

- ☐ Thermodynamics deals chiefly with changes in the internal energies of objects. Temperature is a central concept.

☐ The temperature of any object cannot be lowered beyond a certain limiting value. This value is taken to be zero on the **Kelvin temperature scale**, the temperature scale most widely used in physics. Room temperature on the Kelvin scale is about 300 kelvins (or 300 K).

18–3 The Zeroth Law of Thermodynamics

☐ Temperature is intimately related to the idea of **thermal equilibrium**. If two objects that are not in thermal equilibrium with each other are allowed to exchange energy, they do so and some or all of their macroscopic properties change. When the properties stop changing, there is no longer a net flow of energy and the objects are in thermal equilibrium with each other. They are then at the same temperature.

☐ If two objects are in thermal equilibrium, all their macroscopic properties except temperature might have different values. Suppose the systems are gases. When they are in thermal equilibrium, their pressures may be different, their volumes may be different, their particle numbers may be different, and their internal energies may be different. But their temperatures are the same.

☐ The zeroth law of thermodynamics legitimizes the temperature as a property of a body in thermal equilibrium with another. The law is: If bodies A and B are each in thermal equilibrium with a third body C, then they are in thermal equilibrium with each other. Suppose the law were not valid and suppose further that body A and body B are in thermal equilibrium and therefore at the same temperature. If body C is in equilibrium with A but not with B, then the temperature of C is not a well defined quantity and cannot be considered a property of C.

☐ An important consequence of the zeroth law is that it allows us to select some object for use as a thermometer. Suppose that when the thermometer is in thermal equilibrium with body A it is also in thermal equilibrium with body B. Then, because the law is valid, we know that A and B have the same temperature.

18–4 Measuring Temperature

☐ In general, temperature is measured by measuring some property of an object, called a thermometer. For an ordinary mercury thermometer, the height of the mercury column is measured. When a constant-volume gas thermometer is used, a bulb containing a gas is put in contact with the object whose temperature is to be measured and the pressure of the gas in the bulb is measured. A reservoir of fluid connected to the barometer is used to ensure the gas in the bulb always has the same volume, no matter what its temperature.

☐ The temperature in kelvins of a constant-volume gas thermometer is taken to be proportional to the pressure: $T = Cp$, where the constant of proportionality C is chosen so $T = 273.16$ K at the triple point of water. Thus, $T = T_3(p/p_3)$, where T_3 is the triple-point temperature and p_3 is the triple-point pressure. Actually the ratio p/p_3 must be evaluated in the limit as the amount of gas in the bulb approaches zero. In this limit, the measured value of T does not depend on the type of gas used.

18–5 The Celsius and Fahrenheit Scales

☐ Three temperature scales are commonly used: Kelvin (or absolute), Celsius, and Fahrenheit. A constant-volume gas thermometer gives the temperature on the Kelvin scale. The Celsius temperature T_C is given in terms of the Kelvin-scale temperature T by $T_C = T - 273.15°$. The Fahrenheit temperature T_F is given in terms of the Celsius temperature by $T_F = \frac{9}{5} T_C + 32°$. A Celsius degree is the same as a kelvin and is 9/5 times as large as a Fahrenheit degree .

18–6 Thermal Expansion

☐ Most solids and liquids expand when the temperature is increased and contract when it is decreased. If the temperature is changed by ΔT, a rod originally of length L changes length by $\Delta L = L \alpha \Delta T$, where α is the **coefficient of linear expansion**. Over small temperature ranges α is essentially independent of temperature.

☐ When the temperature changes by ΔT, the length of every line in an isotropic material changes by the same fraction: $\Delta L / L = \alpha \Delta T$. The fractional change in the length of a scratch on the surface of an isotropic solid is also the same. The fractional change in the diameter D of a hole in an object is given by $\Delta D / D = \alpha \Delta T$.

☐ When the temperature changes by ΔT, the area A of a face of a solid changes by the fraction $\Delta A / A = 2\alpha \Delta T$ and the volume V of the solid changes by the fraction $\Delta V / V = 3\alpha \Delta T$, where α is the coefficient of linear expansion. For a fluid, the fractional change in volume is $\Delta V / V = \beta \Delta T$, where β is the **coefficient of volume expansion**.

☐ Water near 4°C and a few other substances decrease in volume when the temperature is increased. These materials have negative coefficients of thermal expansion in the temperature range for which such a contraction occurs.

18–7 Temperature and Heat

☐ Work and heat are alternate ways of changing the internal energy of a system. **Heat is the energy that flows between a system and its environment because they are at different temperatures.** Thermodynamics deals only with work that changes the internal energy, not with work that changes the motion of a system as a whole.

☐ A sign convention is adopted for work and heat. Work W is taken to be positive when it is done *by* the system and heat Q is taken to be positive when energy *enters* the system.

☐ The SI unit of heat is the joule. Other units in common use and their SI equivalents are the calorie (1 cal = 4.186 J), the British thermal unit (1 Btu = 1055 J), the calorie (1 cal = 4.186 J), and the nutritional calorie (1 Cal = 4186 J).

18–8 The Absorption of Heat by Solids and Liquids

☐ The heat capacity relates the energy transferred as heat into or out of a system to the change in temperature of the system. If during some process energy Q is absorbed as heat and the temperature increases by a small increment ΔT, then the heat capacity of the system for that process is given by $Q / \Delta T$.

☐ The heat capacity depends on the kind and amount of material in the system. A related property that depends only on the kind of material is the specific heat c, defined by $Q =$

$mc\,\Delta T$, where m is the mass of the system. Another is the molar specific heat C, defined by $Q = nC\,\Delta T$, where n is the number of moles in the system. A mole is 6.02×10^{23} elementary units (molecules for a gas). This number is called Avogadro's number and is denoted by N_A.

☐ The heat capacity depends on the process by which energy is transferred as heat. It is different for constant volume processes and constant pressure processes, for example.

☐ When two objects, initially at different temperatures, are placed in thermal contact with each other and the composite is isolated from its surroundings, the hotter substance cools, and the cooler substance warms until they reach the same temperature. The energy leaving the hotter object has the same magnitude as the energy entering the cooler object. The algebraic relationship is $Q_A + Q_B = 0$. If object A has mass m_A, specific heat c_A, and initial temperature T_A, then $Q_A = m_A c_A(T_f - T_A)$, where T_f is the final temperature. If object B has mass m_B, specific heat c_B, and initial temperature T_B, then $Q_B = m_B c_B(T_f - T_B)$. The equation $m_A c_A(T_f - T_A) + m_B c_B(T_f - T_B) = 0$ can be solved for one of the quantities.

☐ A substance exchanges energy as heat with its environment when it changes phase (melts, freezes, vaporizes, or condenses), even though the temperature remains constant during a phase change. The magnitude of the energy accompanying a phase change of a system with mass m is given by $|Q| = mL$, where L is the **heat of transformation**. Q is positive (energy absorbed) for melting and vaporization; it is negative (energy emitted) for freezing and condensing. The **heat of fusion** L_F is the energy per unit mass transferred during freezing and melting. The **heat of vaporization** L_V is the energy per unit mass transferred during boiling or condensing.

18–9   A Closer Look at Heat and Work

☐ The **thermodynamic state** of a gas can be described by giving the values of three thermodynamic variables: pressure, volume, and temperature. The state is changed by doing (positive or negative) work on the gas or by transferring energy as heat between the gas and its environment.

☐ Thermodynamics deals with the work done *by* the system, rather than with the work done *on* the system. These are the negatives of each other. When the volume of a gas changes from V_i to V_f, the work that is done *by* the gas is given by the integral of the pressure p:

$$W = \int_{V_i}^{V_f} p\,dV.$$

This expression is valid only if the process is carried out so the gas is nearly in thermal equilibrium at all times. Only then is the pressure well defined throughout the process. W is positive if $V_f > V_i$ and is negative if $V_f < V_i$.

☐ The process can be plotted as a curve on a graph with p as the vertical axis and V as the horizontal axis. Then the work W is the area under the curve.

☐ The work done on the system is different for different processes, represented as different functional dependencies of p on V and plotted as different curves on a p-V diagram. To calculate the work, you must know how the pressure varies as the volume changes.

18–10 The First Law of Thermodynamics

☐ The **first law of thermodynamics** postulates the existence of an internal energy E_{int} and states that its change as the system goes from any initial thermal equilibrium state to any final thermal equilibrium state is independent of how the change is brought about. It then equates the change in the internal energy to $Q - W$, evaluated for any path between the initial and final states. For an infinitesimal change in state, $dE_{int} = dQ - dW$. Positive work done by the system and energy leaving the system as heat (dW positive and dQ negative) both tend to decrease the internal energy.

☐ When a system undergoes a change from one equilibrium state to another, the work done by the system and the energy absorbed by the system as heat depend on the process and may be different for different processes. Changes in the internal energy, like changes in temperature, pressure, and volume, do not depend on the process. At intermediate stages of the process the system might not even be in thermal equilibrium.

18–11 Some Special Cases of the First Law of Thermodynamics

☐ If a system undergoes a change of state at constant volume, then the work done by it is zero and the change in internal energy is related to the energy absorbed as heat by $\Delta E_{int} = Q$. When energy Q is absorbed as heat at constant volume, the internal energy increases by exactly that amount.

☐ If the change of state is **adiabatic**, then the energy absorbed as heat is zero and the change in internal energy is related to the work by $\Delta E_{int} = -W$. When work W is done by the system in an adiabatic process, the internal energy decreases by exactly that amount.

☐ If the process is cyclic, so the initial and final states are the same, then the change in internal energy is zero and the work done by the system is related to the energy absorbed as heat by $W = Q$.

☐ The adiabatic free expansion of a gas is a process for which the system is not in thermal equilibrium during intermediate stages. Initially a partition confines the gas to one part of its container, which is thermally insulated. When the partition is removed, the gas expands into the other part. For this process, $W = 0$ and $Q = 0$. According to the first law, $\Delta E_{int} = 0$.

18–12 Heat Transfer Mechanisms

☐ There are three mechanisms by which energy flows as heat from one place to another:

conduction: Energy is passed from atom to atom in an object during atomic collisions. Example: energy is conducted as heat from the hot to the cold end of a metal rod.

convection: Energy is carried from one place to another by the flow of a fluid. Example: air above a hot radiator is warmed and rises, carrying energy toward the ceiling.

radiation: A hot object emits electromagnetic radiation, which is absorbed by a cooler object. Example: the Earth absorbs radiation from the sun.

☐ If energy Q is uniformly transferred in time t, then the **rate of heat transfer** P over that time is given by $P = Q/t$.

□ Consider a homogeneous bar of material of length L, with one end held at temperature T_H (hot) and the other held at temperature T_C (cold). The temperature in the bar varies from point to point along its length. The rate of heat flow through the bar is proportional to the temperature difference of its ends, proportional to its cross-sectional area A, and inversely proportional to its length L. The constant of proportionality k is called the **thermal conductivity** of the material. Mathematically,

$$P_{\text{cond}} = kA\frac{T_H - T_C}{L}.$$

The thermal conductivity is essentially independent of the temperature over small temperature ranges.

□ Building materials are often characterized by their **thermal resistances** or R values rather than their thermal conductivities. For a slab of material, the R value is related to the thermal conductivity k by $R = L/k$, where L is the thickness of the slab.

□ Objects radiate electromagnetic energy at a rate given by

$$P_{\text{rad}} = \sigma\epsilon AT^4,$$

where A is the area of the object's surface, T is its temperature on the Kelvin scale, ϵ is the emissivity of the surface, and σ is the Stefan-Boltzmann constant ($5.6703 \times 10^{-8}\,\text{W/m}^2\cdot\text{K}^4$), a constant of nature. The emissivity unitless and has value between 0 and 1, depending on properties of the surface.

□ Objects absorb electromagentic energy their environments at the rate

$$P_{\text{abs}} = \sigma\epsilon AT_{\text{env}}^4,$$

where T_{env} is the temperature of the environment on the Kelvin scale.

□ The net rate of electromagnetic energy exchange is given by

$$P_{\text{net}} = P_{\text{abs}} - P_{\text{rad}} = \sigma\epsilon A(T_{\text{env}}^4 - T^4).$$

Hints for Questions

1 The thermometer with the greatest number of degrees between the freezing and boiling points of water has the smallest degree. Similarly the thermometer with the least number of degrees between those two points has the largest degree.

[Ans: Z, X, Y]

3 If energy Q is required to melt mass m of the material, then its heat of fusion is $L_F = Q/m$.

[Ans: B, then A and C tie]

5 Work is done by the gas during the horizontal and curved portions of the cycle. For the gas to do positive work the cycle should be traversed from left to right over the upper of these two portions. According to the first law of thermodynamics $\Delta E_{\text{int}} = Q - W$, where ΔE_{int} is the change in the internal energy, W is the work done by the gas, and Q is the energy absorbed by the gas as heat. The change in the internal energy of the gas is zero over a cycle, so $Q = W$.

[Ans: (a) both clockwise; (b) both clockwise]

7 (a) Look at Eq. 18–37. For a given temperature difference across the composite the rate of energy conduction does not depend on the order in which the materials are placed.

(b) The temperature difference across material 1 depends on the rate of energy conduction through it, its cross-sectional area, and its thermal conductivity.

[Ans: (a) all tie; (b) all tie]

9 (a) For a change in temperature of ΔT the energy extracted as heat is $Q = mc\,\Delta T$, where m is the mass and c is the specific heat. Thus over the time interval Δt, $Q/\Delta t = mc\,\Delta T/\Delta t$. Solve for $\Delta T/\Delta t$.

(b) The freezing point is the temperature associated with the horizontal portion of the graph.

(c) Look at the slope of the curve before the freezing point is reached.

(d) Look at the slope of the curve after the freezing point is reached.

(e) If the heat of fusion is great more energy per unit mass must be extracted to freeze the material.

[Ans: (a) greater; (b) 1, 2, 3; (c) 1, 3, 2; (d) 1, 2, 3; (e) 2, 3, 1]

11 (a) Once the water and what was originally ice reach the same temperature their temperatures remain constant.

(b) The horizontal lines on the graphs indicate the points where freezing or melting occurs. Look at the graphs to see if the freezing or melting occurs at a temperature above or below the final temperature.

(c) If the temperature of what was originally ice reaches the melting point and then continues to rise, the substance ends up fully melted. If the temperature of what was originally water reaches the melting point and then continues to fall, the substance ends up completely frozen.

[Ans: (a) f, because the temperature cannot drop once the melting point is reached; (b) b and c end at the freezing point, d ends above the freezing point, e ends below the freezing point; (c) in b the liquid partly freezes and no ice melts; in c no liquid freezes and the ice fully melts; in e the liquid fully freezes and no ice melts]

Hints for Problems

7 Write $T_X = aT + b$, where T_X is the temperature on the new scale, T is the temperature on the Kelvin scale, and a and b are constants. Use the given data points to calculate a and b, then substitute $T = 340\,\mathrm{K}$.

[Ans: $-92.1°\mathrm{X}$]

13 The change in the surface area is $\Delta A = 2A\alpha\,\Delta T$, where A is the original surface area, ΔT is the change in temperature (on the Kelvin or Celsius scale), and α is the coefficient of linear expansion for brass. See Table 18–2 for the value of α. A cube has 6 faces, each with an area equal to the square of an edge length.

[Ans: $11\,\mathrm{cm}^2$]

17 The diameter of the rod at temperature T is $D_{\text{rod}} = D_{\text{rod, 0}} + D_{\text{rod, 0}}\alpha_S(T - T_0)$, where $D_{\text{rod, 0}}$ is the diameter at temperature T_0. The interior diameter of the brass ring at temperature T is $D_{\text{ring}} = D_{\text{ring, 0}} + D_{\text{ring, 0}}\alpha_B(T - T_0)$, where $D_{\text{ring, 0}}$ is the diameter at temperature T_0. Here α_S is the coefficient of linear expansion for steel and α_B is the coefficient of linear expansion for brass. Set D_{rod} equal to D_{ring} and solve for T.

$\left[\text{Ans: } 360°\text{C}\,\right]$

19 Use $V = V_0 + V_0\beta_{\text{liquid}}\,\Delta T$ to compute the new volume V of the liquid and $A = A_0 + 2A\alpha_{\text{glass}}\Delta T$ to compute the new cross-sectional area of the tube. Here V_0 is the original volume of fluid and A_0 is the original cross-sectional area of the tube. The height h of the fluid after the temperature increase can be found from $Ah = V$. You will need to use $V_0 = A_0 L/2$, where L is the length of the tube.

$\left[\text{Ans: } 0.13\,\text{mm}\,\right]$

29 The power output P of the heater is the energy per unit time absorbed by the water. The energy is $Q = mc\,\Delta T$, where m is the mass of the water, c is the specific heat of water, and ΔT is the temperature change. Use $m = \rho V$, where ρ is the density of water and V is the volume of water, to compute the mass. Solve for the time Δt required for $mc\,\Delta T$ to equal $P\,\Delta t$.

$\left[\text{Ans: } 3.0\,\text{min}\,\right]$

31 The steam is converted to water, giving up energy $m_S L_V$, where m_S is the mass of steam and L_V is the heat of vaporization of water. The temperature of the water is then reduced from T_{S0} ($= 100°\text{C}$) to T ($= 50°\text{C}$), giving up energy $m_S c(T_0 - T)$, where c is the specific heat of water. The ice melts, absorbing energy $m_I L_F$, where m_I is the mass of ice and L_F is the heat of fusion of water. Then the temperature of the resulting water is raised from T_{I0} ($= 0°\text{C}$) to T, absorbing energy $m_I c(T - T_{I0})$. The energy given up by the substance that originally was steam must equal the energy absorbed by the substance that was originally ice. Solve for m_S.

$\left[\text{Ans: } 33\,\text{g}\,\right]$

37 Let m be the mass of the tea. The energy removed from the tea is $mc(T_i - T_f)$, where c is the specific heat of water. Suppose that not all of the ice melts. Then the energy absorbed by the ice is $(m - m_f)L_F$, where L_F is the heat of fusion of water. Furthermore, $T_f = 0°\text{C}$. Equate the two energies and solve for m_f. If you get a value that is larger than m it means that the assumption that all the ice melts is not valid. In that event the temperature of the melted ice increases to T_f and the energy absorbed by the substance that was originally ice is $mL_F + mc(T_f - 0°\text{C})$. Solve the energy equation for T_f.

$\left[\text{Ans: (a) } 5.3°\text{C; (b) } 0; \text{(c) } 0°\text{C; (d) } 69\,\text{g}\,\right]$

47 According to the first law of thermodynamics the change in the internal energy of the gas as it goes from c to a via d is $\Delta E_{\text{int, }c\to a} = Q_{c\to d} + Q_{d\to a} - W_{c\to d} - W_{d\to a}$. Since the internal energy depends only on the state of the gas $\Delta E_{\text{int, }c\to a} = -\Delta E_{\text{int, }a\to c}$. You should recognize that $W_{d\to a} = 0$ since the volume of the gas does not change over this portion of the path. All other quantities are given.

$\left[\text{Ans: } 60\,\text{J}\,\right]$

55 When the rods are welded together end to end the rate of energy conduction is

$$P_{cond} = \frac{kA}{2L}\Delta T,$$

where k is the thermal conductivity, A is the cross-sectional area of each rod, L is the length of each rod, and ΔT is the difference in temperature of the ends of the composite. When they are welded together side by side the length is L and the cross-sectional area is $2A$, so the rate of energy conduction is

$$P_{cond} = \frac{2kA}{L}\Delta T.$$

Note that this is 4 times as large as previously.

[Ans: 0.50 min]

59 According to Eq. 18–37 the rate of energy conduction through the composite is given by

$$P_{cond} = \frac{A\Delta T}{\sum L_i/k_i},$$

where A is the cross-sectional area, ΔT is the temperature difference of the ends, k_i is the thermal conductivity of layer i, and L_i is the thickness of layer i. The rate of energy conduction is the same throughout the composite. Through layer 4 it is

$$P_{cond} = \frac{k_4 A}{L_4}\Delta T_4,$$

where ΔT_4 is the energy difference of the ends of layer 4. Equate the two expressions for the rate of energy conduction and solve for ΔT_4. Note that the area cancels from the equation. Since you know the temperature at one end of layer 4 you can calculate the temperature at the other end.

[Ans: −4.2°C]

67 Take the surface area to be the sum of two parts. One is the surface area of the rounded surface of a cylinder with a length equal to the distance from your feet to your shoulders and a circumference equal to the length of your belt. The other is the surface area of a sphere with a diameter equal to the diameter of your head. The total surface area is $2\pi R_c \ell + (4\pi/3)R_s^3$, where R_c is the radius of the cylinder and R_s is the radius of the sphere. The net radiation loss to the room is $P = \sigma\epsilon A(T_{room}^4 - T^4)$, where σ is the Stefan-Boltzmann constant, ϵ is the emissivity, A is the surface area, T_{room} is the temperature of the room on the Kelvin scale, and T is the skin temperature on the Kelvin scale. To solve (b) take the area to be half the area used in (a) and replace the room temperature with the sky temperature. To solve (c) again take the area to half that used in (a) and replace the room temperature with the snow temperature.

[Ans: (a) 90 W; (b) 2.3×10^2 W; (c) 3.3×10^2 W]

73 The energy required to lift an object of mass m a distance h is mgh and the total energy required to lift it N times is $E = Nmgh$. Solve for N. Covert $E = 3500$ Cal to joules. See Appendix C and be careful to distinguish calories from Calories.

[Ans: 1.87×10^4; (b) 10.4 h]

75 You can obtain an approximate answer using the equation for linear expansion. Ask by how much the radius would shrink if the temperature decreases by ΔT and use $\Delta R = R\alpha\,\Delta T$, where R is the present radius and α is the coefficient of linear expansion. The coefficient of linear expansion is one-third the coefficient of volume expansion.

$\left[\text{Ans: } 1.7 \times 10^2 \text{ km}\right]$

83 The change in internal energy is zero for a cyclic process, so the change along path abc is the negative of the change along path ca. That is, it is $\Delta E_{\text{int}} = +160\,\text{J}$. The first law of thermodynamics is $\Delta E_{\text{int}} = Q - W$. Solve for W. The volume does not change for the bc portion of the cycle, so no work is done on that portion and the work for path ab is the same as that for path abc.

$\left[\text{Ans: (a) } 80\,\text{J; (b) } 80\,\text{J}\right]$

91 The rate with which energy is conducted through the plate is given by $P_{\text{cond}} = (kA/L)\,\Delta T$, where k is the thermal conductivity of iron (see Table 18–6), A is the area of a disk face (πR^2, where R is the radius), L is its thickness, and ΔT is the difference in temperature across the plate. The energy conducted in time t is $E = P_{\text{cond}}t$.

$\left[\text{Ans: } 20\,\text{MJ}\right]$

97 Each linear dimension of the plate is changes according to $\Delta L = L\alpha\,\Delta T$, where L is the length before the temperature change, α is the coefficient of linear expansion, and ΔT is the temperature change. Before the temperature increase the area of the plate is $L_1 L_2$, where L_1 and L_2 are the edge lengths. after the increase the area is $(L_1 + L_1\alpha\,\Delta T)(L_2 + L_2\alpha\,\Delta T)$. To compute the change in area, carry out the multiplication, subtract the original volume, and retain terms that are proportional to ΔT. You should find that $\Delta A = 2A\alpha\,\Delta T$.

$\left[\text{Ans: } 2.16 \times 10^{-5}\,\text{m}^2\right]$

101 The amount of work is equal to the amount of energy need to melt the ice. That is, it is mL_F, where m is the mass of the ice and L_F is its heat of fusion.

Quiz

Some questions might have more than one correct answer.

1. If two objects at the same temperature are brought into contact with each other
   A. net energy may be transferred from one to the other but their volumes do not change
   B. net energy may be transferred from one to the other but their internal energies do not change
   C. net energy may be transferred from one to the other but their shapes do not change
   D. net energy may be transferred from one to the other and their temperatures change
   E. no net energy is transferred from one to the other

2. The temperature of the triple point of water is about

    A. 273.16 K
    B. 0 K
    C. 32° F
    D. 0° F
    E. 0.01° C

3. Let X_3 be the value at the triple point of water of some property of an object that is used as a thermometer. The temperature of the object when the property has the value X is

    A. $(273.15)X_3/X$
    B. $(273.15)X/X_3$
    C. $(273.15)(X - X_3)$
    D. $(273.15)(X_3 - X)$
    E. $(273.15)X$

4. A cube is made of material with a coefficient α of linear expansion and has edges of length L at temperature T_0. If the temperature is changed by a small amount ΔT the area of a cube face changes by

    A. $\Delta A = L^2 \alpha \Delta T$
    B. $\Delta A = 2L^2 \alpha \Delta T$
    C. $\Delta A = 3L^2 \alpha \Delta T$
    D. $\Delta A = L\alpha \Delta T$
    E. $\Delta A = 2L\alpha \Delta T$

5. A straight-line scratch is made on the square surface of a metal object with coefficient of linear expansion is α. If the length of the scratch at temperature T_0 is d and the length of a edge of the surface is L, the length of the scratch when the temperature is $T + \Delta T$ is

    A. d
    B. $d - \alpha L \Delta T$
    C. $d + \alpha L \Delta T$
    D. $d - \alpha d \Delta T$
    E. $d + \alpha d \Delta T$

6. Which of the following statements are true?

    A. net energy is never transferred as heat between two objects at the same temperature
    B. energy may be transferred as heat from one part of an object to another part, provided the two parts are at different temperatures
    C. energy in the form of electromagnetic radiation may be transferred as heat
    D. energy in the form of molecular kinetic energy may be transferred as heat
    E. every object contains a certain amount of heat energy

7. The internal energy of an object might increase if

   A. the object is at a higher temperature than its surroundings, is not isolated from its surroundings, and no net work is done
   B. the object is at a lower temperature than its surroundings, is not isolated from its surroundings, and no net work is done
   C. its surroundings do positive work on it
   D. it does positive work on its surroundings
   E. the work it does is equal to the energy it absorbs as heat

8. If an object undergoes any adiabatic process

   A. its internal energy does not change
   B. its internal energy increases only if the object does work on its surroundings
   C. its internal energy increases only if its surroundings do work on it
   D. its internal energy might increase because it receives energy as heat from its surroundings
   E. its internal energy might increase because it gives up energy as heat to its surroundings

9. Over one cycle of any cyclic process

   A. no net work is done
   B. no net energy is transferred as heat
   C. the internal energy of the object undergoing the process does not change
   D. the internal energy of the object undergoing the process is the same at the end as at the beginning
   E. the energy taken in as heat by the object is equal to the work it does

10. Substance A has a specific heat c and substance B has a specific heat $2c$, both independent of temperature. Object A consists of 4 kg of substance A and object B consists of 2 kg of substance B. Compared to the heat required to raise the temperature of object B by ΔT, the energy required to raise the temperature of object A by the same amount is

    A. twice as much
    B. four times as much
    C. half as much
    D. one-fourth as much
    E. the same

11. A system undergoes a process for which it begins and ends in the same thermodynamic state. For this process

    A. the volume of the system is the same at the beginning and end
    B. the pressure of the system is the same at the beginning and end
    C. the temperature of the system is the same at the beginning and end
    D. the internal energy of the system is the same at the beginning and end
    E. the net work done by the system is zero

Answers: (1) E; (2) A, E; (3) B; (4) B; (5) E; (6) A, B, C, D; (7) B, C; (8) C; (9)D; (10) E; (11) A, B, C, D

Chapter 19
THE KINETIC THEORY OF GASES

Kinetic theory is used to relate macroscopic quantities such as pressure, temperature, and internal energy to the energies and momenta of the particles that comprise a gas. Here you will gain an understanding of the relationship between the microscopic and macroscopic descriptions of a gas. The example used throughout this chapter and the next is known as an ideal gas. Pay careful attention to its properties, particularly its molar heat capacities. Also learn about the average distance a molecule in a gas travels between collisions with other molecules and about the distribution of molecular speeds in a gas.

Important Concepts

☐ Avogadro's number

☐ ideal gas

☐ ideal gas law

☐ root-mean-square speed

☐ translational energy
    of an ideal gas

☐ mean free path

☐ Maxwell speed distribution

☐ molar heat capacities
    of ideal gases

☐ equipartition of energy

☐ isothermal process

Overview

19–2 Avogadro's Number

☐ The quantity of matter in a system of identical particles might be given as the number of particles or as the number of moles. The number of molecules in a mole is Avogadro's number ($N_A = 6.02 \times 10^{23}$). If a gas contains N molecules, then it contains N/N_A moles. If the mass of one mole is M, then the mass of one molecule is $m = M/N_A$. A gas containing N molecules has a total mass of Nm and a gas containing n moles has a total mass of nM.

19–3 Ideal Gases

☐ The molecules of an **ideal gas** are point-like, taking up a negligible portion of the volume, and they interact only in elastic collisions of very short duration. The potential energy of their interactions is negligible compared to their kinetic energies. Real gases become more nearly ideal as the number of molecules per unit volume decreases.

☐ The **ideal gas law** (or ideal gas equation of state) expresses the relationship between the absolute temperature T, pressure p, volume V, and number of moles n for an ideal gas:

$$pV = nRT.$$

R is the universal gas constant ($8.31\,\text{J/mol} \cdot \text{K}$).

☐ An **isothermal process** is one during which the temperature remains constant. For an ideal gas, $p = nRT/V$ and the work it does during an isothermal change in volume is

$$W = \int_{V_i}^{V_f} p(V)\,dV = \int_{V_i}^{V_f} \frac{nRT}{V}\,dV = nRT \ln \frac{V_f}{V_i}.$$

☐ The work done in a constant-pressure process for which the volume changes from V_i to V_f is $W = p(V_f - V_i)$. The work done during a constant-volume process is zero.

19–4 Pressure, Temperature, and RMS Speed

☐ The pressure in a gas is intimately related to the average of the squares of the speeds of the molecules:

$$p = \frac{nM(v^2)_{\text{avg}}}{3V},$$

where n is the number of moles, M is the molar mass, V is the volume, and $(v^2)_{\text{avg}}$ is the average value of the squares of the molecular speeds.

☐ This expression for the pressure can be derived by calculating the average force per unit area exerted by the molecules of a gas on the container walls. Assume that at each collision between a molecule and a wall the normal component of the molecule's momentum is reversed. Compute the change in momentum per unit time to find the average force exerted by the molecule on the wall. When this is divided by the area of the wall, the result is the contribution of the molecule to the pressure.

☐ The **root-mean-square speed** v_{rms} of the molecules in a gas is defined by $v_{\text{rms}} = \sqrt{(v^2)_{\text{avg}}}$. Thus, $p = nMv_{\text{rms}}^2/3V$. Since $pV = nRT$,

$$v_{\text{rms}} = \sqrt{\frac{3RT}{M}}.$$

v_{rms} depends only on the temperature and the molar mass. It remains the same if the temperature does not change.

19–5 Translational Kinetic Energy

☐ For an ideal gas, the average translational kinetic energy per molecule is proportional to the temperature. In terms of the universal gas constant R and Avogadro's number N_A, the exact relationship is $K_{\text{avg}} = \frac{1}{2}mv_{\text{rms}}^2 = 3RT/2N_A$. The quantity $k = R/N_A$ is called the **Boltzmann constant** and has the value $k = 1.38 \times 10^{-23}$ J/K. In terms of k and T,

$$K_{\text{avg}} = \tfrac{3}{2}kT.$$

☐ The molecules of two ideal gases in thermal equilibrium with each other have exactly the same average translational kinetic energy. The root-mean-square speeds, however, are different if the molecular masses are different. The average translational kinetic energy does not change if the temperature does not change.

19–6 Mean Free Path

☐ The **mean free path** λ of a molecule in a gas is the average distance it travels between collisions with other molecules. It can be calculated by considering the number of other molecules that are encountered by any given molecule as it moves through the gas. The result is

$$\lambda = \frac{1}{\sqrt{2}\pi d^2 N/V},$$

where d is the diameter of a molecule and N is the number of molecules in a volume V of the gas.

☐ The number of collisions suffered by a gas molecule per unit time is given by its average speed divided by its mean free path.

19–7 The Distribution of Molecular Speeds

☐ Not all molecules in a gas have the same speed. The distribution of speeds is described by the **Maxwell speed distribution** $P(v)$, defined so that $P(v)\,dv$ gives the fraction of molecules with speeds in the range from v to $v+dv$. If a gas at temperature T has molar mass M, then the Maxwell speed distribution is given by

$$P(v) = 4\pi \left(\frac{M}{2\pi RT}\right)^{3/2} v^2\, e^{-Mv^2/2RT}.$$

The function is graphed in Fig. 19–7 of the text. Notice that at low temperatures the function has a sharp, high peak while at high temperatures the peak is low and broad.

☐ To find the number of molecules with speed between v_1 and v_2, evaluate the definite integral

$$N(v_1, v_2) = \int_{v_1}^{v_2} P(v)\,dv.$$

If the interval $v_2 - v_1$ is small, you can approximate N by $P(v_1)(v_2 - v_1)$.

☐ The Maxwell speed distribution can be used to calculate the most probable speed, the average speed, and the root-mean-square speed. The **most probable speed** v_p is the one for which $P(v)$ is maximum. In terms of the molar mass and temperature, it is given by

$$v_p = \sqrt{\frac{2RT}{M}}.$$

The **average speed** is given by

$$v_{\text{avg}} = \int_0^\infty v P(v)\,dv = \sqrt{\frac{8RT}{\pi M}},$$

where the explicit expression for the Maxwell distribution was substituted for $P(v)$ and the integral was evaluated. The **mean-square speed** is given by

$$v^2{}_{\text{avg}} = \int_0^\infty v^2 P(v)\,dv = \frac{3RT}{M},$$

and the root-mean-square speed v_{rms} is the square root of this:

$$v_{\text{rms}} = \sqrt{\frac{3RT}{M}}.$$

☐ Note that all three speeds (v_p, v_{avg}, and v_{rms}) depend on the temperature as well as on the molar mass. When the temperature increases, they all increase. For a given gas at a given temperature, the greatest of the three characteristic speeds is v_{rms} and the smallest is v_P.

19–8 The Molar Specific Heats of an Ideal Gas

☐ For a *monatomic* ideal gas, the internal energy is just the total kinetic energy of the molecules. Since the average kinetic energy of a molecule is $K_{\text{avg}} = \frac{3}{2}kT$, the internal energy of n moles of an ideal gas is given by $E_{\text{int}} = \frac{3}{2}nN_A kT = \frac{3}{2}nRT$. The internal energy of an ideal gas depends only on the temperature and the amount of gas. For other systems, the internal energy may also depend on the pressure or other macroscopic quantities.

☐ Suppose that the temperature of a monatomic ideal gas is changed by ΔT while the volume is held constant. Then, the internal energy changes by $\Delta E_{\text{int}} = \frac{3}{2}nR\,\Delta T$. Since no work is done, the entire increase or decrease in internal energy is due to energy absorbed or emitted as heat and $Q = \frac{3}{2}nR\,\Delta T$. The heat capacity at constant volume is $Q/\Delta T = 3nR/2$, and the molar specific heat at constant volume is $C_V = 3R/2$ (= $12.5\,\text{J/mol} \cdot \text{K}$). It does not depend on temperature. In this chapter and the next, C is used to represent a *molar* specific heat, not a heat capacity.

☐ You should understand that $\Delta E_{\text{int}} = nC_V\Delta T$ for any process, no matter if it involves heat or not and no matter if the volume changes or not. However, Q is equal to $nC_V\Delta T$ only for a constant volume process, during which no work is done. If, for example, work W is done by n moles of a monatomic ideal gas in thermal isolation (no heat transfer), then the change in internal energy is $\Delta E_{\text{int}} = -W$ and the change in temperature is $\Delta T = -W/nC_V$.

☐ The molar specific heat at constant pressure C_p is related to the molar specific heat at constant volume by $C_p = C_V + R$. More energy is absorbed as heat at constant pressure than at constant volume for the same rise in temperature. Since the internal energy depends only on the temperature, its change is the same. The additional energy absorbed goes into the work done by the gas during the process.

19–9 Degrees of Freedom and Molar Specific Heats

☐ The **equipartition theorem** states that the internal energy is equally divided among the degrees of freedom of a system, with the energy associated with each degree of freedom being $\frac{1}{2}kT$ per molecule or $\frac{1}{2}RT$ per mole. One degree of freedom is associated with each independent energy term. For example, three degrees of freedom are associated with the translational kinetic energy of a molecule since it can move in any of three independent directions. Other degrees of freedom are associated with the rotational motions of the molecules.

☐ For a monatomic ideal gas, there are three degrees of freedom for each molecule, the molar specific heat at constant volume is $3R/2$, and the molar specific heat at constant pressure is $5R/2$. For a diatomic ideal gas, there are five degrees of freedom for each molecule (three translational and two rotational), the molar specific heat at constant volume is $5R/2$, and the molar specific heat at constant pressure is $7R/2$. For a polyatomic ideal gas, there are six degrees of freedom for each molecule (three translational and three rotational), the molar specific heat at constant volume is $3R$ and the molar specific heat at constant pressure is $4R$.

☐ In some cases the rotation of diatomic or polyatomic molecules may be inhibited for some reason. Then no energy is associated with the rotational degrees of freedom and the molar specific heats are the same as those for a monatomic gas.

19–10 A Hint of Quantum Theory

☐ Classical theory predicts that the molar heat capacities of ideal gases should be independent of the temperature. In fact, they are not. If quantization of energy is taken into account, a decrease in the heat capacities with decreasing temperature is correctly predicted. According to quantum theory, the chance that the energy of a molecule is high diminishes rapidly as the temperature decreases, and this leads to a decrease in the heat capacities at lower temperatures.

☐ Atoms in diatomic and polyatomic molecules may also vibrate and the vibration modes may contribute to the heat capacities. Except at the highest temperatures, however, these are also frozen out.

19–11 The Adiabatic Expansion of an Ideal Gas

☐ As an ideal gas undergoes an adiabatic process, passing through equilibrium states, the quantity pV^γ remains constant. Here γ is the ratio C_p/C_V. If p_i and V_i are the pressure and volume of the initial state and p_f and V_f are the pressure and volume of the final state then, $p_i V_i^\gamma = p_f V_f^\gamma$.

☐ If two of the three quantities p, V, and T are given for one state and one of them is given for another state, connected to the first by an adiabatic process, then pV^γ = constant and $pV = nRT$ can be used to find the value of the thermodynamic variable that has not been given.

☐ In a free expansion of an ideal gas from volume V_i to volume V_f the gas does no work and absorbs no energy as heat. Thus the internal energy and temperature do not change. The pressure, however, does change and $p_i V_i = p_f V_f$. The gas is in equilibrium states only at the beginning and end of the expansion.

Hints for Questions

1 Changes in the temperature of an ideal gas are proportional to changes in internal energy. These are given by the first law of thermodynamics: $\Delta E_{\text{int}} = Q - W$ or, since work done on the system is the negative of work done by the system, by $\Delta E_{\text{int}} = Q + W_{\text{on}}$.

[Ans: d, then and b tie, then c]

3 Since the change in temperature is the same for the two situations, the change in internal energy must be the same. Use the first law of thermodynamics: $\Delta E_{\text{int}} = Q - W$. W is zero for the first situation.

[Ans: 20 J]

5 The change in internal energy is given by $\Delta E_{int} = nC_V \Delta T$, where n is the number of moles of gas, C_V is the molar specific heat, and ΔT is the change in temperature. For a constant-volume process $\Delta T = Q/nC_V$ and for a constant-pressure process $\Delta T = Q/nC_p$. Recall that C_p is greater than C_p.

[Ans: constant-volume process]

7 Consider first the constant-pressure processes. According to the first law of thermodynamics the work done by the gas is $W = Q - \Delta E_{int}$, where Q is the energy absorbed as heat and ΔE_{int} is the change in internal energy. Now $\Delta E_{int} = nC_V \Delta T$ and $\Delta T = Q/nC_p$, where n is the number of moles, C_V and C_p are the molar specific heats, and ΔT is the change in temperature. This means $\Delta E_{int} = (C_V/C_p)Q$ and $W = (C_p - C_V)Q/C_p = (R/C_p)Q$. The smaller the molar specific heat for constant-pressure processes the greater the work done by the gas. Recall that C_p is greater for a diatomic gas than for a monatomic gas.

Now consider the constant-volume processes. The ideal gas law gives $\Delta p = (nR/V)\Delta T$. Use $\Delta T = Q/C_V$ to show that the greater the molar specific heat C_V the smaller the pressure change. Recall that C_V is greater for a diatomic gas than for a monatomic gas.

The molecules are rotating if C_p is greater than C_V.

[Ans: (a) 3; (b) 1; (c) 4; (d) 2; (e) yes]

9 (a) The work done during a process is the area under the p-V curve associated with the process.

(b) The change in the internal energy is given by $\Delta E_{int} = nC_V \Delta T$, where n is the number of moles, C_V is the molar specific heat for constant-volume processes, and ΔT is the change in temperature.

[Ans: (a) 1, 2, 3, 4; (b) 1, 2, 3]

Hints for Problems

7 Use the ideal gas law: $pV = nRT$, where p is the pressure, V is the volume, n is the number of moles, and T is the temperature on the Kelvin scale. In part (a) you are given p, V, and T (which you must convert from degrees Celsius to kelvins) and you calculate n. In part (b) You are given p and T (which you convert) and you calculate V using the value you found in part (a) for n.

[Ans: (a) 106 mol; (b) 0.892 m³]

9 Use the ideal gas law. The partial pressure for gas 1 is $p_1 = n_1 RT/V$ and the partial pressure for gas 2 is $p_2 = n_2 RT/V$. Here n_1 is the number of moles of gas 1, n_2 is the number of moles of gas 2, T is the temperature, and V is the volume of the container. The total pressure is $p = p_1 + p_2$. You want to calculate $p_2/(p_1 + p_2)$.

[Ans: 0.2]

15 Let n_{Ai} be the number of moles of gas initially in container A, p_{Ai} be the initial pressure, V_A be the volume and T_A be the temperature on the Kelvin scale. According to the ideal gas law $n_{Ai} = p_{Ai}V_A/RT_A$. Let n_{Bi} be the number of moles of gas initially in container A, p_{Bi} be the initial pressure, V_B be the volume, and T_B be the temperature on the Kelvin scale. Then the ideal gas law gives $n_{Bi} = p_{Bi}V_B/RT_B$. Let n_{Af} and n_{Bf} be the numbers of moles in the containers after the pressures equalize and let p be the final pressure. Then $n_{Af} = pV_A/RT_A$ and $n_{Bf} = pV_B/RT_B$. No molecules are lost, so $n_{Ai} + n_{Bi} = n_{Af} + n_{Bf}$. Substitute for the numbers of moles and use $V_B = 4V_A$, then solve for p.

$\left[\text{Ans: } 2.0 \times 10^5 \, \text{Pa}\right]$

25 (a) Use $\epsilon = L_V/N$, where L_V is the heat of vaporization and N is the number of molecules per gram. Divide the molar mass of water by Avogadro's number to obtain the mass of an atom. The reciprocal of the mass in grams is the number of molecules per gram.
(b) The average translational kinetic energy is $K_{avg} = \frac{3}{2}kT$.

$\left[\text{Ans: (a) } 1.61 \times 10^{-20} \, \text{cal} \ (6.76 \times 10^{-20} \, \text{J}); \text{ (b) } 10.7\right]$

35 (a) and (b) The rms speed of molecules in a gas is given by $v_{rms} = \sqrt{3RT/M}$, where T is the temperature on the Kelvin scale and M is the molar mass of the gas. The speed required for escape from Earth's gravitational pull is $v = \sqrt{2gr_E}$, where g is the acceleration due to gravity at Earth's surface and r_E (= 6.37×10^6 m) is the radius of Earth.
Equate the expressions for the speeds to obtain $\sqrt{3RT/M} = \sqrt{2gr_E}$. Solve for T. The molar mass of hydrogen is 2.02×10^{-3} kg/mol and the molar mass of oxygen is 32.0×10^{-3} kg/mol.
(c) and (d) Replace g in the expression for T by $0.16g$ and repeat the calculation.

$\left[\text{Ans: (a) } 1.0 \times 10^4 \, \text{K}; \text{ (b) } 1.6 \times 10^5 \, \text{K}; \text{ (c) } 4.4 \times 10^2 \, \text{K}; \text{ (d) } 7.0 \times 10^3 \, \text{K}; \text{ (e) no; (f) yes}\right]$

37 (a) The rms speed is given by $v_{rms} = \sqrt{3RT/M}$. The molar mass of hydrogen is 2.02×10^{-3} kg/mol.
(b) When the surfaces of the spheres that represent an H_2 molecule and an Ar atom are touching, the distance between their centers is the sum of their radii.
(c) Since the argon atoms are essentially at rest, in time t the hydrogen atom collides with all the argon atoms in a cylinder of radius d and length vt, where v is its speed. That is, the number of collisions is $\pi d^2 vtN/V$, where N/V is the concentration of argon atoms.

$\left[\text{Ans: (a) } 7.0 \, \text{km/s}; \text{ (b) } 2.0 \times 10^{-8} \, \text{cm}; \text{ (c) } 3.5 \times 10^{10} \, \text{collisions/s}\right]$

45 The heat capacity of N argon atoms, each with mass m, is $mN\alpha$. It is also nC_V, where n is the number of moles and C_V is the molar specific heat. Take n to be 1 mol and N to be Avogadro's number N_A. Since argon is monatomic $C_V = 3R/2$. Solve for m. The molar mass is mN_A.

$\left[\text{Ans: (a) } 6.6 \times 10^{-26} \, \text{kg}; \text{ (b) } 40 \, \text{g/mol}\right]$

47 (a) The work is given by the integral $W = \int p \, dV$, where p is the pressure and dV is a volume element. According to the ideal gas law $p = nRT/V$, so $W = (nRT/V) \, dV$. Take the volume to be $V = V_0 + \alpha t$ and the temperature to be $T = T_0 + \beta t$, where t is the time. Find the values of V_0, T_0, α, and β so that the expressions for V and T give correct values for $t = 0$ (when the process begins) and for $t = 2.00$ h. Substitute the expressions for V, T, and $dV = \alpha dt$ into the equation for the work to obtain

$$W = \int_0^{2.00 \, \text{h}} \frac{nR(T_0 + \beta t)}{V_0 + \alpha t} \alpha \, dt.$$

Evaluate the integral.
(b) According to the first law of thermodynamics the energy absorbed as heat is $Q = \Delta E_{int} + W$, where ΔE_{int} is the change in the internal energy. ΔE_{int} is computed using $\Delta E_{int} =$

$nC_V \Delta T$, where C_V is the molar specific heat for constant volume processes and ΔT is the change in temperature.

(c) The molar specific heat for the process is $Q/n\Delta T$.

(d), (e), and (f) The work done by the gas during the isothermal portion of the process is $W = nRT \ln(V_f/V_i)$, where V_i is the initial volume and V_f is the final volume. The temperature here is the initial temperature. The work done during the constant-volume portion is zero. The energy absorbed as heat and the molar specific heat can be computed in the same manner as in parts (b) and (c).

[Ans: (a) 7.72×10^4 J; (b) 5.46×10^4 J; (c) 5.17 J/mol·K; (d) 4.32×10^4 J; (e) $8,86 \times 10^4$ J; (f) 8.38 J/mol·K]

<u>53</u> If p_i is the initial pressure, V_i is the initial volume, p_f is the final pressure, and V_f is the final volume, then since the process is adiabatic, $p_i V_i^\gamma = p_f V_f^\gamma$. Solve for the final pressure. Then use the ideal gas law $p_i V_i = nRT_i$ to find the number of moles and use $p_f V_f = nRT_f$ to find the final temperature.

[Ans: (a) 14 atm; (b) 6.2×10^2 K]

<u>57</u> The change in the internal energy is the same for the two paths. In particular it is $Q - W$, where Q is the net energy input as heat and W is the net work done by the gas. Add up the heat input and the work for the segments of path 1, then calculate $Q - W$. Along an isotherm the work done is equal to the heat input and along an adiabat the heat input is zero.

[Ans: -20 J]

<u>67</u> (a) The bulk modulus is defined by $B = -V(dp/dV)$, where p is the pressure and V is the volume. For an adiabatic process $pV^\gamma = $ constant, where γ is the ratio of the molar specific heat at constant pressure to the molar specific heat at constant volume: $\gamma = C_p/C_V$. Differentiate $pV^\gamma = C$ with respect to V to find that $dp/dV = -\gamma/V$, then multiply by V.

(b) The speed of sound is $v = \sqrt{B/\rho}$, where ρ is the density of the gas. See Section 17–3. The gas density is $\rho = nM/V$, where n is the number of moles, M is the molar mass, and V is the volume. According to the ideal gas law $n/V = p/RT$, where T is the temperature and R is the universal gas constant. Thus $\rho = pM/RT$. Use this to substitute for ρ and $B = \gamma p$ to substitute for B in the expression for the speed of wound.

<u>69</u> The distance between nodes of a standing wave is $\lambda/2$, where λ is the wavelength. The speed of sound in the tube is $v = \lambda f$, where f is the frequency. Use the value for v in the result of Problem 67, $v = \sqrt{\gamma RT/M}$, and solve for γ. Use Table 19–3 to calculate γ ($= C_p/C_V$) for the various types of ideal gases and compare the results with your answer to part (a).

[Ans: (a) 1.37; (b) diatomic]

<u>73</u> According to the first law of thermodynamics the change in the internal energy is $\Delta E_{int} = Q - W$, where Q is the energy transferred as heat and W is the work done by the gas. The problem statement gives $Q = -210$ J. You need to compute the work. For a constant pressure process this is $W = p(V_f - V_i)$, where p is the pressure, V_i is the initial volume, and V_f is the final volume. Use the ideal gas law to show that T/V has the same value for the initial and final states.

[Ans: (a) -60 J; (b) 90 K]

<u>77</u> According to the ideal gas law the temperature is $T = pV/nR$, where p is the pressure, V is the volume, n is the number of moles, and R is the universal gas constant. In (a) use the value of p and V for state 1 and in (b) use the values for state 2. Convert pressures from atmospheres to pascals and volumes from cubic centimeters to cubic meters. The change in the internal energy is $\Delta E_{int} = nC_V\,\Delta T$, where C_V is the molar specific heat for constant volume processes. You must compute the temperature of state 3.

[Ans: (a) 122 K; (b) 365 K; (c) 0]

<u>87</u> For this process $p = \alpha V$, where α is a constant. Use the ideal gas law to show that $\alpha = nRT_1/V_1$. The work done by the gas is $W = \int p\,dV$ from V_1 to $2V_1$. The change in the internal energy is $\Delta E_{int} = nC_V\,\Delta T$, where C_V is the molar specific heat for constant volume processes and is $(3/2)nR$ for a monatomic ideal gas. The energy transferred as heat is $Q = \Delta E_{int} + W$. The molar specific heat for the process is $Q/n\,\Delta T$.

[Ans: (a) 1.5; (b) 4.5; (c) 6; (d) 2]

Quiz

Some questions might have more than one correct answer.

1. The pressure in an ideal gas is proportional to

   A. the number of molecules in the gas
   B. the volume of the gas
   C. the temperature of the gas on the Kelvin scale
   D. the temperature of the gas on the Celsius scale
   E. the number of molecules per unit volume of gas

2. The pressure in an ideal gas is proportional to

   A. the number of molecules in the gas
   B. the volume of the gas
   C. the mass of a molecule
   D. the average velocity of a molecule
   E. the average of the square of the speed of a molecule

3. No work is done by an ideal gas during

   A. an adiabatic process
   B. an isothermal process
   C. a constant-pressure process
   D. a constant-volume process
   E. a cyclic process

4. No energy is exchanged as heat by an ideal gas during

    A. an adiabatic process
    B. an isothermal process
    C. a constant-pressure process
    D. a constant-volume process
    E. a cyclic process

5. The internal energy of an ideal gas is proportional to

    A. the number of molecules in the gas
    B. the volume of the gas
    C. the pressure in the gas
    D. the temperature on the Kelvin scale
    E. the temperature on the Celsius scale

6. The molar specific heat of an ideal gas for constant-volume processes is proportional to

    A. the number of molecules in the gas
    B. the volume of gas
    C. the temperature on the Kelvin scale
    D. the pressure in the gas
    E. the number of degrees of freedom associated with a molecule of the gas

7. The molar specific heat of an ideal gas for constant-pressure processes

    A. is greater than the specific heat for constant-volume processes
    B. is the same for monatomic and diatomic gases
    C. when multiplied by the change in temperature and the number of moles of gas, gives the change in the internal energy of the gas for all processes
    D. depends on the temperature of the gas
    E. depends on the molar mass of the molecules in the gas

8. The molar specific of an ideal gas for constant-volume processes is C_V and the molar specific heat for constant-pressure processes is C_p. The gas contains n moles. Whenever the gas does work W and absorbs energy Q as heat its temperature changes by

    A. $\Delta T = W/nC_V$
    B. $\Delta T = Q/nC_V$
    C. $\Delta T = Q/nC_p$
    D. $\Delta T = (Q - W)/nC_V$
    E. none of the above

9. n moles of a monatomic ideal gas and n moles of a diatomic ideal gas undergo the processes described below. Order the processes according to the magnitude of the energy exchanged by the gas with its environment as heat, least to greatest.

    I. the temperature of the monatomic gas increases by ΔT at constant volume
    II. the temperature of the monatomic gas increases by $2\Delta T$ at constant volume
    III. the temperature of the monatomic gas increases by ΔT at constant pressure
    IV. the temperature of the diatomic gas increases by ΔT at constant volume

A. I, II, III, IV
B. I, III, IV, II
C. I, III and IV tied, II
D. III, I and II tied, IV
E. II, I and IV tied, III

10. Which of the following quantities change in the adiabatic free expansion of an ideal gas?

A. the volume of the gas
B. the pressure of the gas
C. the temperature of the gas
D. the internal energy of the gas
E. the average kinetic energy of a molecule of the gas

11. The same number of moles of a monatomic ideal gas and a diatomic atomic gas both start in states with pressure p_i and volume V_i, then undergo processes so they both end in states with pressure p_f and volume V_f. Which of the following quantities are the same for the two gases?

A. their final temperatures
B. the changes in their internal energies
C. the work done by them
D. the energy transferred to or from them as heat
E. their specific heats for the processes

12. A monatomic ideal gas ($\gamma = 1.67$) is in a state with pressure p_1, volume V_i, and temperature T_i. Its volume is then doubled by means of one of the three processes listed below. Rank the process in order of the work done by the gas, least to greatest.

     I. isothermal
    II. constant pressure
   III. adiabatic

  A. I, II, III
  B. I, III, II
  C. II, I, III
  D. II, III, I
  E. III, II, I

Answers: (1) A, C, E; (2) A, E; (3) D; (4) A; (5) A, D; (6) E; (7) A; (8) D; (9) C; (10) A, B; (11) A, B; (12) E

Chapter 20
ENTROPY AND THE SECOND LAW
OF THERMODYNAMICS

Here you study the second law of thermodynamics and a closely associated property of any macroscopic system, its entropy. Understand the two definitions: in terms of heat and temperature and in terms of microstates of the system. Pay careful attention to calculations of entropy changes during various processes. Learn to distinguish between reversible and irreversible processes and understand how entropy behaves for each case. The second law has many far-reaching consequences; we can imagine many phenomena that do not violate any other laws of physics, including the conservation laws, but, nevertheless, do not occur. The second law gives the reason.

Important Concepts

- ☐ reversible process
- ☐ irreversible process
- ☐ quasi-static process
- ☐ entropy
- ☐ state variable
- ☐ second law of thermodynamics

- ☐ heat engine
- ☐ thermal efficiency
- ☐ refrigerator
- ☐ coefficient of performance
- ☐ microstate
- ☐ multiplicity of microstates

Overview

20–2 Irreversible Processes and Entropy

☐ There are many phenomena that occur in one direction but not in the other: energy never flows as heat from the cold end of a rod to the hot end, one part of an isolated object never spontaneously becomes hot while the other parts become cold, all of the air molecules in a room never spontaneously congregate in one corner. None of these phenomena are precluded by Newton's second law or the conservation laws. They are precluded by the **second law of thermodynamics**.

20–3 Change in Entropy

☐ A **quasi-static process** is one that is carried out so slowly that the system is essentially in thermodynamic equilibrium throughout. A quasi-static process without friction and turbulence is **reversible**; that is, it can be run backward with the signs of the heat and work changing but not their magnitudes. A process that is not reversible is said to be an **irreversible process**.

☐ If a system receives energy dQ as heat as it goes from an initial equilibrium state to a nearby equilibrium state, then the **entropy** S of the system changes by $dS = dQ/T$, where T is the temperature. This defines entropy. If the state changes reversibly from some initial equilibrium state i to some final equilibrium state f, the change in entropy can be computed by evaluating an integral:

$$\Delta S = S_f - S_i = \int_i^f \frac{dQ}{T}.$$

The process connecting the initial and final states must be reversible for this equation to be valid.

☐ Entropy is a **state variable**. This means that whenever a system is in the same thermodynamic state, it has the same entropy, just as it has the same temperature, pressure, volume, and internal energy. The net change in any of these variables over a cycle is zero.

☐ To evaluate the integral for ΔS, you must pick some *reversible* process that connects the initial and final states but, because entropy is a state variable, it need not be the process actually carried out on the system. Here are some examples:

1. Change in phase. Suppose mass m of a substance with heat of fusion L_F melts at temperature T. For this process, T is constant and $\Delta S = Q/T$. Since $Q = mL_F$,

$$\Delta S = \frac{mL_F}{T}.$$

The change in entropy on freezing is $\Delta S = -mL_F/T$. The entropy of the substance increases on melting and decreases on freezing.

2. Constant volume process. Suppose the molar specific heat at constant volume C_V for a certain gas is independent of T, V, and p. If an n-mole sample of this gas undergoes an infinitesimal reversible change of state at constant volume, so its temperature changes by dT, then the energy absorbed as heat is $dQ = nC_V\,dT$ and the change in entropy is $dS = dQ/T = (nC_V/T)\,dT$. For a finite change in temperature, from T_1 to T_2, say, the change in entropy is the integral of this expression:

$$\Delta S = \int_{T_i}^{T_f} \frac{nC_V}{T}\,dT = nC_V \ln \frac{T_f}{T_i}.$$

3. Constant pressure process. Suppose the molar specific heat at constant pressure C_p for a certain gas is independent of T, V, and p. If an n-mole sample of this gas undergoes an infinitesimal reversible change of state at constant pressure, so its temperature changes by dT, then the energy absorbed as heat is $dQ = nC_p\,dT$ and the change in entropy is $dS = dQ/T = (nC_p/T)\,dT$. For a finite change in temperature, from T_i to T_f, the change in entropy is the integral of this expression:

$$\Delta S = \int_{T_i}^{T_f} \frac{nC_p}{T}\,dT = nC_p \ln \frac{T_f}{T_i}.$$

4. Isothermal process. If an ideal gas undergoes an infinitesimal reversible change of state at constant temperature, then $dE_{\text{int}} = 0$ and $dQ = dW = p\,dV$. Thus $dS = (p/T)\,dV$ and

$\Delta S = \int (p/T) \, dV$. To evaluate this integral, p must be known as a function of volume and temperature. The equation of state gives this information. For n moles of an ideal gas $p/T = nR/V$ and

$$\Delta S = \int_{V_i}^{V_f} \frac{nR}{V} \, dV = nR \ln \frac{V_f}{V_i} \, ,$$

where V_i is the initial volume and V_f is the final volume.

☐ Consider the adiabatic free expansion of an ideal gas from volume V_i to volume V_f, an irreversible process. No energy is received as heat, no work is done, the change in the internal energy is zero, and the change in the temperature is zero. The entropy change can be calculated using a reversible isothermal expansion from V_i to V_f. It is $\Delta S = nR \ln(V_f/V_i)$.

20–4 The Second Law of Thermodynamics

☐ The second law of thermodynamics is expressed in terms of entropy changes: In any thermo-dynamic process that proceeds from an initial equilibrium state to a final equilibrium state, the total entropy of the system and its environment either remains unchanged or increases. If the process is reversible, the entropy is constant; if the process is irreversible, the entropy increases. In equation form, $dS \geq 0$.

☐ You must consider the *total* entropy of the system *and* its environment. Many processes lower the entropy of a system but, according to the second law, they must then increase the entropy of the environment.

20–5 Entropy in the Real World: Engines

☐ A heat engine takes energy as heat from a high-temperature thermal reservoir, converts some of it to work, and rejects the remainder to a low-temperature thermal reservoir.

☐ A heat engine makes use of a working substance whose thermodynamic state varies during the process but that periodically returns to the same state. The same sequence of steps is performed on it during every cycle.

☐ The simplest example is the **Carnot heat engine**, which takes the working substance through a special series of processes: an isothermal expansion in contact with a high-temperature thermal reservoir, followed by an adiabatic expansion, then an isothermal compression in contact with a low-temperature thermal reservoir, and finally an adiabatic compression back to the original state. All of the processes are reversible. Fig. 20–8 illustrates the cycle and shows the p-V diagram.

☐ If the Carnot heat engine receives energy of magnitude $|Q_H|$ from a high-temperature thermal reservoir and rejects energy of magnitude $|Q_C|$ to a low-temperature reservoir, then according to the first law of thermodynamics, it does work $W = |Q_H| - |Q_C|$ over a cycle. Note that because the engine works in cycles the change in the internal energy is zero.

☐ The **efficiency** of a Carnot heat engine is given by

$$\epsilon_C = \frac{W}{|Q_H|} = \frac{|Q_H| - |Q_C|}{|Q_H|} .$$

This is the fraction of the energy input ($|Q_H|$) that is obtained as useful output (W).

□ A *perfect* heat engine takes in energy as heat and does an identical amount of work. No heat is rejected. Since $W = |Q_H|$ for a perfect heat engine, its efficiency is 1. Such an engine does not violate the first law of thermodynamics but it does violate the second law. Over a cycle, the entropy of the high-temperature reservoir decreases by $|Q_H|/T_H$, the entropy of the low-temperature reservoir increases by $|Q_C|/T_C$, and the entropy of the working substance does not change. Thus the change in the total entropy of the engine and its environment is

$$\Delta S = \frac{|Q_C|}{T_C} - \frac{|Q_H|}{T_H}.$$

According to the second law, this cannot be negative. That means $|Q_C|$ must be at least as great as $|Q_H|T_C/T_H$. In particular, it cannot be zero, so $|W|$ cannot be equal to $|Q_H|$ and the efficiency of a heat engine cannot be 1. The first law tells us that in a cycle the energy input as heat is turned into work and energy output as heat. The second law tells us it cannot be turned completely to work; some must be output as heat.

□ A reversible engine is said to be *ideal*. The total entropy of such an engine and the reservoirs does not change. For a Carnot heat engine this means $|Q_C|/T_C = |Q_H|/T_H$ and the efficiency is

$$\epsilon_C = \frac{T_H - T_C}{T_H}.$$

All Carnot engines operating between the same temperatures have the same efficiency, regardless of the working substance and this efficiency is the greatest possible for the temperatures involved.

20–6 Entropy in the Real World: Refrigerators

□ A refrigerator takes in energy as heat from a low-temperature reservoir and ejects energy as heat to a high-temperature reservoir. Work must be done on the working substance of the refrigerator to accomplish this.

□ Consider a Carnot refrigerator, for which the steps are the same as for a Carnot engine except that they are performed in the opposite direction. Suppose that during a cycle a Carnot refrigerator takes in energy of magnitude $|Q_C|$ at a low-temperature reservoir and rejects energy of magnitude $|Q_H|$ at a high-temperature reservoir. Work of magnitude $|W| = |Q_H| - |Q_C|$ is done on the working substance of the refrigerator. The **coefficient of performance** is given by

$$K_C = \frac{|Q_C|}{|W|} = \frac{|Q_C|}{|Q_H| - |Q_C|}.$$

□ For a *perfect* Carnot refrigerator $|W| = 0$ and K is infinite. A perfect Carnot refrigerator does not violate the first law of thermodynamics but it does violate the second law and cannot exist. Over a cycle, the change in the total entropy of the refrigerator and reservoirs is

$$\Delta S = \frac{|Q_H|}{T_H} - \frac{|Q_C|}{T_C}.$$

According to the second law, this cannot be negative. So $|Q_H|$ cannot be less than the quantity $|Q_C|T_H/T_C$ and the work done on the refrigerator cannot be less than $|Q_C|(T_H -$

Work must be done to cyclically absorb energy as heat from a low-temperature reservoir and reject it to a high-temperature reservoir. The statement does NOT preclude energy flowing as heat from a high-temperature reservoir to a low-temperature reservoir without work being done. This is a natural occurrence.

☐ An ideal refrigerator is one that is reversible. Over a cycle, the change in the total entropy of the refrigerator and reservoirs is zero. For a Carnot refrigerator this means

$$\frac{|Q_H|}{T_H} = \frac{|Q_C|}{T_C}$$

and

$$K_C = \frac{T_C}{T_H - T_C}.$$

The coefficients of performance of all Carnot refrigerators operating between the same two temperatures are the same. Other refrigerators have smaller coefficients.

20–7   The Efficiencies of Real Engines

☐ Real engines are not reversible. They are not quasi-static and, in addition, there are energy losses to friction and turbulence. Their efficiencies are less than the efficiency of a reversible engine operating between the same temperatures.

20–8   A Statistical View of Entropy

☐ There are many different arrangements of atoms that lead to the same thermodynamic properties. All arrangements with the same properties are said to have the same configuration. Each arrangement is called a **microstate**. A different number of microstates may be associated with different configurations.

☐ The fundamental hypothesis of statistical mechanics is that every microstate, consistent with the values of p, V, and T, is equally probable. Thus, a system with a large number of particles will, with overwhelming probability, be in the configuration with the greatest number of microstates.

☐ The entropy of a system in a given configuration is related to the number of microstates associated with that configuration. If W is the number of microstates, then the relationship is $S = k \ln W$, where k is the Boltzmann constant.

Hints for Questions

1 The change in entropy is given by $\Delta S = nR \ln(V_f/V_i)$, where n is the number of moles of gas, V_i is the initial volume and v_f is the final volume.

   [Ans: a and c tie, then b and d tie]

3 For a constant-volume process $dQ = nC_V \, dT$, where n is the number of moles, C_V is the molar specific heat for constant-volume processes and dT is an infinitesimal change in temperature. The change in entropy is $\Delta S = nC_V \ln(T_f/T_i)$, where T_i is the initial temperature and T_f is the final temperature. For a constant pressure process $\Delta S = nC_p \ln(T_f/T_i)$, where

C_p is the molar specific heat for constant-pressure processes. Recall that C_p is greater than C_V.

[Ans: b, a, c, d]

5 The energy absorbed as heat is zero for every interval during the compression.

[Ans: unchanged]

7 Since the change in the internal energy is zero over a cycle the first law of thermodynamics tells us that W must equal $Q_H + Q_L$. Over a cycle the entropy of the working substance in the engine is zero. The change in the entropy of the reservoirs is $-(Q_H/T_H) - (Q_L/T_L)$. This must be greater than zero.

[Ans: A: first; B: first and second; C: second; D: neither]

9 A Carnot engine and its reservoirs form an isolated system and the processes involved are reversible. A real engine and its reservoirs also form an isolated system but the processes involved are irreversible. The energy input to the low temperature reservoir of a perfect engine is zero so the entropy change of the system consisting of the engine and its reservoirs is $-Q_H/T_H$, where Q_H is the energy absorbed as heat from the high-temperature reservoir by the engine and T_H is the temperature of that reservoir.

[Ans: (a) same; (b) increase; (c) decrease]

11 The number of microstates is given by $N!/(n!)^2$, where N is the total number of atoms and n is the number in each half. Estimate the value using the Stirling approximation: $N! \approx N \ln N - N$.

[Ans: much more than a year (actually more than the age of the universe)]

Hints for Problems

5 The energy absorbed as heat is $Q = mc\,\Delta T$, where m is the mass of the block, c is the specific heat of copper, and ΔT is the change in temperature on the Kelvin scale. The change in entropy is $\Delta S = \int (mc/T)\,dT$.

[Ans: (a) 5.79×10^4 J; (b) 173 J/K]

9 The change in entropy of a block is $\Delta S = \int (mc/T)\,dT = mc\ln(T_f/T_i)$, where m is the mass of the block, c is the specific heat, T_i is the initial temperature, and T_f is the final temperature. Since the blocks are identical the final temperature is halfway between the two initial temperatures. To find the value of mc, consider the irreversible process of Fig. 20–5. The energy absorbed as heat is related to the change in temperature by $Q = mc\,\Delta T$. Solve for mc.

Now go back to the reversible process of Fig. 20–6. Since the block and its reservoir are isolated and the process is reversible the change in the entropy of the reservoir is the negative of the change in the entropy of the block.

[Ans: (a) -710 mJ/K; (b) $+710$ mJ/K; (c) $+723$ mJ/K; (d) -723 mJ/K; (e) $+13$ mJ/K; (f) 0]

11 (a) Since the temperature is the same for states 1 and 2 the ideal gas law gives $p_1 V_1 = p_2 V_2$. Use $V_2 = 3.00 V_1$.

(b) States 1 and 3 are on the same adiabat, so $p_1 V_1^{\gamma} = p_3 V_3^{\gamma}$. For an ideal diatomic gas $\gamma = 1.40$.

(c) Since $p_2 V_2 = nRT_1$ and $p_3 V_3 = nRT_3$, $T_3/T_1 = p_3/p_2$.

(d) The work is given by $W = \int p\,dV = nRT_1 \int (1/V)\,dV$.

(e) and (f) The internal energy does not change during an isothermal process involving an ideal gas, so $Q = W$.

(g) Since the process is isothermal and reversible, the entropy change is $\Delta S = Q/T_1$.

(h) The volume does not change.

(i) and (j) Use the first law of thermodynamics: $\Delta E_{int} = Q - W$. The change in the internal energy is $\Delta E_{int} = nC_V(T_3 - T_2)$, where C_V ($= 5R/2$) is the molar specific heat for constant-volume processes.

(k) Consider a reversible constant-volume process from state 2 to state 3. The incremental energy absorbed as heat is $dQ = nC_V\,dT$, so the change in entropy is $\Delta S = \int (nC_V/T)\,dT$.

(l) According to the first law of thermodynamics $\Delta E_{int} = -W$ since the process is adiabatic. The change in the internal energy is $\Delta E_{int} = nC_V(T_1 - T_3)$.

(m) The process is adiabatic.

(n) See part (l).

(o) The process is adiabatic and reversible.

$\big[$Ans: (a) 0.333; (b) 0.215; (c) 0.644; (d) 1.10; (e) 1.10; (f) 0; (g) 1.10; (h) 0; (i) -0.889; (j) -0.889; (k) -1.10; (l) -0.889; (m) 0; (n) 0.889; (o) 0 $\big]$

13 (a) The energy absorbed by the lower temperature block equals the energy emitted by the higher temperature block. Thus $m_C c_C (T_C - T) = m_L c_L (T - T_L)$, where m_C is the mass of the copper block, T_C is its initial temperature, c_C is the specific heat of copper, m_L is the mass of the lead block, T_L is its initial temperature, c_L is the specific heat of lead, and T is the final temperature of the blocks. Solve for T.

(b) The two blocks form a system that does no work on its environment and absorbs no energy as heat from its environment.

(c) Consider a constant-volume process that takes a block from its initial temperature to its final temperature. The change in entropy is $\Delta S = \int (mc/T)\,dT$.

$\big[$Ans: (a) 320 K; (b) 0; (c) $+1.72$ J/K $\big]$

23 The efficiency of a Carnot engine is given by $\epsilon_C = 1 - (T_L/T_H)$, where T_L is the temperature of the low-temperature reservoir and T_H is the temperature of the high-temperature reservoir, both on the Kelvin scale. Solve for T_H for each of the given efficiencies, then calculate the difference.

$\big[$Ans: 97 K $\big]$

39 The coefficient of performance of a Carnot refrigerator is $K_C = T_L/(T_H - T_L)$, where T_H is the temperature of the high-temperature reservoir and T_L is the temperature of the low-temperature reservoir, both on the Kelvin scale. Multiply K_C by 0.27 to obtain the coefficient of performance of the air conditioner. The work done by the motor in one cycle is $W = |Q_L|/K$, where Q_L is the energy withdrawn from the low-temperature reservoir as heat in one cycle. Divide by the time Δt of a cycle. $W/\Delta t$ is the power of the motor and

$|Q_L|/\Delta t$ is the cooling capacity. Convert the number given in Btu per hour to horsepower. See Appendix D.

[Ans: 0.25 hp]

43 The possible configurations in the form (n_1, n_2) are: $(1, 7)$, $(2, 6)$, $(3, 5)$, $(4, 4)$, $(3, 5)$, $(2, 6)$, and $(1, 7)$. The corresponding multiplicities are calculated using $W = 8!/(n_1! n_2!)$ and entropies are calculated using $S = k \ln W$, where k is the Boltzmann constant.

53 (a) The ice cube gains energy $m_i L_f$ as it melts and $m_i c_i (T_f - 0°C)$ as its temperature increases from $0°C$ to the final equilibrium temperature T_f. Here m_i is the mass of the ice, LF is the heat of fusion of water, and c_i is the specific heat of ice. The original water loses energy $m_w c_w (T_f - 80.0°C)$ as its temperature drops from $80.0°C$ to the final equilibrium temperature. These two energies must be equal. Solve for T_f. Needed data can be found in Tables 18–3 and 18–4.

(b), (c), (d), and (e) The change in entropy of the ice as it melts is $m_i L_f / T_m$, where T_m is the melting temperature on the Kelvin scale. To find the change in entropy as the resulting water warms to the equilibrium temperature replace the actual process with a reversible process for which the energy gained when the temperature increases by the infinitesimal dT is $dQ = mc_w \, dT$. Evaluate the integral $\int (mc_w/T) \, dT$ from the melting point to the final temperature. Perform a similar calculation for the original water as it cools. The total entropy change for the ice-water system is the sum of the individual entropy changes.

[Ans: (a) 66.5°C; (b) 14.6 J/K; (c) 11.0 J/K; (d) −21.2 J/K; (e) 4.39 J/K]

63 The process is at constant volume, so the gas does no work and, according to the first law of thermodynamics, the energy gained as heat is equal to the change in the internal energy. If the temperature increases by the infinitesimal dT the energy gained is $dQ = mC_V \, dT$, where C_V is the molar specific heat for constant volume processes. Evaluate the integral $\int (dQ/T$ from the initial to the final temperature. The molar specific heat for a monatomic ideal gas is $(3/2)R$, where R is the universal gas constant.

[Ans: +3.59 J/K]

67 The efficiency of a Carnot engine is given by $\epsilon = (T_H - T_L)/T_H$, where T_H is the temperature of the high temperature reservoir and T_L is the temperature of the low temperature reservoir. A change of ΔT_L produces the change $\Delta \epsilon = -(\Delta T_L)/T_H$ in the efficiency.

[Ans: −40 K]

71 The coefficient of performance of a refrigerator is defined by $K = |Q_L|/(|Q_H| - |Q_L|)$, where $|Q_L|$ is the magnitude of the energy taken as heat from the high temperature reservoir (the room) and $|Q_L|$ is the energy delivered as heat to the low temperature reservoir. The work done by the refrigerator is $W = |Q_H| - |Q_L|$, so $K = |Q_L|/W$.

[Ans: (a) 3.73; (b) 710 J]

Quiz

Some questions might have more than one correct answer.

1. The entropy of a system
   A. might decrease for some processes
   B. depends on the process used to go from some initial to some final equilibrium state
   C. does not decrease for any process if the system is closed
   D. does not change for any reversible process if the system is closed
   E. does not increase for any process if the system is closed

2. For any adiabatic process that a system might undergo its entropy

   A. increases
   B. decreases
   C. does not change
   D. might increase, decrease or not change, depending on the process
   E. does none of the above

3. When an ideal gas undergoes a reversible constant-pressure process the change in its entropy is
   A. proportional to the volume change
   B. proportional to the temperature change
   C. proportional to the specific heat for constant-pressure processes
   D. positive if the temperature increases and negative if the temperature decreases
   E. zero

4. When an ideal gas undergoes an irreversible process for which the initial and final temperatures are the same, the change in its entropy is
   A. greater than the change for a reversible isothermal process from the same initial state to the same final state
   B. less than the change for a reversible isothermal process from the same initial state to the same final state
   C. the same as the change for a reversible isothermal process from the same initial state to the same final state
   D. zero
   E. definitely positive

5. n moles of a gas with a constant molar specific heat C_p for constant-pressure processes undergoes a change in temperature from T_i to T_f (on the Kelvin scale) at constant pressure. The change in its entropy is
   A. $\Delta S = nC_p(T_f - T_i)$
   B. $\Delta S = nC_p(T_i - T_f)$
   C. $\Delta S = nC_p \ln(T_i/T_f)$
   D. $\Delta S = nC_p \ln(T_f/T_i)$
   E. $\Delta S = nC_p(T_f/T_i)$

6. n moles of a monatomic ideal gas undergo a change in temperature from T_i to T_f (on the Kelvin scale), during which the pressure p changes in proportion to the volume V. That is, $p = \alpha V$, where α is a constant. The change in entropy of the gas is

   A. $\frac{1}{2}nR\ln(T_f/T_i)$
   B. $\frac{3}{2}nR\ln(T_f/T_i)$
   C. $\frac{5}{2}nR\ln(T_f/T_i)$
   D. $2nR\ln(T_f/T_i)$
   E. $nR\ln(T_f/T_i)$

7. The pressure of n moles of a diatomic ideal gas is changed from p_i to p_f while the volume V remains constant. The change in entropy of the gas is

   A. $(3/2)nR\ln(p_f/p_i)$
   B. $(5/2)nR\ln(p_i/p_f)$
   C. $(5/2)nR\ln(p_f/p_i)$
   D. $(3/2)(p_f - p_i)V$
   E. $(3/2)(p_f - p_i)/V$

8. A Carnot heat engine

   A. is reversible
   B. takes in energy as heat from a thermal reservoir and does an equal amount of work
   C. stops working when its internal energy is exhausted
   D. stops working when its entropy is exhausted
   E. is the least efficient of all possible heat engines

9. The efficiency of a Carnot heat engine depends on

   A. the temperatures of the thermal reservoirs
   B. the type of material (gas, liquid, or solid) in the engine
   C. the quantity of material in the engine
   D. the time to complete one cycle
   E. the specific heat of the material in the engine

10. If a (hypothetical) refrigerator takes energy as heat from a low-temperature thermal reservoir and delivers it to a high-temperature thermal reservoir without any work being done on it

   A. the total entropy of the refrigerator and reservoirs decreases and the refrigerator violates the second law of thermodynamics
   B. the total entropy of the refrigerator and reservoirs increases and the refrigerator violates the second law of thermodynamics
   C. the coefficient of performance is zero
   D. the coefficient of performance is infinite
   E. the second law of thermodynamics is not violated

Answers: (1) A, C, D; (2) D; (3) C, D; (4) C; (5) D; (6) D; (7) C; (8) A; (9) A; (10) A, D

Chapter 21
ELECTRIC CHARGE

You start your study of electromagnetism with Coulomb's law, which describes the force that one stationary charged particle exerts on another. Pay attention to both the magnitude and direction of the force and learn how to calculate the net force when more than one charge acts.

Important Concepts

- [ ] electric charge
- [ ] conductor
- [ ] insulator

- [ ] Coulomb's law
- [ ] quantization of charge
- [ ] conservation of charge

Overview

21–2 Electric Charge

- [ ] **Electric charge** is a property possessed by some particles and, because they have charge, these particles exert electric forces on each other. The SI unit for charge is the coulomb (abbreviated C).

- [ ] There are two types of charge, called positive and negative. Charges of the same type repel each other; charges of different type attract each other. In many of the equations of electromagnetism, a positive number is substituted for the value of a positive charge and a negative number is substituted for the value of a negative charge.

- [ ] The charges on an electron and on a proton have exactly the same magnitude (1.60×10^{-19} C) but opposite signs. An electron is negative; a proton is positive.

- [ ] All macroscopic materials contain enormous numbers of negatively charged electrons and positively charged protons but if the number of electrons in an object equals the number of protons, then the net charge is zero and the object is said to be **neutral** (or uncharged). If there are more electrons than protons, the object has a net negative charge; if there are more protons than electrons, it has a net positive charge. In either case, it said to be **charged**. The net charge is computed by algebraically adding the charges of all particles in the object, taking their signs into account. Normally, macroscopic bodies are neutral.

- [ ] Objects may be given net charges by rubbing them together. When a glass rod is rubbed with silk, the rod becomes positively charged. When a plastic rod is rubbed with fur, the rod becomes negatively charged.

21–3 Conductors and Insulators

☐ The *electrons* in materials move and are transferred to or from other objects; the atomic nuclei (containing protons) are nearly immobile. Materials are often classified according to the freedom with which their electrons can move. Electrons in an **insulator** are not free to move; any charge placed on an insulator remains where it is placed. Scraping or rubbing is required to transfer charge to or from an insulator. **Conductors** contain many electrons that are free to move and easily flow between touching conductors.

☐ All metals, like copper, silver, and aluminum, are conductors. Rubber and mica are good insulators.

☐ A neutral conductor is attracted to a charged rod because the electrons within the inductor are redistributed, making some regions positively charged and others negatively charged. The forces on the negative and positive regions are in opposite directions but the force is greater on the region nearer to the charged object.

☐ Charged objects also attract neutral insulators but the force is much smaller than the force of attraction for a conductor. Under the influence of the external charge, electrons in an insulator move slightly from their normal orbits so the center of the negative charge is slightly displaced from the center of the positive charge. This results in a net force.

21–4 Coulomb's Law

☐ Suppose two point particles, one with charge q_1 and the other with charge q_2, are a distance r apart. The magnitude of the force exerted by either of the charges on the other is given by

$$F = \frac{1}{4\pi\epsilon_0} \frac{|q_1||q_2|}{r^2}.$$

The constant of nature ϵ_0 (= 8.85×10^{-12} C^2/N · m^2) is called the **permittivity constant**. The factor $k = 1/4\pi\epsilon_0$ has the value 8.99×10^9 N · m^2/C. The magnitude of the force is proportional to the reciprocal of the square of the distance between the particles and is also proportional to the product of the magnitudes of their charges.

☐ The direction of the electric force of one charged particle on another is along the line that joins the particles. If the charges have the same sign, the force on either particle is away from the other particle; if they have different signs, the force is toward the other particle. Electric forces between two charged particles obey Newton's third law: they are equal in magnitude and opposite in direction.

☐ When more than two charged particles are present, the force on any one of them is the *vector* sum of the forces due to the others. Each force is computed using Coulomb's law. This is the *principle of superposition* for electric forces.

☐ Just as for gravitational forces there are two shell theorems for electrical forces. They are:

> The electrical force between a uniformly charged spherical shell and a point charged particle *outside* the shell is as if the charge of the shell were concentrated at its center.

> The electrical force between a uniformly charged spherical shell and a charged point particle anywhere *inside* the shell is zero.

21–5 Charge is Quantized

☐ The charge on all particles detected so far is a positive or negative multiple of a fundamental unit of charge e (1.60×10^{-19} C).

21–6 Charge is Conserved

☐ The algebraic sum of the charges (including signs) on all particles in a closed system remains the same. This is true when the particles do not change identities, as in atomic physics and chemistry. It is also true in the subatomic world, where particles can disappear and others can appear. For example, an electron and positron can annihilate each other, leaving two uncharged quanta of light (photons). The charge on the electron is $-e$ and the charge on the positron is $+e$, so the total charge is zero, both before and after the annihilation event.

Hints for Questions

1 Use the shell theorems. The magnitude of the force is proportional to the product of the charge on the particle and the charge inside a sphere of radius d, divided by d^2. For (a) the charge inside a sphere of radius d is 0. For (b) it is $-4Q$ and for (c) it is $+8Q$.

[Ans: b and c tie, then a (zero)]

3 The charges on the two given particles must have opposite signs so their forces on the electron are in opposite directions. The charge on the particle that is further from the electron must have a greater magnitude so the magnitude of the two forces are equal.

[Ans: a and b]

5 Number the particles 1, 2, 3, 4, and 5 from left to right. The magnitudes of the forces of 1 and 5 are equal, as are the forces of 2 and 4. In some situations the forces of 1 and 5 are in opposite directions and sum to zero. In other situations the forces are in the same direction. Furthermore, the magnitudes of the forces of 1 and 5 are less than the magnitudes of the forces of 2 and 4.

[Ans: 3, 1, 2, 4 (zero)]

7 You can pair most of the particles around the perimeter so the forces of two particles of a pair sum to zero. The net force is then the same as the force of the particle with charge $+3q$ at the center of the left side of the square.

[Ans: $6kq^2/r^2$, leftward]

9 Draw vector arrows to represent the forces of the two lower particles on the upper particle. They all have the same length but their directions are either toward or away from one of the lower particles. In some situations the x components sum to zero and in other situations the y components sum to zero.

[Ans: (a) same; (b) less than; (c) cancel; (d) add; (e) adding components; (f) positive y direction; (g) negative y direction; (h) positive x direction; (i) negative x direction]

Hints for Problems

1 (a) The forces of the particles on each other have the same magnitude. Newton's second law tells you that $m_1 a_1 = m_2 a_2$, where m_1 and m_2 are the masses of the particles and a_1 and a_2 are the magnitudes of their accelerations. Solve for m_2.

(b) Use Newton's second law and Coulomb's law to obtain $m_1 a_1 = (1/4\pi\epsilon_0)q^2/r^2$, where q is the charge on either particle and r is their initial separation.

$$\left[\text{Ans: (a) } 4.9 \times 10^{-7}\,\text{kg; (b) } 7.1 \times 10^{-11}\,\text{C}\right]$$

7 Calculate the x and y components of the forces of particles 1, 2, and 4 on particle 3. Add the x components to find the x component of the net force and add the y components to find the y component of the net force. Use Coulomb's law to calculate the magnitude of each force. The charges of particles 1 and 3 have the same sign so the force of 1 is in the negative y direction. The charges of particles 2 and 3 have opposite signs so the force of 2 has positive x and y components. The distance between these particles is $\sqrt{2}a$ and the force makes an angle of $45°$ with the positive x direction. The charges of particles 3 and 4 have opposite signs so the force of 4 is in the negative x direction.

$$\left[\text{Ans: (a) } 0.17\,\text{N; (b) } -0.046\,\text{N}\right]$$

13 Assume the charge on particle 3 is positive. It is then repulsed by particle 1 and attracted by particle 2. Since the charge on particle 2 has a greater magnitude than the charge on particle 1, particle 3 must be the left of particle 1. Furthermore, it must be on the extension of the line that joins particles 1 and 2. Otherwise the forces of those particles would not be along the same line and they would not sum to zero. Suppose particle 3 is a distance d from particle 1. Then the magnitude of the force on it is $(q_3/4\pi\epsilon_0)[(q_1/d^2) + (q_2/(L+d)^2)]$. Set this equal to zero and solve for d.

$$\left[\text{Ans: (a) } 14\,\text{cm; (b) } 0\right]$$

19 Divide the shell into concentric shells of infinitesimal thickness dr. The shell with radius r has volume $dV = 4\pi r^2\,dr$ and contains charge $dq = \rho\,dV = 4\pi r^2\rho\,dr$. Integrate this expression. The lower limit is the inner radius of the original shell and the upper limit is its outer radius.

$$\left[\text{Ans: } 3.8 \times 10^{-8}\,\text{C}\right]$$

25 The current is the charge that is intercepted by Earth's surface per unit time. If N is the number of protons that hit each square meter of the surface per second, then the current is $i = NAe$, where A is the area of the surface, given by $A = 4\pi R^2$. Look up the radius R of Earth in Appendix C.

$$\left[\text{Ans: } 122\,\text{mA}\right]$$

27 When particle 3 is far away the only force on particle 2 is the force of particle 1. This has a magnitude of $1.5 \times 10^{-25}\,\text{N}$ and is in the positive x direction. When particle 3 is at $x = 0.4\,\text{m}$, its force on particle 2 has the same magnitude as the force of particle 1 but it is in the negative x direction. Thus $(1/4\pi\epsilon_0)q_2q_3/(0.4\,\text{m})^2 = 1.5 \times 10^{-25}\,\text{N}$. Solve for q_2.

$$\left[\text{Ans: } +13e\right]$$

33 You want to vectorially sum the six forces. Consider the charges on the x axis. The net force due to them is proportional to $(q_1/d^2) - (q_3/d^2) - (q_4/4d^2)$. Replace q_3 with $q_1/2.00$ and q_4

with $2.00q_1$ and evaluate the expression. Then consider the charges on the y axis. The net force due to them is proportional to $(q_2/d^2) - (q_5/d^2) - (q_6/4d^2)$. Replace q_5 with $q_2/2.00$ and q_6 with $2.00q_2$ and evaluate the expression.

[Ans: 0]

39 Draw a diagram. You want to vectorially sum the two forces on particle 3. If Q_2 is positive the net force is in the positive x direction and you need to compute only that component. Use Coulomb's law to find the magnitude of each individual force, then multiply by the cosine of the appropriate angle. The Pythagorean theorem gives the distance between particle 1 or particle 2 and the particle 3 and a little trigonometry gives the angle that the line joining particle 1 or particle 2 with particle 3 makes with the x axis. If Q_2 is negative the net force on particle 3 is in the negative y direction but the same method can be used to compute its value.

[Ans: (a) $(0.829 \, \text{N})\hat{i}$; (b) $(-0.621 \, \text{N})\hat{j}$]

45 Vectorially sum the three forces on particle 1, set the result equal to zero, and solve for q/Q. Use Coulomb's law to write an expression for each of the individual forces, then multiply by the sine or cosine of the appropriate angle to find the components. The length of a diagonal across a square is $\sqrt{2}a$, where a is the length of an edge. All angles are $45°$.

[Ans: 0.707]

49 You want to vectorially sum the two forces on particle 3. Particle 1 attracts particle 3 and particle 2 repels particle 3. In (a) and (b) particle 3 is to the right of particle 2, so the forces are in opposite directions. Use Coulomb's law to compute the magnitudes of the forces, then add them with appropriate signs. If the force on particle 3 is zero that particle must be on the x axis since otherwise the two forces would not be along the same line and could not sum to zero. Particle 3 cannot be between particles 1 and 2 since then the forces would be in the same direction and could not sum to zero. Since the charge on particle 1 is greater in magnitude than the charge on particle 2, particle 3 must be closer to particle 2 than to particle 1. Thus particle 3 is on the x axis to the right of particle 2. Write an expression for the net force on particle 3 in terms of its coordinate x and solve for the value of x that makes the net force zero.

[Ans: (a) $(89.9 \, \text{N})\hat{i}$; (b) $(-2.50 \, \text{N})\hat{i}$; (c) 68.3 cm; (d) 0]

59 If q_e is the charge on the electron and q_p is the charge on the proton, then $(q_p - |q_e|)/e = 1.0 \times 10^{-6}$. Compute $\Delta q = q_p - |q_e|$. The charge on one copper penny is $29N \, \Delta q$, where N is the number of copper atoms in a penny. The force of one penny on another has magnitude $F = (1/4\pi\epsilon_0)(\Delta q)^2/d^2$, where d is their separation.

[Ans: $1.7 \times 10^8 \, \text{N}$]

65 Use Newton's law of gravitation and Coulomb's law. Set the expressions for the magnitudes of the gravitational and electrostatic force equal to each other and solve for the charge. You will need to look up the masses of the Moon and Earth in Appendix C. The number of hydrogen atoms required is $N = q/e$, where q is the charge found in part (a). The mass of hydrogen required is Nm, where m is the mass of a hydrogen atom (essentially the same as the mass of a proton).

[Ans: (a) $5.7 \times 10^{13} \, \text{C}$; (b) cancels out; (c) $6.0 \times 10^5 \, \text{kg}$]

<u>67</u> When one ball loses its charge there is no electrostatic force between the balls and the balls swing down and touch. Charge is transferred and since the balls are identical each receives $q/2$. In the expression developed in Problem 66 replace q with $q/2$ and calculate x. The value of q found in Problem 66 should have been $2.4 \times 10^{-8}\,\text{C}$.

$\big[$Ans: (b) 3.1 cm $\big]$

Quiz

Some questions might have more than one correct answer.

1. Which of the following statements are true?
   A. particles with oppositely signed charges repel each other
   B. electrostatic forces are parallel to the line that joins the interacting charged particles
   C. electrostatic forces obey Newton's third law
   D. as the distance between two charged particles increases the force of each on the other increases
   E. most particles found in nature are positively charged

2. Electrically neutral objects
   A. have no charged particles
   B. have equal numbers of negatively and positively charged particles
   C. cannot exert electrical forces on charged particles
   D. do not exist in nature
   E. can remain neutral only for short times

3. Which of the following statements are true?
   A. many of the electrons of a conductor are free to travel throughout the conductor
   B. many of the positively charged ions of a conductor are free to travel throughout the conductor
   C. none of the electrons of an insulator are free to travel throughout the insulator
   D. none of the positively charged ions of an insulator are free to travel throughout the insulator
   E. none of the above are true

4. If a charged particle is brought close to a neutral object, the object is
   A. repelled from the particle if it is a conductor and attracted if it is an insulator
   B. repelled from the particle if it is an insulator and attracted if it is a conductor
   C. repelled from the particle no matter if it is a conductor or an insulator
   D. attracted toward the particle no matter if it is a conductor or an insulator
   E. not attracted to or repelled from the particle

5. A positively charged particle is brought near, but not touching, an isolated neutral conductor. You then touch the conductor with your hand and then remove your hand. The conductor is then

   A. positively charged
   B. negatively charged
   C. neutral

6. A positively charged particle is brought near, but not touching, an isolated neutral conductor. While you are touching the conductor with your hand, the particle is removed. The conductor is then

   A. positively charged
   B. negatively charged
   C. neutral

7. A positively charged particle is placed between two neutral conducting spheres. The spheres

   A. do not exert electrical forces on each other
   B. attract each other
   C. repel each other

8. Three identical charged particles are placed at the corners of an equilateral triangle. The net electrical force on any one of the particles is

   A. parallel to one of the triangle sides that meet at the particle
   B. perpendicular to one of the triangle sides that meet at the triangle
   C. parallel to the triangle side opposite the particle
   D. perpendicular to the triangle side opposite the particle
   E. zero

9. When two identical charged particles are a distance d apart, the force of either one on the other has magnitude F. If the distance is doubled and the charge on one of the particles is doubled the magnitude of the force is

   A. F
   B. $2F$
   C. $4F$
   D. $F/2$
   E. $F/4$

10. Two small insulating spheres have identical charge q. When $-4\,C$ of charge is added to one of the spheres, the spheres attract each other with the same magnitude force that they previously repelled each other. What is the value of q?

    A. $2\,C$
    B. $4\,C$
    C. $6\,C$
    D. $-2\,C$
    E. $-4\,C$

11. Three metal spheres are identical except that sphere A is neutral, sphere B has charge Q and sphere C has charge $-3Q$. Sphere C is touch to sphere B and, after it is removed sphere A is touch to sphere B. The charge on sphere A is then

   A. 0
   B. Q
   C. $-Q$
   D. $Q/2$
   E. $-Q/2$

Answers: (1) B, C; (2) B; (3) A, C, D; (4) D; (5) A; (6) C; (7) C; (8) D; (9) D; (10) A; (11) E

Chapter 22
ELECTRIC FIELDS

The idea of an electric field is introduced and used to describe electrical interactions between charges. It is fundamental to our understanding of electromagnetic phenomena and is used extensively throughout the rest of this course.

Important Concepts

- ☐ field
- ☐ electric field
- ☐ electric field lines
- ☐ electric field of a point charge
- ☐ linear charge density

- ☐ surface charge density
- ☐ electric dipole
- ☐ electric field of a dipole
- ☐ torque on a dipole
- ☐ potential energy of a dipole

Overview

22–1   What is Physics?

☐ A **field** gives some property of a region as a function of position. Temperature, for example, is a scalar field and the particle velocity as a function of position in a fluid is a vector field.

☐ **Electric fields** are associated with electric charges. The electric field associated with a charge pervades all space and exerts a force on any other charge in it.

22–2   The Electric Field

☐ The electric field at any point in space is defined as the electric force per unit test charge on a stationary positive test charge placed at that point, in the limit as the test charge becomes vanishingly small. If the test charge is q_0 and $\vec{F}$ is the electric force on it, then the electric field at the position of the test charge is $\vec{E} = \vec{F}/q_0$. $\vec{E}$ is the total field created by all charges other than the test charge. An electric field is associated with the charges that create it and exists regardless of the presence of a test charge. The SI unit of an electric field is a newton per coulomb (N/C).

☐ The limit as q_0 becomes vanishingly small must be included in the definition to minimize the influence of the test charge on the charged particles creating the field being measured.

☐ If a collection of stationary charged particles creates an electric field $\vec{E}$ at the position of another particle, with charge q, then the force exerted by the field on that particle is given by $\vec{F} = q\vec{E}$.

22–3 Electric Field Lines

☐ **Electric field lines** (also called lines of force) graphically depict both the direction and magnitude of an electric field. At any point the field is tangent to the line through that point. The lines are more concentrated in regions of high electric field than in regions of low electric field. In fact, the number of lines per unit area passing through a small area perpendicular to the lines is proportional to the magnitude of the field. Arrows are placed on field lines to indicate the direction of the field but electric field lines themselves are *not* vectors. Electric field lines emanate from positive charge and terminate on negative charge.

☐ To see the electric field lines for some charge distributions, look carefully at Figs. 22–2, 22–3(c), 22–4, and 22–5 of the text. For each figure notice the directions of the lines, where they are close together, and where they are far apart.

22–4 The Electric Field Due to a Point Charge

☐ The magnitude of the electric field at a point a distance r from a single point particle with charge q is given by $E = q/4\pi\epsilon_0 r^2$. The field is along the line that joins the particle q and the point. If q is positive, it points away from the particle; if q is negative, it points toward the particle.

☐ If more than one point particle is responsible for the electric field, the total field is the *vector* sum of the fields due to the individual particles.

22–5 The Electric Field Due to an Electric Dipole

☐ An **electric dipole** consists of a particle with positive charge $+q$ and a particle with negative charge $-q$ (of equal magnitude), separated by a distance d. It is characterized by a dipole moment $\vec{p}$, a vector with magnitude given by $p = qd$. The direction of $\vec{p}$ is along the line joining the particles, from the negative particle toward the positive particle. The field lines of a dipole are shown in Fig. 22–5.

☐ In Fig. 22–8, the z axis is in the direction of the dipole moment and the origin is midway between the particles. The magnitude of the field produced at P by the positively charged particle is $E_+ = q/4\pi\epsilon_0(z - d/2)^2$ and the magnitude of the field produced there by the negatively charged particle is $E_- = q/4\pi\epsilon_0(z + d/2)^2$. The first field is in the positive z direction and the second is in the negative z direction. The total field is

$$E = \frac{q}{4\pi\epsilon_0}\left[\frac{1}{(z - d/2)^2} - \frac{1}{(z + d/2)^2}\right].$$

It is in the positive z direction.

☐ Usually we are interested in situations for which $z \gg d$. The binomial theorem can be used to expand $(z - d/2)^2$ and $(z + d/2)^2$. When the only terms retained are those that are proportional to d, the expression for the field becomes $E = p/2\pi\epsilon_0 z^3$. The field is inversely proportional to z^3, not z^2. It is proportional to the dipole moment, not the charge and particle separation separately

22-6 The Electric Field Due to a Line of Charge

☐ A continuous distribution of charge is characterized by a **charge density**. If the charge is on a line (either straight or curved), the appropriate charge density is the **linear charge density**, denoted by λ and defined so that the charge dq in an infinitesimal segment of the line with length ds is $dq = \lambda \, ds$. If the charge is distributed on a surface, the appropriate charge density is the **surface charge density**, denoted by σ and defined so that the charge dq in an infinitesimal element of area dA is $dq = \sigma \, dA$. If the charge is distributed on throughout a volume, the appropriate charge density is the **volume charge density**, denoted by ρ and defined so that the charge dq in an infinitesimal element of volume dV is $dq = \rho \, dV$. For a *uniform* charge distribution, the appropriate charge density is the same everywhere along the line, on the surface, or in the volume.

☐ To calculate the electric field produced by a uniform ring of charge at a point along its axis, take the z axis to coincide with the axis of the ring and place the origin at the center of the ring. See Fig. 22–10 of the text. Two diametrically opposite elements of the ring produce fields with horizontal components that are the negatives of each other and that sum to zero. The total field is thus in the z direction. An infinitesimal segment ds of the ring has charge $dq = \lambda \, ds$, where λ is the linear charge density. The magnitude of the field produced by the segment at P is $E = (1/4\pi\epsilon_0)(\lambda/r^2) \, ds$, where r is the distance from the segment to P. It is given by $r = (z^2 + R^2)^{1/2}$, where R is the radius of the ring. To obtain the z component, multiply the magnitude of the field by the cosine of the angle θ between the field and the z axis. This is given by $\cos\theta = z/r = z/(z^2 + R^2)^{1/2}$. The total field is found by integrating around the ring. All quantities in the integrand have the same value for all segments of the ring and ds integrates to $2\pi R$. Thus

$$E = \frac{z\lambda(2\pi R)}{4\pi\epsilon_0(z^2 + R^2)^{3/2}} = \frac{qz}{4\pi\epsilon_0(z^2 + R^2)^{3/2}} \, .$$

22-7 The Electric Field Due to a Charged Disk

☐ To calculate the field produced by a uniform disk of charge, the disk is divided into rings, each with infinitesimal width dr. The area of a ring of radius r is $dA = 2\pi r \, dr$ and if σ is the area charge density, then $dq = 2\pi r \sigma \, dr$ is the charge in such a ring. The field it produces at a point on the axis a distance z from the ring center is along the axis and has the magnitude

$$dE = \frac{z\sigma(2\pi r)}{4\pi\epsilon_0(z^2 + R^2)^{3/2}} \, dr \, .$$

To find the total field, this expression is integrated from $r = 0$ to $r = R$. The result is:

$$E = \frac{\sigma}{2\epsilon_0} \left(1 - \frac{z}{\sqrt{z^2 + R^2}} \right) \, .$$

☐ An expression for the field produced by a uniform plane of charge can be found by finding the limit as R becomes large without bound. If $R \gg z$, then $z/\sqrt{z^2 + R^2}$ is much smaller than 1 and $E = \sigma/2\epsilon_0$.

☐ At points near a ring, disk, or plane, superposition produces total fields that are quite different from each other and quite different from the field of a point charge. At points far from a charge distribution, however, the electric field tends to become like that of a point particle with charge equal to the net charge in the distribution.

22–8 A Point Charge in an Electric Field

☐ If an electric force is the only force acting on a charged particle, then Newton's second law takes the form $q\vec{E} = m\vec{a}$, where $\vec{a}$ is the acceleration of the particle and m is its mass. If you know the initial position and velocity of the charge, then you can, in principle, predict its subsequent motion. If the field is uniform, use the kinematic equations for constant acceleration.

22–9 A Dipole in an Electric Field

☐ The net force on a dipole in a uniform electric field is zero because the individual forces on the two charges have the same magnitude and are opposite in direction.

☐ A uniform electric field exerts a torque on a dipole. If $\vec{p}$ is the dipole moment and $\vec{E}$ is the electric field, then the torque is $\vec{\tau} = \vec{p} \times \vec{E}$. This torque tends to rotate the dipole so that $\vec{p}$ becomes more closely aligned with the field.

☐ When an electric dipole rotates in an electric field $\vec{E}$, the field does work on it. If the angle between the dipole moment and field changes from θ_0 to θ, the work done by the field is

$$ W = \int_{\theta_0}^{\theta} \tau \, d\theta = - \int_{\theta_0}^{\theta} pE \sin\theta \, d\theta = pE(\cos\theta - \cos\theta_0). $$

The sign of the work is negative if the angle increases and positive if the angle decreases.

☐ When a dipole rotates in an electric field, the potential energy changes by $\Delta U = -W = -pE(\cos\theta - \cos\theta_0)$. The potential energy for any orientation can be written $U = -pE\cos\theta = -\vec{p} \cdot \vec{E}$. It is a maximum when $\theta = 180°$ and is a minimum when $\theta = 0$.

Hints for Questions

1 The momentum is the greatest for the situation in which the electric field is greatest and for this situation the field lines are closest together in the region between A and B. Similarly, the momentum is least for the situation in which the field lines are farthest apart.

[Ans: a, b, c]

3 Draw arrows with their tails at the central point to represent the electric fields of the four particles. The fields of the inner particles have the same magnitude, as do the fields of the outer particles and the fields of the outer particles are weaker than the fields of the inner particles. Be careful about directions.

[Ans: 2, 4, 3, 1 (zero)]

5 Draw arrows with their tails at point P to represent the electric fields of the two charged particles. The magnitudes depend only on the magnitudes of the charges and the distance of the particles from P. The electric field of a positively charged particle is away from the particle and the electric field of a negatively charged particle is toward the particle.

[Ans: (a) yes; (b) toward; (c) no (the fields are not parallel to each other; (d) cancel; (e) add; (f) adding components; (g) negative y direction]

7 The electric field of a rod that spans a single quadrant is along the line from the origin to the center of the rod. It is toward the rod if the rod is negative and away from the rod if the rod is positive. Vectorially add the field of the rods that are present in each situation.

[Ans: e, b, then a and c tie, then d (zero)]

9 Consider a ring of charge with infinitesimal width. The electric field it produces at P is along the z axis and the greater the radius of the ring the smaller the magnitude of the field. The disk in (b) has the same charge as the disk in (a) but much of it is in rings of greater radius. The ring in (c) also has the same charge but all of it is in rings of greater radius than those of (a) and much of it is in rings of greater radius than those of (b).

[Ans: a, b, c]

11 The potential energy is given by $U = -pE\cos\theta$, where θ is the angle between the dipole moment and the electric field. As the angle increases from 0 the cosine decreases from 1 to 0 (at 90°), then becomes negative and increases in magnitude until it reaches -1 (at 180°). The magnitude of the torque is given by $pE\sin\theta$. As the angle increases from 0 the sine increases from 0 to 1 (at 90°), then decreases to 0 (at 180°).

[Ans: (a) 4, 3, 1, 2; (b) 3, then 1 and 4 tie, then 2]

Hints for Problems

1 Use the shell theorems. All field lines are radial. The electric field is zero inside the inner shell. Between the shells the field is the same as that of a point particle with charge q_1, located at the center of the cavity. Outside the outer shell the field is the same as that of a point particle with charge $q_1 - q_2$, located at the center of the cavity.

13 The particles are equidistance from P and their charges have the same magnitude. The y components of their fields sum to zero and their x components are the same, so you need calculate only the x component of one of the fields, then double it. The x component of the field of either particle is given by $E_x = -(1/4\pi\epsilon_0)qx/r^3$, where x is the coordinate of the particle and r is its distance from P.

[Ans: (a) 1.38×10^{-10} N/C; (b) negative x direction]

15 The fields of the particles all have the same magnitude: $E = (1/4\pi\epsilon_0)/r^2$. Calculate the x and y components, then sum the x components to find the x components of the net field and sum the y components to find the y component of the net electric field. The magnitude of the net field is the square root of the sum of the squares of its components and the tangent of the angle the field makes with the positive x direction is the y component divided by the x component. Be careful of the signs when you compute the components.

[Ans: (a) 3.93×10^{-6} N/C; (b) $-76.4°$]

19 At P the electric fields of the two charged particles have the same magnitude, which is given by $E = (1/4\pi\epsilon_0)q/(r^2 + d^2/4)$. Their x components sum to zero and their y components have the same value: $E_y = -E\sin\theta$, where θ is the angle between the x axis and the line from either particle to P. Use $\sin\theta = (d/2)/\sqrt{r^2 + d^2/4}$ to substitute for $\sin\theta$. Double the result to take both particles into account, then use the binomial theorem to find the expression for the net field for $r \gg d$.

[Ans: (a) $qd/4\pi\epsilon_0 r^3$; (b) negative y direction]

25 At P the electric fields of the two quarter-circles have the same magnitude. Sample Problem 22–4 shows you how to obtain an expression for this magnitude. The field of the upper arc points downward at 45° to the positive x direction and the field of the lower arc points downward at 45° to the negative x direction. The x components of the field sum to zero. Add the y components.

[Ans: (a) 20.6 N/C; (b) negative y direction, at $-90°$ to the positive x direction]

29 Symmetry tells us that the horizontal component of the net electric field at P is zero. Divide the rod into sections of infinitesimal width dx and treat each section as a point particle with charge $dq = \lambda\, dx$, where λ (= q/L) is the linear charge density of the rod. Put the origin at the center of the rod. The magnitude of the field produced at P by the section at x is $dE = (1/4\pi\epsilon_0)(\lambda\, dx)/r^2$ and its vertical component is $dE\sin\theta$. Here r is the distance from the section to P and θ is the angle between the line from the section to P and the positive x direction. Substitute $r = \sqrt{x^2 + R^2}$ and $\sin\theta = R/r = R/\sqrt{x^2 + R^2}$ and integrate over the length of the rod.

[Ans: a) 12.4 N/C; (b) positive y direction]

33 Use Eq. 22–26 to find the electric field produced by the disk. Think of the ring as the superposition of a disk of radius $R/2$ and surface charge density $-\sigma$ and the original disk (with surface charge density σ). Add the electric fields of these two disks to obtain the electric field of the ring.

[Ans: 28%]

53 The potential energy of an electric dipole is given by $U = -\vec{p}\cdot\vec{E} = -pE\cos\theta$, where $\vec{p}$ is the dipole moment, $\vec{E}$ is the electric field, and θ is the angle between the dipole moment and the electric field. The work required of an external agent is the change in the potential energy.

[Ans: 1.22×10^{-23} J]

61 Vectorially sum the individual fields. Notice that the fields of q_1 and q_5 cancel since the charges have the same sign, are the same distance from the center of the square, and are on opposite sides of the center. Look for other pairs of fields that cancel, then carry out the vector addition of the remaining fields.

[Ans: $(1.08 \times 10^{-5}$ N/C$\hat{i}$]

<u>67</u> Follow the procedure given in Sample Problem 22–4. Divide the arc into infinitesimal segments of length ds and take the charge in a segment to be $dq = \lambda\, ds$, where λ is the linear charge density (the charge per unit length). The magnitude of the field produced by a segment at the center of curvature is given by $dE = (1/4\pi\epsilon_0)(\lambda/R^2)\, ds$. Symmetry tells you that the total field is along the line that joins the center of the arc and the center of curvature, so you need to find only one component. Multiply the magnitude of the field due to a segment by the cosine or sine of the appropriate angle and integrate over the arc. It is convenient to convert the integration variable from ds to $R\, d\theta$, where θ is the angle you used to find the field component.

$\left[\text{Ans: } 5.39\,\text{N/C}\,\right]$

<u>75</u> The gravitational force is given by mg, where m is the mass of the drop. According to Table 14–1 the density of water is $\rho = 998\,\text{kg/m}^3$. Use $m = (4\pi/3)R^3\rho$ to compute m. The net charge on the drop is $Q = F/E$, where F is the magnitude of the electrostatic force and E is the magnitude of the electric field, and the number of excess electrons is $N = Q/e$.

$\left[\text{Ans: (a) } 8.87 \times 10^{-15}\,\text{N; (b) } 120\,\right]$

<u>77</u> Take the string to be stretched from $x = 0$ to $x = L$ and divide it into infinitesimal segments of length dx. A segment contains charge $dq = \lambda\, dx$, where λ is the linear charge density. The field produced at $x = a$ by the segment at x is $dE = (1/4\pi\epsilon_0)(\lambda\, dx)/(a - x)^2$. Integrate from $x = 0$ to $x = L$.

$\left[\text{Ans: } 61\,\text{N/C}\,\right]$

<u>85</u> The magnitude of the electrostatic force must equal the magnitude of the gravitational force, which is the weight of the sphere. The magnitude of the electrostatic force is given by QE, where Q is the magnitude of the net charge on the sphere and E is the magnitude of the electric field. The electrostatic force must be upward. That is, it is opposite the direction of the electric field.

$\left[\text{Ans: (a) } -0.029\,\text{C; (b) repulsive forces would explode the sphere}\right]$

<u>91</u> Draw a free body diagram of the small sphere when it has swung out from the vertical through an angle θ. The forces on it are the force of gravity, the tension force of the thread, and the electrostatic force. This last force is upward in part (a) and downward in part (b). Calculate the net torque on the sphere, about the point of attachment of the upper end of the thread and set this equal to $I\alpha$, where α is the angular acceleration of the pendulum, and $I\ (= mL^2)$ is the rotational inertia of the pendulum. Use the small-angle approximation by replacing $\sin\theta$ with θ (in radians). You should get an equation of the form $\alpha = -C\theta$, where C is a constant that depends on m, L, and E. The angular frequency is $\omega = \sqrt{C}$ and the period is $T = 2\pi/\omega$.

$\left[\text{Ans: (a) } 2\pi\sqrt{L/|g - qE/m|}; \text{ (b) } 2\pi\sqrt{L/(g + qE/m)}\,\right]$

```
Quiz
```

Some questions might have more than one correct answer.

1. At any point in space the electric field produced by a positively charged point particle
   A. is directed from the point toward the particle
   B. is directed from the point away from the particle
   C. has a magnitude that is inversely proportional to the distance from the particle to the point
   D. has a magnitude that is inversely proportional to the square of the distance from the particle to the point
   E. obeys Newton's third law

2. For any electric field the electric field line through any point in space
   A. is tangent to the electric field at that point
   B. might cross another electric field line
   C. is a vector
   D. is the path of a charged particle placed at the point
   E. might have an end at the point even if there is no particle there

3. Two particles with identical positive charge are situated on the x axis, equidistant from the origin. The electric field at a point on the y axis is
   A. in the positive y direction no matter where the point is on the y axis
   B. in the negative y direction if the y coordinate of the point is negative
   C. in the positive y direction if the y coordinate of the point is negative
   D. in the positive y direction if the y coordinate of the point is positive
   E. parallel to the x axis no matter where the point is on the y axis

4. A particle with a large positive charge is placed at a point on the x axis with a negative x coordinate and a particle with a small negative charge is placed at a point on the x axis with a positive x coordinate. The electric field is zero
   A. at a point on the x axis between the particles
   B. at a point on the x axis with an x coordinate that is more positive than that of the negative charge
   C. at a point on the x axis with an x coordinate that is more negative than that of the positive charge
   D. at a point off the x axis
   E. at no point

5. Three particles, with charges of $-4\,C$, $+6\,C$, and $-5\,C$ are positioned near the origin of a coordinate system. At any point that is far away from the charges

A. the electric field lines are nearly along directed lines that are radially outward from the origin

B. the electric field lines are nearly along directed lines that are radially inward toward the origin

C. the electric field is like that of a single particle with a charge of $-3\,C$, located at the origin

D. the electric field is like that of a single particle with a charge of $+15\,C$, located at the origin

E. none of the above are true

6. Suppose charge is distributed with volume charge density ρ and you want to find the electric field at some point P. Let r be the distance from a point within the charge distribution to P and let ϕ be the angle that the line from that point to P makes with the x axis. (The integrals given below are volume integrals over the charge distribution.) Then

A. the quantity $\rho\,dV$ is the charge in the infinitesimal volume dV

B. the quantity $\rho\,dV$ can be treated as the charge of a point particle

C. the magnitude of the electric field at P is given by $E = \dfrac{1}{4\pi\epsilon_0}\displaystyle\int \dfrac{\rho}{r^2}\,dV$

D. the x component of the electric field at P is given by $E_x = \dfrac{1}{4\pi\epsilon_0}\displaystyle\int \dfrac{\rho}{r^2}\,dV$

E. the x component of the electric field at P is given by $E_x = \dfrac{1}{4\pi\epsilon_0}\displaystyle\int \dfrac{\rho\cos\phi}{r^2}\,dV$

7. An electric dipole

A. does not produce a net electric field because it is neutral

B. produces a net electric field because the two charged particles are not at the same place

C. produces a net electric field that is everywhere parallel to the dipole moment

D. produces an electric field with a magnitude far away that is proportional to $1/r^2$, where r is the distance from the dipole

E. produces an electric field with a magnitude far away that is proportional to $1/r^3$, where r is the distance from the dipole

8. The dipole moment of an electric dipole

A. has a magnitude that is the product of the total charge in the dipole and the distance between charges

B. has a magnitude that is the product of the a positive charge in the dipole and the distance between charges

C. is directed from the negatively charged particle toward the positively charged particle

D. is directed from the positively charged particle toward the negatively charged particle

E. is a scalar

9. If no other force except the electric force acts on a negatively charged particle
    A. the particle moves in the direction of the electric field at its position
    B. the particle moves in the direction opposite that of the electric field at its position
    C. the acceleration of the particle is in the direction of the electric field at its position
    D. the acceleration of the particle is in the direction opposite that of the electric field at its position
    E. the acceleration of the particle is perpendicular to the direction of the electric field at its position

10. Two identical particles have the same positive charge. The electric field produced by them is zero
    A. at the point midway between them
    B. at a point between them that is closer to one particle than to the other
    C. at a point that is equidistance from the particles, a short distance from them, but is not on the line that runs through them
    D. at a point that is on the line that runs through the particles and is not between them
    E. nowhere except infinitely far from both particles

Answers: (1) B, D; (2) A; (3) B, D; (4) B; (5) B, C; (6) A, B, E; (7) B, E; (8) B, C; (9) D; (10) E

Chapter 23
GAUSS' LAW

Gauss' law is one of the four fundamental laws of electromagnetism. It relates an electric field to the charged particles that create it and is, therefore, closely akin to Coulomb's law. Unlike Coulomb's law, however, it is valid when the charges are moving, even at relativistic speeds. The central concept for an understanding of Gauss' law is that of electric flux, a quantity that is proportional to the number of field lines penetrating a given surface. Pay careful attention to the definition of flux and learn how to compute it for various fields and surfaces. Then learn how to use Gauss' law to compute the charge in any region if the electric field is known on the boundary and also how to use the law to compute the electric field in certain highly symmetric situations.

Important Concepts

☐ flux

☐ electric flux

☐ Gauss' law

☐ Gaussian surface

Overview

23–1   What is Physics?

☐ To use Gauss' law, you consider a *closed* surface, called a **Gaussian surface**. It can be a real surface, such as the boundary of a material object, or an imaginary surface, constructed only for the purpose of applying the law. When Gauss' law is used to solve for the electric field, the Gaussian surface should reflect the symmetry of the problem: a sphere for a point charged particles or a spherical conductor, a cylinder for a wire or a cylindrical conductor.

☐ Gauss' law gives the relationship between the *normal* component of the electric field at a Gaussian surface and the electric charge enclosed by the surface. This component is directly related to the **flux** of the electric field.

23–2   Flux

☐ For a moving fluid, the flux through an area is the volume of fluid that crosses the area per unit time per unit area. If the fluid velocity vector $\vec{v}$ makes the angle θ with the normal to the surface, the flux is $\Phi = (v \cos \theta)A$, where A is the area.

☐ $\Phi = 0$ if the velocity is parallel to the surface ($\theta = 90°$). No fluid moves through the surface then. Φ has its maximum value if the velocity is perpendicular to the surface ($\theta = 0$).

23–3 Flux of an Electric Field

☐ Although the electric field does not represent a velocity, a flux can be associated with it. If the field is uniform, the electric flux through a plane area A is given by $\Phi = (E \cos \theta) A$, where θ is the angle between the field and the normal to the area. If the field is not uniform, the flux is defined by the integral $\Phi = \int (E \cos \theta) \, dA$. The surface is divided into a large number of small area elements, the flux through each element is calculated, and the results are summed.

☐ If the *vector* area element $d\vec{A}$ is defined to be in the direction of the normal to the surface, the definition of electric flux can be written $\Phi = \int \vec{E} \cdot d\vec{A}$.

☐ The vector area element $d\vec{A}$ may be in either of the two directions that are perpendicular to the surface (up or down for a horizontal surface). Which one is used determines the sign but not the magnitude of the flux. If the surface is closed, like a Gaussian surface, $d\vec{A}$ is always chosen to point *outward* and the definition of electric flux is written $\Phi = \oint \vec{E} \cdot d\vec{A}$, with a small circle on the integral sign to indicate that the surface is closed.

☐ An area element contributes to the electric flux only if the field pierces the element. In fact, the flux through a surface is proportional to the number of field lines that penetrate the element. Field lines that cross a closed surface from inside to outside make a positive contribution to Φ while lines that cross the surface from outside to inside make a negative contribution. Since some parts of a closed surface may make positive contributions to the flux while other parts make negative contributions, the total flux through a surface may vanish even though an electric field exists at every point on the surface.

☐ Remember that the number of electric field lines per unit area through an infinitesimal area dA, perpendicular to the field, is proportional to the magnitude of the field. If the vector $d\vec{A}$ is in the same direction as the field, the number of lines that penetrate it is proportional to $E \, dA$. If the area is rotated so $d\vec{A}$ makes the angle θ with the field, the number of lines that penetrate the area decreases to $E \cos \theta \, dA$.

23–4 Gauss' Law

☐ Gauss' law states that the total electric flux through any closed surface is proportional to the total charge enclosed by the surface. Mathematically,

$$\epsilon_0 \oint \vec{E} \cdot d\vec{A} = q_{enc} ,$$

where q_{enc} is the total charge enclosed, the sum of the charge (including sign) on all particles within the surface.

☐ The law should not surprise you since the magnitude of the electric field and the number of field lines associated with any charge are both proportional to the charge. All lines from a single positive charge within a closed surface penetrate the surface from inside to outside, so the total flux is positive and proportional to the charge. All lines associated with a single negative charge within a closed surface penetrate from outside to inside, so the total flux is negative and again proportional to the charge. Some lines associated with a charge outside a surface penetrate the surface, but those that do, penetrate twice, once from outside to inside and once from inside to outside, so this charge does not contribute to the total flux through the surface.

23-5 Gauss' Law and Coulomb's Law

☐ Gauss' law can be used to find an expression for the electric field of a point charge. Imagine a sphere of radius r with a positive point charge q at its center. Since the electric field is radially outward from the charge, the normal component of the field at any point on the surface of the sphere is the same as the magnitude E of the field. Furthermore, the magnitude E is uniform on the surface, so the total flux through the sphere is $\Phi = 4\pi r^2 E$. Equate this to q/ϵ_0 and obtain $E = q/4\pi\epsilon_0 r^2$, in agreement with Coulomb's law.

☐ If the electric field is given at all points on a surface, Gauss' law can be used to calculate the net charge enclosed by the surface. If the charge is known, the law can be used to find the total electric flux through a surface. If the charge distribution is highly symmetric, so a symmetry argument can be used to show that $E\cos\theta$ has the same value at all points on a surface, then Gauss' law can be used to solve for the electric field at points on the surface.

23-6 A Charged Isolated Conductor

☐ A conductor contains a large number of electrons that are free to move throughout the material. As a consequence the electric field vanishes at all points in the interior of a conductor in electrostatic equilibrium, with all charge stationary. If this were not true, the electric field would exert a force on the electrons in the conductor and they would move until the force on each of them is zero. $\vec{E} = 0$ inside a conductor even when excess charge is placed on it or when an external field is applied to it. In the interior, the external field is canceled by the field produced by the redistributed charge on the surface. By way of contrast, an electrostatic field might exist inside an insulator.

☐ In an electrostatic situation, any excess charge on a conductor must reside on its surface; there can be no net charge in its interior. Imagine a Gaussian surface that is completely within the conductor. The electric field is zero at every point on the surface, so the total flux through the surface is zero and, according to Gauss' law, the net charge enclosed by the surface is zero. Any net charge on the conductor must lie outside the Gaussian surface. Since this result is true for *every* Gaussian surface that can be drawn completely within the conductor, no matter how close to its surface, we conclude that any excess charge must be on the surface of the conductor. If the object is an insulator, the field may not be zero inside and excess charge may be distributed throughout its volume.

☐ If a conductor has a cavity (completely surrounded by the conductor), then the charge on the inside surface of the conductor (its boundary with the cavity) must be the negative of the charge in the cavity. According to Gauss' law, the sum of the charge in the cavity and on the inside surface must be zero. If there is charge Q in the cavity, then there is charge $-Q$ on the inside surface of the conductor. The charge on the inside and outside surfaces must sum to give the total charge on the conductor. If, for example, the total charge on the conductor is zero, then the charge on its outside surface is $+Q$.

☐ The magnitude of the electric field at any point just outside the surface of a conductor is directly proportional to the surface charge density at the corresponding point on the surface.

The exact relationship is $E = \sigma/\epsilon_0$, where σ is the area charge density. This result follows directly from Gauss' law, applied to a small closed surface that is partially inside and partially outside the conductor.

23–7 Applying Gauss' Law: Cylindrical Symmetry

☐ Successful use of Gauss' law to solve for the electric field depends greatly on choosing the right Gaussian surface. First, it must pass through the point where you want the value of the field. Second, either the normal component of the electric field must have constant magnitude over the entire surface or else it must have constant magnitude over part of the surface and the flux through the other parts must be zero. A symmetry argument should be made to justify the use of the Gaussian surface you have chosen.

☐ If a long, straight, cylindrical rod has positive charge distributed uniformly along it, symmetry tells us that the electric field is radially outward from the rod and has the same magnitude at points that are the same distance from the rod.

☐ To calculate the electric field a distance r from a uniformly charged rod, use a cylindrical Gaussian surface with radius r and length h, concentric with the rod. The field has the same magnitude at all points on the rounded portion of the cylinder and at all these points it is normal to the surface. Over the rounded portion of the cylinder, $\int \vec{E} \cdot d\vec{A} = 2\pi E r h$. The electric field is parallel to the ends of the cylinder (perpendicular to $d\vec{A}$), so $\int \vec{E} \cdot d\vec{A} = 0$ for these portions of the Gaussian surface. Thus $\oint \vec{E} \cdot d\vec{A} = 2\pi E r h$. The charge enclosed is $q = \lambda h$, where λ is the linear charge density of the rod. Gauss' law becomes $2\pi \epsilon_0 E r h = \lambda h$, so

$$E = \frac{\lambda}{2\pi \epsilon_0 r} .$$

23–8 Applying Gauss' Law: Planar Symmetry

☐ If an infinite sheet has a uniform, positive surface charge density σ, symmetry tells us that the electric field points perpendicularly away from the sheet and that the magnitude of the field is uniform over any plane that is parallel to the sheet. Take the Gaussian surface to be a cylinder with its ends parallel to the sheet and its axis perpendicular to the sheet. The sheet cuts through the midpoint of the cylinder's length, so the field has the same magnitude on both ends of the cylinder. The flux through the rounded portion of the cylinder is zero because the field is parallel to this portion of the surface (perpendicular to $d\vec{A}$); the flux through either of the cylinder ends is $\int \vec{E} \cdot d\vec{A} = EA$, where A is the area of an end. The charge enclosed by the Gaussian surface is $q = \sigma A$, so Gauss' law becomes $2\epsilon_0 E A = \sigma A$ and

$$E = \frac{\sigma}{2\epsilon_0} .$$

☐ You might think this result is inconsistent with the expression for the magnitude of the electric field just outside a conductor, $E = \sigma/\epsilon_0$. It is not. Consider an infinite plane conducting sheet with a uniform charge density σ on one surface. This charge produces an electric field with magnitude $\sigma/2\epsilon_0$, just like any other large uniform sheet of charge. The field exists on both sides of the sheet, in the interior of the conductor as well as in the exterior, and on each side it points away from the surface if σ is positive. Another electric field must be present, produced perhaps by charge on another portion of the conductor or by external charge. In the interior of the conductor, the second field exactly cancels the field due to the charge layer to produce a total field of zero and, in the exterior, it augments the field due to the charge layer to produce a total field with magnitude σ/ϵ_0.

23–9 Applying Gauss' Law: Spherical Symmetry

☐ If a spherical shell is uniformly charged, we use a Gaussian surface in the form of a sphere, concentric with the shell. The electric field is normal to the surface and its magnitude is uniform over the surface. The flux through the surface is $\Phi = 4\pi r^2 E$, where r is the radius of the Gaussian sphere. If the Gaussian surface is inside the shell, the charge enclosed is zero, so the electric field is $E = 0$ at points inside the shell. If the Gaussian surface is outside the shell, the charge enclosed is the total charge q on the shell, so the electric field is $E = q/4\pi\epsilon_0 r^2$ at points outside the shell.

☐ Consider a uniform sphere of charge with radius R. The field inside the sphere, a distance r (with $r < R$) from its center, is $E = q'/4\pi\epsilon_0 r^2$, where q' is the charge enclosed by a sphere of radius r. If the volume charge density is ρ, then $q' = (4\pi/3)\rho r^3$ and

$$E = \frac{\rho r}{3\epsilon_0}.$$

If q is the total charge in the sphere, then $\rho = 3q/4\pi R^3$ and

$$E = \frac{qr}{4\pi\epsilon_0 R^3}.$$

Hints for Questions

1 Evaluate $\Phi = \vec{E} \cdot \vec{A} = E_x A_x + E_y A_y + E_z A_z$.

[Ans: (a) $8\,\mathrm{N} \cdot \mathrm{m}^2/\mathrm{C}$; (b) 0]

3 Since the electric field is radial and its magnitude is uniform over any of the Gaussian surfaces, Gauss's law yields $4\pi r^2 E = q_{\mathrm{enc}}/\epsilon_0$, where r is the radius of the Gaussian surface and q_{enc} is the net charge inside the surface. Thus the magnitude of the field on a Gaussian surface is proportional to q_{enc}/r^2.

[Ans: all tie]

5 Since the electric field is radial and its magnitude is uniform over any of the Gaussian surfaces, Gauss's law yields $2\pi r L E = q_{\mathrm{enc}}/\epsilon_0$, where r is the radius of the Gaussian surface and q_{enc} is the net charge inside the surface. Thus the magnitude of the field on a Gaussian surface is proportional to q_{enc}/r.

[Ans: all tie]

7 The magnitude of the electric field is given by $(1/2\epsilon_0)|\sigma_+ - \sigma_-|$, the magnitude of the force on the electron is $F = eE$, and the magnitude of the acceleration of the electron is $a = F/m$. Thus the magnitude of the acceleration is proportional to $|\sigma_+ - \sigma_-|$.

[Ans: all tie]

9 The electrostatic field is zero inside a conductor. The magnitude of the field outside the spherical shell is proportional to the net charge enclosed by the shell.

[Ans: (a) all tie ($E = 0$); (b) all tie]

Hints for Problems

5 Use Gauss' law in the form $\Phi = q_{enc}/\epsilon_0$, where Φ is the net flux through the surface of the cube and q_{enc} is the net charge inside the cube.

[Ans: $2.0 \times 10^5 \, \text{N} \cdot \text{m}^2/\text{C}$]

7 Think of the proton as being at the center of a cube with edge length d. According to Gauss's law the net electric flux through the surface of the cube is e/ϵ_0. Since the proton is at the center one-sixth of the net flux is through each face of the cube.

[Ans: $3.01 \, \text{nN} \cdot \text{m}^2/\text{C}$]

17 The surface charge density is the net charge on the surface divided by the surface area ($4\pi R^2$, where R is the radius). The magnitude of the electric field is given by $E = \sigma/\epsilon_0$, where σ is the surface charge density.

[Ans: (a) $4.5 \times 10^{-7} \, \text{C/m}^2$; (b) $5.1 \times 10^4 \, \text{N/C}$]

23 (a) The charge on the drum is the product of its surface charge density and the surface area of its curved surface: $q = \sigma 2\pi R L$, where σ is the charge density ($2.0 \, \mu\text{C/m}^2$ from Problem 16), R is its radius, and L is its length.

(b) The magnitude of the electric field at the drum surface is given by $E = \lambda/2\pi\epsilon_0 R$, where λ is the linear charge density. Since $\lambda = q/L$, $E = q/2\pi\epsilon_0 R L$. Thus $q_2/R_2 L_2 = q_1/R_1 L_1$, where the subscripts 1 refer to the old values and the subscripts 2 refer to the new values. Solve for q_2.

[Ans: (a) $0.32 \, \mu\text{C}$; (b) $0.14 \, \mu\text{C}$]

25 The radial component E of the electric field of a long cylindrical shell is given by $E = \lambda/2\pi\epsilon_0 r$, where λ is the linear charge density and r is the distance from the cylindrical axis. For part (a) the point is between the shells so $\lambda = 5.0 \times 10^{-6} \, \text{C/m}$ and for part (b) the point is outside both shells so $\lambda = 5.0 \times 10^{-6} \, \text{C/m} - 7.0 \times 10^{-6} \, \text{C/m} = -2.0 \times 10^{-6} \, \text{C/m}$. If E is positive the field is radially outward and if E is negative the field is radially inward.

[Ans: (a) $2.3 \times 10^6 \, \text{N/C}$; (b) outward; (c) $4.5 \times 10^5 \, \text{N/C}$; (d) inward]

27 The electric field at a distance r from the cylindrical axis is given by $E = \lambda/2\pi r$, where λ is the linear charge density inside the Gaussian cylinder of radius r. For part (a) the point is outside both the rod and the cylindrical shell, so $\lambda = (Q_1 + Q_2)/L$. For part (b) the point is outside the rod but inside the shell, so $\lambda = Q_1/L$. The field is radially outward if E is positive and radially inward if E is negative. There can be no net charge inside a Gaussian cylinder that is completely within the shell, so the charge on the rod and the interior surface of the shell must sum to zero. The charge on the interior and exterior surfaces of the shell must sum to Q_2.

[Ans: (a) $0.214 \, \text{N/C}$; (b) inward; (c) $0.855 \, \text{N/C}$; (d) outward; (e) $-3.4 \times 10^{-12} \, \text{C}$; (f) $-3.40 \times 10^{-12} \, \text{C}$]

29 The magnitude of the electric field a distance r from the cylinder axis is given by $E = \lambda/2\epsilon_0 r$, where λ is the liner charge density inside a Gaussian cylinder of radius r, concentric with the wire and shell. Now $\lambda = \lambda_{\text{wire}} + \lambda_{\text{shell}}$, where λ_{wire} is the linear charge density on the wire and λ_{shell} is the linear charge density on the shell. For the field to be zero outside the shell $\lambda_{\text{shell}} = -\lambda_{\text{wire}}$. Since the surface area of the curved surface of the shell is $2\pi RL$, where R is the radius of the shell and L is its length, the surface charge density of the shell is $\sigma_{\text{shell}} = \lambda_{\text{shell}} L/2\pi RL = \lambda_{\text{shell}}/2\pi R$.

$\left[\text{Ans: } 3.8 \times 10^{-8}\,\text{C/m}^2\right]$

33 The magnitude of the electric field of a large plate with surface charge density σ is $E = |\sigma|/2\epsilon_0$. If σ is positive the field points away from the plate and if σ is negative it points toward the plate. Thus the x component of the field is $\sigma/2\epsilon_0$ to the right of the plate and $-\sigma/2\epsilon_0$ to the left. Write equations for the x component of the field between plates 1 and 2, between plates 2 and 3, and outside plate 3, then solve these for σ_2/σ_2.

$\left[\text{Ans: } -1.5\right]$

41 Symmetry tells us that the electric field at $x = 0$ is zero. If it were not which way would it point? Use Gauss' law to find the electric field at other points. Use a Gaussian surface in the form of a rectangular solid with two of the faces having area A. These faces are parallel to the plates and one of them is at $x = 0$. The cross section in the plane of the figure is a rectangle with sides of length L and x. There is electric flux through the right face only and its value is EA. For parts (a), (b), and (c) the charge enclosed is ρAx. For part (d) the charge enclosed is $\rho Ad/2$.

$\left[\text{Ans: (a) } 0; \text{ (b) } 1.31\,\mu\text{N/C}; \text{ (c) } 3.08\,\mu\text{N/C}; \text{ (d) } 3.08\,\mu\text{N/C}\right]$

45 Use Gauss' law with a spherical Gaussian surface in the form of a sphere with radius r. The electric flux through the surface is $4\pi r^2 E$, where E is the magnitude of the electric field. For part (a) only the charge on the smaller shell is enclosed by the Gaussian surface and for part (b) the charge on both shells is enclosed.

$\left[\text{Ans: (a) } 2.50 \times 10^4\,\text{N/C}; \text{ (b) } 1.35 \times 10^4\,\text{N/C}\right]$

51 To find the total charge in the sphere integrate the volume charge density over the volume of the sphere. Divide the sphere into spherical shells of infinitesimal thickness dr. The volume of the shell with radius r is $4\pi r^2\,dr$.

To find the electric field use Gauss' law with a Gaussian surface in the form of a sphere with radius r, concentric with the sphere of charge. Symmetry tells us that electric field is normal to the surface and that its magnitude is uniform over the surface. Thus the electric flux through the surface is $4\pi r^2 E$, where E is the radial component of the electric field. To find the charge enclosed by the Gaussian surface integrate the volume charge density over the volume of a sphere of radius r.

$\left[\text{Ans: (a) } 7.78\,\text{fC}; \text{ (b) } 0; \text{ (c) } 5.58\,\text{mN/C}; \text{ (d) } 22.3\,\text{mN/C}\right]$

53 Use a spherical Gaussian surface with radius r, concentric with the charge distribution. The electric field is radially outward, normal to the surface, and has a uniform magnitude over the surface, so the electric flux through the surface is $4\pi r^2 E$, where E is the magnitude of the electric field a distance r from the center of the charge distribution. The charge enclosed by the Gaussian surface is

$$q_{\text{enc}} = \int_0^r 4\pi(r')^2 \rho \, dr'$$

Thus

$$4\pi r^2 \epsilon_0 E = \int_0^r 4\pi(r')^2 \rho \, dr'.$$

Differentiate with respect to r and solve for ρ.

$\left[\text{Ans: } 6\epsilon_0 K r^3\,\right]$

<u>65</u> Use a Gaussian surface in the form of a cube that is centered at the origin and has faces parallel to the xy, xz, and yx planes. To find the electric field at x place the faces that are parallel to the yz plane at $-x$ and $+x$. Symmetry then tells us that the field has the same magnitude on these faces. The field is in the positive x direction at one of these faces and is in the negative x direction at the other, so the net flux through the cube is $\Phi = 2EA$, where A is the area of a face. You must find the net charge enclosed. For part (a) it is only part of the total charge in the slab but for part (b) it is the entire charge of the slap.

$\left[\text{Ans: (a) } 5.4\,\text{N/C; (b) } 6.8\,\text{N/C}\,\right]$

<u>69</u> Gauss' law applied to a uniform spherical distribution of charge leads to the equation $E = q/4\pi\epsilon_0 d^2$ for the magnitude of the electric field a distance d from the center of the distribution. Here q is the net charge inside a Gaussian sphere of radius d, centered at the center of the distribution. For part (a) the entire charge of the distribution is inside the Gaussian sphere while for part 9b) only part of it is. The ratio of the charge inside to the total charge is the same as the ratio of the volume of the sphere of radius d is the volume of the charge distribution. That is, it is $(d/r)^3$.

$\left[\text{Ans: (a) } 15.0\,\text{N/C; (b) } 25.3\,\text{N/C}\,\right]$

<u>71</u> Sum the individual fluxes through the six faces, with appropriate signs. The net flux is $\Phi = (-1 + 2 - 3 + 4 - 5 + 6) \times 10^3\,\text{N} \cdot \text{m}^2/\text{C}$. According to Gauss' law the charge inside the die is $q = \epsilon_0 \Phi$.

$\left[\text{Ans: } 26.6\,\text{nC}\,\right]$

<u>73</u> Since the field is uniform and inward at the flat base, the flux through the flat base is $-EA$, where A is the area of the base, which is a circle of radius R. The net flux of a uniform field through a closed surface that contains no net charge is zero. Thus the sum of the flux through the base and the flux through the curved surface is zero.

$\left[\text{Ans: (a) } -2.53 \times 10^{-2}\,\text{N} \cdot \text{m}^2/\text{C; (b) } +2.53 \times 10^{-2}\,\text{N} \cdot \text{m}^2/\text{C}\,\right]$

<u>79</u> Use Gauss' law with a spherical Gaussian surface of radius r that is concentric with the shell. The electric field is radial and has uniform magnitude over the surface, so the integral in the law is $\oint \vec{E} \cdot d\vec{A} = 4\pi r^2 E$, where E is the radial component of the field. In part (a) the charge enclosed is $+5.00\,\text{C}$ and in part (d) it is $(5.00\,\text{pC}) + (-3.00\,\text{pC}) = 2.00\,\text{pC}$. In part (c) the point is inside the shell. If E is positive the field is outward and if E is negative it is inward.

$\left[\text{Ans: (a) } 0.180\,\text{N/C; (b) outward; (c) } 0; \text{(d) } 4.50\,\text{mN/C}\,\right]$

<u>83</u> The total field is the superposition of the field produced by the positive point particle at the center of the atom and the uniform distribution of negatively charged particles around it.

The field due to the central charge is radially outward and has magnitude $E_+ = Ze/4\pi\epsilon_0 r^2$. To find the field of the negatively charged particles use a spherical Gaussian surface with radius r, centered on the positive particle. The field is radially inward and uniform over the Gaussian surface, so the integral in Gauss' law is $\oint \vec{E} \cdot d\vec{A} = 4\pi r^2 E$, where E is the radial component of the field. The charge enclosed is not the total charge of the negative distribution but only the charge that is enclosed by the Gaussian surface. This is proportional to the volume enclosed by the Gaussian surface.

Quiz

Some questions might have more than one correct answer.

1. The total electric flux through any surface

   A. depends on the normal component of the electric field at every point on the surface
   B. depends on the tangential component of the electric field at every point on the surface
   C. might be zero even when the electric field is not zero at any point on the surface
   D. is always positive
   E. is a valid concept only if the surface is spherical

2. A plane surface with an area of $2.0\,\mathrm{m}^2$ lies in the xy plane. In the region of the surface the electric field is uniform, has a magnitude of $5.0\,\mathrm{N/C}$, and makes an angle of $30°$ with the positive z axis. The magnitude of the electric flux through the surface is

   A. $1.3\,\mathrm{N \cdot m^2/C}$
   B. $5.0\,\mathrm{N \cdot m^2/C}$
   C. $8.7\,\mathrm{N \cdot m^2/C}$
   D. $10\,\mathrm{N \cdot m^2/C}$
   E. $12\,\mathrm{N \cdot m^2/C}$

3. The length of a side of a certain cube is L. A uniform electric field has a magnitude E and is normal to two of the sides (and tangent to the other four). The net electric flux through the surface of the cube is

   A. EL^2
   B. $2EL^2$
   C. $4EL^2$
   D. $6EL^2$
   E. zero

12. Two large flat conducting plates are parallel to each other. Both are positively charged, one with a surface charge density of σ_1 and other with a larger surface charge density of σ_2. The magnitude of the electric field at a point between the plates is

    A. $(\sigma_1 + \sigma_2)/\epsilon_0$
    B. $|\sigma_1 - \sigma_2|/\epsilon_0$
    C. $(\sigma_1 + \sigma_2)/2\epsilon_0$
    D. $|\sigma_1 - \sigma_2|/2\epsilon_0$
    E. $\sigma_1\sigma_2/\epsilon_0$

Answers: (1) A, C; (2) C; (3) E; (4) D; (5) B; (6) B; (7) A, E; (8) B, D, E; (9) B, D; (10) A, D; (11) A; (12) D

Chapter 24
ELECTRIC POTENTIAL

Because the electric force is conservative, a potential energy is associated with a collection of charged particles. If the particles are released, potential energy is converted to kinetic energy as each particle moves in response to the forces of the other particles. Electric potential is closely related to potential energy and plays a vital role in most succeeding discussions of electricity. Play careful attention to its definition and learn how to compute it for a collection of charged point particles.

Important Concepts

☐ electric potential

☐ equipotential surface

☐ electric potential energy

Overview

24–2   Electric Potential Energy

☐ Since the electric force is conservative, a potential energy is associated with it, just as a potential energy is associated with the gravitational force. If a system of charged particles changes from some initial configuration to some final configuration, the change in the potential energy is given by $\Delta U = U_f - U_i = -W_{if}$, where W_{if} is the work done by electrical forces.

☐ A special case is of great importance: a test charge is moved in the electric field created by other particles, which remain stationary. The change in the potential energy of the system consisting of the test charge and the other particles is the negative of the work done by the electrical forces on the test charge. If the test charge is moved from a reference point (usually far away from all other charged particles) and the potential energy is taken to be zero for the test charge at that point, then the negative of the work is the potential energy for the test charge at its final position.

24–3   Electric Potential

☐ Electric potential and electric potential energy are closely related but they are not the same. To find the electric potential of a collection of charged point particles, a reference point is chosen and the electric potential at that point is set equal to zero. Then, a positive *test* charge, not one of the particles in the collection, is moved from the reference point to any point P and the work done by the electric field on the test charge is calculated. The electric potential at P is the negative of this work divided by the test charge. This is also the change in the electric potential energy per unit test charge of the system consisting of the original collection of particles and the test charge. The particles of the collection must remain in fixed positions as the test charge is moved. If the total charge is finite, the reference point is usually selected to be infinitely far removed from the charged particles. A value for the electric potential is associated with each point in space and exists regardless of whether a test charge is present.

☐ Since electric potential is an energy divided by a charge, its SI unit is J/C. This unit is called a volt (abbreviation: V). The unit of an electric field may be taken to be a volt per meter (V/m), which is the same as a newton per coulomb.

24–4 Equipotential Surfaces

☐ An equipotential surface is a surface (imaginary or real) such that the electric potential has the same value at all points on it. It can be labeled by giving the value of the potential on it. Equipotential surfaces do not cross each other. One and only one goes through any point in space.

☐ The electric field line through any point is perpendicular to the equipotential surface through that point. If the field has a non-vanishing component tangent to a surface, then a potential difference must exist between points on the surface and the surface cannot be an equipotential surface.

☐ The equipotential surfaces of an isolated charged point particle are spheres, centered on the particle. The equipotential surfaces of a uniform field are planes, perpendicular to the field. Equipotential surfaces associated with other charge distributions are more complicated but if the distribution has a net charge, they are nearly spheres far from the distribution. Look at Fig. 24–3 of the text to see the surfaces associated with a dipole, for which the net charge is zero. All conductors are equipotential volumes and their boundaries are equipotential surfaces.

24–5 Calculating the Potential from the Field

☐ When a test charge q_0 moves through an infinitesimal displacement $d\vec{s}$, the work done by the electric field is $dW = q_0\vec{E} \cdot d\vec{s}$ and the change in the potential energy is $dU = -dW = -q_0\vec{E} \cdot d\vec{s}$. The difference in the electric potential for two points i and f is given by

$$V_f - V_i = -\int_i^f \vec{E} \cdot d\vec{s}.$$

The potential at any point is

$$V = -\int_i^f \vec{E} \cdot d\vec{s},$$

if i is taken to be the reference point and the potential is taken to be zero there. These integrals are line integrals: the path is divided into infinitesimal segments, the integrand is evaluated for each segment, and the results are summed. Because the electric field is conservative, every path will give the same value for the potential difference of the given points.

☐ If the field is uniform, as it is outside a large sheet with uniform charge density, then $V_f - V_i = -\vec{E} \cdot \Delta\vec{r}$, where $\Delta\vec{r}$ is the displacement of point f from point i.

☐ Roughly speaking, the electric field points from a region where the potential is high toward a region where the potential is low. A positively charged particle is accelerated toward a low potential region; a negatively charged particle is accelerated toward a high potential region.

24–6 Potential Due to a Point Charge

☐ As a test charge q_0 moves from far away to a point a distance r from a single isolated point particle with charge q, the work done by the electric field on the charge is $W = -qq_0/4\pi\epsilon_0 r$ and the change in the potential energy of the two-charge system is $\Delta U = +qq_0/4\pi\epsilon_0 r$. If the potential is taken to be zero at infinity, then its value at the final position of the test charge is

$$V = \frac{q}{4\pi\epsilon_0 r} \, .$$

This expression is also valid if q is negative. The potential then has a negative value.

24–7 Potential Due to a Group of Point Charges

☐ Electric potential is a scalar, so the potential of a collection of charged point particles is the algebraic sum of the individual potentials of the particles in the collection.

☐ Suppose the collection consists of particles with charges q_1, q_2, and q_3. The electric potential at some point P is

$$V = \frac{1}{4\pi\epsilon_0} \left(\frac{q_1}{r_1} + \frac{q_2}{r_2} + \frac{q_3}{r_3} \right) ,$$

where r_1 is the distance from the particle with charge q_1 to P, r_2 is the distance from the particle with charge q_2 to P, and r_3 is the distance from the particle with charge q_3 to P. When you substitute values for the charges, be sure you include the appropriate signs.

24–8 Potential Due to an Electric Dipole

☐ An electric dipole consists of a positively charged and a negatively charged particle with the same magnitude charge q, separated by a distance d. The potential at any point can be found by summing the potentials of the two particles. Let $\vec{r}$ be the vector from the midpoint of the line joining the particles to the point and suppose this line makes the angle θ with the dipole moment $\vec{p}$. If the particles are close together compared to r, the potential is $V = p\cos\theta/4\pi\epsilon_0 r^2$.

24–9 Potential Due to a Continuous Charge Distribution

☐ Charge may be distributed continuously along a line, on a surface, or throughout a volume. If it is, divide the region into infinitesimal elements, each containing charge dq, and sum (integrate) the potential due to the regions. If an infinitesimal region is a distance r from point P, then the contribution of that region to the potential at P is $dV = dq/4\pi\epsilon_0 r$ and the potential due to the whole distribution is

$$V = \frac{1}{4\pi\epsilon_0} \int \frac{dq}{r} \, .$$

In practice, dq is replaced by $\lambda\, ds$ for a line distribution and by $\sigma\, dA$ for a surface distribution. Here λ is a linear charge density and σ is a surface charge density.

□ Consider a uniform line of charge that extends from $x = 0$ to $x = L$, as shown in Fig. 24–12. If the linear charge density is λ, then the charge in an infinitesimal segment dx is $dq = \lambda\,dx$. If the coordinate of the segment is x and you are calculating the potential at a point that is a perpendicular distance d from the end of the line at $x = 0$, then the distance from the segment to the point is $r = (x^2 + d^2)^{1/2}$. The potential at the point due to the segment is

$$dV = \frac{1}{4\pi\epsilon_0} \frac{\lambda}{(x^2 + d^2)^{1/2}}\,dx$$

and the potential due to the entire line is

$$V = \frac{\lambda}{4\pi\epsilon_0} \int_0^L \frac{dx}{(x^2 + d^2)^{1/2}} = \frac{\lambda}{4\pi\epsilon_0} \ln\left(x + \sqrt{x^2 + D^2}\right)\bigg|_0^L$$

$$= \frac{\lambda}{4\pi\epsilon_0} \ln\left(\frac{L + (L^2 + d^2)^{1/2}}{d}\right).$$

See integral 17 of Appendix E.

□ To calculate the potential produced by a uniform disk of charge at a point on the axis a distance z above the disk center, first consider a ring with radius R' and width dR'. The area of the ring is $dA = 2\pi R'\,dR'$ and if σ is the surface charge density, the charge contained in the ring is $dq = 2\pi\sigma R'\,dR'$. All charge in the ring is the same distance from the point: $r = (z^2 + R'^2)^{1/2}$. Thus the potential produced by the ring at z is

$$dV = \frac{1}{4\pi\epsilon_0} \frac{\sigma(2\pi R')}{(z^2 + R'^2)^{1/2}}\,dR'$$

and the total potential due to the disk is given by

$$V = \frac{\sigma}{2\epsilon_0} \int_0^R \frac{dR'}{(z^2 + R'^2)^{1/2}} = \frac{\sigma}{2\epsilon_0}\left(\sqrt{z^2 + R^2} - z\right).$$

24–10 Calculating the Field from the Potential

□ If the electric potential is a known function of position in a region of space, then the components of the electric field $\vec{E}$ can be found:

$$E_x = -\frac{\partial V}{\partial x}, \qquad E_y = -\frac{\partial V}{\partial y}, \qquad \text{and} \qquad E_z = -\frac{\partial V}{\partial z}.$$

When you differentiate with respect to x to find E_x, you treat y and z as constants. Similar statements hold for E_y and E_z.

24–11 Electric Potential Energy of a System of Point Charges

□ Suppose particles with charge q_1 and q_2 start a distance r_i apart and move so they end a distance r_f apart. Then, the work done by the electric field is

$$W_{if} = -\frac{q_1 q_2}{4\pi\epsilon_0}\left[\frac{1}{r_f} - \frac{1}{r_i}\right]$$

and the change in the potential energy of the two-particle system is

$$\Delta U = +\frac{q_1 q_2}{4\pi\epsilon_0}\left[\frac{1}{r_f} - \frac{1}{r_i}\right].$$

The result is the same if either particle remains stationary while the other moves or if both move. All that matters is their initial and final separations.

☐ If the total charge in the collection is finite, the potential energy is usually taken to be zero for infinite particle separation. Let the initial separation r_i become large without bound and replace r_f with r. Then,

$$U = \frac{q_1 q_2}{4\pi\epsilon_0 r}$$

gives the potential energy when the particle separation is r. This expression is valid no matter what the signs of the charges. If they have the same sign, the potential energy is positive; it decreases if the particle separation becomes greater and increases if the particle separation becomes less. If the two charges have opposite signs, the potential energy is negative; it increases (becomes less negative) if the particle separation becomes greater and decreases (becomes more negative) if the particle separation becomes less.

☐ To calculate the potential energy of a collection of charged point particles, sum the potential energies of all *pairs* of particles. If the system consists of four particles, for example, add the potential energies of particle 1 and 2, 1 and 3, 1 and 4, 2 and 3, 2 and 4, and 3 and 4.

☐ The electric potential energy of a system is the work that must be done by an external agent to assemble the collection, bringing the particles from infinite separation to their final positions. The particles are at rest at the beginning and end of the process, so there is no change in kinetic energy.

☐ If no external agent acts on a system of charges, its mechanical energy is conserved: during any interval $\Delta K + \Delta U = 0$. If the charged particles are released from some initial configuration, their mutual attractions and repulsions cause them to move. The potential energy of the system will decrease as it is converted to kinetic energy.

24–12   Electric Potential of a Charged Isolated Conductor

☐ All points in the interior and on the surface of an isolated conductor have the same electric potential, once equilibrium is established. This follows immediately from the definition of the potential difference between two points and the condition $\vec{E} = 0$ inside a conductor in equilibrium.

☐ Consider a conducting sphere of radius R, with charge Q uniformly distributed on its surface. If r is the distance from the sphere center to a point outside the sphere, the electric potential there is given by $V = Q/4\pi\epsilon_0 r$. If r is the distance from the sphere center to a point inside the sphere, the potential there is given by $V = Q/4\pi\epsilon_0 R$. The potential inside does not depend on r and has the same value as the potential at the surface. The zero of potential was taken to be at $r = \infty$.

Hints for Questions

<u>1</u> For the electric potential to be zero anywhere the particles must have charges with opposite signs. The zero of the potential is then closer to the particle with the smaller magnitude charge. Recall that the magnitude of the electric field of a point particle is inversely proportional to the square of the distance from the particle while the magnitude of the electric potential is inversely proportional to the distance itself.

[Ans: (a) 1 and 2; (b) none; (c) no; (d) 1 and 2, yes; 3 and 4, no]

<u>3</u> Since all the particles are the same distance from the origin the electric potential there is proportional to the net charge in the arrangement.

[Ans: b, then a, c, and d tie]

<u>5</u> The magnitude of the electric field is given by $\Delta V/\Delta x$, where ΔV is the difference in electric potential of two equipotential lines that are separated by Δx. The field points from a region of high electric potential toward a region of lower electric potential.

[Ans: (a) 1, then 2 and 3 tie; (b) 3]

<u>7</u> Since the separations are all the same, the electric potential energy is proportional to the product of the charges. It is positive if the charges have the same sign and negative if they have opposite signs. As the separation increases the magnitude of the potential energy decreases. Thus the potential energy itself decreases for charges of the same sign and increases for charges of opposite sign.

[Ans: (a) 3 and 4 tie, then 1 and 2 tie; (b) 1 and 2, increase; 3 and 4 decrease]

<u>9</u> After the move from A to B the neighborhood of the particle with charge q still contains the same charge at the same distance from that particle. The work done by the field of the other two charged particles is the negative of the change in the potential energy and the work done by your force is the same as the change in the potential energy. The same is true for the move from B to C.

[Ans: (a) 0; (b) 0; (c) 0; (d) all three quantities are still 0]

Hints for Problems

<u>7</u> The magnitude of the electric field is given by $E = \sigma/2\epsilon_0$, where σ is the surface charge density of the sheet. If the x axis is perpendicular to the sheet, then the electric potential is given by $V = -Ex + C$, where C is a constant. The work done by the electric field is $W = -q\Delta V$, where ΔV is the difference in electric potential between the beginning and ending points of the particle's path.

[Ans: (a) 1.87×10^{-21} J; (b) -11.7 mV]

<u>9</u> The magnitude of the electric field a distance r from the sphere center, inside the sphere, is $E = qr/4\pi\epsilon_0 R^3$. Integrate the negative of this from $r = 0$ to $r = 1.45$ cm in part (a) and to $r = R$ in part (b).

[Ans: (a) -0.268 mV; (b) -0.681 mV]

<u>13</u> The electric potential is given by $V = (1/4\pi\epsilon_0)[(q_1/|x|) + (q_2/|x - d|)]$. For points to the left of the origin this becomes $V = (1/4\pi\epsilon_0)[-(q_1/x) - (q_2/(d - x))]$ and for points between

the particles it becomes $V = (1/4\pi\epsilon_0)[(q_1/x)+(q_2/(d-x))]$. Find the values of x for which $V = 0$. Treat each region separately.

[Ans: (a) 6.0 cm; (b) -12.0 cm]

19 The electric potential due to an electric dipole is given by $V = (1/4\pi\epsilon_0)(p/r)\cos\theta$, where p is the magnitude of the dipole moment and the angle θ is measured from the dipole axis. Here $\theta = 0$.

[Ans: 16.3 μV]

21 Consider only the right-hand half of the rod. Divide it into infinitesimal strips of width dx. The strip at x is a distance $\sqrt{x^2 + d^2}$ from P and can be treated as a point charged particle. Its charge is $dq = \lambda\,dx$, where λ is its linear charge density, and the potential it produces at P is $dV = (1/4\pi\epsilon_0)(\lambda/\sqrt{x^2 + d^2})\,dx$. Integrate from $x = 0$ to $x = L/2$. See integral 17 in Appendix 17. To obtain the answer for the full rod in either part (a) or (b), add the contributions of the two halves of the rod.

[Ans: (a) 24.3 mV; (b) 0]

27 All the charge on any of the rods is equidistant from the origin so the electric potential produced at the origin by any rod is $V = (1/4\pi\epsilon_0)q/R$, where Q is the charge on the rod and R is the radius of the rod. Add the contributions of the three rods.

[Ans: 13 kV]

29 Divide the rod into infinitesimal strips of width dx. The strip at x is a distance $x + D$ from P_1 and can be treated as a point charged particle. Its charge is $dq = \lambda\,dx = cx\,dx$, where λ is the linear charge density, and the potential it produces at P is $dV = (1/4\pi\epsilon_0)[cx/(x+D)]\,dx$. Integrate from $x = 0$ to $x = L$. See integral 21 of Appendix E.

[Ans: 18.6 mV]

35 The force on the electron is $\vec{F} = -e\vec{E}$, where $\vec{E}$ is the electric field at the position of the electron. The electric field components are $E_x = -\partial V/\partial x$, $E_y = -\partial V/\partial y$, and $E_z = -\partial V/\partial z$. The first two partial derivatives are the slopes of the graphs.

[Ans: $(-4.0 \times 10^{-16}\,\text{N})\hat{\imath} + (1.6 \times 10^{-16}\,\text{N})\hat{\jmath}$]

43 Use conservation of mechanical energy. If U_1 is the initial potential energy, U_2 is the final potential energy, and v_2 is the final speed, then $U_2 + \frac{1}{2}mv_2^2 = U_1$. The initial potential energy is given by $U_1 = q^2/4\pi\epsilon_0 r_1$ and the final potential energy is given by $U_2 = q^2/4\pi\epsilon_0 r_2$.

[Ans: 2.5 km/s]

51 Use conservation of mechanical energy. The potential energy when the positron is at some coordinate x is $U = eV$, where values of V can be read from the graph. When the positron is at $x = 0$ the potential energy is zero, so the mechanical energy is the same as the positron's kinetic energy there: it is $\frac{1}{2}mv^2$, where v is the speed of the positron as it enters the field. If the mechanical energy is greater than the potential energy when the positron is at $x = 50$ cm, then the positron will continue in the positive x direction and will exit the field at $x = 50$ cm. If the mechanical energy is less than the potential energy with the positron at $x = 50$ cm then the direction of motion of the positron will reverse at some point before $x = 50$ cm and the positron will exit the field at $x = 0$.

[Ans: (a) 0; (b) 1.0×10^7 m/s]

57 The magnitude of the electric field at the sphere's surface is $E = \sigma/\epsilon_0$, where σ is the surface charge density ($Q/4\pi R^2$, where Q is the charge on the sphere and R is radius of the sphere). Since the electric potential is $V = Q/4\pi\epsilon_0 R$, you can use $V = ER$ to compute it.

[Ans: (a) 12 kV/m; (b) 1.8 kV; (c) 5.8 cm]

59 Use Gauss' law in the form $\Phi = q/\epsilon_0$ to find the charge q on the sphere, then use $V = q/4\pi\epsilon_0 r$ to find the electric potential (relative to that at infinity) a distance r from the sphere center.

[Ans: 3.71×10^4 V]

65 Treat the ring as the superposition of two disks, one with radius R and uniform surface charge density σ and the other with radius r and surface charge density $-\sigma$. The electric potential for the first disk given by Eq. 24–37: $V = (\sigma/2\epsilon_0)(\sqrt{z^2 + R^2} - z)$. For the second disk substitute r for R and $-\sigma$ for σ.

[Ans: 10.3 mV]

69 Use conservation of energy. The initial potential energy of the two-particle system is $qQ/4\pi\epsilon_0 r_i$, where r_i is the initial separation of the particles. The final potential energy is $qQ/4\pi\epsilon_0 r_f$, where r_f is their final separation. These expressions are value no matter what the signs of q and Q. The initial kinetic energy is zero and the final kinetic energy is unknown.

[Ans: (a) 0.90 J; (b) 4.5 J]

79 The work done by an external agent is equal to the change in the potential energy of the three-particle system. Let L be the separation of the two fixed electrons. The initial potential energy is $e^2/4\pi\epsilon_0 L$ and the final potential energy is $3e^2/4\pi\epsilon_0 L$.

[Ans: 240 kV]

81 Both particles are positively charged, so their electric potentials (relative to that at infinity) are both positive. The electric field is zero at some point on the z axis between the particles. If x is the coordinate of the point, then $q_1/x^2 = q_2/(d - x)^2$. Solve for x.

[Ans: (a) at no point; (b) 0.41 m]

85 To find the total charge sum the charges in the two regions. Each is the product of a surface charge density and an area. The area of the first region is $\pi(R/2)^2$ and the area of the second is $\pi R^2 - \pi(R/2)^2$. To find the electric potential superpose the potentials of three disks: one with radius $R/2$ and surface charge density σ_1 ($= 1.50 \times 10^{-6}$ C/m^2), the second with radius R and surface charge density σ_2 ($= 8.00 \times 10^{-7}$ C/m^2), and the third with radius $R/2$ and surface charge density $-\sigma_2$. The superposition of the second and third disks produces the ring that extends from $R/2$ to R. The potential of a disk is given by Eq. 24–37: $V = (\sigma/2\epsilon_0)(\sqrt{z^2 + r^2} - z)$, where r is the radius of the disk and σ is its surface charge density.

[Ans: (a) 1.48 nC; (b) 795 V]

95 There are three pairs of particles and the particles in each pair are separated by L ($= 2.82 \times 10^{-15}$ m). The potential energy of a pair is $q^2/4\pi\epsilon_0 L$, where $q = -e/3$. Sum the energies of the pairs.

[Ans: (a) 2.72×10^{-14} J; (b) 3.02×10^{-31} kg, about 1/3 of accepted value]

103 (a) The electric potential at the point with coordinates x and y is

$$V = \frac{q_1}{4\pi\epsilon_0 \sqrt{x^2 + y^2}} + \frac{q_2}{4\pi\epsilon_0 \sqrt{(x-d)^2 + y^2}}.$$

Put $V = 0$ and algebraically manipulate the resulting equation until you put it into the form $(y - y_c)^2 + (x - x_c)^2 = R^2$. This is the equation of a circle with radius R and center at x_c, y_c.

(b) You can show that the $v = 5\,\text{V}$ equipotential line is not a circle by finding where it crosses the x axis (put $y = 0$ and solve for x) and where it crosses the y axis (put $x = 0$ and solve for y). Compare the two values.

[Ans: (a) $-4.8\,\text{nm}$; (b) $8.1\,\text{nm}$; (c) no]

105 If there are N electrons on a grain then the charge of a grain is $-Ne$ and the electric potential at its surface is $V = -Ne/4\pi\epsilon_0 R$, where R is the radius of the grain. Solve for N.

[Ans: 2.8×10^5]

109 The electric potential a distance r from a charged point particle is given by $V = q/4\pi\epsilon_0 r$, where q is the charge of the particle. All points that are the same distance from the particle are at the same potential. The separation dr of equipotential surfaces that differ in potential by dV is

$$dr = \frac{1}{dV/dr} dV.$$

The separation is uniform if dV/dr is independent of r.

[Ans: (a) spherical, centered on q, radius $4.5\,\text{m}$; (b) no]

Quiz

Some questions might have more than one correct answer.

1. Which of the following statements are true?
    A. only changes in the electric potential as the charges creating it move have physical significance, not the beginning and ending values themselves
    B. only differences in the values of the electric potential at different places have physical significance, not the values themselves
    C. the change in the electric potential energy of a collection of charged particles is the product of the net charge and the change in the electric potential
    D. the electric potential is valid only if a test charge is present
    E. the electric potential at any point is the algebraic sum of the electric potentials produced at that point by all charges

2. The electric potential of a charged point particle, relative to the electric potential far away, is proportional to

    A. the distance from the particle
    B. the square of the distance from the particle
    C. the reciprocal of the distance from the particle
    D. the reciprocal of the square of the distance from the particle
    E. the square root of the distance from the particle

3. Two particles, each with charge q, are each at a vertex of an equilateral triangle with sides of length L. The electric potential at the third vertex, relative to the electric potential far away, is

    A. $q/4\pi\epsilon_0 L^2$
    B. $2q/4\pi\epsilon_0 L$
    C. $q^2/4\pi\epsilon_0 L$
    D. $q^2/4\pi\epsilon_0 L^2$
    E. $(2q/4\pi\epsilon_0 L^2)\sin 60°$

4. Equipotential surfaces

    A. might intersect
    B. are always spheres
    C. are always planes
    D. are perpendicular to electric field liens
    E. are tangent to electric field lines

5. The electric potential

    A. has the same value at all points inside a conductor, no matter what the net charge on the conductor
    B. has the same value at any two points that are the same distance from a charged point particle if there is no other charge
    C. has the same value at all points outside a uniformly charged nonconducting disk
    D. has the same value at all points inside a uniformly charged spherical insulator
    E. has the same value at all points on a plane that is parallel to a uniformly charged plane sheet

6. At points far from an electric dipole, the electric potential due to the dipole, relative to the electric potential at infinity, is proportional to

    A. the distance from the particle
    B. the square of the distance from the dipole
    C. the reciprocal of the distance from the dipole
    D. the reciprocal of the square of the distance from the dipole
    E. the square root of the distance from the dipole

7. The difference between the values of the electric potential at any two points can be found by

    A. differentiating the electric field with respect to the coordinates of the points
    B. differentiating the electric field with respect to the charge in the region
    C. integrating the perpendicular component of the electric field along any path between the points
    D. integrating the tangential component of the electric field along any path between the points
    E. multiplying the electric field by the distance between the points

8. A spherical conductor of radius R has charge Q uniformly distributed on its surface. Relative to the electric potential far away, the electric potential at a point inside the conductor a distance r from its center is

    A. zero
    B. $Q/4\pi\epsilon_0 r$
    C. $Q/4\pi\epsilon_0 r^2$
    D. $Qr/4\pi\epsilon_0$
    E. $Q/4\pi\epsilon_0 R$

9. The electric potential energy of two charged point particles

    A. increases as their separation increases if the charges have the same sign
    B. increases as their separation increases if the charges have opposite signs
    C. decreases as their separation increases if the charges have the same sign
    D. decreases as their separation increases if the charges have opposite signs
    E. does not change as their separation increases

10. A particle with charge q is released from rest at a point where the electric potential (due to all other particles) is V_1. When it gets to a point where the electric potential is V_2 its kinetic energy is

    A. $q(V_2 - V_1)$
    B. $q(V_1 - V_2)$
    C. $(V_2 - V_1)$
    D. $(V_1 - V_2)$
    E. $(V_2 - V_1)/q$

11. Two conducting spheres are far apart. The smaller sphere has radius R_1 and charge q_1. The larger sphere has radius R_2 and charge q_2, which is less than q_1. They are then connected by a conducting wire. Afterwards

    A. their charges are equal
    B. the electric potentials at their surfaces are equal
    C. the sum of their charges is equal to $q_1 + q_2$
    D. their surface charge densities are equal
    E. the electric field at their surfaces are equal

12. An external agent brings a particle with charge Q from far away, where the electric potential is zero, to the center of a fixed charge distribution, where the electric potential is V. The particle starts and ends at rest. Which of the following statements are true?

   A. the work done by the agent is $+QV$
   B. the work done by the agent is $-QV$
   C. the work done by the electric field of the fixed charge distribution is $+QV$
   D. the work done by the electric field of the fixed charge distribution is $-QV$
   E. the potential energy of the system consisting of the fixed charge distribution and the particle changes by $+QV$

Answers: (1) A, B, E; (2) C; (3) B; (4) D; (5) A, B, E; (6) D; (7) D; (8) E; (9) B, C; (10) A; (11) B, C; (12) A, D, E

Chapter 25
CAPACITANCE

Capacitors are electrical devices that are used to store charge and electrical energy and, as you will see in a later chapter, are important for the generation of electromagnetic oscillations. This chapter is an excellent review of the principles you have learned in previous chapters: you will make extensive use of Gauss' law and the concepts of electric potential and electric potential energy.

Important Concepts

- ☐ capacitor
- ☐ capacitance
- ☐ equivalent capacitance

- ☐ energy in a capacitor
- ☐ electric energy density
- ☐ dielectric

Overview

25–2 Capacitance

☐ A **capacitor** is simply two conductors, isolated from each other. Each conductor is called a **plate**. The symbol used to represent a capacitor in an electrical circuit is ⊣⊢.

☐ In normal use, one plate holds positive charge and the other holds negative charge of the same magnitude. The charge creates an electric field, pointing roughly from the positive plate toward the negative plate, and because an electric field exists in the region between the plates, the positive plate is at a higher electric potential than the negative.

☐ The potential difference of the plates and the charge on either one are proportional to each other. If the magnitude of the potential difference is V and the magnitude of the charge on either plate is q, then $q = CV$ defines the **capacitance** C of the capacitor. The capacitance is independent of both q and V. It does, however, depend on the geometry of the capacitor. When a capacitor is charged by transferring charge from one plate to the other, the capacitance is a measure of how much is transferred for a given potential difference.

☐ A capacitor can be charged by connecting one plate to the positive terminal of a battery and the other plate to the negative terminal. The battery pulls electrons from one plate and deposits them on the other until the potential difference between the plates is the same as the potential difference between the battery terminals.

☐ The SI unit for capacitance is a coulomb/volt. This unit is called a farad and is abbreviated F. Microfarad (abbreviated μF) and picofarad (abbreviated pF) capacitors are commonly used in electronic circuits.

25-3 Calculating the Capacitance

☐ To calculate capacitance, first imagine charge q is placed on one plate and charge $-q$ is placed on the other. Use Gauss' law to find an expression for the electric field in the region between the plates. The field at every point is proportional to q. Use $V = -\int \vec{E} \cdot d\vec{s}$ to compute potential difference V of the plates. V is also proportional to q. Finally, use $q = CV$ to find an expression for the capacitance.

☐ Suppose a capacitor consists of two parallel metal plates, each of area A, separated by a distance d. If charge q is paced on one and charge $-q$ is placed on the other, the electric field between the plates is $E = q/\epsilon_0 A$, the potential difference of the plates is $V = Ed = qd/\epsilon_0 A$, and the capacitance is $C = q/V = \epsilon_0 A/d$. The capacitance depends on the permittivity constant ϵ_0, the plate area A, and the plate separation d.

☐ Suppose a capacitor consists of two concentric cylindrical shells with length L, the inner one with radius a and the outer one with radius b. Charge q is placed on the inner cylinder and charge $-q$ is placed on the outer cylinder. The electric field between the cylinders, a distance r from the axis, is $E = q/2\pi\epsilon_0 Lr$, the potential difference of the cylinders is $V = (q/2\pi\epsilon_0 L)\ln(b/a)$, and the capacitance is $C = 2\pi\epsilon_0 L/\ln(b/a)$. C depends only on ϵ_0, a, b, and L.

☐ Suppose a capacitor consists of two concentric spherical shells, with radii a and b. Charge q is placed on the inner shell and charge $-q$ is placed on the outer shell. The electric field between the shells, a distance r from the sphere centers, is $E = q/4\pi\epsilon_0 r^2$, the potential difference of the shells is $V = q(b-a)/4\pi\epsilon_0 ab$, and the capacitance is $C = 4\pi\epsilon_0 ab/(b-a)$.

☐ Capacitance is also defined for a single conductor. The other plate is assumed to be infinitely far away. To find an expression for the capacitance, imagine charge q is on the conductor, find an expression for the electric field outside the conductor, calculate the potential V of the conductor relative to the potential at infinity, and use $q = CV$ to find the capacitance. For example, the capacitance of a spherical conductor of radius R is $C = 4\pi\epsilon_0 R$.

25-4 Capacitors in Parallel and in Series

☐ The potential difference is the same for all capacitors in a parallel connection; the charge is the same for all capacitors in a series connection. If neither the potential difference nor the charge is the same, then the capacitors do not form either a parallel or series combination.

☐ The capacitors in either a series or a parallel connection can be replaced by a single capacitor with capacitance C_{eq} such that when the potential difference is the same as the potential difference across the original combination then the charge transferred is the same as the total charge transferred for the original combination. If the individual capacitances are C_1 and C_2, then the value of C_{eq} for a parallel combination is $C_{eq} = C_1 + C_2$ and the value of C_{eq} for a series combination is $C_{eq} = C_1 C_2/(C_1 + C_2)$.

☐ For a parallel combination, the equivalent capacitance is greater than the greatest capacitance in the combination; for a series combination, the equivalent capacitance is less than the smallest capacitance in the combination.

25–5 Energy Stored in an Electric Field

☐ As a capacitor is being charged, energy must be supplied by an external source, such as a battery, and energy is stored by the capacitor. You may think of the stored energy in either one of two ways: as the potential energy of the charge on the plates or as an energy associated with the electric field produced by the charges.

☐ Suppose that, at one stage in the charging process, the positive plate of a capacitor has charge q' and the potential difference across the plates is V'. If an additional infinitesimal charge dq' is taken from the negative plate and placed on the positive plate, the energy is increased by $dU = V' dq'$ or, what is the same, by $dU = (q'/C) dq'$, where C is the capacitance. This expression is integrated from 0 to the final charge q to obtain $U = q^2/2C$. Since $q = CV$ it is also given by $U = CV^2/2$.

☐ The energy density (or energy per unit volume) in a parallel plate capacitor is given by $u = \frac{1}{2}\epsilon_0 E^2$, where E is the magnitude of the electric field between the plates. This expression is quite generally valid for any electric field, not only those in capacitors. Most fields are functions of position so a volume integral must be evaluated to calculate the energy required to produce the field: $U = \frac{1}{2}\epsilon_0 \int E^2 \, dV$.

☐ If, after charging a capacitor, the plates are connected by a conducting wire, then electrons flow from the negative to the positive plate until both plates are neutral and the electric field vanishes. The stored energy is converted to kinetic energy of motion and eventually to internal energy in resistive elements of the circuit.

25–6 Capacitor with a Dielectric

☐ If the space between the plates of a capacitor is filled with insulating material, the capacitance is greater than if the space is a vacuum. If the space is completely filled, the capacitance is given by $C = \kappa C_0$, where C_0 is the capacitance of the unfilled capacitor and κ is the **dielectric constant** for the insulator. This last quantity is a property of the insulator, is unitless, and is always greater than unity.

☐ The energy stored in a capacitor with a dielectric is $U = \frac{1}{2}q^2/C = \frac{1}{2}CV^2$, where q is the magnitude of the charge on either plate and V is the potential difference. As a dielectric is inserted between the plates of a capacitor with the potential difference held constant (by a battery, say), both the charge on the positive plate and the stored energy increase. For the potential difference to remain the same, the battery must transfer charge as the dielectric is inserted. As a dielectric is inserted with the capacitor in isolation, so the charge cannot change, both the potential difference and the stored energy deceases. In either case, if the dielectric is inserted by an external agent, the difference in energy is associated with the work done by the agent and battery, if present.

25–7 Dielectrics: An Atomic View

☐ **Polarization** of the dielectric by the electric field causes an increase in capacitance. If the dielectric is composed of polar molecules, they rotate to align with the electric field between the plates. If the dielectric is composed of non-polar molecules, dipoles are induced on the molecules by the field.

☐ Electric dipoles in a polarized dielectric produce an electric field that is directed opposite to the field produced by the charge on the conducting plates. Thus the total field is weaker than the field produced by charge on the plates alone. It is, in fact, weaker by the factor κ. That is, if $\vec{E}_0$ is the field produced by the charge on the plates, then the total field is given by $E = E_0/\kappa$. If the electric field produced by charge on the plates is uniform, as it essentially is between the plates of a parallel plate capacitor, then the dipole field and the total field are also uniform.

☐ Since the electric field for a given charge on the plates is weaker if the capacitor is filled with a dielectric than if it is not, the potential difference is less when the dielectric is present. Since $q = CV$, this means the capacitance is larger when the dielectric is present.

25–8 Dielectrics and Gauss' Law

☐ For a parallel plate capacitor, the effect of a polarized dielectric is exactly the same as a uniform distribution of positive charge q' on the dielectric surface nearest the negative plate and a uniform distribution of negative charge $-q'$ on the dielectric surface nearest the positive plate. If the dielectric constant of the dielectric is κ and the charge on the positive plate is q, then $q' = (\kappa - 1)q/\kappa$. If the plates have area A, then Gauss' law can be used to show that the electric field due to the charge on the plates is $q/\epsilon_0 A$, the field due to the dipoles is $q'/\epsilon_0 A = (\kappa - 1)q/\kappa\epsilon_0 A$, and the total field is $E = (q/\epsilon_0 A) - [(\kappa - 1)q/\kappa\epsilon_0 A] = q/\kappa\epsilon_0 A$.

Hints for Questions

1 The capacitance is the slope of the graph of the charge as a function of the potential difference and since the capacitors are all parallel-plate capacitors the capacitance is also given by $\epsilon_0 A/d$, where A is the plate area and d is the plate separation.

[Ans: a, 2; b, 1; c, 3]

3 The charge on any capacitors is the same as the charge on any other and is the same as the charge on the equivalent capacitor. The potential difference across a capacitor is the charge divided by the capacitance. The net charge through the meter during charging is the same as the charge on any one of the capacitors.

[Ans: (a) $V/3$; (b) $CV/3$; (c) $CV/3$]

5 Two capacitors are in series if their charges are the same and are in parallel if their potential differences are the same.

[Ans: (a) no; (b) yes; (c) all tie]

7 After the second capacitor is added the potential difference across each capacitor is the same as the potential difference across the battery. The charge on the capacitor is the product of the potential difference and the capacitance. The equivalent capacitance of two capacitors in parallel is the sum of the capacitances.

[Ans: (a) same; (b) same; (c) more; (d) more]

9 Charge is conserved, so the total charge on the two capacitors is $9q$ for each circuit. After the switch is closed the potential differences across the capacitors are the same. These are given by the charge divided by the capacitance.

[Ans: (a) 2; (b) 3; (c) 1]

<u>11</u> When the dielectric slab is inserted the capacitance becomes κC_0, where κ is the dielectric constant of the slab and C_0 is capacitance before the slab is inserted. κ is greater than 1. The charge on each capacitor is the same as the charge on the equivalent capacitor. Figure out if the equivalent capacitance increases, decreases, or remains the same. The potential difference across each capacitor is the charge divided by the capacitance. The stored energy is half the product of the square of the charge and the reciprocal of the capacitance.

[Ans: (a) increase; (b) increase; (c) decrease; (d) decrease; (e) capacitance: same, charge: increase, potential difference: increase, potential energy: increase]

Hints for Problems

<u>1</u> The final potential difference across the capacitor is the same as the potential difference across the battery. The charge that passes through the switch is the same as the final charge on the capacitor, which is the product of the capacitance and the potential difference.

[Ans: 3.0 mC]

<u>9</u> C_1 and C_2 are in parallel and the combination is in series with C_3. First find the equivalent capacitance C_{12} of C_1 and C_2, then the equivalent capacitance of C_{12} and C_3.

[Ans: 3.16 μF]

<u>11</u> First find the original potential difference across C_1. The original charge on C_1 is $q_1 = C_{123}V$, where C_{123} is the equivalent capacitance of the circuit, and the potential difference is $V_1 = q_1/C_{12}$, where C_{12} is the equivalent capacitance of C_1 and C_2. After breakdown the potential difference across C_1 is V. If ΔV_1 is the change in the potential difference across C_1 then $\Delta q_1 = C_1 \Delta V_1$ is the change in its charge.

[Ans: (a) 790 μC; (b) 78.9 V]

<u>15</u> C_3 and C_5 are in series and this combination is in parallel with C_2 and C_4. C_1 and C_6 are in parallel and this combination is in series with the combination of C_3, C_5, C_2, and C_4. The charge stored by the equivalent capacitor is $C_{eq}V$. Find the charge and potential difference of any of the capacitors by remembering that the potential differences are the same across two capacitors in parallel and that the charges are the same for two capacitors in series. In addition, the potential difference for a parallel connection is the same as the potential difference across the equivalent capacitor and the charge on each capacitor of a series connection is the same as the charge for the equivalent capacitor.

[Ans: (a) 3.00 μF; (b) 60 μC; (c) 10 V; (d) 30.0 μC; (e) 10 V; (f) 20.0 μC; (g) 5.00 V; (h) 20.0 μC]

<u>19</u> The total charge that passes through point a is the sum of the charges on capacitors 3 and 4. The total charge that passes through point b is the charge on capacitor 4. The potential difference across capacitor 4 is the charge through b divided by C_4 and this is the same as the potential difference across capacitor 3. You now know the charge and potential difference of capacitor 3 and so can compute its capacitance. The equivalent capacitance of the circuit is

the charge through a divided by the potential difference across the battery. Write an expression for the equivalent capacitance of the circuit in terms of the individual capacitances and solve for C_1.

[Ans: (a) 4.0 μF; (b) 2.0 μF]

23 If only switch 1 is closed then C_1 and C_3 are in series and C_2 and C_4 are also in series. The two combinations are in parallel. If both switches are closed then C_1 and C_2 are in parallel and C_3 and C_4 are also in parallel. The two combinations are in series. Remember that the potential differences are the same across two capacitors in parallel and that this potential difference is the same as the potential difference across the equivalent capacitor. Also remember that the charges are the same for two capacitors in series and that this charge is the same as the charge on the equivalent capacitor.

[Ans: (a) 9.00 μC; (b) 16.0 μC; (c) 9.00 μC; (d) 16.0 μC; (e) 8.40 μC; (f) 16.8 μC; (g) 10.8 μC; (h) 14.4 μC]

25 The energy stored in a capacitor is $\frac{1}{2}CV^2$, where C is its capacitance and V is the potential difference across its plates. You need to convert 10 kW $\cdot$ h to joules.

[Ans: 72 F]

29 The energy density of an electric field of magnitude E is $\epsilon_0 E^2/2$ and the magnitude of the field of an electron a distance r from the particle is $E = (1/4\pi\epsilon_0)e/r^2$.

[Ans: (a) 9.16×10^{-18} J/m^3; (b) 9.16×10^{-6} J/m^3; (c) 9.16×10^6 J/m^3; (d) 9.16×10^{18} J/m^3; (e) ∞]

33 The three capacitors each carry the same charge, so the one with the smallest capacitance has the greatest potential difference. Take this to be 100 V. Calculate the charge on this capacitor (and hence on each of the others) and then the potential differences across the other capacitors. The potential difference between points A and B is the sum of the potential differences across the capacitors.

[Ans: (a) 190 V; (b) 95 mJ]

41 The magnitude of the electric field between the plates is $E = V/d$, where V is the potential difference and d is the plate separation. Solve for V. The energy stored is $\frac{1}{2}CV^2$, where C is the capacitance. The capacitance of a parallel-plate capacitor is given by $C = \epsilon_0 A/d$, where A is the area of a plate.

[Ans: 66 μJ]

49 The capacitance of the capacitor before the slab is inserted is $C = \epsilon_0 A/d$, where A is a plate area and d is the plate separation. You may think of the capacitor after the slab is inserted as two capacitors in series. One has capacitance $\epsilon_0 A/(d - b)$ and the other has capacitance $\kappa\epsilon_0 A/b$, where b is the thickness of the slab. The charge on the capacitor before the slab is inserted is $q = CV$, where V is the potential difference across the battery and, since the battery is then disconnected, no charge leaves the capacitor as the slab is inserted. The electric field between a plate and the dielectric is given by $E = q/\epsilon_0$ and the electric field in the dielectric is given by E/κ. The potential difference across the plates is the charge on a plate divided by the capacitance. The work done in inserting the slab is the change in the energy stored by the capacitor, calculated using $U = \frac{1}{2}q^2/C$.

[Ans: (a) 89 pF; (b) 0.12 nF; (c) 11 nC; (d) 11 nC; (e) 10 kV/m; (f) 2.1 kV/m; (g) 88 V; (h)

$-0.17\,\mu\text{J}\,\big]$

51 The energy stored in a capacitor is $\frac{1}{2}CV^2$, where C is the capacitance and V is the potential difference across its plates. These capacitors are identical and are connected in series. This means that the potential difference across the plates is the same for all of them and is equal to V_{battery}/n, where V_{battery} is the potential difference across the battery terminals. The total energy is $\frac{1}{2}nC(V_{\text{battery}}/n)^2$.

$\big[$Ans: 4 $\big]$

57 (a) Try a series connection of n capacitors. The equivalent capacitance is C/n, where C is the capacitance of one of capacitors. The combination can withstand $n(200\,\text{V})$.
(b) Try connecting m of the series combinations of part (a) in parallel. The equivalent capacitance is mC_n, where C_n is the equivalent capacitance of n capacitors in series. The combination can withstand the same potential difference as in part (a).

$\big[$Ans: (a) five capacitors in series; (b) one possible answer: three rows in parallel, each row containing five capacitors in series $\big]$

61 The upper two capacitors are in parallel. Replace them with their equivalent capacitor, with capacitance C_{12}, say. This equivalent capacitor is in series with the bottom capacitor. replace them with their equivalent capacitor, with capacitance C_{123}, say. The charge on C_{123} is the same as the charge on the lower capacitor and is $q = C_{123}V$. The potential difference across the bottom capacitor is q/C, where C is its capacitance.

$\big[$Ans: (a) 24.0 μC; (b) 6.00 V $\big]$

65 (a) Use $q = CV$. The potential difference across capacitor 1 is 10.0 V.
(b) Reduce the lower portion of the circuit to an equivalent capacitor with the battery across its plates. Find the charge on the equivalent capacitor. This the same as the charge on the combination of three capacitors in the lower right corner, considered as a single equivalent capacitor. Since you know the charge you can find the potential difference. This is also the potential difference across the two capacitors in series in that corner, so you can find the charge on their equivalent capacitor. That is the same as the charge on capacitor 2.

$\big[$Ans: (a) 100 μC; (b) 20.0 μC $\big]$

67 There are $n-1$ identical capacitors connected in parallel, each of which has capacitance $C = \epsilon_0 A/d$, where A is the area of overlap and d is the plate separation. For maximum capacitance the plates have maximum overlap.

$\big[$Ans: 2.28 pF $\big]$

77 The energy stored in a capacitor is given by $U = \frac{1}{2}CV^2$, where C is the capacitance and V is the potential difference across the plates. After the potential difference is increased by ΔV the energy is $U + \Delta U = \frac{1}{2}C(V + \Delta V)^2$. Find an expression for $(\Delta U)/U$ in terms of $(\Delta V)/V$ and solve for $(\Delta V)/V$.

$\big[$Ans: 4.9% $\big]$

83 Think of this as two parallel-plate capacitors, each with a plate separation of $(d-b)/2$. The stored energy is $CV^2/2$. Use $C = A\epsilon_0/d$ for the capacitance before the slab is in place and $C = A\epsilon_0/(d-b)$ for the capacitance after it is in place. The work done is the change in the stored energy. Since the charge on the plates changes work is done both by the battery and by the agent placing the slab. The work done by the battery is $(\Delta q)V$, where Δq is the change in the capacitor charge. Write $\Delta U = W_{battery} + W_{agent}$ and calculate W_{agent}. If it is positive the slab was pushed in; if it is negative the slab is sucked in and the agent had to pull back on it.

[Ans: (a) 0.708 pF; (b) 0.600; (c) 1.02×10^{-9} J; (d) sucked in]

Quiz

Some questions might have more than one correct answer.

1. If the potential difference between the two plates of a capacitor is doubled
   A. the capacitance is doubled
   B. the capacitance is halved
   C. the total charge on the two plates is doubled
   D. the total charge on the two plates is halved
   E. the charge on each of the plates is doubled

2. The capacitance of a parallel-plate capacitor is proportional to
   A. the area of a plate
   B. the reciprocal of the area of a plate
   C. the plate separation
   D. the reciprocal of the plate separation
   E. the charge on the positive plate

3. The capacitances of two capacitors are $2\,\mu F$ and $4\,\mu F$. They are wired in parallel and a potential difference of 6 V is applied across the combination. As a result
   A. the potential difference across the plates of the 2-μF capacitor is 6 V
   B. the potential difference across the plates of the 4-μF capacitor is 6 V
   C. the charge on the positive plate of the 2-μF capacitor is $12\,\mu C$
   D. the charge on the positive plate of the 2-μF capacitor is $2\,\mu C$
   E. the charge on the positive plate of the 2-μF capacitor is $1\,\mu, C$

4. The capacitances of two capacitors are $2\,\mu F$ and $4\,\mu F$. They are wired in series and a potential difference of 6 V is applied across the combination. As a result
   A. the potential difference across the plates of the 2-μF capacitor is 6 V
   B. the potential difference across the plates of the 4-μF capacitor is 4 V
   C. the charge on the positive plate of the 2-μF capacitor is $12\,\mu C$
   D. the charge on the positive plate of the 2-μF capacitor is $8\,\mu C$
   E. the charge on the positive plate of the 4-μF capacitor is $8\,\mu C$

5. The electric field in a charged capacitor is produced by
    A. the battery used to transfer charge from one plate to the other
    B. charge in the wires that connect the capacitor to the battery
    C. charge on the capacitor plates
    D. charge that is external to both the capacitor and the battery
    E. none of the above

6. The electrical energy stored in a charged capacitor is proportional to
    A. the charge on the positive plate
    B. the square of the charge on the positive plate
    C. the potential difference of the plates
    D. the square of the potential difference between the plates
    E. the reciprocal of the square of the potential difference between the plates

7. A parallel-plate capacitor is wired to a battery. If the plate separation is increased
    A. the electrical energy stored in the capacitor increases if the battery is connected while the plate separation is increased
    B. the electrical energy stored in the capacitor decreases if the battery is connected while the plate separation is increased
    C. the electrical energy stored in the capacitor increases if the battery is disconnected while the plate separation is increased
    D. the electrical energy stored in the capacitor decreases if the battery is disconnected while the plate separation is increased
    E. the electrical energy stored in the capacitor does not change

8. The energy stored in an electrostatic field is
    A. equal to the work that must be done by an agent to bring the charges that produce the field from well-separated positions to their final positions
    B. equal to the volume integral of the electrical energy density over all space
    C. zero
    D. proportional to the volume integral of the electric field over all space
    E. proportional to the volume integral of the square of the electric field over all space

9. Two parallel-plate capacitors have the same plate area and the same plate separation but one has a dielectric between the plates and the other has a vacuum. The same potential difference is applied across the plates. As a result,
    A. the capacitor with the dielectric has the greater charge on its positive plate
    B. the two capacitors have the same charge on their positive plates
    C. the capacitor with the dielectric has the greater stored electrical energy
    D. the two capacitors have the same stored electrical energy
    E. the capacitor with the dielectric has the greater magnitude electric field between its plates

10. Two parallel-plate capacitors have the same plate area and the same plate separation but one has a dielectric between the plates and the other has a vacuum. The capacitors have the same charge. As a result,

    A. the capacitor with the dielectric has the smaller potential difference across its plates
    B. the two capacitors have the same potential difference across their plates
    C. the capacitor with the dielectric has the smaller stored electrical energy
    D. the two capacitors have the same stored electrical energy
    E. the capacitor with the dielectric has the smaller magnitude electric field between its plates

Answers: (1) E; (2) A, D; (3) A, B, C; (4) D, E; (5) C; (6) D; (7) B; (8)A, B, E; (9) A, C; (10) A, C, E

Chapter 26
CURRENT AND RESISTANCE

Here you begin the study of electric current, the flow of charged particles. Pay close attention to the definitions of current and current density and understand how they depend on the concentration and speed of the particles. For most materials, a current is established and maintained only when an electric field is present. You will learn about the properties of materials that determine the current for any given field.

Important Concepts

☐ electric current
☐ current density
☐ resistance
☐ resistor
☐ resistivity
☐ conductivity

☐ temperature coefficient
   of resistivity
☐ mean free time
☐ energy dissipated by a resistor
☐ semiconductor
☐ superconductor

Overview

26–2   Electric Current

☐ When a material object, such as a wire, contains a current, it is usually electrons that flow, not atomic nuclei. However, current from the sun consists of both electrons and protons and cosmic rays are chiefly highly energetic protons.

☐ Positively and negatively charged particles moving in the same direction result in currents in opposite directions. If the two currents have the same magnitude, the net current is zero. There is no net transport of charge. Thus, a neutral moving object consists of moving electrons and protons but does not constitute an electric current. The random motion of electrons in a piece of metal do not constitute a current because at any instant just as many are moving in any one direction as there are moving in the opposite direction. Again, there is no net transport of charge. When a battery is connected across the metal, an organized motion, opposite the electric field, is superimposed on the random motion of the electrons. Charge is now transported and there is a current.

☐ The current i through any cross-sectional area of a conducting wire is given by

$$i = \frac{dq}{dt},$$

where dq is the net charge that passes through the area in the time interval dt. Although current is a scalar, it is assigned a direction. Positive charge moving to the right and negative charge moving to the left are both currents to the right. For steady flow of charge through a conductor, the current is the same for all cross sections, no matter how the cross-sectional area differs along the wire, because equal charge passes every cross section in equal times.

☐ The SI unit for current is the coulomb/second. This unit is called an ampere and is abbreviated A.

26–3 Current Density

☐ Current is associated with an area, like the cross-sectional area of a conductor. On the other hand, **current density** $\vec{J}$ is a related quantity that can be associated with each *point* in a conductor. At any point, it is the current per unit area through an area at the point, oriented perpendicularly to the particle flow, in the limit as the area shrinks to the point. Current density is a vector. If positively charged particle are flowing, $\vec{J}$ is in the direction of their velocity; if negatively charged particles are flowing, $\vec{J}$ is opposite the direction of their velocity.

☐ Consider any surface within a current-carrying conductor and let $d\vec{A}$ be an infinitesimal vector area, with direction perpendicular to the surface. If $\vec{J}$ is the current density, then the current i through the surface is given by the integral over the surface:

$$i = \int \vec{J} \cdot d\vec{A}.$$

The scalar product indicates that only the component of $\vec{J}$ normal to the surface contributes to the current through the surface. If the surface is a plane with area A and is perpendicular to the particle velocity and if the current density is uniform over the surface, then $i = JA$.

☐ For a collection of particles with uniform concentration n, the current density is given by the vector relationship

$$\vec{J} = en\vec{v}_d,$$

where e is the charge on any one of the particles and $\vec{v}_d$ is the average velocity of the particles (the drift velocity). This expression can be derived by considering the number of particles per unit time that pass through an area perpendicular to their velocity. $\vec{J}$ and $\vec{v}_d$ are in opposite directions for negatively charged particles.

☐ The velocity of an electron can be considered to be the sum of two velocities, one of which changes direction often as the electron collides with atoms. When all electrons are taken into account, this velocity averages to zero and so it does not contribute to the current. The second component, the drift velocity, is much smaller in magnitude than the first. This is the velocity that enters the expression for the current density, not the total velocity.

☐ In an ordinary conductor, an electric field is required to produce a net drift and hence an electric current. For electrons, the direction of the drift velocity is opposite the direction of the electric field. This means the current is in the same direction as the field and it is directed from a region of high electric potential toward a region of lower electric potential.

☐ That an electric field can be maintained in a conductor does not contradict the statement made earlier that the electrostatic field in a conductor is zero. A conductor containing a current is not in electrostatic equilibrium.

26–4 Resistance and Resistivity

☐ **Resistance** is a measure of the current generated in a given conductor by a given potential difference. To determine resistance, a potential difference V is applied between two points in the material and the current i is measured. The resistance for that material and those points of application is defined by

$$R = \frac{V}{i}.$$

The resistance depends on where the current leads are attached, as well as on the material and its geometry. Resistance has an SI unit of volt/ampere, a unit that is called an ohm and is abbreviated Ω. In drawing an electrical circuit, an element whose function is to provide resistance, called a **resistor**, is indicated by the symbol —⋀⋀—.

☐ Resistance is intimately related to a property of the material called its **resistivity**. If, at some point in the material the electric field is $\vec{E}$ and the current density is $\vec{J}$, then the resistivity ρ at that point is defined by $\vec{E} = \rho \vec{J}$. The resistivities of the materials we consider are the same at every point in the material and are the same for every orientation of the electric field. The SI unit for resistivity is the ohm·meter ($\Omega \cdot$ m).

☐ Typical semiconductors have resistivities that are greater than those of metals by factors of 10^5 to 10^{11} and insulators have resistivities that are greater than those of metals by factors of 10^{18} to 10^{24}. The **conductivity** σ of any substance is related to the resistivity by $\sigma = 1/\rho$.

☐ Let L be the length of a homogeneous wire with uniform cross section, A be its cross-sectional area, and ρ be its resistivity. If one end of the wire is held at potential 0 and the other is held at potential V, the electric field in the wire is given by $E = V/L$. If the current is i then, since the current density is uniform, it is given by $J = i/A$. Substitute $E = V/L$ and $J = i/A$ into $E = \rho J$ and solve for V/i $(= R)$. The result is $R = \rho L/A$. The resistance depends on a property of the material (ρ) and on the geometry of the sample (L and A).

☐ The resistivity of a metal increases with increasing temperature, chiefly because electrons suffer more collisions per unit time at high temperatures than at low temperatures. The temperature dependence is characterized by a quantity α, called the **temperature coefficient of resistivity**, which is a measure of the deviation of the resistivity from its value at a reference temperature. Let ρ_0 be the resistivity at the reference temperature T_0 and let ρ be the resistivity at a nearby temperature T. Then, in terms of α,

$$\rho - \rho_0 = \rho_0 \alpha (T - T_0).$$

Temperatures are given in kelvins or degrees Celsius.

☐ Table 26–1 lists values for some materials. Notice that α is negative for semiconductors. For them, the resistivity decreases as the temperature increases because the concentration n of nearly free electrons increases rapidly with temperature. For metals near room temperature, n is essentially independent of temperature.

26–5 Ohm's Law

☐ Ohm's law describes an important characteristic of the resistance of certain samples, called ohmic samples (or devices). It is: For an ohmic device, the current is proportional to the potential difference. The equation $V = iR$ defines the resistance R and so holds for every sample. For an ohmic sample, V is a *linear* function of i or, what is the same, R does not depend on V or i. In addition, for ohmic samples a reversal of the potential difference simply reverses the direction of the current without changing its magnitude. For some non-ohmic samples, such as a *pn* junction diode, a reversal of the potential difference not only changes the direction of the current but also its magnitude.

26–6   A Microscopic View of Ohm's Law

☐ For a substance that obeys Ohm's law, the resistivity is independent of the electric field and the current density is therefore proportional to the electric field. This implies that the drift velocity is also proportional to the field. Since an electric field accelerates charges, you should be surprised.

☐ To understand why Ohm's law is valid for some materials, consider the collection of free electrons in a conducting sample. These electrons are accelerated by the electric field applied to the sample and suffer collisions with atoms of the material. On average, the effect of a collision is to stop the drift of an electron so that, as far as drift is concerned, an electron starts from rest at a collision and is accelerated by the field until the next collision. The collision stops it and the process is repeated. The time between collisions, averaged over all electrons, is called the **mean free time** and is designated by τ. Because collisions occur at random times, τ is independent of the time; its value is the same no matter when the averaging is done.

☐ The magnitude of the acceleration is $a = qE/m$, where E is the magnitude of the electric field and m is the mass of an electron, so the average electron drift speed is given by $v_d = a\tau = e\tau E/m$. Although each electron accelerates between collisions, v_d does not vary with time if the field is constant. Also note that v_d is proportional to E if τ does not depend on E. The drift speed is such a small fraction of the electron's speed that the electric field does not appreciably alter its speed and, therefore, does not significantly change the mean time between collisions.

☐ The current density is given by $J = env_d = (e^2 n\tau/m)E$. This expression immediately gives $\rho = m/e^2 n\tau$ for the resistivity. If τ is independent of E, the resistivity is also independent of E and the material obeys Ohm's law.

☐ A long mean free time leads to a small resistivity because an electron is accelerated for a long time before it is stopped by a collision. The drift velocity is great so the resistivity is small. Similarly, a short mean free time implies a small drift velocity and a large resistivity.

☐ Because the mean time between collisions is determined to a large extent by thermal vibrations of the atoms, it is temperature dependent. As the temperature increases, an electron suffers more collisions per unit time, so τ becomes shorter as the temperature increases.

26–7   Power in Electric Circuits

☐ If current i passes through a potential difference V, from high to low potential, the moving charges lose potential energy at a rate given by $P = dU/dt = iV$. For a resistor, the energy is transferred to atoms of the material in collisions. The result is an increase in the internal energy of the resistor and is usually accompanied by an increase in temperature. The phenomenon finds practical application in toasters and electrical heaters.

☐ Since $V = iR$ for a resistor, the expression for the rate of energy **dissipation** can be written as $P = i^2R$ or as $P = V^2/R$.

26–8 Semiconductors

☐ The resistivities of semiconductors are greater than the resistivity of a typical metal but less than the resistivity of a typical insulator. The concentration of nearly free electrons is intermediate between the concentration for a conductor and the concentration for an insulator. The nearly free electron concentration in a semiconductor can be adjusted by the addition of certain impurities.

☐ The resistivity of a semiconductor, unlike that of a metal, decreases with increasing temperature. Although the mean free time decreases with increasing temperature, like that of a metal, the number of free electrons in the current increases dramatically as the temperature increases.

☐ The special properties of semiconductors make them useful in a wide variety of electronic devices.

26–9 Superconductors

☐ When the temperature of a superconductor is lowered below what is called its critical temperature, its resistivity becomes zero. Only certain materials become superconducting but they are of great technological value because they can carry large currents without heating. Electromagnets, motors, and magnetic field detectors make use of superconducting wires. Materials are being sought that turn superconducting at room temperatures and above, although none have been found as yet.

Hints for Questions

1 The net charge is the integral of the current with respect to time and so is the area under the curve of the current as a function of time.

[Ans: a, b, and c all tie, then d (zero)]

3 The resistance of a wire is proportional to its length and to the reciprocal of its cross-sectional area.

[Ans: b, a, c]

5 The electric field is given by the potential difference divided by the length. The current density is given by the electric field divided by the resistivity (which is the same for all the rods). The drift speed is proportional to the current density and to the reciprocal of the free electron concentration (which is the same for all the rods).

[Ans: (a) 1 and 2 tie, then 3; (b) 1 and 2 tie, then 3; (c) 1 and 2 tie, then 3]

7 The rate of generation of thermal energy is given by $P = V^2/R$, where V is the potential difference and R is the electrical resistance. The electrical resistance is proportional to the resistivity and to the length. It is also proportional to the reciprocal of the cross-sectional area but that is the same for all the wires.

[Ans: C, A, B]

9 Since the sections are connected in series the current is the same in all of them. The current density is the current divided by the cross-sectional area and the electric field is the product of the resistivity and the current density.

[Ans: (a) all tie; (b) B, C, A; (c) B, C, A]

Hints for Problems

1 The charge that passes through any cross section is the product of the current and time. It is also the product of the number of electrons that pass a cross section and the charge on an electron.

[Ans: (a) $1.1\,\mathrm{kC}$; (b) 7.5×10^{21}]

7 The cross-sectional area of wire is given by $A = \pi r^2$, where r is its radius. The magnitude of the current density is $J = i/A$. Solve for r and double the result to obtain the diameter.

[Ans: $0.38\,\mathrm{mm}$]

11 The current is the area integral of the current density over the cross-sectional area of the wire. An infinitesimal element at a distance r from the center has area $dA = 2\pi r\, dr$, so the current is $\int 2\pi r J\, dr$, where the limits of integration are 0 and R. To answer part (c) notice that the current density increases with r in part (a) and decreases with r in part (b).

[Ans: (a) $1.33\,\mathrm{A}$; (b) $0.666\,\mathrm{A}$; (c) J_a]

13 The conductivity is given by $\sigma = J/E$, where J is the magnitude of the current density and E is the magnitude of the electric field in the wire. The magnitude of the current density is the current divided by the cross-sectional area and the magnitude of the electric field is the potential difference divided by the length.

[Ans: $2.0 \times 10^6\,(\Omega \cdot \mathrm{m})^{-1}$]

25 Assume that changes with temperature of the length and cross-sectional area of the filament are negligible. Then the resistance changes because the resistivity changes. Use $R - R_0 = R_0 \alpha(T - T_0)$, where T_0 is room temperature, R_0 is the room temperature resistance, T is the operating temperature, R is the operating resistance, and α is the temperature coefficient of resistivity for tungsten (see Table 26–1). The operating resistance is the potential difference divided by the current under operating conditions.

[Ans: $1.9 \times 10^3\,^\circ\mathrm{C}$]

29 The current is the potential difference divided by the resistance. The magnitude of the current density is the current divided by the cross-sectional area. To calculate the drift speed v_d use $J = nev_d$, where J is the magnitude of the current density and n is the free-electron concentration. The magnitude of the electric field is the potential difference divided by the front-to-rear length.

[Ans: (a) $38.3\,\mathrm{mA}$; (b) $109\,\mathrm{A/m^2}$; (c) $1.28\,\mathrm{cm/s}$; (d) $227\,\mathrm{V/m}$]

31 Since the strands are in parallel the total current is the sum of the currents in the individual strands. Since the strands are identical and the potential differences are the same each strand carries the same current. The end-to-end potential difference is the product of the current in a strand and the resistance of a strand. The electrical resistance of the cable is the potential difference divided by the total current. (You might also use the expression for the resistance of a collection of resistors in parallel.)

$\left[\text{Ans: } 6.00\,\text{mA; (b) } 1.59 \times 10^{-8}\,\text{V; (c) } 21.2\,\text{n}\Omega\,\right]$

37 The rate of energy dissipation is given by $P = V^2/R$, where V is the potential difference across the resistor and R its resistance. For the same resistor connected to two different batteries the ratio of the dissipation rates is $P_1/P_2 = V_1^2/V_2^2$.

$\left[\text{Ans: } 0.135\,\text{W}\,\right]$

41 The rate of energy dissipation is given by $P = V^2/R$, where V is the potential difference across the wire and R is its resistance. The resistance of a wire is given by $R = \rho L/A$, where ρ is the resistivity, L is the length of the wire, and A is its cross-sectional area. Solve for L.

$\left[\text{Ans: (a) } 5.85\,\text{m; (b) } 10.4\,\text{m}\,\right]$

43 Multiply the power of the bulb by the duration of a month in hours and divide by 1000 to get the energy dissipated in kilowatt·hours. Finally, multiply by the cost of a kilowatt·hour to get the total cost. Use $P = V^2/R$, where P is the power of the bulb, V is the potential difference, and R is the resistance, to calculate the resistance of the bulb and $V = iR$ to calculate the current i.

$\left[\text{Ans: (a) \$4.46\,US; (b) } 144\,\Omega\text{; (c) } 0.833\,\text{A}\,\right]$

49 The electrical resistance of a wire is given by $R = \rho L/A$, where ρ is the resistivity, L is the length, and A is the cross-sectional area. The magnitude of the current density is $J = i/A$, where i is the current.

$\left[\text{Ans: (a) } 64\,\Omega\text{; (b) } 0.25\,\right]$

53 The upper terminal of the battery is the positive terminal and the battery pushes electrons internally from the positive to the negative terminal. The magnitude of the work done on the electron by the electric field in the strip is given by eV, where V is the potential difference across the strip. This the same as the potential difference across the battery terminals. All of the energy acquired by the electron is converted to thermal energy in collisions with atoms of the strip.

$\left[\text{Ans: (a) upward in the strip; (b) } 12\,\text{eV; (c) } 12\,\text{eV}\,\right]$

59 According to Ohm's law the current is $i = V/R$, where V is the applied potential difference and R is the resistance of the wire. The resistance is given by $R = \rho L/A$, where ρ is the resistivity of copper (see Table 26–1), L is the length of the wire, and A is its cross-sectional area. The magnitude of the current density is $J = i/A$ (assuming the distribution of current is uniform over a cross-section). The electric field $\vec{E}$ and current density $\vec{J}$ are related by $\vec{E} = \rho\vec{J}$. The rate of production of thermal energy in a resistor is $P = i^2R$ (or V^2/R).

$\left[\text{Ans: (a) } 1.74\,\text{A; (b) } 2.15\,\text{MA/m}^2\text{; (c) } 36.3\,\text{mV/m; (d) } 2.09\,\text{W}\,\right]$

65 The rate of dissipation of energy in the device is $P = iV$, where i is the current and V is the potential difference. The charge that passes through the device in time t is $q = it$.

[Ans: 28.8 kC]

73 The current density $\vec{J}$ is related to the electric field $\vec{E}$ by $\vec{E} = \rho\vec{J}$, where ρ is the resistivity. The magnitude of the electric field is $E = V/L$, where V is the potential difference across the length of the wire and L is the length of the wire. The field points from the high potential end toward the low potential end.

[Ans: (a) $1.5 \times 10^7 \, \text{A/m}^2$; (b) toward]

81 According to Ohm's law the potential difference is iR, where i is the current and R is the resistance of a length of wire equal to the caterpillar's length. The resistance is $R = \rho L/A$, where ρ is the resistivity (given in Table 26–1), L is the length of the caterpillar, and A is the cross-sectional area of the wire. Electrons move from regions of low electric potential toward regions of high electric potential. The magnitude of the current density is $J = i/A$ and is also nev_d, where n is the number of charge carriers per unit volume and v_d is the drift speed.

[Ans: (a) 0.38 mV; (b) negative; (c) 3 min 58 s]

Quiz

Some questions might have more than one correct answer.

1. When an electric field is introduced into an ionized gas 3.4×10^{17} electrons/s stream in one direction through an aperture and 7.9×10^{17} protons/s stream through the aperture in the opposite direction. The charge on an electron is -1.60×10^{-19} C and the charge on a proton is $+1.60 \times 10^{-19}$ C The total current through the aperture is

    A. 54 mA
    B. 72 mA
    C. 130 mA
    D. 180 mA
    E. 200 mA

2. A wire with radius R and length L carries current i. If the current density is uniform, its magnitude is

    A. $4\pi R^2 Li$
    B. $(4\pi/3)R^2 Li$
    C. $i/2\pi RL$
    D. $i/\pi RL$
    E. $i/\pi R^2$

3. The current density in a metal is proportional to

A. the average electron speed
B. the electron drift velocity
C. the concentration of electrons
D. the concentration of protons
E. the mass of an electron

4. The conductivity of a metal is

A. the reciprocal of the resistivity
B. the current density per unit electric field
C. the electric field per unit current density
D. equal to the resistance of the material
E. proportional to the reciprocal of the electric field magnitude

5. Which of the following statements are true?

A. the resistivities of metals increase with increasing temperature
B. the resistivities of metals decrease with increasing temperature
C. the resistivities of pure semiconductors increase with increasing temperature
D. the resistivities of pure semiconductors decrease with increasing temperature
E. the resistivities of most materials are independent of temperature

6. Which of the following statements are true?

A. the electrical resistance of every object is independent of the potential difference across it
B. the electrical resistance of an object depends on where the current enters and leaves the object
C. Ohm's law holds for all objects
D. Ohm's law does not hold for pn junctions
E. for Ohm's law to hold the resistivity must be independent of the electric field

7. Three wires made of the same material have the following lengths L and radii r.

    I. $L = L_0$, $r = r_0$
    II. $L = 2L_0$, $r = 2r_0$
    III. $L = 4L_0$, $r = 2r_0$

Order them according to their resistances, least to greatest.
A. I, then II and III tied
B. II, then I and III tied
C. I, II, III
D. III, I, II
E. III, then I and II tied

8. Three wires are made of the same material and have the same diameters. When the following potential differences V are applied to their ends a current i, given below, results:

    I. $V = V_0$, $i = i_0$
    II. $V = 2V_0$, $i = 4i_0$
    III. $V = 4V_0$, $i = 2i_0$

Order them according to their lengths, least to greatest.
A. I, II, III
B. III, II, I
C. I, III, II
D. III, I, II
E. II, I, III

9. If the potential difference across a resistor is doubled

A. the current in it is halved
B. the current in it is doubled
C. the rate with which it converts electrical energy to internal energy is doubled
D. the rate with which it converts electrical energy to internal energy is four times as great
E. its resistance is doubled

10. If the potential difference across a metal wire is doubled the drift speed of electrons in the wire

A. is doubled
B. is four times as great
C. is half as great
D. is one-fourth as great
E. is not changed

11. Around room temperature the resistivity of a pure metal is proportional to

A. the number of electrons per unit volume
B. the reciprocal of the number of electrons per unit volume
C. the mass of an electron
D. the average time between collisions between electrons and ions
E. the reciprocal of the average time between collisions between electrons and ions

Answers: (1) D; (2) E; (3) A, B, C; (4) A, B; (5) A, D; (6) B, D, E; (7) B; (8) E; (9) B, D; (10) A; (11) B, C, E

Chapter 27
CIRCUITS

The concept of emf is introduced in this chapter. Emf devices are used to maintain potential differences and to drive currents in electrical circuits. You will use this concept, along with those of electric potential, current, resistance, and capacitance, learned earlier, to solve both simple and complicated circuit problems. You will also learn about energy balance in electrical circuits.

Important Concepts

- ☐ emf device
- ☐ emf
- ☐ battery
- ☐ internal resistance
- ☐ energy in a circuit
- ☐ Kirchhoff loop rule
- ☐ Kirchhoff junction rule

- ☐ resistors in series
- ☐ resistors in parallel
- ☐ voltmeter
- ☐ ammeter
- ☐ charging a capacitor
- ☐ discharging a capacitor
- ☐ capacitive time constant

Overview

27–2 "Pumping" Charges

☐ A potential difference must be maintained for a current to exist in a resistor. An **emf device**, such as a battery, produces the required potential difference. It "pumps" charge from one of its terminals to the other.

27–3 Work, Energy, and Emf

☐ An emf device performs two functions: it maintains a potential difference and it moves charge from one terminal to the other inside the device. An ideal emf device contains only a seat of emf; a real emf device, such as a **battery**, also contains **internal resistance**.

☐ In electrical circuits, an ideal emf device is indicated by the symbol $\overset{\circ\rightarrow}{\dashv\vdash}$. The arrow with the small circle at its tail points from the negative terminal toward the positive terminal. An emf tends to drive current in the direction of the arrow but, for any circuit, the actual direction of the current through an emf device depends on other elements in the circuit.

☐ An **emf**, denoted by $\mathcal{E}$, is defined in terms of the work it does as it transports charge. If it does work dW on charge dq, then $\mathcal{E} = dW/dq$. The rate at which an emf does work is $dW/dt = (dq/dt)\mathcal{E} = i\mathcal{E}$. Emf has an SI unit of volt.

☐ An emf device does positive work on the charge passing through it if the current is from the negative to the positive terminal inside the device and negative work if the current is in the other direction. This is true whether the current consists of positive or negative charges.

☐ Positive work done by an emf device results in a decrease in the store of energy of the device, chemical energy in the case of a battery. Negative work results in an increase in the store of energy of the device. If the device is a battery, then in the first case it is *discharging* and in the second it is *charging*.

☐ The potential difference across an ideal emf device with emf $\mathcal{E}$ is $\mathcal{E}$, with the positive terminal being at the higher potential. This statement is true no matter what the direction of the current through the device.

27–4   Calculating the Current in a Single-Loop Circuit

☐ Charge flows if a resistor is connected across an emf device. In steady state, the rate with which the emf does work on the charge must equal the rate with which electrical energy is converted to internal energy in the resistor: $i\mathcal{E} = i^2 R$. Thus the current is $i = \mathcal{E}/R$.

☐ Circuits in steady state obey the **Kirchhoff loop rule**: The algebraic sum of the changes in potential encountered in a complete traversal of the circuit must be zero. To apply the rule, pick a direction for the current in the loop, then traverse the circuit. If you cross an emf from the negative to the positive terminal, the change in potential is $+\mathcal{E}$. If you traverse it in the other direction, the change is $-\mathcal{E}$. If you traverse a resistor in the direction of the current arrow, the change in potential is $-iR$; if you traverse it in the other direction the change is $+iR$. These expressions are valid even if you picked the wrong direction for the current arrow; i then has a negative value.

☐ For a single-loop circuit consisting of an ideal emf device with emf $\mathcal{E}$ and a resistor with resistance R, the loop rule gives $\mathcal{E} - iR = 0$, in agreement with the energy equation.

27–5   Other Single-Loop Circuits

☐ A real battery contains an internal resistance in addition to a seat of emf. Although the internal resistance is distributed inside the battery, you may think of it as being in series with an ideal emf.

☐ Suppose a single-loop circuit consists of a battery with emf $\mathcal{E}$ and internal resistance r, connected to an external resistance R. Then the loop rule gives $\mathcal{E} - ir - iR = 0$ and the current in the circuit is $i = \mathcal{E}/(r+R)$. Study Sample Problem 27–1 of the text for an example of a circuit with two batteries. Note that the battery with the larger emf is discharging and determines the direction of the current while the battery with the smaller emf is charging.

☐ The current is the same for all resistors in a series combination. Resistors in series can be replaced by a single resistor with resistance R_{eq}, such that the current is the same as the total current into the original combination when the potential difference is the same as the potential difference across the original combination. For a series combination of N resistors,

$$R_{eq} = \sum_{i=1}^{N} R_i$$

27–6 Potential Differences Between Two Points

☐ Once the current in a circuit is determined, its value can be used to calculate the potential difference across any of the circuit elements. The potential difference across a resistor, for example, is iR. The potential difference across two or more circuit elements in series is the sum of the individual potential differences.

27–7 Multiloop Circuits

☐ To analyze multiloop circuits, the **Kirchhoff junction rule** is needed in addition to the loop rule. The rule is: At any junction, the sum of the currents entering the junction is equal to the sum of the currents leaving. If, for example, four wires join at a junction, with i_1 and i_2 going into the junction and i_3 and i_4 coming out, then $i_1 + i_2 = i_3 + i_4$. The form of the junction equation does not depend on the directions of the actual currents, only on the directions of the current arrows.

☐ A junction is a point where three or more wires are joined. A branch is a portion of a circuit with junctions at its ends and none between. You must place a current arrow in each branch and label the arrows in different branches with different current symbols. In general, for any circuit with N junctions there will be $N - 1$ independent junction equations. The total number of independent equations equals the number of branches so the number of independent loop equations you will use equals the number of branches minus the number of junction equations. In addition, each current symbol must appear at least once in a loop equation. The loop and junction equations can be solved simultaneously for the currents in all the branches.

☐ The potential difference is the same for all the resistors in a parallel combination and the current is the same for all resistors in a series combination. If neither the potential difference nor the current is the same, then the resistors do not form either a parallel or series combination.

☐ A parallel combination of resistors can be replaced by a single resistor with resistance R_{eq}, such that the current is the same as the total current into the original combination when the potential difference is the same as the potential difference across the original combination. For a parallel combination of N resistors

$$\frac{1}{R_{eq}} = \sum_{i=1}^{N} \frac{1}{R_i} .$$

27–8 The Ammeter and the Voltmeter

☐ An **ammeter** measures the current in a branch of a circuit. The circuit is broken and the meter is inserted, so the current in the branch goes through the meter. An ammeter has a resistance and this alters the current it is measuring. To minimize this effect, the resistance of the ammeter should be very small.

☐ A **voltmeter** measures the potential difference between two points in a circuit. The connection is made so the meter is in *parallel* with a portion of the circuit. Voltmeters draw current and thereby change the value of the potential difference they are measuring. To minimize this effect, the resistance of the meter should be very large.

☐ A capacitor is charged by connecting it in series with a resistor and an emf to form a single-loop circuit. The loop rule gives $\mathcal{E} - iR - (q/C) = 0$. Here the direction of the current arrow was chosen to be in the direction of the emf and q represents the charge on the capacitor plate to which the current arrow points. This means the current and charge are related by $i = dq/dt$ and the equation satisfied by the charge is

$$\mathcal{E} - R\frac{dq}{dt} - \frac{q}{C} = 0.$$

☐ If the charging process is started at time $t = 0$ (by throwing a switch then, for example), the charge is given by

$$q(t) = C\mathcal{E}\left(1 - e^{-t/RC}\right)$$

and the current is given by

$$i(t) = \frac{dq}{dt} = \left(\frac{\mathcal{E}}{R}\right)e^{-t/RC}.$$

The solution obeys the initial condition $q(0) = 0$. These functions are graphed in Fig. 27–16.

☐ According to this result, the current just after the switch is closed, at $t = 0$, is given by $i(0) = \mathcal{E}/R$, just as if there were no capacitor. The results also predict that after a long time the charge on the capacitor is $q(\infty) = C\mathcal{E}$ and the current is $i(\infty) = 0$. The capacitor is then fully charged.

☐ The quantity $\tau = RC$ is called the **capacitive time constant** of the circuit. It has units of seconds and controls the time for the charge on the capacitor to reach any given value. If τ is made longer (by increasing C or R), more time is taken for the capacitor to reach any given fraction of its full charge $C\mathcal{E}$.

☐ A capacitor C, initially with charge q_0, is discharged by connecting it to a resistor R. The loop equation is

$$R\frac{dq}{dt} + \frac{q}{C} = 0$$

and its solution is

$$q(t) = q_0\, e^{-t/RC}.$$

After a long time, q becomes zero. The current as a function of time is

$$i(t) = \frac{dq}{dt} = -\frac{q_0}{RC}e^{-t/RC}.$$

Initially, the current is $i(0) = q_0/RC$; after a long time, it is zero.

Hints for Questions

1 The potential difference across a segment is $V = iR$, where R is the resistance of the segment and i is the current in the segment. The magnitude of the electric field in the segment is

$E = V/L$, where L is the length of the segment. The electric field is in the direction of the current.

[Ans: (a) b and d tie, then a, c, and e tie; (b) b, d, then a, c, and e tie; (c) in the positive x direction]

3 Two resistors are in series if the currents in them are the same. There can be no circuit junction between them. Two resistors are in parallel if the potential differences across them are the same. The equivalent resistance is $\mathcal{E}/i$, where $\mathcal{E}$ is the emf of the battery and i is the current in the battery.

[Ans: (a) no; (b) yes; (c) all tie]

5 When R_2 is added the potential difference across R_2 is the same as the potential difference across R_1 and is the same as the potential difference was before the resistor was added. The current in R_1 is $\mathcal{E}/R_1$ and the current in R_2 is $\mathcal{E}/R_2$, where $\mathcal{E}$ is the emf of the battery.

[Ans: (a) same; (b) same; (c) less; (d) more]

7 The current in the battery is given by $i = \mathcal{E}/R_{eq}$, where $\mathcal{E}$ is the emf of the battery and R_{eq} is the equivalent resistance of the circuit. Remember that the equivalent resistance of a series connection is greater than any of the individual resistances and the equivalent resistance of a parallel connection is less than any of the individual resistances.

[Ans: parallel, R_2, R_1, series]

9 After R_2 is added the sum of the potential differences across the resistors is the same as the potential difference across R_1 before R_2 was added. The current in R_1 is V_1/R_1, where V_1 is the potential difference across R_1. The equivalent resistance of a series connection is greater than any of the individual resistances.

[Ans: (a) less; (b) less; (c) more]

11 The capacitive time constant is given by $\tau = RC$, where R is the resistance of the resistor and C is the capacitance of the capacitor. The current at $t = 0$ is q_0/RC and the slope of the current versus time curve at $t = 0$ is q_0/R^2C^2, where q_0 is the initial charge on the capacitor. Assume the capacitor is fully charged when the switch is thrown. Then $q_0 = C\mathcal{E}$, where $\mathcal{E}$ is the emf of the battery.

[Ans: 1, c; 2, a; 3, d; 4, b]

Hints for Problems

3 The energy that can be delivered by the battery is the product of its charge and emf. The energy required is the product of the rate of delivery and the duration of operation. Equate the two expressions for the energy and solve for the duration of operation.

[Ans: 14.4 h]

7 Since the battery is being charged, the potential difference across its terminals is given by $\mathcal{E} + ir$, where $\mathcal{E}$ is its emf and r is its internal resistance. Energy is dissipated at a rate that is given by i^2r and is being converted to chemical energy at a rate of $\mathcal{E}i$. When the battery is being discharged the potential difference across its terminals is given by $\mathcal{E} - ir$.

[Ans: (a) 14 V; (b) 1.0×10^2 W; (c) 6.0×10^2 W; (d) 10 V, 1.0×10^2 W]

11 Write the loop equations for the two circuits. Use one of them to eliminate the emf of the battery from the other, then solve for the resistance.

$$\big[\text{Ans: } 8.0\,\Omega\,\big]$$

17 R_1 and R_2 are in parallel and their equivalent resistor is in series with R_3.

$$\big[\text{Ans: } 4.50\,\Omega\,\big]$$

23 Write three loop and three junction equations, then solve them simultaneously for the emf of the battery. The easiest way to proceed is to use the loop containing R_4 and R_6, the loop containing R_3 and R_4, and the loop containing R_1 and R_3.

$$\big[\text{Ans: } 48.3\,\text{V}\,\big]$$

25 The dissipation rate in R_3 is given by $i_3^2 R_3$, where i_3 is the current in R_3. Use the loop and junction rules to find an expression for i_3 in terms of R_3, then set its derivative with respect to R_3 equal to zero and solve for R_3.

$$\big[\text{Ans: } 1.43\,\Omega\,\big]$$

29 Group the resistors into m series combinations, each containing n resistors. Wire the series combinations in parallel. Now show that $m = n$ if the equivalent resistance is to be equal to the resistance of one of the resistors. The total dissipation rate is the product of the number of resistors (n^2) and the dissipation rate of one of them. Select the smallest value of n for which this is greater than $5.0\,\text{W}$.

$$\big[\text{Ans: } 9\,\big]$$

37 Write one junction and two loop equations, then solve them simultaneously for the current in R_3. Assume the ammeter has zero resistance.

$$\big[\text{Ans: } 0.45\,\text{A}\,\big]$$

41 Write one junction and two loop equations, then solve for the current i in the ammeter (and in R). The voltmeter reading is given by $V' = i(R + R_A)$. The apparent resistance is $R' = V'/i$.

$$\big[\text{Ans: (a) } 55.2\,\text{mA; (b) } 4.86\,\text{V; (c) } 88.0\,\Omega\text{; (d) decrease}\,\big]$$

43 Use the loop equation for the loop containing R_1, R_2, R_s, and R_x. Write an expression for the potential difference $V_b - V_a$ and set it equal to zero. Solve these two equations for R_x.

47 The potential difference across the capacitor is given by $V = q/C = \mathcal{E}(1 - e^{-t/\tau})$, where q is the charge on the capacitor and τ is the capacitive time constant. Solve for τ by taking the natural logarithm of both sides of the equation, then solve $\tau = RC$ for C.

$$\big[\text{Ans: (a) } 2.41\,\mu\text{s; (b) } 161\,\text{pF}\,\big]$$

51 The energy initially stored by the capacitor is given by $q_0^2/2C$, where q_0 is the initial charge. Solve for q_0. At any time t the charge on the capacitor is given by $q = q_0 e^{-t/\tau}$, and the current is given by $i = (q_0/\tau)e^{-t/\tau}$, where $\tau\ (= RC)$ is the capacitive time constant. The potential difference across the capacitor is q/C, the potential difference across the resistor is iR, and the rate of thermal energy production in the resistor is given by $i^2 R$.

$$\big[\text{Ans: (a) } 1.0 \times 10^{-3}\,\text{C; (b) } 1.0 \times 10^{-3}\,\text{A; (c) } (1.0 \times 10^3\,\text{V})\,e^{-t}\text{; (d) } (1.0 \times 10^3\,\text{V})\,e^{-t}\text{; (e) } P = e^{-2t}\,\text{W}\,\big]$$

59 Use the loop rule around either of the two loops that contain the resistor to find the current i in the resistor. Each of the batteries is discharging, so the potential difference across the terminals is $\mathcal{E} - ir$, where $\mathcal{E}$ is the emf of the battery and r is the internal resistance. The power of a battery is $i\mathcal{E}$ and the rate of transfer of energy to thermal energy in a battery is $i^2 r$.

[Ans: (a) 6.0 A; (b) 8.0 V; (c) 60 W; (d) 36 W]

69 The charge on the capacitor at time t is given by $q = C\mathcal{E}\left[1 - e^{-t/\tau}\right]$ and the current is $i = (\mathcal{E}/R)e^{-t/\tau}$, where τ is the capacitive time constant. Solve the first equation for $e^{-t/\tau}$, substitute into the second equation, and evaluate the result.

[Ans: 2.5 A]

71 The work done by the battery on the electron is $e\mathcal{E}$, where $\mathcal{E}$ is the emf of the battery. The power is $i\mathcal{E}$, where i is the current (the charge that passes through per unit time).

[Ans: (a) 12.0 eV; (b) 6.53 W]

75 (a) and (b) Apply the loop rule to the loop that goes through the two lowest resistors, $\mathcal{E}_4$, $\mathcal{E}_1$, and $\mathcal{E}_3$. The only current to enter the equation is i_1.
(c) and (d) Replace the two parallel resistors in the second branch from the bottom with their equivalent resistor, then apply the loop rule to the bottom-most loop.
(e) and (f) The power of the battery is given by $i\mathcal{E}_4$, where i is the current in battery. Use the junction rule to find i in terms of i_1 and i_2. If the current is in the direction of the emf (from the negative to the positive terminal in the interior), the battery is supplying energy. Otherwise it is absorbing energy.

[Ans: (a) 7.50 A; (b) left; (c) 10.0 A; (d) left; (e) 87.5 W; (f) being supplied]

83 Apply the loop rule to the loop containing $\mathcal{E}$, R_4, and R_5 and solve for the current i in the two resistors. The potential difference across R_5 is iR_5.

[Ans: 7.50 V]

99 Replace R_2 and R_3 with their equivalent resistance R_{23}, then calculate the current in the resulting circuit. The potential difference across R_{23} is the same as the potential difference across R_3 of the original circuit and is $V_3 = iR_{23}$. The current in R_3 is V_3/R_3.

[Ans: (a) 1.00 A; (b) 24.0 W]

103 The rate of energy dissipation in the line is given by $i^2 R$, where i is the current. The power supplied by the emf device is $P = i\mathcal{E}$, so $i = P/\mathcal{E}$ and the rate of energy dissipation is $P^2 R/\mathcal{E}^2$. Find the ratio for the two values of $\mathcal{E}$, with P and R the same for the two cases.

[Ans: 1.00×10^{-6}]

111 In the steady state the current in each of the branches with a capacitor is zero and all resistors have the same current i. Use the loop that passes through the resistors and the battery to solve for the current. The potential difference across C_1 is the same as the potential difference across R_1 and is $V_1 = iR_1$. The energy stored in this capacitor is $\frac{1}{2}C_1 V_1^2$. Similarly, the potential difference across C_2 is the same as the potential difference across R_2 and is V_2/R_2. The energy stored in this capacitor is $\frac{1}{2}C_2 V_2^2$.

[Ans: 250 μJ]

Quiz

Some questions might have more than one correct answer.

1. An emf is

    A. the work done on charged particles by any force
    B. the work per unit charge done on charged particles by a non-electrostatic forces
    C. the work done on any particles by non-electrical forces
    D. the potential energy per unit charge of any system of charged particles
    E. the kinetic energy per unit charge of any system of charged particles

2. Which of the following statements are true?

    A. electrical energy increases when the current inside an emf device is from the positive terminal toward the negative terminal
    B. electrical energy increases when the current inside an emf device is from the negative terminal toward the positive terminal
    C. the electric potential difference across the terminals of an emf device without internal resistance is equal to the emf
    D. an emf device increases the current exponentially
    E. none of the above are true

3. A single-loop circuit contains an ideal emf device and a resistor. Which of the following statements are true?

    A. the energy supplied by the emf device equals the energy converted to internal energy in the resistor
    B. the emf device supplies twice as much energy as is converted to internal energy in the resistor
    C. the electric potential difference across the terminals of the emf device equals the electric potential difference across the resistor
    D. the current in the emf device is greater than the current in the resistor
    E. none of the above are true

4. A single-loop circuit contains an ideal emf device with emf $\mathcal{E}$ and a resistor with resistance R. The rate with which the emf device supplies energy to the circuit is

    A. $\mathcal{E}R$
    B. $\mathcal{E}/R$
    C. $\mathcal{E}^2 R$
    D. $\mathcal{E}^2/R$
    E. $\mathcal{E}/R^2$

5. Current i is from left to right in a resistor with resistance R. This means that
   A. the electric potential difference across the resistor is given by iR and the left end is at a higher electric potential than the right end
   B. the electric potential difference across the resistor is given by iR and the right end is at a higher electric potential than the left end
   C. the electric potential difference across the resistor is given by i/R and the left end is at a higher electric potential than the right end
   D. the electric potential difference across the resistor is given by i/R and the right end is at a higher electric potential than the left end
   E. the electric potential difference across the resistor is given by i^2R and the left end is at a higher electric potential than the right end

6. A battery consists of an emf device with emf $\mathcal{E}$ and internal resistance r. If the current is i the electric potential difference across its terminals is
   A. $\mathcal{E}$
   B. $\mathcal{E} - ir$ if the current inside is from the negative terminal toward the positive terminal
   C. $\mathcal{E} - ir$ if the current inside is from the positive terminal toward the negative terminal
   D. $\mathcal{E} + ir$ if the current inside is from the negative terminal toward the positive terminal
   E. $\mathcal{E} + ir$ if the current inside is from the positive terminal toward the negative terminal

7. The resistance of resistor A is twice the resistance of resistor B. If the two resistors are wired in series and a potential difference is maintained across the combination
   A. the current in resistor A is twice the current in resistor B
   B. the current in resistor A is half the current in resistor B
   C. the current in the two resistors is the same
   D. the potential difference across resistor A is twice the potential difference across resistor B
   E. the potential difference across resistor A is half the potential difference across resistor B
   F. the potential difference is the same across the two resistors

8. The resistance of resistor A is twice the resistance of resistor B. If the two resistors are wired in parallel and a potential difference is maintained across the combination
   A. the current in resistor A is twice the current in resistor B
   B. the current in resistor A is half the current in resistor B
   C. the current in the two resistors is the same
   D. the potential difference across resistor A is twice the potential difference across resistor B
   E. the potential difference across resistor A is half the potential difference across resistor B
   F. the potential difference is the same across the two resistors

9. A certain multiloop circuit contains 6 branches and 4 junctions. The values of all resistances and emfs are given. To solve for the currents in the branches you must write

   A. 1 junction equation and 5 loop equations
   B. 2 junction equations and 4 loop equations
   C. 3 junction equations and 3 loop equations
   D. 4 junction equations and 2 loop equations
   E. 4 junction equations and 6 loop equations

10. Consider the following three combinations of resistors:

    I. a single resistor, with resistance R
    II. two resistors, each with resistance R, wired in parallel
    III. two resistors, with resistance R, wired in series

    The same potential difference is maintained across each combination. Rank them according to the rate with which electrical energy is converted to internal energy, least to greatest.

    A. I, II, III
    B. all tied
    C. I, II and III tied
    D. III, I, II
    E. II, III, I

11. A circuit containing a capacitor with capacitance C, a resistor with resistance R, an ideal emf device with emf $\mathcal{E}$, and a switch, all in series, is being used to charge the capacitor. Which of the following statements is true?

    A. just after the switch is closed the potential difference across the resistor is zero
    B. just after the switch is closed the potential difference across the resistor is $\mathcal{E}$
    C. just after the switch is closed the potential difference across the capacitor is zero
    D. just after the switch is closed the potential difference across the capacitor is $\mathcal{E}$
    E. after the switch has been closed for a long time the potential difference across the resistor is zero and the potential difference across the capacitor is $\mathcal{E}$

Answers: (1) B; (20 B, C; (3) A,, C; (4) D; (5) A; (6) B, E; (7) C, D; (8) B, F; (9) C; (10) D; (11) B, C, E

Chapter 28
MAGNETIC FIELDS

Charged particles, when moving, exert magnetic as well as electric forces on each other. Magnetic fields are used to describe magnetic forces and, although the geometry is a little more complicated, the idea is much the same as the idea of using electric fields to describe electric forces. Here you will learn about the force exerted by a magnetic field on a moving charge and on a wire carrying an electric current. In each case, pay particular attention to what determines the magnitude and direction of the force.

Important Concepts

- ☐ magnetic field
- ☐ magnetic force on a charge
- ☐ Hall effect
- ☐ orbit of a charge in a magnetic field

- ☐ magnetic force on a current-carrying wire
- ☐ magnetic torque on a current loop
- ☐ magnetic dipole
- ☐ magnetic dipole moment

Overview

28–2   The Magnetic Field

☐ A moving charged particle creates a **magnetic field** in all of space and this magnetic field exerts a force on any other moving charged particle. Magnetic fields do not exert forces on charges at rest. Many particles, among them the electron, proton, and neutron, have intrinsic magnetic fields associated with them, even if they are at rest.

28–3   The Definition of $\vec{B}$

☐ The force exerted by the magnetic field $\vec{B}$ on a particle with charge q, moving with velocity $\vec{v}$, is given by the vector product

$$\vec{F}_B = q\vec{v} \times \vec{B}\,.$$

The magnitude of $\vec{v} \times \vec{B}$ is given by $vB\sin\phi$, where ϕ is the angle between $\vec{v}$ and $\vec{B}$ when they are drawn with their tails at the same point. The direction of $\vec{v} \times \vec{B}$ is determined by the right-hand rule: curl the fingers of the right hand so they rotate $\vec{v}$ into $\vec{B}$ around the point where their tails meet, through the angle ϕ. The magnetic force is zero if $\vec{v}$ is parallel to $\vec{B}$. It is a maximum if $\vec{v}$ is perpendicular to $\vec{B}$.

☐ A magnetic force is perpendicular to the magnetic field and to the velocity of the particle. Because the force is perpendicular to the velocity, a magnetic field cannot change the speed or kinetic energy of a particle. It can, however, change the direction of motion.

☐ The sign of the charge is important for determining the direction of the force. A positively charged particle and a negatively charged particle moving with the same velocity in the same magnetic field experience magnetic forces in opposite directions. The force on a positively charged particle is in the direction of $\vec{v} \times \vec{B}$ while the force on a negatively charged particle is in the opposite direction.

☐ The magnetic field is defined in terms of the force on a moving positive test charge. First the electric force is found by measuring the force on the test charge when it is at rest. This force is subtracted vectorially from the force on the test charge when it is moving in order to find the magnetic force. Second, the direction of the magnetic field is found by causing the test charge to move in various directions and seeking the direction for which the magnetic force is zero. Lastly, the test charge is given a velocity perpendicular to the field and the magnetic force on it is found. If the charge is q_0, its speed is v, and the magnitude of the force on it is F_B, then the magnitude of the magnetic field is given by $B = F_B/q_0 v$.

☐ The SI unit for a magnetic field is the tesla and is abbreviated T: $1\,\text{T} = 1\,\text{N/A} \cdot \text{m}$. Another unit in common use is the gauss: $1\,\text{T} = 10^4$ gauss.

☐ Magnetic field lines are drawn so that at any point the field is tangent to the field line through that point and so that the magnitude of the field is proportional to the number of lines through a small area perpendicular to the field.

☐ Magnetic field lines form closed loops. For example, they continue through the interior of a magnet. The end of a magnet from which they emerge is called a north pole; the end they enter is called a south pole. If a bar magnet is free to rotate in Earth's magnetic field, its north pole tends to point roughly toward the north geographic pole.

28–4 Crossed Fields: Discovery of the Electron

☐ If both a magnetic field $\vec{B}$ and an electric field $\vec{E}$ act simultaneously on a particle with charge q moving with velocity $\vec{v}$, the total force on the particle is given by

$$\vec{F} = q\vec{E} + q\vec{v} \times \vec{B}.$$

An electric force can be used to balance a magnetic force on a charged particle, but the magnitude and direction of the electric field depend on the velocity of the particle. If a charged particle is in a magnetic field $\vec{B}$ and has velocity $\vec{v}$, then the total force on it is zero if $\vec{E} = -\vec{v} \times \vec{B}$. This result does not depend on either the sign or magnitude of the charge. The balancing electric field is perpendicular to both the magnetic field and the velocity of the particle. An important special case occurs if the velocity of the particle is perpendicular to the magnetic field. Then the magnitude of the balancing electric field is $E = vB$.

☐ Crossed electric and magnetic fields can be used to measure the mass-to-charge ratio m/q of an electron (or other charged particle). A beam of electrons, emitted from a hot filament, is formed by accelerating them in an electric field and passing them through a slit. They then enter a region of crossed electric and magnetic fields. The beam hits a fluorescent screen and produces a spot where it hits. The steps of the measurement are:

1. The fields are turned off and the position of the spot is noted.

2.   Only the electric field $\vec{E}$ is turned on and the deflection y of the beam at the edge of the deflecting plates is computed from the deflection of the spot on the screen. It is given by $y = qEL^2/2mv^2$, where L is the length of the plates and v is the speed of the particles.

3.   The magnetic field is now adjusted until the spot returns to its original position. Then the magnetic and electric forces cancel and the speed is $v = E/B$. Substitute this expression into the equation for y and solve for m/q. The result is $m/q = B^2L^2/2yE$. All of the quantities on the right side can be determined experimentally.

28–5   Crossed Fields: The Hall Effect

☐   A Hall effect experiment can be used to obtain the sign and concentration of charged particles in a current. A magnetic field, perpendicular to the current, pushes charged particles to one side of the sample. These particles create a transverse electric field and when there are enough of them, the electric force on particles in the current balances the magnetic force on them. Then no more particles collect along the side. The direction of the transverse electric field and, hence, the sign of the charge on the particles can be found with a voltmeter. The concentration of particles in the current is given by $n = iB/qV\ell$, where i is the current, B is the magnetic field, V is the transverse potential difference, and ℓ is the thickness of the sample.

28–6   A Circulating Charged Particle

☐   Since a magnetic force is always perpendicular to the velocity of the particle on which it acts, it can be used to hold a charged particle in a circular orbit. All that is required is to fire the particle perpendicularly into a uniform field. Suppose a region of space contains a uniform magnetic field with magnitude B and a particle with charge of magnitude q is given a velocity $\vec{v}$ perpendicular to the field. Equate the magnitude of the magnetic force (qvB) to the product of the mass and acceleration (mv^2/r) and solve for the radius r of the orbit. The result is $r = mv/qB$.

☐   The time taken by the particle to go once around its orbit does not depend on its speed because the radius of the orbit and hence the distance traveled are both proportional to the speed. Since the circumference of the orbit is $2\pi R$, the period is $T = 2\pi r/v = 2\pi m/qB$. The number of times the particle goes around per unit time, or the frequency of the motion, is given by $f = 1/T = qB/2\pi m$. The angular frequency is given by $\omega = 2\pi f = qB/m$.

☐   If the particle velocity is not perpendicular to the field, but has both parallel and perpendicular components, the trajectory is a helix. The radius is determined by the velocity component in the plane perpendicular to the field and the pitch is determined by the velocity component parallel to the field.

28–7   Cyclotrons and Synchrotrons

☐   A cyclotron is diagramed in Fig. 28–15. A uniform magnetic field is everywhere perpendicular to the dees and an electric field is in the gap between them. A charged particle enters the cyclotron near the center. The magnetic field causes the particle to travel in a circular arc. The electric field causes the speed of the particle to increase and when it does, the particle moves to an orbit of larger radius. The electric field must reverse direction twice each period if it is to increase the speed of the particle during every traversal of the gap. This is fairly easy to do since the time between reversals does not change as long as the particle speed is significantly less than the speed of light.

☐ At speeds near the speed of light, the period does depend on the particle speed and the interval between reversals of the electric field must be decreased as the particle speeds up. Synchrotrons are accelerators that change the frequency of the electric field so its reversals remain in step with the circulating particles. The magnetic field is also varied so the radius of the orbit remains the same.

28–8 Magnetic Force on a Current-Carrying Wire

☐ A current is a collection of moving charged particles and so experiences a force when a magnetic field is applied. If the current is in a wire, the force is transmitted to the wire itself.

☐ To calculate the force on a wire carrying current i, divide the wire into infinitesimal segments, calculate the force on each segment, then vectorially sum the forces. Let dL be the length of an infinitesimal segment of wire with cross-sectional area A. If n is the concentration of charge carriers and each carrier has charge q, then the total charge in the segment is given by $qnA\,dL$. The magnetic force on the segment is the product of this and the force on a single charged particle. That is, $d\vec{F}_B = qnA\vec{v}_d \times \vec{B}\,dL$, where $\vec{v}_d$ is the drift velocity. Notice that the combination $qnAv_d$ appears in the expression for the magnitude of the force. This is the current i. If we take the vector $d\vec{L}$ to be in the direction of $q\vec{v}_d$, then the force is given by $d\vec{F}_B = i\,d\vec{L} \times \vec{B}$.

☐ For a finite wire, the resultant force is given by the integral

$$\vec{F}_B = i \int d\vec{L} \times \vec{B}.$$

For a straight wire of length L in a uniform field $\vec{B}$, the magnitude of the force is $F_B = i\vec{L} \times \vec{B}$. Since $\int d\vec{L} = 0$ for a closed loop, the magnetic force on a closed loop of wire in a uniform magnetic field is zero.

28–9 Torque on a Current Loop

☐ Although a uniform magnetic field does not exert a net force on a current-carrying loop, it may exert a torque: the center of mass of the loop does not accelerate in a uniform field but if the magnetic torque is not balanced, the loop has an angular acceleration about the center of mass.

☐ Consider a rectangular loop of wire carrying current i, in a uniform magnetic field $\vec{B}$. If the sides of the loop have lengths a and b, then the magnetic torque (about any axis of rotation) is given by $\tau = iab\sin\theta$, where θ is the angle between the magnetic field and the normal to the plane of the loop. If the loop has N turns of wire, the torque is $\tau = NiAB\sin\theta$, where $A\,(=ab)$ is the area of the loop.

☐ Let $\vec{n}$ be a unit vector that is normal to the loop, in the direction of the thumb when the fingers of the right hand curl around the loop in the direction of the current. Then the magnetic torque is in the direction of $\vec{n} \times \vec{B}$.

☐ The torque has maximum magnitude when $\theta = 90°$; the normal to the loop is perpendicular to the field and the loop itself is parallel to the field. The torque is zero when $\theta = 0$; the normal to the loop is in the direction of the field and the loop itself is perpendicular to the field. The torque exerted by the magnetic field tends to rotate the normal toward the direction of the field.

28–10   The Magnetic Dipole

☐ The torque exerted by a uniform field on any planar current loop is easily expressed in terms of the **magnetic dipole moment** $\vec{\mu}$ of the loop. For a loop with area A, having N turns and carrying current i, the magnitude of the dipole moment is given by $\mu = NiA$. The direction of the dipole moment is determined by the right-hand rule: Curl the fingers of the right hand around the loop in the direction of the current. Then the thumb points in the direction of the dipole moment.

☐ The torque exerted by a field $\vec{B}$ on a loop with magnetic dipole moment $\vec{\mu}$ is given by $\vec{\tau} = \vec{\mu} \times \vec{B}$. Its magnitude is given by $\tau = \mu B \sin \theta$, where θ is the angle between $\vec{\mu}$ and $\vec{B}$ when they are drawn with their tails at the same point. The direction of the torque is given by the right-hand rule for vector products.

☐ Although magnetic fields are not conservative and a potential energy cannot be associated with a charged particle in a magnetic field, a potential energy can be associated with a magnetic dipole $\vec{\mu}$ in a magnetic field $\vec{B}$. It is given by $U = -\vec{\mu} \cdot \vec{B}$. The potential energy is a minimum when the dipole moment is aligned with the magnetic field. It is a maximum when the dipole moment is anti aligned with the field. In both these cases, the plane of the loop is perpendicular to the field.

☐ Suppose a dipole moment initially makes the angle θ_i with the magnetic field, but then rotates so it makes the angle θ_f. During this rotation, the work done on the dipole by the field is given by $W = -(U_f - U_i) = \mu B(\cos \theta_f - \cos \theta_i)$.

Hints for Questions

1 Since the magnetic force is given by the vector product of the particle velocity and the magnetic field it must be perpendicular to the directions of these two quantities.

[Ans: (a) no ; (b) yes; (c) no]

3 The magnetic force is small if the particle speed is small and is large if the particle speed is large.

[Ans: (a) $\vec{F}_E$; (b) $\vec{F}_B$]

5 The magnetic force must point toward the center of the path. It is always perpendicular to the particle velocity and so cannot change the speed of the particle. Recall that the period is independent of the particle speed and that the radius of the orbit is proportional to the particle speed.

[Ans: (a) negative; (b) equal; (c) equal; (d) half-circle]

7 The radius of the orbit is proportional to the reciprocal of the magnetic field magnitude, the period is proportional to the magnetic field magnitude, and the magnetic field does not change the speed of the particle. Because the particle is negatively charged the direction of the magnetic force on it is opposite to the direction of the vector product of the particle velocity and the field. It must point from the particle toward the center of the path. The period is proportional to the reciprocal of the magnetic field magnitude.

[Ans: (a) $\vec{B}_1$; (b) $\vec{B}_1$ into page, $\vec{B}_2$ out of page; (c) less]

9 First determine the direction of the magnetic force on the electron when it is between each set of plates. Then determine the direction of the electric field so that the electric force is in the opposite direction. The electric field points from the plate at the higher electric potential toward the plate at the lower electric potential.

[Ans: (a) upper plate; (b) lower plate; (c) out of page]

Hints for Problems

5 Use the equation $\vec{F}_B = -e\vec{v} \times \vec{B}$ for the magnetic force. Equate the z component of the left side to the z component of the right side and solve for B_x.

[Ans: −2.0 T]

7 Take the magnetic field to be perpendicular to both the electric field and the particle velocity. Since the net force on the electron is zero the magnitudes of the fields are related by $E = vB$, where E is the magnitude of the electric field, B is the magnitude of the magnetic field, and v is the speed of the electron. The magnitude of the electric field is $E = V/d$, where V is the potential difference between the plates and d is the plate separation. The electric field points from the positive plate toward the negative plate. Choose the direction of the magnetic field so that the direction of the magnetic force is opposite to the direction of the electric force.

[Ans: −0.267 mT]

11 A transverse electric field is produced by the electrons that are moved to the edges of the strip by the magnetic field, as in the Hall effect. The net force on electrons in the strip is zero, so $E = vB$, where E is the magnitude of the transverse electric field, B is the magnitude of the magnetic field, and v is the speed of the strip. The magnitude of the transverse electric field is V/d, where V is the potential difference across the strip and d is the width of the strip.

[Ans: 38.2 cm/s]

13 Since the net force on an electron in the solid is zero, the electric field is given by $\vec{E} = -\vec{v} \times \vec{B}$, where $\vec{v}$ is the velocity of the solid and $\vec{B}$ is the magnetic field. The potential difference across the solid is Ed, where d is the width of the solid along the direction of the electric field.

[Ans: (a) $(-600 \, \text{mV/m})\hat{k}$; (b) 1.20 V]

19 The frequency is given by $f = eB/2\pi m$ and the orbit radius by $r = mv/eB$, where B is the magnitude of the magnetic field, m is the mass of an electron, and v is the speed of an electron. The kinetic energy of the electron can be used to compute its speed.

[Ans: (a) 0.978 MHz; (b) 96.4 cm]

21 The magnitude of the magnetic force is $F_B = ev_\perp B$, where $v_\perp$ is the component of the particle velocity perpendicular to the magnetic field and B is the magnitude of the field. The pitch is $p = v_\parallel T$, where $v_\parallel$ is the component of the particle velocity along the direction of the field and T is the period of the motion. The period is $T = 2\pi m/eB$.

[Ans: 65.3 m/s]

<u>25</u> According to Sample Problem 28–3 the magnitude of the magnetic field is $B = \sqrt{8Vm/qx^2}$, where V is the accelerating potential difference, m is the mass of a uranium atom and q is its charge, and x is the diameter of the orbit in the magnetic field. See Fig. 28–14. The total mass M that passes through in an hour is given by $M = Nm$, where N is the number of ions that pass through in that time. The total charge that passes through per hour is given by $Q = Nq$ and the current is $i = Q/t$, where t is one hour (converted to seconds).

[Ans: (a) 495 mT; (b) 22.7 mA; (c) 8.17 MJ]

<u>27</u> Linear momentum is conserved in the decay and the particles have the same mass, so they move away in opposite directions with the same speed. They collide after they have gone halfway around their circular orbits. This occurs after a time equal to half a period of the motion. The period is given by $T = 2\pi m/eB$.

[Ans: (a) 5.06 ns]

<u>31</u> The kinetic energy increase with each pass is $\Delta K = eV$, where V is the accelerating potential difference. After N passes its kinetic energy is $K = N\,\Delta K$. Use $K = \frac{1}{2}mv^2$ to compute the speed v after 100 passes and after 101 passes, then use $r = mv/eB$ to compute the radius of the orbit after each of those passes.

[Ans: (a) 200 eV; (b) 20.0 keV; (c) 0.499%]

<u>35</u> The magnetic force on the wire must be upward and equal in magnitude to the weight of the wire. The magnetic force is given by $i\vec{L} \times \vec{B}$, where i is the current, $\vec{B}$ is the magnetic field, and L is the length of the wire. The vector $\vec{L}$ is in the direction of the current.

[Ans: (a) 467 mA; (b) right]

<u>43</u> For part (a) the current in the galvanometer should be 1.62 mA when the potential difference across the galvanometer-resistor combination is 1.00 V. Calculate the equivalent resistance of the combination, then use the equation for either a series or parallel combination to find the auxiliary resistance. (Only one will work.) For part (b) the current in the galvanometer should be 1.62 mA when the total current in the galvanometer-resistor combination is 50.0 mA.

[Ans: (a) 542 Ω; (b) series; (c) 2.52 Ω; (d) parallel]

<u>53</u> The magnitude of the magnetic dipole moment is the product of the loop area (πR^2, where R is the radius) and the current in the loop. Multiply this by the unit vector in the direction of the dipole moment to find the vector moment. The torque is the vector product of the dipole moment and the magnetic field. The magnetic potential energy is the negative of the scalar product of the dipole moment and the field.

[Ans: (a) $(-9.7 \times 10^{-4}\,\text{N} \cdot \text{m})\hat{\imath} - (7.2 \times 10^{-4}\,\text{N} \cdot \text{m})\hat{\jmath} + (8.0 \times 10^{-4}\,\text{N} \cdot \text{m})\hat{k}$; (b) $-6.0 \times 10^{-4}\,\text{J}$]

<u>57</u> The electron moves with constant speed along the direction of the magnetic field while at the same time it undergoes uniform circular motion around the field direction. It next crosses the line though the injection point at a time equal to the period of rotation. Find the component of

the velocity along the field direction and multiply by the period, which is given by $2\pi m/eB$, where m is the mass of the electron and B is the magnitude of the magnetic field.

$\left[\text{Ans: } 0.53\,\text{m}\,\right]$

59 The magnitude of the magnetic force is evB, where v is the speed of the ion and B is magnitude of the magnetic field. The acceleration of the ion is v^2/r, where r is the radius of the orbit. Newton's second law gives $evB = mv^2/r$, where m is the mass of the ion. The period is $T = 2\pi r/v$. Use this to substitute for v in the second law equation, then solve for m. Use $1\,\text{u} = 1.661 \times 10^{-27}\,\text{kg}$ to convert to atomic mass units.

$\left[\text{Ans: } 127\,\text{u}\,\right]$

65 Take the magnetic field to be $\vec{B} = B_x\,\hat{\text{i}} + B_x\,\hat{\text{j}} + B_z\,\hat{\text{k}}$ and find the components of $q\vec{v} \times \vec{B}$ in terms of B_x and B_z. Equate each component to the same component of the force and solve for B_x and B_z.

$\left[\text{Ans: } -(3.0\,\text{T})\,\hat{\text{i}} - (3.0\,\text{T})\,\hat{\text{j}} - (4.0\,\text{T})\,\hat{\text{k}}\,\right]$

69 The net electromagnetic force is $\vec{F} = q(\vec{E} + \vec{v} \times \vec{B})$, where $\vec{E}$ is the electric field, $\vec{v}$ is the particle velocity, $\vec{B}$ is the magnetic field, and q is particle's charge. See Section 3-8 to see how to evaluate the vector product in terms of components.

$\left[\text{Ans: } (0.80\,\text{mN})\,\hat{\text{k}}\,\right]$

77 The torque is given by $\vec{\tau} = \vec{\mu} \times \vec{B}$, where $\vec{\mu}$ is the magnetic dipole moment of the coil and $\vec{B}$ is the magnetic field. The magnitude of the torque is $\mu = iA$, where i is the current in the coil and A is its area. The torque is perpendicular to the magnetic field and so is either along the line toward the 20 min mark or along the line toward the 50 min mark. Use the right-hand rule for magnetic dipole moments to find the direction of $\vec{\mu}$, then use the right-hand rule for vector products to find the direction of the torque.

$\left[\text{Ans: (a) } 20\,\text{min}; \text{ (b) } 5.9 \times 10^{-2}\,\text{N} \cdot \text{m}\,\right]$

83 The x component of the velocity is constant. The y and z components are those of a particle in uniform circular motion with angular frequency $\omega = eB/m$. That is, they are proportional to the sine or cosine of ωt. The expression for the velocity must reduce to $v_x\,\hat{\text{i}} + v_{0y}\,\hat{\text{j}}$ for time $t = 0$.

$\left[\text{Ans: } \vec{v} = v_{0x}\,\hat{\text{i}} + v_{0y}\cos(\omega t)\,\hat{\text{j}} - v_{0y}\sin(\omega t)\,\hat{\text{k}}, \text{ where } \omega = eB/m\,\right]$

87 The magnetic force on the wire is given by $\vec{F} = i\int d\vec{s} \times \vec{B}$, an integral along the wire. $\vec{B}$ is the magnetic field and $d\vec{s}$ is an infinitesimal vector with length ds and directed in the direction of the current. Since $\vec{B}$ is uniform it may be factored from the integral. What is the value of $\int d\vec{s}$ for a closed loop? Think of the integral as a vector sum of a large number of small vectors that are place tail to head and form a closed figure.

$\left[\text{Ans: (b) no}\,\right]$

```
Quiz
```

Some questions might have more than one correct answer.

1. Consider the force $\vec{F}$ of the magnetic field $\vec{B}$ on a positively charged particle moving with velocity $\vec{v}$. Which of the following statements are true?

   A. the magnetic field is perpendicular to the velocity of the particle
   B. the magnetic field is parallel to the velocity of the particle
   C. the magnetic force on the particle is perpendicular to the magnetic field
   D. the magnetic force on the particle is perpendicular to the velocity of the particle
   E. the magnetic force on the particle is parallel to the velocity of the particle

2. The force of a magnetic field on a moving charged particle is proportional to

   A. the speed of the particle
   B. the reciprocal of the speed of the particle
   C. the magnitude of the magnetic field
   D. the angle between the magnetic field and the velocity of the particle
   E. the sine of the angle between the magnetic field and the velocity of the particle
   F. the cosine of the angle between the magnetic field and the velocity of the particle

3. Magnetic field lines

   A. start at north poles and end at south poles
   B. start at south poles and end at north poles
   C. form closed loops
   D. may intersect each other
   E. are always uniformly spaced

4. A certain region of space contains a uniform electric field $\vec{E}$ and a uniform magnetic field $\vec{B}$ that are perpendicular to each other. An electron is shot into this region and passes straight through without a change in speed. This means that

   A. the electron's velocity was perpendicular to the electric field
   B. the electron's velocity was perpendicular to the magnetic field
   C. the electron's speed must have been E/B
   D. the electron's speed must have been B/E
   E. the electric field must have been perpendicular to the magnetic field
   F. none of the above

5. A certain region of space contains a uniform magnetic field and no electric field. An electron enters this region with a velocity that is perpendicular to the magnetic field. As a result the electron moves

    A. with constant speed around a circle with a radius that is proportional to the magnetic field magnitude

    B. with constant speed around a circle with a radius that is proportional to the reciprocal of the magnetic field magnitude

    C. with constant speed around a circle with a radius that is proportional to the electron's speed

    D. with constant speed around a circle with a radius that is proportional to the reciprocal of the electron's speed

    E. along a straight line

6. A charged particle moves around a circle in a uniform magnetic field. The magnetic field supplies the necessary centripetal force. The time for the particle to go around once depends on

    A. the magnitude of the magnetic field

    B. the particle's speed

    C. the charge to mass ratio of the particle

    D. the radius of the circular path

    E. none of the above

7. The force of a uniform magnetic field on a current-carrying wire, not necessarily lying along a straight line, is proportional to

    A. the current in the wire

    B. the length of the wire

    C. the distance between the end points of the wire

    D. the magnitude of the magnetic field

    E. none of the above

8. A uniform magnetic field exerts a torque on a closed loop of current-carrying wire that is

    A. in a plane that is perpendicular to the magnetic field

    B. in any plane that is not perpendicular to the magnetic field

    C. in a plane that is parallel to the magnetic field

    D. in any plane that is not parallel to the magnetic field

    E. none of the above

9. A uniform magnetic field exerts a torque on a closed loop of current-carrying wire that is proportional to

   A. current in the loop
   B. area enclosed by the loop
   C. the magnitude of the magnetic field
   D. the reciprocal of the current in the loop
   E. the reciprocal of the area enclosed by the loop
   F. the reciprocal of the magnitude of the magnetic field

10. The magnetic dipole moment of a closed loop of current-carrying wire

    A. is in the plane of the loop
    B. is perpendicular to the plane of the loop
    C. has a magnitude that is proportional to the magnitude of the magnetic field
    D. has a magnitude that is proportional to the area enclosed by the loop
    E. has a magnitude that is proportional to the current in the loop

11. The torque of a uniform magnetic field on a magnetic dipole is

    A. parallel to the dipole moment
    B. parallel to the magnetic field
    C. perpendicular to the dipole moment
    D. perpendicular to the magnetic field
    E. none of the above

12. The energy of a magnetic dipole in a magnetic field has its greatest value when

    A. it is in the same direction as the field
    B. it is opposite the field
    C. it is perpendicular to the field
    D. it makes an angle of 45° with the field
    E. when it is in none of the orientations given above

13. The magnitude of the torque of a magnetic field on a magnetic dipole has its greatest value when the dipole moment

    A. is in the same direction as the field
    B. is opposite the field
    C. is perpendicular to the field
    D. makes an angle of 45° with the field
    E. it is in none of the orientations given above

14. You hold a neutral metal rod while wearing insulating gloves, with one hand at each end. Starting with your arms outstretched in front of you pull the rod rapidly toward you in a vertically downward magnetic field. As a result

    A. an electric field points from right to left in the rod
    B. an electric field points from left to right in the rod
    C. the left end of the rod is charged negatively compared to the right end
    D. the right end of the rod is charged negatively compared to the left end
    E. none of the above occur

Answers: (1) C, D; (2) A, C, E; (3) C; (4) A, C, E; (5) B, C; (6) A; (7) A, C, D; (8) B, C, E; (9) A, B, C; (10) B, D, E; (11) C, D; (12) B; (13) C; (14) A, C

Chapter 29
MAGNETIC FIELDS DUE TO CURRENTS

Learn to calculate the magnetic field produced by a current using two techniques, one based on the Biot-Savart law and the other based on Ampere's law. When using the first, you will sum the fields produced by infinitesimal current elements. Carefully note how the directions of the current and the position vector from a current element to the field point influence the direction of the field. Ampere's law relates the integral of the tangential component of the field around a closed path to the net current through the path. Pay attention to the role played by symmetry when you use this law to find the field.

Important Concepts

- [ ] Biot-Savart law
- [ ] permeability constant
- [ ] magnetic field of a
     long straight wire
- [ ] magnetic field of an arc

- [ ] Ampere's law
- [ ] solenoid
- [ ] toroid
- [ ] magnetic field of a dipole

Overview

29–2 Calculating the Magnetic Field Due to a Current

- [ ] The **Biot-Savart law** is a prescription for calculating the magnetic field of a current. To calculate the magnetic field produced at any point P by a wire carrying a current i, first choose an infinitesimal element of the wire, with length $d\vec{s}$ and in the same direction as the current. Draw the displacement vector $\vec{r}$ from the selected current element to P. The field produced at P by the current element is given by

$$dR = \left(\frac{\mu_0}{4\pi}\right) \frac{i \, d\vec{s} \times \vec{r}}{r^3}.$$

  Note the *vector product*. The field of the element is perpendicular to both $d\vec{s}$ and $\vec{r}$. To find the total field at P, sum (integrate) the contributions from all elements of the wire.

- [ ] The constant μ_0 is called the **permeability constant** and its value is exactly $4\pi \times 10^{-7}$ T·m/A. Do not confuse the symbol with that for the magnitude of a magnetic dipole moment (μ without a subscript).

- [ ] In Section 29–2 of the text, the Biot-Savart law is used to find an expression for the magnitude of the magnetic field produced by a long straight wire carrying current i. Each infinitesimal element of the wire produces a field in the same direction so the magnitude of the total field is the sum of the magnitudes of the fields produced by all the elements. Go over the calculation carefully. The result is

$$B = \frac{\mu_0 i}{2\pi r},$$

where r is the perpendicular distance from the wire to the field point.

☐ The magnetic field lines around a long straight wire are circles centered on the wire. The field has the same magnitude at all points on any given circle. The direction is determined by a right-hand rule: If the thumb of the right hand points in the direction of the current, then the fingers curl around the wire in the direction of the magnetic field lines.

☐ You should also know how to compute the magnetic field produced at its center by a circular arc of current. Consider the wire shown on the right. It has radius R, subtends the angle ϕ, and carries current i. Other wires, not shown, carry the current into and out of the arc. Since the values of i and R are the same for all segments and the fields at C due to all segments are in the same direction (into the page), the magnitude of the total field at C is given by

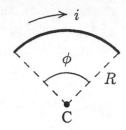

$$B = \frac{\mu_0}{4\pi} \int \frac{i}{R^2}\, ds = \frac{\mu_0 i s}{4\pi R^2} = \frac{\mu_0 i \phi}{4\pi R},$$

where s is the length of the arc. The relationship $s = R\phi$, with ϕ in radians, was used. For a semicircle, $\phi = \pi$ and $B = \mu_0 i/4R$ and for a complete circle, $\phi = 2\pi$ and $B = \mu_0 i/2R$.

29–3 Force Between Two Parallel Currents

☐ The expression for the field of a long straight wire can be combined with the expression for the magnetic force on a wire, developed in the last chapter, to find an expression for the force exerted by two parallel wires on each other. The magnetic field produced by one wire at a point on the other wire is perpendicular to the second wire. If the wires are separated by a distance d and carry currents i_a and i_b, then the magnitude of the force per unit length of one on the other is given by $F/L = \mu_0 i_a i_b/2\pi d$. If the currents are in the same direction, the wires attract each other; if the currents are in opposite directions, they repel each other. The forces of the wires on each other obey Newton's third law: they are equal in magnitude and opposite in direction.

29–4 Ampere's Law

☐ Ampere's law tells us that the integral of the tangential component of the magnetic field around any closed path is equal to the product of μ_0 and the net current that pierces the loop. Mathematically for any closed path,

$$\oint \vec{B} \cdot d\vec{s} = \mu_0 i,$$

where $d\vec{s}$ is an infinitesimal displacement vector, tangent to the path. The integral on the left side is a path integral around a *closed* path, called an Amperian path or Amperian loop. The current i on the right side is the net current through a surface that is bounded by the path. For example, if the path is formed by the edges of a page, then i is the net current through the page. The Amperian path need not be the boundary of any physical surface and, in fact, may be purely imaginary. The surface need not be a portion of a plane.

☐ To apply Ampere's law, choose a direction (clockwise or counterclockwise) to be used in evaluating the integral on the left side. The choice is immaterial but it must be made since it determines the direction of $d\vec{s}$. If the tangential component of the field is in the direction of $d\vec{s}$, then the integral is positive; if it is in the opposite direction, the integral is negative. Now, curl the fingers of your right hand around the loop in the direction chosen. Your thumb will point in the direction of positive current. Examine each current through the surface and algebraically sum them. If a current arrow is in the direction your thumb pointed, it enters the sum with a positive sign; if it is in the opposite direction it enters with a negative sign.

☐ A current outside the path produces a magnetic field at every point on the path but the integral $\oint \vec{B} \cdot d\vec{s}$ of the field it produces vanishes, so it is not included on the right side of the equation.

☐ Only the tangential component of the magnetic field enters Ampere's law. When the law is used to calculate the magnetic field of a current distribution, the path is taken, if possible, to be along a field line. Then, the integral reduces to $\int B\, ds$. If, in addition, the magnitude of the field is constant along the path, then the integral is Bs, where s is the total length of the path.

☐ Ampere's law can be used to find an expression for the magnetic field produced by a long straight wire carrying current i. The Amperian path used is a circle in a plane perpendicular to the wire and centered at the wire. Since the magnetic field is tangent to the circle and has the same magnitude at all points around the circle, the integral $\oint \vec{B} \cdot d\vec{s}$ is $2\pi r B$, where r is the radius of the circle. Ampere's law gives $2\pi r B = \mu_0 i$, so $B = \mu_0 i / 2\pi r$, as before.

☐ Ampere's law can also be used to find the magnetic field *inside* a long straight wire. Suppose a cylindrical wire of radius R carries a current that is uniformly distributed over its cross section. Take the Amperian path to be a circle of radius r (with $r < R$). The magnetic field is tangent to the circle and has constant magnitude around the circle, so $\oint \vec{B} \cdot d\vec{s} = 2\pi r B$. Not all the current in the wire goes through the Amperian circle. In fact, the fraction through the Amperian circle is the ratio of the circle area to the wire area: r^2 / R^2. Thus the right side of the Ampere's law equation is $\mu_0 i r^2 / R^2$. Ampere's law gives $2\pi r B = \mu_0 i r^2 / R^2$, so $B = \mu_0 i r / 2\pi R^2$.

☐ The magnetic field at the center of the wire is zero. It increases linearly with distance from the wire center until it reaches its maximum value at the surface of the wire ($r = R$). This value is $B = \mu_0 i / 2\pi R$.

29–5 Solenoids and Toroids

☐ Ampere's law can be applied to a **solenoid**, a cylinder tightly wrapped with a thin wire. For an ideal solenoid (long and tightly wrapped), the magnetic field outside is negligible and the field inside is uniform and is parallel to the axis.

☐ As the Amperian path take a rectangle with one side of length h, inside the solenoid and parallel to its axis, and the opposite side outside the solenoid. The first of these sides contributes Bh to the left side of the Ampere's law equation and the other three sides of the rectangle contribute zero. Thus $\oint \vec{B} \cdot d\vec{s} = Bh$. If the solenoid has n turns of wire per unit length, then the number of turns that pass through the surface bounded by the Amperian path is nh. Each carries current i so the right side of the Ampere's law equation is $\mu_0 nhi$. Thus $Bh = \mu_0 nhi$, so $B = \mu_0 ni$.

☐ Ampere's law can also be applied to a toroid, with a core shaped like a doughnut and wrapped with a wire, like a solenoid bent so its ends join. The magnetic field is confined to the interior of the core and the field lines are concentric circles centered at the center of the hole. The Amperian path is a circle of radius r, inside the core and centered at the center of the hole. The integral $\oint \vec{B} \cdot d\vec{s}$ is $2\pi r B$ and if there are N turns of wire, each carrying current i, the total current through the surface bounded by the path is Ni. Thus the Ampere's law equation is $2\pi r B = \mu_0 Ni$ and $B = \mu_0 Ni/2\pi r$. The field is zero inside the hole and outside the toroid.

29–6 A Current-Carrying Coil as a Magnetic Dipole

☐ The Biot-Savart law can be used to show that the magnetic field produced by a circular loop of wire of radius R and carrying current i, at a point on its axis of symmetry a distance z from its center, is

$$B(z) = \frac{\mu_0 i R^2}{2(R^2 + z^2)^{3/2}}.$$

For points far away from the loop ($z \gg R$), the expression becomes $B(z) = \mu_0 i R^2/2z^3$, which can be written in terms of the dipole moment $\vec{\mu}$ of the loop: $B(z) = \mu_0 \mu/2\pi z^3$. $\vec{B}$ is in the direction of $\vec{\mu}$ for points above the loop (positive z) and in the opposite direction for points below the loop (z negative). This expression is valid for the magnetic field of *any* plane loop, regardless of its shape, for points far away along the axis defined by the dipole moment.

Hints for Questions

1 The fields of the two wires must be in opposite directions at P. Since P is to the left of both wires, this means that the currents must be in opposite directions. The magnitudes of the fields must be the same. Recall that the magnitude decreases with distance from the wire.

[Ans: (a) into; (b) greater]

3 The field at the center of a circular arc is proportional to the angle subtended by the arc and to the reciprocal of the arc radius. Note that in some cases the fields due to the two arcs are in the same direction while in other cases they are in opposite directions.

[Ans: c, a, b]

5 The magnitude of the force of one wire on another is proportional to the reciprocal of the distance between the wires. If the two currents are in the same direction it is a force of attraction along a line joining the wires and if the currents are in opposite directions it is a force of repulsion along that line.

[Ans: (a) 1, 3, 2; (b) less]

__7__ According to Ampere's law the integral is equal to $\mu_0 i_{enc}$, where i_{enc} is the current through the surface bounded by the Amperian loop. If the Amperian loop is inside the wire, $i_{enc} = (r/R)^2 i$, where r is the radius of the Amperian loop, R is the radius of the wire, and i is the total current in the wire. If the Amperian loop is outside the wire, $i_{enc} = i$.

[Ans: c and d tie, then b, a]

__9__ According to Ampere's law the integral is equal to $\mu_0 i_{enc}$, where i_{enc} is the current through the surface bounded by the Amperian loop. Sum the currents through each Amperian loop. Note that some currents are positive and others are negative.

[Ans: b, a, d, c (zero)]

Hints for Problems

__9__ At P the magnitudes of the two magnetic fields of the wires must be the same. The magnitude of the field a distance r from a long straight current-carrying wire is given by $B = \mu_0 i/2\pi r$, where i is the current. Set the two magnitudes equal to each other and solve for the current in wire 2. The two fields must be in opposite directions. When you point the thumb of your right hand in the direction of the current your fingers curl around the wire in the direction of the field lines.

[Ans: 4.3 A (b) out]

__11__ Since the velocity of the proton is perpendicular to the magnetic field, the magnitude of the force on the proton is given by $F = evB$, where v is the speed of the proton and B is the magnitude of the field. The magnitude of the field is given by $B = \mu_0 i/2\pi r$, where i is the current in the wire and r is the distance of the proton from the wire. The direction of the field is given by the fingers of your right hand when your thumb is along the wire in the direction of the current. The direction of the force on the proton is the direction of the vector product $\vec{v} \times \vec{B}$.

[Ans: $(-7.76 \times 10^{-23}\,\text{N})\hat{\imath}$]

__21__ The magnetic field produced by a circular loop of current-carrying wire at its center is given by $\mu_0 i/2R$, where i is the current in the loop and R is the radius of the loop. The field is out of the page for the current direction shown. The magnitude of the field produced by a long straight current-carrying wire is given by $\mu_0 i/2\pi r$, where i is the current and r is the distance from the wire. Add the two fields vectorially. The positive z direction is out of the page.

[Ans: (a) $(253\,\text{nT})\hat{k}$; (b) $(192\,\text{nT})\hat{\imath} + (61.2\,\text{nT})\hat{k}$]

__25__ The magnitude of the magnetic field produced by a straight wire of length L at a point that is a distance R from an end of the wire on a line that is perpendicular to the wire is given by

$$B = \frac{\mu_0 i}{4\pi R} \frac{L}{\sqrt{L^2 + R^2}}.$$

(See problem 19.) To find the direction of the field point the thumb of your right hand along the wire in the direction of the current. Your fingers then curl around the wire in the direction of the field lines. Vectorially add the field of the six wires.

$\left[\text{Ans: (a) } 20\,\mu\text{T; (b) into }\right]$

29 To a good approximation the wires may be considered to be long. The magnitude of the force of any one of them on another is given by $F = \mu_0 i_a i_b L/2\pi d$, where i_a and i_b are the currents in the wires, d is the separation of the wires, and L is the length of a wire. If the currents are in the same direction, the force is one of attraction along a line joining the wires and if the currents are in opposite directions, the force is one of repulsion along such a line. To find the net force on one of the wires sum the forces of the other wires on it.

$\left[\text{Ans: (a) } (469\,\mu\text{N})\,\hat{\jmath}; \text{ (b) } (188\,\mu\text{N})\,\hat{\jmath}; \text{ (c) } 0; \text{ (d) } (-188\,\mu\text{N})\,\hat{\jmath}; \text{ (e) } (-469\,\mu\text{N})\,\hat{\jmath}\right]$

31 The magnitude of the force per unit length of a long straight current-carrying wire on a parallel long straight current-carrying wire is given by $F/L = \mu_0 i_a i_b/2\pi d$, where i_a and i_b are the currents in the wires and d is the separation of the wires. If the currents are in the same direction, the force is one of attraction along a line joining the wires and if the currents are in opposite directions, the force is one of repulsion along such a line. To find the net force on wire 4 sum the forces of the other wires on it.

$\left[\text{Ans: } (-125\,\mu\text{N/m})\,\hat{\imath} + (41.7\,\mu\text{N/m})\,\hat{\jmath}\right]$

33 The forces on the two sides that are perpendicular to the long wire have the same magnitude and opposite directions, so they sum to zero. The long wire attracts the loop side that is closest to it and repels the other loop side. The magnitude of the force on either of these sides is given by $\mu_0 i_W i_L L/2\pi d$, where i_W is the current in the long wire, i_L is the current in the loop, and d is the distance from the loop side to the long wire. Vectorially sum the two forces.

$\left[\text{Ans: } (3.20\,\text{mN})\,\hat{\jmath}\right]$

37 Use Ampere's law with an Amperian loop that is a circle of radius r, concentric with the wire. The law gives $B2\pi r = \mu_0 i_{\text{enc}}$, where B is the magnitude of the magnetic field a distance r from the central axis of the wire and i_{enc} is the current through the loop. If $r < a$, then $i_{\text{enc}} = (r/a)^2 i$, where i is the total current in the wire. If $r > a$, then $I_{\text{enc}} = i$.

$\left[\text{Ans: (a) } 0; \text{ (b) } 0.850\,\text{mT; (c) } 1.70\,\text{mT; (d) } 0.850\,\text{mT}\right]$

39 Use ampere's law with an Amperian loop that is circle of radius r, concentric with the wire. The law gives $B2\pi r = \mu_0 i_{\text{enc}}$, where B is the magnitude of the magnetic field a distance r from the central axis of the wire and i_{enc} is the current through the loop. A circular ring with radius r and infinitesimal width dr has area $2\pi r\,dr$ and the current through such a ring is $2\pi r J\,dr$. Thus $i_{\text{enc}} = \int_0^r 2\pi r' J\,dr'$.

$\left[\text{Ans: (a) } 0; \text{ (b) } 0.10\,\mu\text{T; (c) } 0.40\,\mu\text{T}\right]$

41 The magnitude of the magnetic field inside a solenoid is given by $B = \mu_0 n i$, where n is the number of turns per unit length of the solenoid and i is the current in the solenoid.

$\left[\text{Ans: } 0.30\,\text{mT}\right]$

51 The magnitude of the magnetic field of a circular arc of wire at its center is given by $\mu_0 i\phi/4\pi R$, where i is the current in the arc, ϕ is the angle in radians subtended by the arc, and R is the radius of the arc. The fields of both arcs are into the page. The fields at P of the straight sections are both zero since they are along lines that pass through P.

$\left[\text{Ans: (a) } 0.497\,\mu\text{T; (b) into; (c) } 1.06\,\text{mA}\cdot\text{m}^2\text{; (d) into }\right]$

57 You should be able to show that the sum of the magnetic fields produced at the center by the circular arcs is zero and the field of each of the straight wires with extensions through the center is also zero. That leaves the two straight wires that are not along lines through the center. Use the Biot-Savart law to show that the field of a semi-infinite wire at a point that is a distance R from the wire along a perpendicular line through its end is $\mu_0 i/4\pi R$, where i is the current in the wire. The magnitude of the field a distance R from a long straight wire is $B = \mu_0/2\pi R$.

63 Use Ampere's law to find an expression for the magnetic field inside a wire a distance r from its center and another expression for the magnetic field outside the wire a distance r from its center. Set the first expression equal to the field magnitude given for $r = 4.0\,\text{mm}$ and the second equal to the field magnitude given for $r = 10\,\text{mm}$. These equations can be solved simultaneously for the radius of the wire. The other unknown is the current in the wire.

$\left[\text{Ans: } 5.3\,\text{mm }\right]$

67 Use Ampere's law with a cylindrical Amperian path that has a radius r and is concentric with the cylindrical conductor. The integral on the left side of the Ampere's law equation is $\oint \vec{B} \cdot d\vec{s} = 2\pi r B$. In (a) the enclosed current is $24\,\text{A}$. In (b) it is $24\,\text{A} - (24\,\text{A})f$, where f is the fraction of the cross-sectional area of the cylindrical shell that is within the Amperian path. In (c) it is zero.

$\left[\text{Ans: (a) } 4.8\,\text{mT; (b) } 0.93\,\text{mT; (c) } 0 \right]$

73 Use Ampere's law with a cylindrical Amperian path that has a radius r $(= 5.0\,\text{mm})$ and is concentric with the conducting surface and wire. The integral on the left side of the Ampere's law equation is $\oint \vec{B} \cdot d\vec{s} = 2\pi r B$ and the current enclosed is the sum of the currents in the wire and on the conducting surface. They might be in opposite directions.

$\left[\text{Ans: (a) } 5.0\,\text{mA; (b) downward }\right]$

79 There is a radial line outward from the wire at points along which the magnetic field of the wire is opposite in direction to the uniform field. You want the distance along this line to the point where the field of the wire has the same magnitude as the uniform field. The magnetic field a distance r from a wire is $B_{\text{wire}} = \mu_0 i/2\pi r$, where i is the current in the wire.

$\left[\text{Ans: } 4.0\,\text{mm }\right]$

81 Use the Biot-Savart law to show that the magnetic field a distance d along the perpendicular bisector of a wire with length a is $\mu_0 i/\pi\sqrt{2}a$, where i is the current in the wire. Set $d = a/a$ and sum the fields of the four wires that make up the square.

87 Draw the wire horizontally across your paper and select a direction for the current. The magnitude of the magnetic field a distance r from the wire is $B = \mu_0 i/2\pi r$, where i is the current in the wire. Pick a point above or below the wire and position the electron there. Use the right-hand rule to determine the direction of the field at the electron, then use $\vec{F} = -e\vec{v} \times \vec{B}$, where $\vec{v}$ is the electron's velocity, to determine the force. Use the right-hand rule for vector products to find the direction of the force.

[Ans: (a) 3.2×10^{-16} N; (b) 3.2×10^{-16} N; (c) 0]

Quiz

Some questions might have more than one correct answer.

1. Let $i\,d\vec{s}$ be an infinitesimal current element of length ds, carrying current i. Suppose you want to calculate the x component of the magnetic field it produces at a point that has the position vector $\vec{r}$ relative to the current element. Suppose further that the angle between $\vec{r}$ and $d\vec{s}$ is ϕ and that the vector product $d\vec{s} \times \vec{r}$ makes the angle θ with the x axis. The x component of the field is then

   A. $\dfrac{\mu_0 i\,ds}{4\pi r^2} \cos\phi \cos\theta$

   B. $\dfrac{\mu_0 i\,ds}{4\pi r^2} \cos\phi \sin\theta$

   C. $\dfrac{\mu_0 i\,ds}{4\pi r^2} \sin\phi \cos\theta$

   D. $\dfrac{\mu_0 i\,ds}{4\pi r^2} \sin\phi \sin\theta$

   E. $\dfrac{\mu_0 i\,ds}{4\pi r^2} \sin\phi$

2. Let $i\,d\vec{s}$ be an infinitesimal current element of length ds, carrying current i, and consider the magnetic field it produces at a point that has the position vector $\vec{r}$ relative to the current element. The direction of this field is

   A. parallel to $d\vec{s}$
   B. parallel to $\vec{r}$
   C. perpendicular to $d\vec{s}$
   D. perpendicular to $\vec{r}$
   E. perpendicular to $d\vec{s} \times \vec{r}$

3. The magnetic field outside a long straight current-carrying wire

   A. has a magnitude that is proportional to the distance from the wire
   B. has a magnitude that is proportional to the reciprocal of the distance from the wire
   C. has a magnitude that is proportional to the reciprocal of the square of the distance from the wire
   D. is tangent to a circle that is centered on the wire
   E. is along a line that extends radially outward from the wire

4. The magnetic field inside a long straight wire with current uniformly distributed over its cross section

   A. is parallel to the central axis of the wire
   B. is tangent to circles centered on the central axis of the wire
   C. has a magnitude that is proportional to the reciprocal of the distance from the central axis
   D. has a magnitude that is proportional to reciprocal of the square of the distance from the central axis
   E. has a magnitude that is proportional to the distance from the central axis

5. Two long straight wires are parallel to each other and each carries current i in the same direction. If the distance between the wires is d the magnetic field at the point that is midway between them is

   A. zero
   B. $\mu_0 i/\pi d$
   C. $2\mu_0 i/\pi d$
   D. $4\mu_0 i/\pi d$
   E. $8\mu_0 i/\pi d$

6. The magnetic field produced by a circular arc of current-carrying wire at its center is proportional to

   A. the arc radius
   B. the reciprocal of the arc radius
   C. the current
   D. the angle subtended by the arc at its center
   E. the reciprocal of the angle subtended by the arc at its center

7. Two long straight wires are parallel to each other and each carries the same current i. The magnetic forces of the wires on each other

   A. are forces of attraction if the currents are in the same direction
   B. are forces of repulsion if the currents are in the same direction
   C. are equal in magnitude and opposite in direction
   D. have magnitudes that are proportional to the current
   E. have magnitudes that are proportional to the square of the current

8. Three wires penetrate the side of a building. At the same instant one of them carries a 4-A current into the building, the second carries a 2-A current into the building, and the third carries a 5-A current out of the building. The magnitude of the integral $\oint \vec{B} \cdot d\vec{s}$ of the magnetic field $\vec{B}$ around the periphery of the building side is (in SI units)

   A. zero
   B. $1\,\mu_0$
   C. $3\,\mu_0$
   D. $7\,\mu_0$
   E. $11\,\mu_0$

9. Consider a region of space in which there is a uniform magnetic field $\vec{B}$. An Amperian path is in the form of a square with two sides parallel to the field and two sides perpendicular to the field. If the length of a side is a, the magnitude of the integral $\oint \vec{B} \cdot d\vec{s}$ around the path is

   A. zero
   B. Ba
   C. 2Ba
   D. 3Ba
   E. 4Ba

10. The magnetic field produced by current in a solenoid

    A. is parallel to the solenoid axis nearly everywhere inside and is nearly zero everywhere outside
    B. is parallel to the current both inside and outside
    C. is zero inside and parallel to the solenoid axis outside
    D. on the inside has a magnitude that is proportional to the number of turns per unit length
    E. is nearly uniform on the inside

Answers: (1) C; (2) C, D; (3) B, D; (4) B, E; (5) A; (6) B, D; (7) A, C, E; (8) B; (9) A; (10) A, D, E

Chapter 30
INDUCTION AND INDUCTANCE

A changing magnetic flux through any area induces an emf around the boundary and, if the boundary is conducting, also induces a current. Faraday's law gives the relationship between the emf and the rate of change of the flux. An emf can be generated by changing the magnetic field or by moving a physical object through a magnetic field. Concentrate on how to calculate the magnetic flux and the emf in each case. A non-conservative electric field is induced by a changing magnetic field and this field is responsible for the emf when the magnetic field is changing. You will also learn about inductive circuit elements, which produce emfs via Faraday's law when their currents are changing and which store energy in magnetic fields. Pay attention to the definition of inductance and learn to calculate its value for solenoids and toroids. Learn about the influence of inductance on the current in a circuit. Also learn how to calculate the energy stored in an inductor.

Important Concepts

☐ magnetic flux

☐ Faraday's law

☐ induced emf

☐ induced current

☐ Lenz's law

☐ inductance

☐ flux linkage

☐ inductor

☐ *RL* circuit

☐ magnetic energy

☐ magnetic energy density

☐ mutual inductance

Overview

30–2 Two Experiments

☐ When a current-carrying loop of wire is placed in a magnetic field, the field exerts a torque on the loop and the loop rotates. The reverse also happens. If the loop initially has no current but is then rotated in a magnetic field, current is induced in it.

☐ If a magnet is moved toward a closed loop of wire, current is induced in the wire. When the magnet stops moving, the current stops. As the magnet is withdrawn, current is again induced. The direction of the current is reversed when the direction of motion of the magnet is reversed.

☐ If two loops of wire are close to each other and a changing current is produced in one, then current is induced in the second loop.

☐ In each of these cases, a changing magnetic flux through a closed loop induces an emf around the loop and this emf generates a current. The emf is induced only when the magnetic field through the loop is changing or else when the loop is moving in a magnetic field.

30–3 Faraday's Law of Induction

☐ The **magnetic flux** through a surface is defined by the integral

$$\Phi_B = \int \vec{B} \cdot d\vec{A}$$

over the surface. Here $d\vec{A}$ is an infinitesimal vector area, normal to the surface. If the field is uniform over the surface, then $\Phi_B = BA\cos\theta$, where θ is the angle between the field and the normal to the surface.

☐ The magnetic flux through any area is proportional to the number of magnetic field lines through that area. Recall that the number of field lines through a small area perpendicular to the field is proportional to the magnitude of the field and that the number of lines through *any* small area is proportional to the component B_n along a normal to the area.

☐ The SI unit of magnetic flux is weber and is abbreviated Wb. A weber is equivalent to a tesla·meter squared.

☐ Faraday's law tells us that whenever the flux through any area is changing with time, then an emf is generated around the boundary of that area. Symbolically, the law is

$$\mathcal{E} = -\frac{d\Phi_B}{dt},$$

where $\mathcal{E}$ is the induced emf. The induced emf is distributed around the boundary and is not localized as is the emf of a battery. Its direction is specified as clockwise or counterclockwise, for example. The negative sign that appears in Faraday's law is closely related to the direction of the emf.

☐ The direction of $d\vec{A}$ is ambiguous since two directions are normal to any surface. Φ_B has the same magnitude for both choices but is positive for one and negative for the other. You may choose either but you may find thinking about Faraday's law a bit easier if you pick the one for which the angle between $\vec{B}$ and $d\vec{A}$ is less than 90° and Φ_B is positive. If you point the thumb of your right hand in the direction chosen for $d\vec{A}$, then your fingers will curl around the boundary in the direction of positive emf. If $\mathcal{E}$, calculated using Faraday's law including the minus sign, turns out to be positive, then the emf is in the direction of your fingers. If it turns out to be negative, then it is in the opposite direction. If the boundary of the region is conducting and no other emf's are present, the induced current will be in the direction of the induced emf.

☐ Suppose a coil of wire consists of N identical turns, like a solenoid, with the same magnetic field through each turn. If Φ_B is the flux through each turn, then the total emf induced around the coil is given by $\mathcal{E} = -N\,d\Phi_B/dt$.

☐ The flux through the interior of a loop can be changed by changing the magnitude or direction of the magnetic field, by changing the area of the loop, or by changing the orientation of the loop. No matter how the emf is generated, it represents work done on a unit test charge as the test charge is carried around the loop.

30–4 Lenz's Law

☐ Lenz's law provides another way to determine the direction of an induced emf. Imagine that the boundary of an area is a conducting wire and that an externally produced magnetic field penetrates the area. When the field changes, a current is induced in the wire and the induced current also produces a magnetic field. According to Lenz's law, the sign of the flux of the induced current is the same as that of the externally produced flux if that flux is decreasing and is opposite that of the externally produced flux if that flux is increasing. In the first case, the magnetic field of the induced current is roughly in the same direction as the externally produced field; in the second case, it is roughly in the opposite direction.

☐ Once you have determined the direction of the magnetic field produced by the induced current, you can determine the direction of the current itself and hence that of the emf. Very near any segment of the loop, the field is quite similar to the field of a long straight wire; the lines are nearly circles around the segment. Use the right-hand rule explained in the last chapter: curl your fingers around the segment so they point in the direction of the field in the interior of the loop. That is, they should curl upward through the loop if the field is upward through the loop and they should curl downward if the field is downward. Your thumb will then point along the segment in the direction of the current. This is also the direction of the induced emf.

30–5 Induction and Energy Transfers

☐ An emf is also generated if all or part of the loop moves in a manner that changes the flux through it. If a loop is moved through a uniform field, the flux through it does not change and no emf is generated. If, however, the field is not uniform, then the flux changes as the loop moves and an emf is generated. Find an expression for the flux as a function of the position of the loop, then differentiate it with respect to time to obtain $d\Phi_B/dt$. The emf clearly depends on the velocity of the loop. Either Lenz's law or the sign convention associated with Faraday's law can be used to find the direction of the emf.

☐ If part of a loop moves in such a way that the area changes, an emf is generated even if the field is uniform. The flux through the loop is a function of time because the area is a function of time. If the field is uniform and perpendicular to the area, the emf is given by $\mathcal{E} = -B\,dA/dt$. It is generated only along the moving part, not around the whole loop.

☐ Suppose a closed loop is pulled with constant speed out of the region of a uniform magnetic field, perpendicular to the loop, as shown on the right. At any instant, x is the length of the loop within the region of the field and the flux through the loop is $\Phi_B = BLx$. The emf generated around it is $\mathcal{E} = -BL\,dx/dt = -BLv$, where v is the speed of the loop. The emf is proportional to the speed.

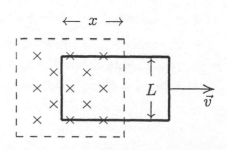

☐ Part or all of the loop being considered may be imaginary. An emf is generated along an isolated rod moving in a magnetic field, for example. To calculate it, you may complete the loop with imaginary lines. If the rod is conducting, electrons move in the direction opposite to the emf when the emf first appears. They collect at one end of the rod, leaving the other end positively charged, and as a result, an electric field exists along the rod. Very quickly the field prevents further build-up of charge and the current becomes zero. Do not confuse the electric field created by charge at the ends of a moving rod with an electric field associated with a time-varying magnetic field.

☐ An emf is also generated around a loop that is changing orientation so the angle between its normal and the magnetic field is changing with time. Suppose a loop with area A rotates in a uniform magnetic field, perpendicular to the axis of rotation. The flux through the loop is given by $\Phi_B = BA \cos \theta$, where θ is the angle between the normal to the plane of the loop and the magnetic field. If θ is changing according to $\theta = \omega t$, then $\mathcal{E} = -d\Phi_B/dt = \omega AB \sin(\omega t)$.

☐ When a current is induced, energy is transferred, via the emf, to the moving charges. Recall that the rate with which an emf $\mathcal{E}$ does work on a current i is given by $P = \mathcal{E}i$. When the emf is associated with a changing magnetic field, the energy comes from the agent changing the field; when the emf is motional, the energy comes from the agent moving the loop or from the kinetic energy of the loop. In either case, the energy is dissipated in the resistance of the loop.

☐ Consider the example above in which a rectangular loop moves with constant speed out of a region with a uniform magnetic field and recall that the magnitude of the emf generated is given by $\mathcal{E} = BLv$. If the resistance of the loop is R, then the current is given by $i = \mathcal{E}/R = BLv/R$, so the emf transfers energy at the rate $P = B^2 L^2 v^2/R$.

☐ This is precisely the rate with which the external agent does work on the loop to keep it moving at a constant speed. The external field $\vec{B}$ exerts a net force of magnitude $F = iBL = B^2 L^2 v/R$ on the loop. The direction of the current is clockwise in the diagram and the magnetic field is into the page so the magnetic force on the moving loop is directed to the left. If the loop is to move with constant velocity, an external agent must apply a force of equal magnitude but in the opposite direction, to the right. Since the loop is moving to the right with speed v, the rate with which the agent does work is $P = Fv = B^2 L^2 v^2/R$.

☐ If the agent stops pulling, the magnetic field exerts a force that slows the moving loop and eventually stops it. The kinetic energy of the loop is then dissipated in its resistance. This is the basis of magnetic braking.

30–6 Induced Electric Fields

☐ An electric field is always associated with a changing magnetic field and this electric field is responsible for the induced emf. The relationship between the electric field $\vec{E}$ and the emf around a closed loop is given by the integral

$$\mathcal{E} = \oint \vec{E} \cdot d\vec{s},$$

where $d\vec{s}$ is an infinitesimal displacement vector. The integral is zero for a conservative field, such as the electrostatic field produced by charges at rest. The electric field induced by a changing magnetic field, however, is non-conservative and the integral is not zero. For a changing magnetic field, Faraday's law becomes

$$\oint \vec{E} \cdot d\vec{s} = -\frac{d\Phi_B}{dt}.$$

☐ Suppose a cylindrical region of space contains a uniform magnetic field, directed along the axis of the cylinder, and that the field is zero outside the region. If the magnetic field changes with time, the lines of the electric field it induces are circles, concentric with the cylinder.

☐ First, consider a point inside the cylinder, a distance r from the center. The magnetic flux through a circle of radius r is $\pi r^2 B$. The circle coincides with an electric field line and the electric field has uniform magnitude around the circle, so the emf is $\mathcal{E} = 2\pi r E$. Faraday's law gives $2\pi r E = -\pi r^2\, dB/dt$, so $E = -\frac{1}{2} r\, dB/dt$.

☐ Now consider a point outside the cylinder. The electric field lines are again circles, concentric with the cylinder. Since the region of the magnetic field extends only a distance R from the cylinder center, the magnetic flux through a circle of radius r is $\pi R^2 B$, where R is the radius of the cylinder. Since the emf is again $\mathcal{E} = 2\pi r E$, Faraday's law gives $2\pi r E = -\pi R^2\, dB/dt$ and $E = -(R^2/2r)\, dB/dt$.

30–7   Inductors and Inductance

☐ Current in a circuit produces a magnetic field and magnetic flux through the circuit. If the circuit consists of N turns and the flux is the same through all of them, then its **inductance** L is defined by

$$L = \frac{N\Phi_B}{i},$$

where i is the current that produces flux Φ_B through each turn. Since Φ_B is proportional to i, L does not depend on the current or flux. It does depend on the geometry of the circuit.

☐ The SI unit of inductance is called the henry (abbreviated H). The quantity $N\Phi_B$ is called the **flux linkage**.

☐ For an ideal solenoid of length ℓ and cross-sectional area A, with n turns per unit length and carrying current i, the magnetic field in the interior is given by $B = \mu_0 in$, the flux through each turn is given by $\Phi_B = BA = \mu_0 inA$, the flux linkage is $N\Phi_B = \mu_0 in^2 A\ell$, and the inductance is $L = n\ell\Phi_B/i = \mu_0 n^2 A\ell$.

☐ Consider a toroid with N turns, a square cross section, inner radius a, and outer radius b. If the current in the toroid is i, the magnetic field a distance r from the center, within the toroid, is $B = \mu_0 iN/2\pi r$ and the flux through a cross section is

$$\Phi_B = \int_a^b B\, dA = \int_a^b \frac{\mu_0 iN}{2\pi r} h\, dr = \frac{\mu_0 iNh}{2\pi} \ln(b/a).$$

The inductance is $L = N\Phi_B/i = (\mu_0 N^2 h/2\pi) \ln(b/a)$.

30–8 Self-Induction

☐ When the current in a circuit changes, the flux changes and an emf is induced in the circuit. If the circuit has inductance L, the induced emf is

$$\mathcal{E}_L = -L\frac{di}{dt}.$$

☐ Every circuit has an inductance, usually small, but there are electrical devices, called **inductors,** that are used expressly to add inductance to a circuit. They usually consist of a coil of wire, like a solenoid. The symbol for an inductor is ◌◌◌◌◌ .

☐ The diagram on the right shows an inductor carrying current i, directed from a to b. The rest of the circuit is not shown. If the current is increasing, then the emf induced in the inductor is from b toward a (a is the positive terminal); if it is decreasing, then the emf is from a toward b (b is the positive terminal). In either case, the potential V_b at point b is given by $V_b = V_a - L\,di/dt$, where V_a is the potential at point a and L is the inductance.

30–9 *RL* Circuits

☐ Consider a circuit consisting of an inductor L, a resistor R, and an emf $\mathcal{E}$. Take the current to be positive in the direction of the emf. The loop rule then gives $\mathcal{E} - iR - L\,di/dt = 0$. Assume the source of emf $\mathcal{E}$ is connected to the circuit at time t = 0, at which time the current is 0. The solution to the loop equation is then

$$i(t) = \frac{\mathcal{E}}{R}\left(1 - e^{-t/\tau_L}\right),$$

where $\tau_L = L/R$ is the **inductive time constant**. This expression predicts $i = 0$ at $t = 0$, but di/dt is not zero then. It also predicts that long after the emf is connected, the current is $i = \mathcal{E}/R$, as if the inductor were not present. The rate at which the current increases to its final value is controlled by the inductive time constant. The larger the time constant, the slower the rate.

☐ The potential difference across the inductor is given by

$$V_L(t) = L\frac{di}{dt} = \mathcal{E}\,e^{-t/\tau_L}$$

and the potential difference across the resistor is given by

$$V_R(t) = iR = \mathcal{E}\left(1 - e^{-t/\tau_L}\right).$$

These functions are graphed in Fig. 30–20.

☐ The current is the least but its rate of change is the greatest at $t = 0$. Then the potential difference across the resistor is zero and the potential difference across the inductor is $\mathcal{E}$. A long time after the emf is connected, the current is $\mathcal{E}/R$ and its rate of change is zero. Then the potential difference across the resistor is $\mathcal{E}$ and the potential difference across the inductor is zero.

☐ Suppose that at time $t = 0$ the circuit has a steady current i_0 and the emf is replaced by a wire. The loop equation is $L\,di/dt + iR = 0$ and its solution is

$$i(t) = i_0\, e^{-t/\tau_L} .$$

The current decreases exponentially to zero. The rate of decrease is controlled by the inductive time constant.

30–10   Energy Stored in a Magnetic Field

☐ Energy must be supplied to build up the magnetic field in an inductor, perhaps by a source of emf. The energy may be considered to be stored in the magnetic field and can be retrieved when the current and field decrease. If current i is in an inductor with inductance L, the energy stored is given by $U_B = \frac{1}{2}Li^2$.

☐ Multiply the loop equation for a series LR circuit by the current to obtain $i\mathcal{E} = i^2R + Li\,di/dt$. The quantity on the left is the rate with which the emf device is supplying energy to the circuit. The first term on the right is the rate with which energy is dissipated in the resistor. The second term on the right is the rate with which energy is being stored in the magnetic field of the inductor.

30–11   Energy Density of a Magnetic Field

☐ The inductance of a solenoid with cross-sectional area A, length ℓ, and n turns per unit length is $L = \mu_0 n^2 A\ell$ and the energy stored is $U_B = \frac{1}{2}Li^2 = \frac{1}{2}\mu_0 n^2 A\ell i^2$, where i is the current. Since the magnetic field in the solenoid is $B = \mu_0 ni$, this can be written $U_B = B^2 A\ell/2\mu_0$. $A\ell$ is the volume of the solenoid, so the energy density (energy per unit volume) is

$$u_B = \frac{B^2}{2\mu_0} .$$

This expression gives the energy density stored in *any* magnetic field, not just the field of a solenoid. The total energy stored in a field is the volume integral of the energy density.

30–12   Mutual Induction

☐ The current in one circuit influences the current in another, although the two are not physically connected. The current in the first circuit produces a magnetic field at all points in space and thus is responsible for a magnetic flux through the second circuit. When the current in the first circuit changes, the flux through the second circuit changes and an emf is induced in that circuit.

☐ If the second circuit has N_2 turns and Φ_{21} is the flux produced through it by the current i_1 in the first circuit, then

$$M_{21} = \frac{N_2\Phi_{21}}{i_1}$$

is the **mutual inductance** of circuit 2 with respect to circuit 1. Similarly, the mutual inductance of circuit 1 with respect to circuit 2 is $M_{12} = N_1\Phi_{12}/i_2$, where N_1 is the number of turns in circuit 1, i_2 is the current in circuit 2, and Φ_{12} is the flux through circuit 1 produced by the current in circuit 2. The two mutual inductances M_{12} and M_{21} are always equal and the subscripts are not needed.

□ When the current i_1 in circuit 1 changes at the rate di_1/dt, the emf induced in circuit 2 is $\mathcal{E}_2 = -M\, di_1/dt$ and when the current i_2 in circuit 2 changes at the rate di_2/dt, the emf induced in circuit 1 is $\mathcal{E}_1 = -M\, di_2/dt$.

Hints for Questions

1 Use Lenz's law. As the conductor expands, the magnetic flux through it increases, so the magnetic field produced by the induced current in the interior of the loop must be opposite the ambient field.

[Ans: out]

3 Use Lenz's law to determine the direction of the emf. In both cases the magnetic flux through the loop is increasing so the magnetic field of the induced current in the interior of the loop is opposite the ambient field. Use the information given for circuit 1 to determine the direction of the ambient field, then determine the direction of the emf induced in circuit 2. The magnitude of the emf is determined by the rate of change of the flux through the circuit.

[Ans: (a) into; (b) counterclockwise; (c) larger]

5 The induced emf is proportional to the rate of change of the z component of the magnetic field. It is clockwise if the field is increasing with time and counterclockwise if the field is decreasing.

[Ans: 1 and 3 tie (clockwise), then 2 and 5 tie (zero), then 4 and 6 tie (counterclockwise)]

7 The potential difference across the resistor is proportional to the current. The current gets to about 63% of its final value in a time equal to the inductive time constant, which is L/R, where L is the inductance and R is the resistance in the circuit.

[Ans: c, b, a]

9 Immediately after the switch is first closed the current in the inductor is zero. A long time after the switch is first closed the currents are constant and the emf of the inductor is zero. Immediately after the switch is reopened the current in the inductor retains its previous value. A long time later the current is zero.

[Ans: (a) more; (b) same; (c) same; (d) same (zero)]

Hints for Problems

3 Use Faraday's law. The magnitude of the emf is equal to the magnitude of the rate of change of the magnetic flux, which is the product of the magnetic field magnitude and the area of the loop. The rate of change of the magnetic field is the slope of the given graph.

[Ans: (a) $-11\,\text{mV}$; (b) 0; (c) $11\,\text{mV}$]

11 The magnetic flux though the coil is the product of the magnitude of the magnetic field, the area of the coil, the number of turns, and the cosine of the angle between the normal to the coil and the magnetic field. This angle, in radians, is given by $2\pi ft$, where f is the frequency of rotation and t is the time. Differentiate the flux with respect to time to find the emf.

[Ans: (b) 0.786 m^2]

13 The current in the circuit is $i = \mathcal{E}/R$, where $\mathcal{E}$ is the induced emf and R is the resistance. According to Faraday's law $\mathcal{E} = -d\Phi_B/dt$, where Φ_B is the magnetic flux. Since $\Phi_B = BA$, where A is the area of the coil, $i = -(A/R)\,dB/dt$. The charge is the integral of the current.

[Ans: 29.5 mC]

15 Use Faraday's law. The area of the circuit is $A_R + A_S\cos(\omega t)$, where A_R is the area of the rectangular portion, A_S is the area of the semicircle, and ω is the angular frequency of rotation. The magnetic flux through the circuit is the product of the area and the magnitude of the magnetic field. Differentiate the flux with respect to time to find the induced emf.

[Ans: (a) 40 Hz; (b) 3.2 mV]

23 Use Faraday's law to find the emf. To calculate the flux divide the loop into strips of length a ($= 2.0$ cm) and infinitesimal width dy. The flux through a strip is $Ba\,dy = 4.0t^2ya\,dy$ and the total flux is the integral of this from $y = 0$ to $y = a$. The emf is the negative of the derivative of the flux with respect to time. Use Lenz's law to find the direction of the emf.

[Ans: (a) 80 μV; (b) clockwise]

31 Use Faraday's law. The area of the circuit is changing. The magnetic flux through the circuit is Bwx, where B is the magnitude of the magnetic field, w is the width of the circuit, and x is the distance of the rod from the right end of the rails. The rate of change of the flux is Bwv, where v ($= dx/dt$) is the speed of the rod. The current in the circuit is the induced emf divided by the resistance. The rate of production of thermal energy is i^2R, where i is the current and R is the resistance. Since the rod moves with constant velocity, the force that must be applied is equal in magnitude to the magnetic force on the rod. The rate with which the external force does work is the product of the force and the speed of the rod.

[Ans: (a) 0.60 V; (b) up; (c) 1.5 A; (d) clockwise; (e) 0.90 W; (f) 0.18 N; (g) 0.90 W]

37 The radius of the magnet is 1.65 cm, so a point 1.6 cm from the central axis is inside the magnet and you should use Eq. 30–25. The magnitude of the magnetic field is $B = B_0 + a\sin(2\pi f)$, where B_0 is the average of the maximum and minimum fields and a is half their difference.

[Ans: 0.15 V/m]

45 The current is the same in the two inductors and the total emf is the sum of their emfs. The equivalent inductance is the total emf divided by the rate of change of the current.

$$\left[\text{Ans: (b) } L_{eq} = \sum_{j=1}^{N} L_j \right]$$

47 L_2 and L_3 are in parallel. Replace them with their equivalent inductor, with inductance L_{23} (see Problem 46). L_1, L_4, and L_{23} are in series. Their equivalent inductance is the equivalent inductance of the circuit (see Problem 45).

[Ans: 59.3 mH]

51 The current is given by $i = i_0 e^{-t/\tau_L}$, where i_0 is the current at time $t = 0$ and τ_L is the inductive time constant. Solve for τ_L by taking the natural logarithm of both sides, then use

$\tau_L = L/R$ to compute R. Here L is the inductance and R is the resistance.

[Ans: $46\,\Omega$]

59 The rate with which energy is stored in the inductor is given by $d(\frac{1}{2}Li^2)/dt = Li\,di/dt$, where L is the inductance, i is the current, and t is the time. The rate of generation of thermal energy is i^2R, where R is the resistance. The current is given by Eq. 30–40: $i = (\mathcal{E}/R)(1 - e^{-t/\tau_L})$, where $\tau_L\ (= L/R)$ is the inductive time constant. Equate the two energy rates, substitute the expression for i, and solve for t. To do this, isolate e^{-t/τ_L} on one side of the equation, then take the natural logarithm of both sides.

[Ans: $25.6\,\text{ms}$]

65 The electric energy density is given by $\frac{1}{2}\epsilon_0 E^2$ and the magnetic energy density is given by $B^2/2\mu_0$. Here E is the magnitude of the electric field and B is the magnitude of the magnetic field. Equate the two energy densities and solve for E.

[Ans: $1.5 \times 10^8\,\text{V/m}$]

73 Use $M = N\Phi_B/i$, where Φ_B is the magnetic flux through the loop, N is the number of turns in the loop, and i is the current in the long straight wire. The magnetic field of the wire is given by $B = \mu_0 i/2\pi r$, where r is the radial distance from the wire. To calculate the flux divide the interior of the loop into strips of length ℓ and infinitesimal width dr. The flux in a strip is $B\ell\,dr = (\mu_0 i/2\pi r)\ell\,dr$. Integrate from $r = a$ to $r = a + b$.

[Ans: $13\,\mu\text{H}$]

75 The magnetic flux through the loop is $\Phi_B = BA = B_0 e^{-t/\tau}\pi r^2$. According to Faraday's law the induced emf is the rate of change of the flux.

[Ans: $(\pi B_0 r^2/\tau)\exp(-t/\tau)$]

79 Since the fuse has zero resistance the current in the fuse is the same as the current in the battery. There is no current in the resistor. Treat this as a single loop circuit as if the resistor branch were not there. Write the loop equation using $L(di/dt)$ for the magnitude of the emf of the inductor. Solve the equation for the current as a function of time, then solve for the time when the current reaches $3.0\,\text{A}$.

[Ans: (a) $1.5\,\text{s}$]

83 (a) Just after the switch is closed the current in the inductor is zero and the circuit can be treated as a single loop containing the two resistors and the battery. Solve the loop equation for the current, then calculate the potential difference across R_1. This must be the same as the potential difference across the inductor and so is equal to $L(dI_L/dt)$, where i_L is the current in the inductor.
(b) Apply the loop rule to the outer loop, containing R_2 and L. Use $L(di_L/dt)$ for the potential difference across the inductor and solve for di_L/dt.
(c) When steady state is reached the currents are constant and the potential difference across the inductor is zero. Again apply the loop rule to the outer loop and solve for the battery current.

[Ans: (a) $400\,\text{A/s}$; (b) $200\,\text{A/s}$; (c) $0.600\,\text{A}$]

93 The magnetic flux through the loop is given by $\Phi_B = BA\cos\theta$, where B is the magnitude of the magnetic field, A is the area of the loop (πr^2), and θ is the angle between the magnetic field and the normal to the loop. According to Faraday's law the emf induced in the loop is given by $\mathcal{E} = d\Phi_B/dt = A(\cos\theta)(dB/dt)$. The current in the loop is $i = \mathcal{E}/R$, where R is the resistance of the loop. The rate of change of the field is $2.0\,\text{T/s}$.

[Ans: $221\,\text{mA}$]

95 The magnetic energy density is given by $u_B = B^2/2\mu_0$, where B is the magnitude of the magnetic field. Multiply this by the volume of the shell. This is the difference in volume between a sphere of radius R and a sphere of radius $R+h$, where R is the radius of Earth and h is the thickness of the shell. To a good approximation the volume is $4\pi R^2 h$ since h is much less than R.

[Ans: (a) $1.0 \times 10^{-3}\,\text{J/m}^3$; (b) $8.4 \times 10^{15}\,\text{J}$]

97 The emf is given by Faraday's law: $\mathcal{E} = -d\Phi_B/dt$. In this case the magnetic field is changing, not the area of the wire loop, so $\mathcal{E} = -A(dB/dt)$. The rate of change of the field is the slope of the graph. The current is $i = \mathcal{E}/R$, where R is the resistance of the loop. Use Lenz's law to find the direction of the emf and current.

[Ans: (a) $0.50\,\text{mA}$; (b) counterclockwise; (c) $0.50\,\text{mA}$; (d) counterclockwise; (e) 0;]

Quiz

Some questions might have more than one correct answer.

1. Suppose a long, straight, current-carrying wire passes perpendicularly through the center of this page. An emf is induced around the boundary of the page if

   A. the current is changing with time
   B. the wire is moving parallel to itself
   C. the angle that the wire makes with the page is changing
   D. the current reverses direction
   E. none of the above are occurring

2. A circular loop is moving in the xy plane through a region where there is a magnetic field. In which of the following situations is an emf generated around the loop?

   A. the magnetic field is uniform and constant
   B. the magnetic field is constant but its z component is different at different places in the xy plane
   C. the magnetic field is constant and its x and z components are uniform but its y component is different at different place in the xy plane
   D. the magnetic field has only an x component but that component is changing with time
   E. the magnetic field is uniform but its z component is changing with time

3. This page is framed by a rectangular loop of wire. You thrust a bar magnet toward the page, with the north pole leading. While the magnet is approaching the page

   A. the magnetic field is into the page and the magnetic flux is increasing
   B. in the interior of the loop the magnetic field of the current induced in the loop is into the page
   C. in the interior of the loop the magnetic field of the current induced in the loop is out of the page
   D. the current and emf induced in the loop are clockwise around the loop
   E. the current and emf induced in the loop are counterclockwise around the loop

4. An airplane is flying north on a horizontal path in a magnetic field that is straight downward. As a result, electrons in the metal body of the plane accumulate

   A. at the front of the plane
   B. at the back of the plane
   C. at the tip of the eastern wing
   D. at the tip of the western wing
   E. nowhere

5. A circular loop of wire lies in the xy plane and is centered at the origin. The magnetic field in the region is in the positive z direction. An emf is generated around the loop as

   A. the loop is rotated about the x axis
   B. the loop is rotated about the y axis
   C. the loop is rotated about the z axis
   D. the loop moves in the positive z direction
   E. the loop is deformed into a square

6. A closed loop of wire has an electrical resistance of $5\,\Omega$. When it is pushed by a constant force of $10\,\mathrm{N}$ through a region with a nonuniform magnetic field it moves in the direction of the force with a constant speed of $2\,\mathrm{m/s}$. Which of the following statements are true?

   A. all the energy supplied by the pushing agent is converted to internal energy
   B. the rate with which the pushing agent does work on the loop is $20\,\mathrm{W}$
   C. the rate with which the pushing agent does work on the loop cannot be computed from the given data
   D. the current induced in the loop is $2\,\mathrm{A}$
   E. the rate of change of magnetic flux through the loop is $10\,\mathrm{Wb/s}$

7. The current is from left to right in an inductor and is decreasing. As a result

   A. the induced emf is from left to right
   B. the induced emf is from right to left
   C. the left end is more positive than the right end
   D. the right end is more positive than the left end
   E. the electric potential is the same at the two ends

8. An inductor, a resistor, an ideal battery, and a switch are wired in a single-loop circuit. Immediately after the switch is closed

A. the current in the circuit is zero
B. the current in the circuit is constant
C. the current in the circuit has its maximum value
D. the electric potential difference across the inductor is equal to the emf of the battery
E. the electric potential difference across the resistor is equal to the emf of the battery

9. An inductor, a resistor, an ideal battery, and a switch are wired in a single-loop circuit. After the switch has been closed for a long time

A. the current in the circuit is zero
B. the current in the circuit is constant
C. the current in the circuit has its maximum value
D. the electric potential difference across the inductor is equal to the emf of the battery
E. the electric potential difference across the resistor is equal to the emf of the battery

10. An inductor, a resistor, an ideal battery, and a switch are wired in a single-loop circuit. When the switch has been closed for a long time, it is opened. The inductive time constant for the circuit is the time required for

A. the current to become zero
B. the current to reach half its maximum value
C. the current to reach about one third of its initial value
D. the induced emf to reach its maximum value
E. the induced emf to reach about one-third of its initial value

11. The magnetic energy stored in an inductor is proportional to

A. the current in the inductor
B. the square of the current in the inductor
C. the reciprocal of the current in the inductor
D. the rate of change of the current in the inductor
E. the square of the rate of change of the current in the inductor

12. A (nonconservative) electric field is induced in any region in which

A. there is changing magnetic flux
B. there is a changing magnetic field
C. the inductive time constant is large
D. the electrical resistance is small
E. there is electrical current

Chapter 32
MAXWELL'S EQUATIONS; MAGNETISM OF MATTER

Maxwell's equations, the four equations that describe the electric and magnetic fields produced by any source, are presented together in this chapter so you can see how they complement each other. One new law and two new concepts are introduced to complete the full set of equations. The new law is Gauss' law for magnetism. Understand both the mathematical statement and the important ramifications of the law. Magnetic dipoles, not monopoles, are the fundamental magnetic entities responsible for the magnetic properties of matter. New concepts deal with magnetic induction and the displacement current.

Magnetic properties of materials arise primarily from the magnetic dipoles that are both intrinsic to electrons and that are associated with their motions. You should understand that the dipole moment of an atom is intimately related to its angular momentum. You should be able to describe the magnetic properties of paramagnetic, diamagnetic, and ferromagnetic materials and you should understand how the behavior of atomic dipoles leads to these properties.

Important Concepts

□ Gauss' law for magnetism

□ induced magnetic field

□ Maxwell's law of induction

□ displacement current

□ Maxwell's equations

□ relationship between
  magnetic dipole moment
  and angular momentum

□ magnetic monopole

□ spin magnetic dipole moment

□ orbital magnetic dipole moment

□ Bohr magneton

□ magnetization

□ diamagnetism

□ paramagnetism

□ Curie's law

□ ferromagnetism

□ Curie temperature

□ hysteresis

□ ferromagnetic domain

Overview

32–2 Gauss' Law for Magnetic Fields

□ If the normal component of the magnetic field is integrated over a *closed* surface (one that completely surrounds a volume), the result is zero no matter what closed surface is chosen. That is,

$$\oint \vec{B} \cdot d\vec{A} = 0$$

for every closed surface. The direction of the infinitesimal element of area $d\vec{A}$ is normal to the surface. This is Gauss' law for magnetism.

☐ The law does not necessarily mean that the magnetic field is zero at any point on the surface, only that the total magnetic flux through any closed surface is zero. Usually the field is essentially outward over some portions and essentially inward over others. Every magnetic field line that enters any region also leaves that region: no field lines start or stop anywhere; they are closed curves.

☐ The law means that **magnetic monopoles** do not exist, or at any rate are so rare that their influence has not been detected. If monopoles existed and if a closed surface surrounded one of them, the right side of the equation would not be zero. Magnetic field lines would start and stop at monopoles so a net magnetic flux would pass through the surface. On the other hand, magnetic dipoles do exist: a charge circulating around a small loop is an example. They do not violate Gauss' law for magnetism and are, in fact, the fundamental sources of magnetism in matter.

32–3 Induced Magnetic Fields

☐ A changing electric field produces a magnetic field. The relationship is

$$\oint \vec{B} \cdot d\vec{s} = \mu_0 \epsilon_0 \frac{d\Phi_E}{dt} .$$

The left side is a path integral around a closed path. Φ_E is the electric flux through the area bounded by the path. In this form, the equation is called the **Maxwell law of induction**. Recall that $\Phi_E = \int \vec{E} \cdot d\vec{A}$, where $d\vec{A}$ is an infinitesimal surface element and is normal to the surface.

☐ A sign convention is associated with the Maxwell induction law. First, pick the direction of $d\vec{A}$ to be used to compute the electric flux. It is usually chosen to make the flux positive. Then, point the thumb of your right hand in the direction of $d\vec{A}$; your fingers then curl around the boundary in the direction of $d\vec{s}$.

☐ Suppose a uniform but time-varying electric field exists in a cylindrical region of radius R and is parallel to the axis of the cylinder. The magnetic field lines are circles that are concentric with the cylinder and the magnetic field is uniform on any one of them. For the field line with radius r, $\oint \vec{B} \cdot d\vec{s} = 2\pi r B$. If the circle is inside the cylinder ($r < R$), then $\Phi_E = \pi r^2 E$ and the Maxwell induction law gives $2\pi r B = \pi r^2 \mu_0 \epsilon_0 \, dE/dt$, so $B = \frac{1}{2}\mu_0 \epsilon_0 r \, dE/dt$. If the circle is outside the cylinder ($r > R$), then $\Phi_E = \pi R^2 E$ and the Maxwell induction law gives $2\pi r B = \pi R^2 \mu_0 \epsilon_0 dE/dt$, so $B = (\mu_0 \epsilon_0 R^2 /r) \, dE/dt$. The magnetic field increases linearly with distance away from the center of the cylinder until the boundary is reached, then it decreases like $1/r$.

☐ Currents also produce magnetic fields, as you learned when you studied Ampere's law. This law and the Maxwell induction law can be combined to yield

$$\oint \vec{B} \cdot d\vec{s} = \mu_0 i + \mu_0 \epsilon_0 \frac{d\Phi_E}{dt} ,$$

which is known as the Ampere-Maxwell law.

32–4 Displacement Current

☐ According to the Ampere-Maxwell law, a changing electric field in a region produces a magnetic field around the boundary of the region, just as if a current passed through the region. In fact, the quantity

$$i_d = \epsilon_0 \frac{d\Phi_E}{dt}$$

is called a **displacement current**. A displacement current is emphatically NOT a true current, which consists of moving charges.

☐ For a cylindrical region of radius R containing a uniform changing electric field parallel to its axis, the displacement current through a circular loop of radius r is $i_d = \epsilon_0 \pi r^2 \, dE/dt$ if $r < R$ and is $i_d = \epsilon_0 \pi R^2 \, dE/dt$ is $r > R$. The direction of the displacement current is the same as the direction of the field if E is increasing and opposite the direction of the field if E is decreasing.

☐ The equations for the magnetic field associated with the cylindrical region considered above are just like the equations for the magnetic field of a long straight wire, developed in Chapter 30, but the true current i is replaced by the displacement current i_d. That is, the field inside the cylinder is $B = \mu_0 i_d (r^2/R^2)/2\pi r$ and the field outside the cylinder is $B = \mu_0 i_d/2\pi r$, where i_d is the total displacement current in the cylinder.

☐ A charging (or discharging) capacitor provides an example of a displacement current. The charge on the plates produces an electric field in the interior and, since the charge is changing, the field is also. Consider a parallel-plate capacitor with plate area A and let $q(t)$ be the charge on the positive plate. Then, the magnitude of the electric field between the plates is $E(t) = q(t)/\epsilon_0 A$. Consider a cross section between the plates with the same area as a plate. The electric flux through the cross section is $\Phi_E = EA$ and the total displacement current through it is $i_d = \epsilon_0 A \, dE/dt = dq/dt$. That is, the total displacement current in the interior is the same as the current in the capacitor wires. The sum of the true and displacement currents is continuous. Although the true current stops at the plates, the displacement current continues into the interior.

32–5 Maxwell's Equations

☐ Here is list of Maxwell's equations, including the displacement current term in the Ampere-Maxwell law:

Gauss' law for electricity: $\oint \vec{E} \cdot d\vec{A} = \dfrac{q}{\epsilon_0}$.

Gauss' law for magnetism: $\oint \vec{B} \cdot d\vec{A} = 0$.

Faraday's law of induction: $\oint \vec{E} \cdot d\vec{s} = -\dfrac{d\Phi_B}{dt}$.

Ampere-Maxwell law: $\oint \vec{B} \cdot d\vec{s} = \mu_0 i + \mu_0 \epsilon_0 \dfrac{d\Phi_E}{dt}$.

☐ The left sides of Gauss' law for electricity and Gauss' law for magnetism contain integrals over closed surfaces. The symbol q on the right side of the law for electricity represents the net charge enclosed by the surface. The zero on the right side of the law for magnetism indicates that magnetic monopoles do not exist. The left sides of Faraday's law and the Ampere-Maxwell law contain integrals around closed paths. On the right side of Faraday's law, Φ_B is the magnetic flux through the surface bounded by the path. On the right side of the Ampere-Maxwell law, i is the net current through the surface bounded by the path and Φ_E is the total electric flux through that surface.

☐ We used Gauss' law for electricity to find the electric field of various charge distributions; we used Faraday's law to find the electric field produced by a changing magnetic field; and we used the Ampere-Maxwell law to find the magnetic field produced by currents and by changing electric fields.

32–6   Magnets

☐ The magnetic dipole is the simplest known magnetic structure. Magnetism arises from the dipole moments of electrons in materials, associated with their spins and orbital motions.

☐ Magnetic field lines exit a bar magnet from the end called the north pole and enter the end called the south pole. The lines continue through the interior of the magnet to form closed loops.

☐ The magnetic field in the exterior of a permanent bar magnet can be closely approximated by the field that would be produced by a positive monopole (a north pole) at one end and a negative monopole (a south pole) at the other but the field does not actually arise from single monopoles but rather from magnetic dipoles associated with electron motion. As proof that the field is not due to monopoles, the magnet can be cut in half with the result that each half produces a field that is closely approximated as a dipole field. This process can be continued to the atomic level with the same result.

☐ The magnetic field of Earth can be approximated by the field of a magnetic dipole with moment $\mu = 8.0 \times 10^{22}$ J/T, located at Earth's center. The dipole moment is not along the axis of rotation but is tilted slightly from that direction. It points from north to south: magnetic field lines leave Earth in the southern hemisphere and enter it in the northern hemisphere. Thus the north geomagnetic pole is actually the south pole of Earth's magnetic dipole.

☐ The direction of Earth's magnetic field at any point is often specified in terms of its **declination** and **inclination**. The declination is the angle between the true geographic north and the horizontal component of the magnetic field. The inclination is the angle between the field and a horizontal plane. If the field has horizontal component B_h and vertical component B_v, then the inclination ϕ_i is given by $\tan \phi_i = B_v/B_h$.

32–7   Magnetism and Electrons

☐ Every electron has an intrinsic angular momentum, often called its spin angular momentum or simply its spin. Only one component, usually taken to be the z component, can be measured at any one time. It is

$$S_z = \pm \frac{h}{4\pi} = \pm 5.2729 \times 10^{-35} \, \text{J} \cdot \text{s} \, ,$$

where h is the Planck constant. A magnetic dipole moment is associated with spin and its z component is

$$\mu_{s,z} = -\frac{e}{m}S_z = \mp\frac{eh}{4\pi m} = \mp 9.27 \times 10^{-24} \, \text{J/T},$$

where m is the electron mass. Since an electron is negatively charged, its dipole moment and spin angular momentum are in opposite directions.

☐ Particle and atomic magnetic moments are often measured in units of the **Bohr magneton** μ_B: $\mu_B = eh/4\pi m$ ($= 9.27 \times 10^{-24}$ J/T). This is the magnitude of the electron spin dipole moment.

☐ A magnetic moment is also associated with the orbital motion of an electron. The z component of the angular momentum is

$$L_{\text{orb},z} = m_\ell \frac{h}{2\pi},$$

where $m_\ell = \pm 1, \pm 2, \ldots, \pm(\text{limit})$ and "limit" is the largest magnitude of m_ℓ. The z component of the dipole moment is

$$\mu_{\text{orb},z} = -m_\ell \frac{eh}{4\pi} = -m_\ell \mu_B.$$

☐ The z component of the total dipole moment is $\mu_z = \mu_{s,z} + \mu_{\text{orb},z}$.

☐ The magnetic energy of an electron in an external magnetic field, in the z direction, is given by $U = -\mu_z B_{\text{ext}}$.

32–8 Magnetic Materials

☐ Three types of magnetic materials (diamagnetic, paramagnetic, and ferromagnetic) are discussed in the following sections. A dipole moment, opposite the applied magnetic field, is induced in all atoms by an externally applied field. This gives rise to diamagnetism. Some atoms are paramagnetic. They have permanent dipole moments but these are randomly oriented in the absence of an external field. When an external field is turned on they tend to align with the field. Ferromagnetic atoms also have permanent dipole moments, but they align with each other even in the absence of an external field.

☐ Paramagnetism and ferromagnetism, when they occur, are much stronger than diamagnetism.

32–9 Diamagnetism

☐ In the absence of an applied field, the atoms of a diamagnetic substance have no magnetic dipole moments, but moments are induced when a field is applied. As an external field is turned on, the orbits of the electrons change so the electrons produce an opposing field, in accordance with Faraday's law. The direction of the magnetization (and the induced dipole moments, on average) is opposite that of the local magnetic field, so the total magnetic field in a diamagnetic material is less in magnitude than the applied field. For most diamagnetic materials, the dipole field is extremely weak.

☐ The effect occurs for all materials, but if the atoms have permanent dipole moments, the effect of their alignment with the field dominates and the material is paramagnetic rather than diamagnetic.

☐ If the external field is not uniform, a diamagnetic material is repelled from a region of greater field toward a region of lesser field.

32–10 Paramagnetism

☐ The **magnetization** of a uniformly magnetized object is its magnetic moment per unit volume. The SI unit of magnetization is an ampere per meter (A/m). If the substance is not uniformly magnetized, the magnetization at a point in the object is the limiting value of the dipole moment per unit volume as the volume shrinks to the point.

☐ Atoms of paramagnetic materials have permanent dipole moments: the dipole moments of the electrons in one of these atoms do not sum to zero. When no external magnetic field is applied, however, the magnetization of the material is zero because the atomic dipole moments are randomly oriented. An external field aligns the atomic moments, and the substance becomes magnetized. When the applied field is removed, the magnetization is quickly reduced to zero by thermal motions, which randomize the moments.

☐ When an external magnetic field is applied, the direction of the field produced by dipoles of the material is in the same direction as the applied field, so the total field in the material is greater in magnitude than the applied field.

☐ According to **Curie's law,** the magnetization at any point in a paramagnetic substance is directly proportional to the magnetic field B at that point and inversely proportional to the absolute temperature T:

$$M = C\left(\frac{B}{T}\right),$$

where the constant of proportionality C is called the **Curie constant** for the material. The relationship is valid only for small values of B/T. The magnetization increases with the applied field because the stronger the field the more atomic dipoles become aligned. The decrease with temperature reflects the randomizing effect of thermal motions.

☐ For large magnetic fields, the magnetization is no longer proportional to the field and, in fact, is less in magnitude than a proportional relationship predicts. For sufficiently high fields, the magnetization is independent of the field. The magnetization is then said to be *saturated*. This occurs when all the dipoles are aligned. If the sample contains N dipoles per unit volume, each with dipole moment μ, then the saturation value of the magnetization is $M_{\max} = N\mu$. See Fig. 32–14 for a graph of the magnetization as a function of the magnetic field.

☐ The potential energy of a dipole $\vec{\mu}$ in a magnetic field $\vec{B}$ is given by $U = -\vec{\mu}\cdot\vec{B}$. The energy changes if the orientation of the dipole changes. Suppose a dipole originally aligned with a field is turned end for end. Its initial potential energy is $U_i = -\mu B$, its final potential energy is $U_f = +\mu B$, and the energy required to turn the dipole is $U_f - U_i = 2\mu B$. To see if the thermal energy of the material is adequate to inhibit magnetization, the change in energy is compared with the mean kinetic energy of the atoms of the material. For an ideal gas of atoms, this is $\frac{3}{2}kT$ at absolute temperature T. At room temperature, the thermal energy is typically several hundred times the magnetic energy and is easily sufficient to randomize the dipole orientations.

32–11 Ferromagnetism

☐ Atoms of ferromagnetic materials have permanent dipole moments but unlike the atomic moments of paramagnetic materials, they *spontaneously* align with each other. Iron is the best known ferromagnetic material.

☐ The spontaneous alignment of dipoles in a ferromagnet is *not* due to the magnetic torque exerted by one magnetic dipole on another. These torques are not sufficiently strong to overcome thermal motions, which tend to randomize the dipole orientations. The source of dipole alignment is in the quantum mechanics of electrons in solids.

☐ For temperatures above its **Curie temperature**, a ferromagnetic substance is paramagnetic. For iron, this temperature is 1043 K.

☐ Ferromagnetic materials exhibit **hysteresis**. This may be demonstrated by plotting the magnetic field B_M, due to the atomic dipoles, as a function of the applied field B_0. See Fig. 32–19. If the substance starts in the unmagnetized state and the external field is increased from zero, B_M is nearly linear in B_0 at first but then at higher applied fields, the slope becomes less. Eventually the magnetization becomes saturated and B_M is constant. When the applied field is reduced, B_M does not follow the same curve downward and, in fact, when $B_0 = 0$, B_M is not zero. A residual magnetism remains: the material is magnetized even though there is no external field.

☐ **Ferromagnetic domains** are responsible for hysteresis. All the dipole moments in a domain are aligned but dipoles in neighboring domains are in different directions. The field B_M is the vector sum of the fields due to all domains. If a ferromagnet is unmagnetized, the vector sum of all the dipole moments is zero. When an external field is applied, domains with dipole moments parallel to the field tend to grow in size while domains with dipole moments in other directions tend to shrink. This amounts to changes in the orientations of dipoles near domain boundaries. In some materials, slight re-orientation of dipoles within a domain may also take place.

☐ Hysteresis comes about because the growth and shrinkage of domains is not reversible. When an external field is applied to an unmagnetized sample and then turned off, the domains do not spontaneously revert to their original sizes.

☐ To measure the magnetic field of the atomic dipoles, the core of a toroid is filled with the ferromagnetic material. If the toroid is narrow and has a large radius, the field of the current i_p in it can be approximated by the field of a solenoid: $B_0 = \mu_0 n i$, where n is the number of turns per unit length. The total field is measured by wrapping a secondary coil of wire around the toroid, measuring the current in it when i is changed, and using Faraday's law to find the total field B. Finally, $B = B_0 + B_M$ is used to compute B_M.

Hints for Questions

1 Use Maxwell's law of induction. Take the vector area to be in the direction of $\vec{E}$. Point the thumb of your right hand in that direction. If your fingers curl in the direction of the field line then E is increasing. If they curl in the opposite direction then E is decreasing.

[Ans: a, decreasing; b, increasing]

3 The displacement current is in the same direction as the true current. Since the capacitor is discharging the left plate is positive and the right plate is negative. To find the direction of the magnetic field at P use the right-hand rule for the magnetic field of a long straight current-carrying wire.

[Ans: (a) rightward; (b) leftward; (c) into]

5 The energy of a magnetic dipole with dipole moment $\vec{\mu}$ in an external magnetic field $\vec{B}_{ext}$ is $U = -\vec{\mu} \cdot \vec{B}_{ext}$. The direction of the spin dipole moment of an electron is opposite that of its spin angular momentum.

[Ans: supplied]

7 To answer part (a) use the right-hand rule for finding the direction of a magnetic dipole moment. Since electrons are negatively charged the current is clockwise, opposite the direction of motion of the electron. If the magnetic dipole moment is in the direction of the magnetic field the dipole is attracted toward regions of greater magnetic field; If the moment is opposite the magnetic field the dipole is repelled from regions of greater magnetic field.

[Ans: (a) all down; (b) 1 up, 2 down, 3 zero]

9 The net magnetic dipole moment of paramagnetic material is in the direction of the external magnetic field. Paramagnetic material is attracted toward regions of greater magnetic field.

[Ans: (a) 1 down, 2 down, 3 up; (b) 1 up, 2 down, 3 zero]

11 Look at Fig. 32–12 and determine what happens to the force if the magnitude of $\vec{B}_{ext}$ increases and if the angle that $\vec{B}_{subext}$ makes with the vertical increases.

[Ans: (a) increase; (b) increase]

Hints for Problems

1 According to Gauss' law for magnetism the net flux through the closed surface is zero. Find the flux through each of the top and bottom faces. The flux curved part of the surface is the negative of their sum. It is outward if it is positive and inward if it is negative.

[Ans: (a) 1.1 mWb; (b) inward]

9 Use the Maxwell law of induction. The magnetic field lines are circles concentric with the boundary of the region and the magnitude of the field is uniform on a line. Thus $\oint \vec{B} \cdot d\vec{s} = 2\pi r B$. Set this equal to $\mu_0 \epsilon_0 \, d\Phi_{Eenc}/dt$ and solve for B. The region of electric flux extends only to $r = R$.

[Ans: (a) 3.54×10^{-17} T; (b) 2.13×10^{-17} T]

11 Use the Maxwell law of induction with a concentric circle of radius r as the integration path. The magnetic field is tangent to the path and its magnitude is uniform around the path, so $\oint \vec{B} \cdot d\vec{s} = 2\pi r B$. The electric flux through the surface bounded by the path is $\int E 2\pi r \, dr$ where the limits of integration are 0 and r for part (a) and are 0 and R for part (b).

[Ans: (a) 3.09×10^{-20} T; (b) 1.67×10^{-20} T]

17 The magnetic field a distance r from the central axis of a uniform distribution of displacement current is given by $B = \mu_0 i_d/2\pi r$, similar to the field of a long straight current-carrying wire. Here i_d is the displacement current through a cross section of radius r. In this case,

$i_d = \pi r^2 J_d$, where J_d is the displacement current density. Use $J_d = \epsilon_0\, dE/dt$ to compute the rate of change of the electric field in the region. (See Problem 15.)

[Ans: (a) $0.63\,\mu$T; (b) 2.3×10^{12} V/m $\cdot$ s]

23 The magnetic field a distance r from the central axis of a uniform distribution of displacement current is given by $B = \mu_o i_d/2\pi r$, where i_d is the displacement current though a cross-section. For part (a) the point is inside the displacement current distribution so you should take i_d to be $\pi r^2 J_d$. For part (b) the point is outside so you should take it to be $\pi R^2 J_d$.

[Ans: (a) 75.4 nT; (b) 67.9 nT]

25 The magnetic field a distance r from the central axis of a uniform distribution of displacement current is given by $B = \mu_o i_d/2\pi r$, where i_d is the displacement current though a cross-section. The displacement current is $\int J_d 2\pi r\, dr$, where the limits of integration are 0 and r for part (a) (the point is inside the displacement current region) and are 0 and R for part (b) (the point is outside that region).

[Ans: (a) 27.9 nT; (b) 15.1 nT]

33 For each of the original levels there is a new level associated with each possible value of m_ℓ. Thus one value of m_ℓ is associated with level E_1 and three values are associated with level E_2. Use $\mu_{\text{orb}\, z} = m_\ell \mu_B$ and $U = -\mu_{\text{orb}\, z} B_{\text{ext}}$ to compute the difference in energy of the levels for which $m_\ell = 0$ and $m_\ell = 1$, *say*.

[Ans: (a) 0; (b) $-1, 0, 1$; (c) 4.64×10^{-24} J]

39 Calculate the largest value of B/T for the experiment. Look at Fig. 32–14 to see if the curve is linear from 0 to that value.

[Ans: yes]

41 The magnitude of the dipole moment is iA where i is the current and A ($= \pi r^2$, where r is the orbit radius) is the area bounded by the electron's path. The current is $e/T = ev/2\pi r$, where T is the period of the motion and v is the speed of the electron. The radius of the orbit is $r = mv/eB$ (see Chapter 28). Make substitutions to write the expression for the dipole moment in terms of v, then use $K = \frac{1}{2}mv^2$ to write it in terms of the kinetic energy K. Since the magnetic force must be inward toward the center of the path you can find the direction of travel for a given field direction and hence can find the direction of the dipole moment. To find the magnetization of the gas, vectorially add the dipole moments per unit volume of the electrons and ions.

[Ans: (b) K_i/B; (b) $-z$; (c) 0.31 kA/m]

45 First find the number of nickel atoms per unit volume. This is the density of nickel divided by the mass of a nickel atom, which is the molar mass divided by the Avogadro constant. The dipole moment is the magnetization divided by the number of atoms per unit volume. You can find the molar mass in Appendix F.

[Ans: 5.15×10^{-24} A $\cdot$ m^2]

47 The magnitude of the magnetic field inside a toroid, a distance r from the center, is given by $B_0 = \mu_0 i_p N_p/2\pi r$, where N_p is the number of turns in the primary and i is the current (see Eq. 29–24). Use the average of the inside and outside radii for r and solve for i_p. The total field is $B = B_0 + B_M$ and the magnetic flux through one turn of the secondary coil is $\Phi_B = BA$, where A is the cross-sectional area of the toroid. According to Faraday's law the emf generated in the secondary is $\mathcal{E} = d\Phi_B/dt$. The current is $i_s = \mathcal{E}/R$, where R is the resistance of the secondary coil. The charge is the integral of the current with respect to time.

[Ans: (a) 0.14 A; (b) 79 μC]

55 Read the graph to find the values of B_{ext}/T for which $M/M_{max} = 0.50$ and 0.90, then solve for T with $B = 2.0$ T.

[Ans: (a) 4 K; (b) 1 K]

61 (a) At every instant the current in the wire equals the total displacement current between the plates, so the maximum current is in the wire is $i_m = 7.60\,\mu$A.
(b) The displacement current is $i_d = \epsilon_0(d\Phi_E/dt)$. Solve for the value of $d\Phi_E/dt$ when i_d has its maximum value.
(c) The magnitude of the electric field between the plates is given by $E = V/d$, where V is the potential difference across the plates. In this case $V = \mathcal{E}$, where $\mathcal{E}$ is the emf of the generator. Thus $\Phi_E = AE = A\mathcal{E}/d$ and $i_d = (\epsilon_0 A/d)(d\mathcal{E}/dt)$. Solve for d.
(d) The point is between the plates. See Sample Problem 1 for the expression for the magnetic field.

[Ans: (a) 7.60 μA; (b) 859 kV · m/s; (c) 3.39 mm; (d) 5.16 pT]

65 Magnetic field lines enter the magnet at the south pole. They are close together near the magnet and further apart away from the magnet. See Fig. 32–4.
(b) The material is paramagnetic, so the field produced by the magnetic moment of the loop in its interior is in the same direction as the magnetic field of the magnet. Near a magnetic dipole the field is in the same direction as the dipole moment.
(c) Use the right-hand rule for magnetic dipoles. If the fingers of your right hand curl around the loop in the direction of the current then your thumb points in the direction of the dipole moment.
(d) The force of a nonuniform magnetic field on a magnetic dipole to toward regions of greater field magnitude.

[Ans: (b) $-x$; (c) counterclockwise; (d) $-x$]

75 See Sample Problem 1 for the equations that give the magnetic field between the plates and outside the plates of a charging capacitor. The maximum field occurs at the boundary of these two regions.

[Ans: (a) 27.5 mm; (b) 110 mm]

Quiz

Some questions might have more than one correct answer.

1. The magnetic flux through one end of a cylinder containing several magnetic dipoles is 5 Wb, outward, and the magnetic flux through the other end is 7 Wb, inward. The magnetic flux though the side of the cylinder

   A. is 12 Wb, outward
   B. is 12 Wb, inward
   C. is 2 Wb, outward
   D. is 2 Wb, inward
   E. cannot be determined from the given data

2. Let Φ_E be the electric flux through a region. As far as producing a magnetic field, the quantity that acts like an electrical current (but is not) is

   A. $\mu_0 \epsilon_0 \dfrac{d\Phi_E}{dt}$
   B. $\mu_0 \epsilon_0 \Phi_E$
   C. $\epsilon_0 \dfrac{d\Phi_E}{dt}$
   D. $\epsilon_0 \Phi_E$
   E. none of the above

3. A magnetic field is induced in any region that

   A. contains a constant electric field
   B. contains a varying electric field
   C. has a varying electric flux through it
   D. is moving
   E. has a cylindrical shape

4. Lines of Earth's magnetic field

   A. leave near the geographic north pole and enter near the geographic south pole
   B. leave near the geographic south pole and enter near the geographic north pole
   C. do not pass through Earth
   D. are constant in time
   E. are uniformly distributed over the surface

5. The orbital magnetic dipole moment of an electron in an atom

   A. is in the same direction as its orbital angular momentum
   B. is opposite in direction to its orbital angular momentum
   C. is perpendicular to its orbital angular momentum
   D. is proportional in magnitude to its orbital angular momentum
   E. is not related to its orbital angular momentum

6. If an electron in an atom has a nonzero orbital angular momentum, its total magnetic dipole moment

    A. is in the same direction as its total angular momentum

    B. is opposite in direction to its total angular momentum

    C. is not parallel to its total angular momentum

    D. is proportional in magnitude to its total angular momentum

    E. is not related to its total angular momentum

7. Atoms in a paramagnetic substance

    A. have permanent magnetic dipole moments, but these moments are not aligned unless an external field is turned on

    B. do not have permanent dipole moments

    C. are responsible for a magnetization that is proportional to the applied external field for all fields

    D. are responsible for a magnetization that is proportional to the applied external field provided the applied field is small in magnitude

    E. produce a magnetic field with a direction that is opposite that of the applied field in the interior of the substance

8. Atoms in a diamagnetic substance

    A. have permanent magnetic dipole moments, but these moments are not aligned unless an external field is turned on

    B. do not have permanent dipole moments, but acquire moments when an external magnetic field is applied

    C. are responsible for a permanent magnetization of the substance

    D. are responsible for a magnetization that is proportional to the applied external field provided the applied field is small in magnitude

    E. produce a magnetic field with a direction that is opposite that of the applied field in the interior of the substance

9. Atoms in a ferromagnetic substance

    A. have permanent magnetic dipole moments which can be at least partially aligned by an external applied field

    B. do not have permanent dipole moments but dipole moments are induced by an external applied magnetic field

    C. are responsible for a magnetization that remains when the applied field is removed

    D. influence the directions of each other's dipole moments by means of their magnetic fields

    E. influence the directions of each other's dipole moments by means of nonmagnetic interactions

Answers: (1) C; (2) C; (3) B; (4) B; (5) B, D; (6) C; (7) A, D; (8) B, D, E; (9) A, C, E

Chapter 33
ELECTROMAGNETIC WAVES

Maxwell's equations predict the possibility of electric and magnetic fields that propagate in free space, with the changing magnetic field producing changes in the electric field, via Faraday's law, and the changing electric field producing changes in the magnetic field, via the displacement current term in the Ampere-Maxwell law. Pay close attention to the relationship between the amplitudes of the fields, to the relationship between their phases, and to the relationship between their directions. Learn about the energy and momentum carried by electromagnetic waves. Also learn how to determine the propagation directions of light that is reflected or refracted at the boundary between two materials.

Important Concepts

- ☐ electromagnetic wave
- ☐ speed of light
- ☐ Poynting vector
- ☐ intensity
- ☐ radiation pressure
- ☐ polarization
- ☐ polarization direction

- ☐ law of Malus
- ☐ wave optics
- ☐ geometrical optics
- ☐ law of reflection
- ☐ law of refraction
- ☐ total internal reflection
- ☐ Brewster's angle

Overview

33–2 Maxwell's Rainbow

☐ Electromagnetic waves exist for all wavelengths and frequencies, from the very large to the very small. Various ranges of wavelengths have been named: gamma radiation, x-ray radiation, ultraviolet radiation, visible radiation, infrared radiation, and microwave radiation. All these electromagnetic radiations are exactly alike except for their frequencies and wavelengths. They all consist of traveling electric and magnetic fields and they all travel in free space with the same speed, the speed of light $c = 2.9979 \times 10^8$ m/s. The ranges do not have sharp boundaries; they blend into each other.

33–3 The Traveling Electromagnetic Wave, Qualitatively

☐ Classically, electromagnetic waves are produced by accelerating charges. Charges at rest or moving with constant velocity do not radiate. In the quantum mechanical picture, electromagnetic radiation is produced when a charge (perhaps an electron in an atom) changes its state or in reactions of fundamental particles.

☐ Figs. 33-33, 33–4, and 33–5 show the electric and magnetic fields of a radiating antenna consisting of straight wires in which the current varies sinusoidally, driven by an RLC circuit. The electric field in the antenna varies sinusoidally and because its current varies sinusoidally, the magnetic field it produces also varies sinusoidally. The fields produced by the antenna travel outward from it with the speed of light.

☐ A traveling electromagnetic wave has the following properties:

    1.   The electric and magnetic fields are perpendicular to each other.

    2.   The wave travels in a direction that is perpendicular to both the electric and magnetic fields.

    3.   The electric and magnetic fields are in phase. Both of them reach their maximum values at the same time and are zero at the same time.

☐ Very far from the source, the wave becomes a plane wave: the electric and magnetic field lines are nearly straight lines. Mathematically, the fields of an electromagnetic wave traveling in the positive x direction are then given by

$$E = E_m \sin(kx - \omega t) \qquad \text{and} \qquad B = B_m \sin(kx - \omega t),$$

where $k = 2\pi/\lambda$, λ is the wavelength, and ω is the angular frequency. These quantities are related by $\omega/k = c$, where c is the speed of light. The phase constants are the same for the two fields.

☐ The electric field of an electromagnetic wave is the source, via the Ampere-Maxwell law, of the magnetic field and the magnetic field is the source, via Faraday's law, of the electric field. Once started, the fields produce each other and travel together through space. When these laws are applied, you find that the amplitudes of the fields are related by

$$\frac{E_m}{B_m} = c$$

and that the wave speed is determined by the permittivity and permeability constants ϵ_0 and μ_0:

$$c = \frac{1}{\sqrt{\mu_0 \epsilon_0}}.$$

33–4   The Traveling Electromagnetic Wave, Quantitatively

☐ For a wave traveling along the x axis, the electric field is different at a point with coordinate x and a point with coordinate $x + dx$, an infinitesimal distance away, because a changing magnetic flux penetrates the region between these points. The magnetic field is different at x and $x + dx$ because a changing electric flux penetrates the region between these points. Differences in the fields at different points can be computed using the Faraday and Ampere-Maxwell laws.

☐ The following diagram shows a small region of space in which a wave is traveling toward the right. The electric fields at two infinitesimally separated points, x and $x + \Delta x$, are shown, as is the magnetic field in the region between. Apply Faraday's law to this situation. To calculate the magnetic flux, take $d\vec{A}$ to be out of the page and to calculate the emf, traverse the path shown in the counterclockwise direction. Clearly, $\Phi_B = Bh\Delta x$ and $\oint \vec{E} \cdot d\vec{s} = h[E(x + \Delta x) - E(x)]$. As Δx becomes small, $[E(x + \Delta x) - E(x)]/\Delta x$ becomes $\partial E/\partial x$ and Faraday's law yields $\partial E/\partial x = -\partial B/\partial t$.

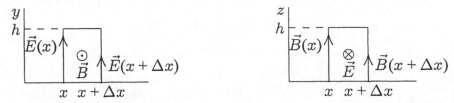

☐ The diagram on the right above shows the view looking up from underneath the previous diagram. The magnetic field at the two points and the electric field between are shown. Apply the Ampere-Maxwell law to this situation. Take $d\vec{A}$ to be into the page and traverse the path in the clockwise direction. Clearly, $\Phi_E = Eh\Delta x$ and $\oint \vec{B} \cdot d\vec{s} = h[B(x) - B(x + \Delta x)]$. As Δx becomes small, $[B(x) - B(x + \Delta x)]/\Delta x$ becomes $-\partial B/\partial x$ and the Ampere-Maxwell law yields $\partial B/\partial x = \mu_0 \epsilon_0 \, \partial E/\partial t$.

☐ Substitute $E = E_m \sin(kx - \omega t)$ and $B = B_m \sin(kx - \omega t)$ into these two equations and show that $kE_m = \omega B_m$ and that $kB_m = \mu_0 \epsilon_0 \omega E_m$. Then use $\omega/k = c$ to show that $E_m = cB_m$ and $c^2 = 1/\mu_0 \epsilon_0$.

33–5 Energy Transport and the Poynting Vector

☐ Electromagnetic waves carry energy. The energy per unit volume associated with an electric field $\vec{E}$ is given by $u_E = \frac{1}{2}\epsilon_0 E^2$ and the energy per unit volume associated with a magnetic field $\vec{B}$ is given by $u_B = B^2/2\mu_0$. You can use $B = E/c$ and $c = 1/\sqrt{\epsilon_0 \mu_0}$ to show that for a plane wave $u_E = u_B = EB/2\mu_0 c$ and that the total energy density is $u = EB/\mu_0 c$.

☐ This energy moves with the wave, with speed c. Consider a region of space with infinitesimal width dx and cross-sectional area A, perpendicular to the direction of travel of a plane electromagnetic wave. The volume of the region is $A\,dx$, so the electromagnetic energy in the region is $dU = uA\,dx$. All this energy will pass through the area A in time $dt = dx/c$. Substitute $dx = c\,dt$ and divide by dt. The energy passing through the area per unit time is $dU/dt = ucA$ and the energy per unit area passing through per unit time is $dU/A\,dt = uc = EB/\mu_0$.

☐ The transport of energy is described in terms of the **Poynting vector** $\vec{S}$, defined by the vector product

$$\vec{S} = \frac{\vec{E} \times \vec{B}}{\mu_0}.$$

Since $\vec{E}$ and $\vec{B}$ are perpendicular to each other, the magnitude of the Poynting vector is given by $S = EB/\mu_0$. The energy passing through a surface of area A, perpendicular to the direction of travel, per unit time, is given by $dU/dt = SA$ and the energy passing through per unit area, per unit time, is given by $dU/A\,dt = S$.

☐ For most sinusoidal electromagnetic waves of interest, the fields oscillate so rapidly that their instantaneous values cannot be detected or else are not of interest. Energies and energy densities are then characterized by their average over a period of oscillation. For a sinusoidal wave, the average value of E^2 is $\frac{1}{2}E_m^2$, where E_m is the amplitude. Since the rms value of E is $E_{rms} = E_m/\sqrt{2}$ the average of E^2 is also E_{rms}^2.

☐ The **intensity** of a wave is the magnitude of the Poynting vector averaged over a period of oscillation:

$$I = \frac{E_m B_m}{2\mu_0}.$$

This can be written as $I = E_m^2/2c\mu_0$ and as $I = cB_m^2/2\mu_0$. For sinusoidal waves, the intensity is often written in terms of the rms value E_{rms} of the electric field: $I = E_{rms}^2/c\mu_0$.

☐ The direction of $\vec{S}$ is the direction in which the wave is traveling. Recall that this direction is intimately connected with the relative signs of the terms kx and ωt in the argument of the trigonometric function that describes the fields. If the wave travels in the negative x direction, you write $E_y(x,t) = E_m \sin(kx + \omega t)$ and $B_z(x,t) = -B_m \sin(kx + \omega t)$. The negative sign appears so that $\vec{E} \times \vec{B}$ is in the negative x direction.

☐ For a source that emits isotropically the energy per unit time that crosses every sphere centered at the source is the same. This means that the intensity a distance r from the source is given by $I = P_s/4\pi r^2$, where P_r is the power of the source.

33–6 Radiation Pressure

☐ Electromagnetic waves also carry momentum. If ΔU is the energy in any small volume, then $\Delta p = \Delta U/c$ is the magnitude of the momentum in that volume. The momentum is in the direction of $\vec{S}$, the direction of propagation. If u is the energy per unit volume, then u/c is the momentum per unit volume and if I is the average energy transported through an area per unit area per unit time (the intensity), then I/c is the average momentum transported through the area per unit area per unit time.

☐ The electric field component of an electromagnetic wave exerts a force on a charged particle and does work on it. If the charge is moving, the magnetic field component also exerts a force. Thus momentum and energy may be transferred to an object when radiation interacts with it.

☐ When electromagnetic radiation is absorbed by a material object, all its energy and momentum are transferred to the object. Consider a plane wave with time averaged energy density u, incident normally on the plane surface of an object. If the area of the surface is A, then, averaged over a period, the energy transferred in time Δt is given by $IA\,\Delta t$ and the momentum transferred is given by $(IA/c)\,\Delta t$. The momentum transferred per unit time is the force of the radiation on the object and the force per unit area is the **radiation pressure**. Thus the radiation pressure is I/c.

☐ If the wave is reflected without loss back along the original path, the energy transferred is zero and the momentum transferred is $2(IA/c)\,\Delta t$. The radiation pressure is now $2I/c$, twice what it would be if the radiation were absorbed.

33–7 Polarization

☐ The electric field of a **linearly polarized** electromagnetic wave is always parallel to the same direction. The **direction of polarization** is parallel to the electric field and the **plane**

of polarization is determined by the directions of propagation and the electric field. For maximum signal, the wires in an electric dipole antenna used to detect polarized waves must be aligned with the polarization direction.

☐ An electromagnetic wave is said to be **unpolarized** if the electric field at any point changes direction randomly and often. Light from an incandescent bulb or from the Sun is unpolarized. The direction of its electric field changes rapidly in a random fashion because the light comes from many different atoms, each of which emit waves in short bursts, with random polarization directions.

☐ Any linearly polarized wave can be treated as the sum of two waves, polarized in any two mutually orthogonal directions that are perpendicular to the direction of propagation. Suppose a wave with an electric field amplitude E_m is traveling along a z axis and is polarized with its electric field along a line that makes an angle θ with the x axis. You can consider the electric field to be the vector sum of two fields, one with amplitude $E_m \cos \theta$, polarized along the x axis and one with amplitude $E_m \sin \theta$, polarized along the y axis.

☐ Polarized radiation can be produced by shining unpolarized radiation through a sheet of Polaroid. These sheets contain certain long-chain molecules that are aligned in the manufacturing process. Radiation with its electric field vector parallel to the chains is preferentially absorbed while radiation with its electric field vector perpendicular to the chains is preferentially transmitted. A line perpendicular to the molecules is said to be along the **polarizing direction** of the sheet.

☐ Suppose that linearly polarized radiation with amplitude E_m and intensity I_m is incident on an ideal polarizing sheet and that its electric field is along a line that makes an angle θ with the polarizing direction. Then the amplitude of the transmitted radiation is given by $E = E_m \cos \theta$ and its intensity is given by $I = I_m \cos^2 \theta$. This is called the **law of Malus**. Note that the transmitted amplitude is the component of the incident amplitude along the polarization direction of the sheet. The transmitted radiation is polarized along the polarizing direction of the sheet.

☐ If unpolarized radiation with intensity I_i is incident on an ideal polarizing sheet, the intensity of the transmitted radiation is given by $I = I_i/2$. The transmitted radiation is polarized along the polarizing direction of the sheet.

33–8 Reflection and Refraction

☐ In **wave optics**, the wave nature of light is used to describe and understand optical phenomena. Wave optics is close to the fundamental principles, Maxwell's equations, and is valid for all optical phenomena outside the quantum realm. When any obstacles to light or any openings through which light passes are large compared to the wavelength of the light, then details of its wave nature are not important. The direction of travel of the light is important. This is the realm of **geometrical optics** (or ray optics).

☐ Geometrical optics uses rays to describe the path of light. A ray is a line in the direction of propagation of a light wave. The propagation direction changes when light is reflected or enters a region where the wave speed is different.

□ When light in one medium encounters a boundary with another region, in which the wave speed is different, some light is reflected back into the region of incidence and some is transmitted into the second region. The diagram on the right shows a ray in medium 1 incident on the boundary with medium 2. It makes the angle θ_1 with the normal to the boundary. The reflected ray makes the angle θ_1' with the normal and the transmitted ray makes the angle θ_2 with the normal. The angle θ_1 is called the **angle of incidence**, θ_1' is called the **angle of reflection**, and θ_2 is called the **angle of refraction**. Reflected and refracted rays all lie in the plane of incidence, defined by the incident ray and the normal to the surface. The **law of reflection** is $\theta_1' = \theta_1$ and the **law of refraction** is $n_2 \sin\theta_2 = n_1 \sin\theta_1$, where n_1 is the **index of refraction** for medium 1 and n_2 is the index of refraction for medium 2. The law of refraction is also known as Snell's law.

□ The index of refraction of a medium is the ratio of the speed of light c in vacuum to the speed of light v in the medium: $n = c/v$.

□ Light incident from a medium with a small index of refraction bends toward the normal as it enters a medium with a higher index of refraction. Light incident from a medium with a large index of refraction bends away from the normal as it enters a medium with a smaller index of refraction.

□ The index of refraction depends on the wavelength of the light. Light rays of different wavelengths are bent through different angles on refraction. This accounts for the bands of colors produced by prisms, for example.

33–9   Total Internal Reflection

□ When **total internal reflection** occurs, incident light is only reflected; it is not refracted into the neighboring region. This can occur when light is incident from a medium with a large index of refraction on a medium with a smaller index of refraction *and* the angle of incidence is greater than a certain critical value denoted by θ_c. Suppose $n_1 > n_2$ and light is incident from medium 1 on the boundary with medium 2. Then $\theta_1 = \theta_c$ if $\theta_2 = 90°$ ($\sin\theta_2 = 1$). The critical angle is given by $\theta_c = \sin^{-1}(n_2/n_1)$.

33–10   Polarization by Reflection

□ If unpolarized light is incident on a boundary between two different materials, both the reflected and refracted waves are partially polarized. For a special angle of incidence, called **Brewster's angle** and denoted by θ_B, the reflected wave is completely polarized. If θ_r is the angle of refraction for incidence at Brewster's angle, then $\theta_B + \theta_r = 90°$. If n_1 is the index of refraction for the medium of incidence and n_2 is the index of refraction for the medium of refraction, then Snell's law leads to $\tan\theta_B = n_2/n_1$. For the special case $n_1 = 1$ (air or vacuum), $\tan\theta_B = n_2$. This is **Brewster's law** for the polarizing angle.

□ For incidence at Brewster's angle, the reflected light is polarized perpendicularly to the plane determined by the incident and reflected rays. Even at Brewster's angle, the *refracted* light contains components with all polarization directions.

Hints for Questions

1 The direction of propagation of an electromagnetic wave is the direction of the vector product $\vec{E} \times \vec{B}$.

[Ans: into]

3 If I_0 is the incident intensity, then the transmitted intensity is $I = I_0/2$ for unpolarized light and is $I = I_0 \cos^2 \theta$ for polarized light with its direction of polarization making the angle θ with the polarizing direction of the sheet.

[Ans: (a) same; (b) increase; (c) decrease]

5 The transmitted intensity is zero if the polarizing directions of the second and third sheets are perpendicular to each other or if the polarizing directions of the third and fourth sheets are perpendicular to each other.

[Ans: 20° and 90°]

7 Use the law of refraction and remember that the sine of an angle increases as the angle increases from zero. The light bends toward the normal to the surface if the index of the medium of refraction is greater than the index of the medium of incidence and bends away from the normal if the index is less.

[Ans: a, b, c]

9 Use the law of refraction and remember that the sine of an angle increases as the angle increases from zero.

[Ans: d, b, a, c]

11 The critical angle for total internal reflection is the smallest angle of incidence θ for which $(1/n) \sin \theta$ is greater than 1. Rays at these angles lie along the boundaries between the shaded and unshaded regions on the diagrams.

[Ans: n_3, n_2, n_1]

Hints for Problems

3 The frequency is given by $f = c/\lambda$ and the period is given by $T = 1/f$, where λ is the wavelength. Find the radius of Earth in Appendix C.

[Ans: (a) 4.7×10^{-3} Hz; (b) 3 min 32 s]

13 The intensity is given by $E_m^2/2c\mu_0$, where E_m is the electric field amplitude. The magnetic and electric field amplitudes are related by $B_m = E_m/c$.

[Ans: (a) 1.03 kV/m; (b) 3.43 μT]

17 The intensity is given by $E_m^2/2c\mu_0$, where E_m is the electric field amplitude. The magnetic and electric field amplitudes are related by $B_m = E_m/c$. Since the source radiates isotropically the energy per unit time passing through any sphere centered at the source is the same and is equal to the power of the source.

21 The radiation pressure on a perfectly absorbing object is given by $p_r = I/c$, where I is the intensity at the object. The intensity a distance r from an isotropically emitting source of power P_s is $I = P_s/4\pi r^2$.

[Ans: 5.9×10^{-8} Pa]

29 The intensity a distance r from a point source is given by $I = P_s/4\pi r^2$, where P_s is the power of the source. Thus the slope of the line on the graph is $P_s/4\pi$.

[Ans: 0.25 kW]

35 The transmitted intensity is $I = I_0/2$, where I_0 is the incident intensity. The intensity is related to the electric field amplitude E_m by $I = E_m^2/2\mu_0 c$. Since the sheet is absorbing the radiation pressure is $p_r = I_a/c$, where I_a is the intensity of the absorbed light. This is $I_a = I_0 - I$.

[Ans: (a) 1.9 V/m; (b) 1.7×10^{-11} Pa]

39 Let I_p be the intensity associated with the portion of the light that is polarized and let I_u be the intensity associated with the portion that is unpolarized. The transmitted intensity is $I_p \cos^2\theta + I_u/2$, where θ is the angle between the direction of polarization of the polarized portion and the polarizing direction of the sheet. It has a maximum of $I_p + I_u/2$ and a minimum of $I_u/2$. Solve for $I_p/(I_p + I_u)$.

[Ans: 0.67]

47 Use the law of refraction. The angle of refraction is $90°$ and the angle of incidence is given by $\tan\theta = L/D$.

[Ans: 1.26]

51 Use the law of refraction at each boundary. For the ray that passes through all the layers the angle of incidence at the exit point of a layer is equal to the angle of refraction at the entrance point. For the boundary between the upper layer and air the angle of incidence is equal to the angle of reflection at the boundary between the two upper layers.

[Ans: (a) 56.9°; (b) 35.3°]

57 The angle of incidence for the light approaching the boundary between materials 2 and 3 is ϕ. This is the critical angle for total internal reflection, so $n_2 \sin\phi = n_3$. Apply the law to the refraction at the interface between materials 1 and 2: $n_1 \sin\theta = n_2 \sin(90° - \phi)$. Solve for θ.

[Ans: (a) 1.39 (b) 28.1°; (c) no]

63 Let θ_2 be the angle of refraction at P, θ_3 be the angle of incidence at Q, and θ_4 (= 90°) be the angle of refraction at Q. The law of refraction applied at point P gives $\sin\theta = n\sin\theta_2$ and applied at point Q gives $n\sin\theta_3 = 1$. With a little geometry you can show that $\theta_3 = 90° - \theta_2$. (The key here is that the normals to the two surfaces are perpendicular to each other.) Use two of these relationships to eliminate θ_2 and θ_3 from the third, then solve for n as a function of θ. Find the value of θ for which n is a maximum. The critical angle for total internal reflection at Q is $\sin\theta_3 = 1/n$. If θ is increased does $\sin\theta_3$ become greater or less than $1/n$? What happens to $\sin\theta_3$ if θ is decreased?

[Ans: (a) $\sqrt{1 + \sin^2\theta}$; (b) $\sqrt{2}$; (c) yes; (d) no]

67 Since the incident light is unpolarized its intensity as it leaves the first sheet is $I_0/2$, where I_0 in the incident intensity. Furthermore it is polarized along the polarizing direction of the sheet. At each sheet the polarization direction of the incident light is the same as the polarizing direction of the previous sheet and the intensity is multiplied by $\cos^2\theta$, where θ is the angle between the polarization direction of the incident light and the polarizing direction of the sheet.

$\left[\text{Ans: } 0.50\,\text{W/m}^2\,\right]$

71 Since the catfish "sees the world" consider a light ray that is incident along the surface of the water and is refracted to the eye of the catfish. If θ is the angle of refraction, then the law of refractions gives $n \sin\theta = 1$. Take n to be 1.33 and solve for θ. A little trigonometry shows that the radius r of circle of sight is related to the depth d of the catfish and the angle of refraction by $\tan\theta = r/d$.

$\left[\text{Ans: (a) } 4.56\,\text{m; (b) increase}\,\right]$

73 To find the direction of travel look at the argument of the trigonometric function in the expression for the magnetic field. The direction of polarization is the direction of the electric field. Recall that the direction of travel is the same as the direction of the vector product $\vec{E} \times \vec{B}$. The intensity is given by $I = E_m^2/2\mu_0 c$, where E_m is the electric field amplitude, which is related to the magnetic field amplitude by $E_m = cB_m$. The wavelength is given by $\lambda = 2\pi c/\omega$, where ω is the angular frequency, which can be read from the given function. Look at Fig. 33-1 to decide on the region of the electromagnetic spectrum.

$\left[\text{Ans: (a) negative } y; \text{ (b) along the } z \text{ axis; (c) } 1.01\,\text{kW/m}^2; \text{ (d) } E_z = (1.20\,\text{kV/m}) \sin[(6.67 \times 10^6\,\text{m}^{-1})y + (2.00 \times 10^{15}\,\text{s}^{-1})t]; \text{ (e) } 942\,\text{nm; (f) infrared}\,\right]$

79 The angle of refraction and the Brewster angle sum to $90°$. Furthermore the tangent of the Brewster angle is the index of refraction for the medium of refraction divided by the index for the medium of incidence.

$\left[\text{Ans: (a) } 1.60; \text{ (b) } 58.0°\,\right]$

83 Solve the result of Problem 53 for ψ. First solve for $(\psi + \phi)/2$, then for ψ. Since the crystals are six-sided the apex angle of the dotted triangle is $60°$.

$\left[\text{Ans: } 22°\,\right]$

87 Look at Fig. 33–19 to find the index of refraction n for each of the two wavelengths, then use $\theta_B = \tan^{-1} n$ to calculate the Brewster angle θ_B.

$\left[\text{Ans: (a) } 55.8°; \text{ (b) } 55.5°\,\right]$

91 Because the sphere is totally absorbing the radiation pressure on it is $p_r = I/c$, where I is the intensity of the radiation at the sphere. The intensity a distance r from a point source that emits isotropically is $P_s/4\pi r^2$, where P_s is the power of the source. The force on the sphere is the product of the radiation pressure and the cross-sectional area of the sphere: $F = p_r 4\pi R^2$, where R is the radius of the sphere.

$\left[\text{Ans: } 1.7 \times 10^{-13}\,\text{N}\,\right]$

101 The magnitudes E and B of the electric and magnetic fields are related by $E = cB$. The direction of travel of the wave is the same as the direction of the vector product $\vec{E} \times \vec{B}$.

$\left[\text{Ans: (a) } 0.33\,\mu\text{T; (b) } -x\,\right]$

107 The light is polarized along the axis that is parallel to the electric field. The direction of travel of the wave can be found by observing the argument of the given trigonometric function, then the direction of the electric field is chosen so that the vector product $\vec{E} \times \vec{B}$ is in the direction of travel. The angular frequency ω can be read from the given function. The frequency is

$\omega/2\pi$. The intensity is given by $E_m^2/2\mu_0 c$, where E_m is the electric field amplitude. Recall that $E_m = cB_M$, where B_m is the magnetic field amplitude.

[Ans: (a) z axis; (b) 7.5×10^{14} Hz; (c) $1.9\,\text{kW/m}^2$]

Quiz

Some questions might have more than one correct answer.

1. All electromagnetic waves have

    A. the same frequency
    B. the same wavelength
    C. the same amplitude
    D. the same speed in a vacuum
    E. the same phase constant

2. An electromagnetic wave consists of

    A. vibrating electrons
    B. vibrating protons
    C. electric and magnetic fields
    D. electric fields alone
    E. magnetic fields alone

3. the direction of travel of an electromagnetic wave is the same as the direction of

    A. the electric field component
    B. the magnetic field component
    C. the vector product $\vec{B} \times \vec{E}$, where $\vec{E}$ is the electric field and $\vec{B}$ is the magnetic field
    D. the vector product $\vec{E} \times \vec{B}$, where $\vec{E}$ is the electric field and $\vec{B}$ is the magnetic field
    E. some other vector

4. For a traveling sinusoidal electromagnetic wave

    A. the electric and magnetic field amplitudes are proportional to each other
    B. the electric and magnetic field amplitudes are reciprocal to each other
    C. the electric and magnetic fields have their maximum values at the same place and time
    D. the magnetic field is zero where the electric field is a maximum
    E. the magnetic field is decreasing where the electric field is increasing

5. Which of the following statements are true?

    A. the expressions for the electric and magnetic field components of a traveling electromagnetic wave satisfy Maxwell's equations
    B. Maxwell's equations predict the speed of electromagnetic waves
    C. Maxwell's equations predict that visible light is an electromagnetic wave
    D. Maxwell's equations predict the ratio of the electric and magnetic field amplitudes
    E. Maxwell's equations predict the wavelength of an electromagnetic wave

6. The energy in a traveling electromagnetic wave
    A. is transported with the speed of the wave
    B. is not carried by the wave
    C. is divided equally between electric and magnetic field energies
    D. is mostly electric field energy
    E. fluctuates sinusoidally

7. The intensity of a traveling sinusoidal electromagnetic wave is
    A. the average over a cycle of the energy per unit area per unit time passing through a (perhaps imaginary) surface that is perpendicular to the direction of travel of the wave
    B. the energy per unit area per unit time passing through a (perhaps imaginary) surface that is perpendicular to the direction of travel of the wave
    C. equal to the magnitude of the vector $\vec{E} \times \vec{B}/\mu_0$
    D. equal to the average over a cycle of EB/μ_0, where E is the magnitude of the electric field component of the wave and B is the magnitude of the magnetic field component
    E. zero because the fields vary sinusoidally

8. The radiation pressure of an electromagnetic wave on a surface of incidence
    A. is twice as great for a totally absorbing surface as for a totally reflecting surface
    B. is twice as great for a totally reflecting surface as for a totally absorbing surface
    C. proportional to the intensity of the wave
    D. occurs because electromagnetic waves carry momentum as well as energy
    E. zero because electromagnetic waves do not carry momentum

9. In a linearly polarized electromagnetic wave
    A. the electric field is parallel to the direction of travel of the wave
    B. the electric field is parallel to the magnetic field
    C. the electric field at any point is parallel to the electric field at any other point
    D. the electric field component travels faster than the magnetic field component
    E. the electric field vector at any point rotates about the direction of travel of the wave

10. Light with an intensity I_0 is incident on a polarizing sheet. The intensity of the transmitted light is
    A. $I_0/2$ in all cases
    B. $I_0/2$ if the incident light is unpolarized
    C. $I_0/2$ if the incident light is polarized with its electric field making an angle of 45° with the polarizing direction of the sheet
    D. I_0 if the incident light is polarized with its electric field making an angle of 90° with the polarizing direction of the sheet
    E. I_0 if the incident light is polarized with its electric field parallel to the polarizing direction of the sheet

11. Light is incident in medium A on the plane surface of medium B. A and B have different indices of refraction. An incident ray makes an angle of 30° with the normal to the surface. Which of the following statements are true?

A. the reflected ray makes an angle less than 30° with the normal to the surface
B. the reflected ray makes an angle of 30° with the normal to the surface
C. the refracted ray makes an angle less than 30° with the normal to the surface if the index of refraction of medium A is greater than the index of refraction of medium B
D. the refracted ray makes an angle less than 30° with the normal to the surface if the index of refraction of medium A is less than the index of refraction of medium B
E. the refracted ray makes an angle of 30° with the normal to the surface

Answers: (1) D; (2) C; (3) D; (4) A, C; (5) A, B, D; (6) A, C; (7) A, D; (8) B, C, D; (9) C; (10) B, C, E; (11) B, D

Chapter 34
IMAGES

The law of reflection is applied to plane and spherical mirrors and the law of refraction is applied to spherical refracting surfaces. Each of these form an image of an object placed in front of it and you will learn to find the position, size, and orientation of the image. You will also apply what you learn to lenses with spherical faces and to systems of lenses, such as those used in telescopes and microscopes. Pay careful attention to ray tracing techniques, which help you to visualize image formation.

Important Concepts

☐ plane mirror

☐ real image

☐ virtual image

☐ object distance

☐ image distance

☐ spherical mirror

☐ focal length of a mirror

☐ focal point of a mirror

☐ lateral magnification

☐ spherical refracting surface

☐ lens

☐ focal length of a lens

☐ focal points of a lens

☐ angular magnification

Overview

34–2　Two Types of Image

☐ Light rays diverge from an image and your brain thinks the image is at the place from which the rays diverge. An image is said to be a **real image** if light actually passes through the image point. It is said to be a **virtual image** if light does not. The image in a plane mirror, for example, appears to come from behind the mirror but it actually diverges from the surface of the mirror. The image is virtual. Real images can be formed on a screen; virtual images cannot.

34–3　Plane Mirrors

☐ After light from an object has been reflected by a plane mirror, it appears to come from an **image** located behind the mirror. The diagram shows a point source (or **object**) P in front of a mirror and two rays emanating from it. The reflected rays are drawn according to the law of reflection. The dotted lines extend the reflected rays to behind the mirror, where they intersect at the image point P'.

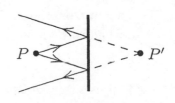

Many other rays from the source to the mirror could have been drawn. All reflected rays, extended backward to the region behind the mirror, intersect at P'. P' is on the line through P that is perpendicular to the mirror and is the same distance behind the mirror as P is in front.

☐ If the source is extended, you may think of each point on it as a point source, emitting light in all directions. An image is formed for each point that sends light to the mirror. To find the image of a straight line, simply find the images of each end and connect them with a straight line. Clearly the image of an extended source is the same size as the source.

☐ The distance from a point source to a mirror is denoted by p. It is positive. The image position is denoted by i. It is positive for real images, negative for virtual images, and its magnitude is the distance from the image to the mirror. For a plane mirror, $i = -p$.

34–4 Spherical Mirrors

☐ The diagrams below show two spherical mirrors, with their centers labeled c and their centers of curvature labeled C. A point source of light S is in front of each mirror and its image is formed at I. F marks the focal point of the mirror. The line through the center of curvature and the center of the mirror is called the **central axis**. Angles are usually measured with respect to this line and distances are measured along it. The object distance is p and the image distance is i.

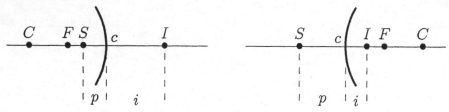

☐ Light from the source is reflected by the mirror. In some cases, reflected rays converge to a point on the source side of the mirror and form a real image. In other cases, the reflected rays diverge as they leave the mirror but they follow lines that pass through a single point behind the mirror. They form a virtual image. Strictly speaking, sharp images are formed only by light whose rays make small angles with the central axis.

☐ Incident rays that are parallel to the central axis of a *concave* mirror, like the mirror on the left above, are reflected so they pass through the focal point, which is a distance $f = r/2$ in front of the mirror. Incident rays that are parallel to the central axis of a *convex mirror*, like the mirror on the right above, are reflected so they diverge, but they appear to come from the focal point, a distance $r/2$ behind the mirror. Here r is the radius of curvature of the mirror.

34–5 Images from Spherical Mirrors

☐ The mirror equation relates the object distance p, the image distance i, and the radius of curvature r. It is

$$\frac{1}{p} + \frac{1}{i} = \frac{2}{r}.$$

This is often written in terms of the **focal length** of the mirror, defined by $f = r/2$:

$$\frac{1}{p} + \frac{1}{i} = \frac{1}{f}.$$

By convention, both r and f are positive if the center of curvature is on the same side of the mirror as the object and negative if the center of curvature is on the opposite side. The image distance i is positive for a real image (on the same side of the mirror as the object) and negative for a virtual image (on the opposite side).

☐ **Ray tracing** provides a graphical way of finding the image of an object that is not on the central axis. Special rays are drawn from the object to the mirror, then the reflected rays are drawn. They intersect at the image. The special rays are shown on the diagram to the right and are described below:

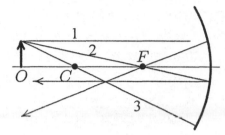

1. An incident ray that is parallel to the central axis is reflected so the reflected ray, perhaps extended backward, is through the focal point.

2. An incident ray that passes through the focal point is reflected so the reflected ray is parallel to the central axis.

3. An incident ray that passes through the center of curvature is reflected so the reflected ray is along the same line.

If the image is real, it is at their intersection. If it is virtual, it is at the intersection of their extensions to the region behind the mirror.

☐ If the mirror is concave and the object is between the focal point and the mirror, then the image is virtual and not inverted. If the object is outside the focal point, it is real and inverted. If the object is more than twice the focal length, from the mirror the image is smaller than the object; otherwise it is larger. If the object is at the focal point, all small-angle reflected rays are parallel to each other and the image is said to be at infinity. If the mirror is convex, the image is virtual and not inverted, no matter where the object is. These statements can be verified easily by tracing rays or by using the mirror equation.

☐ The **lateral magnification** m of a mirror is the ratio of the lateral size (the dimension normal to the central axis) of the image to the lateral size of the object. The lateral magnification of a spherical mirror is given by $m = -i/p$. Values for p and i are substituted with their signs. If m is positive, then the orientations of the object and image are the same; if m is negative, then the image is inverted with respect to the object. Virtual images are not inverted; real images are inverted.

34–6 Spherical Refracting Surfaces

☐ Suppose a spherical surface separates the medium of incidence, with index of refraction n_1, from the medium of refraction, with index of refraction n_2. The central axis is the line through the center of the surface and its center of curvature. If rays from a point source on the central axis make small angles with that axis, then after refraction, they either converge on a real image or else their extensions diverge from a virtual image. If the image is virtual, it is formed in the region containing the source. If the image is real, it is formed on the side into which light is transmitted.

☐ Distances are measured along the central axis from the surface. For a single refracting surface, the object distance p is positive. The image distance i is positive if the image is real and negative if it is virtual.

☐ The law of refraction yields a relationship between the object distance p, the image distance i, the radius of curvature r of the surface, and the indices of refraction for the two sides:

$$\frac{n_1}{p} + \frac{n_2}{i} = \frac{n_2 - n_1}{r} .$$

If the surface is concave with respect to the source, the center of curvature is on the same side of the surface as the object and r is negative. If the surface is convex with respect to the source, the center of curvature is on the opposite side and r is positive.

☐ If $n_2 > n_1$, then a convex surface forms a real image of an object that is far from it and a virtual image of an object that is near it. Details depend on the values of the indices of refraction and the radius of curvature. A concave surface, on the other hand, always forms a virtual image.

34–7   Thin Lenses

☐ A **lens** has two refracting surfaces. The image formed by the first is considered to be the object for the second. If the surrounding medium is a vacuum and the thickness of the lens can be neglected, then the object distance p, image distance i, and focal length f are related by

$$\frac{1}{p} + \frac{1}{i} = \frac{1}{f} .$$

The focal length f is given by

$$\frac{1}{f} = (n - 1) \left(\frac{1}{r_1} - \frac{1}{r_2} \right) .$$

Here r_1 is the radius of curvature of the surface nearer the object, r_2 is the radius of curvature of the surface farther away, and n is the index of refraction of the lens material.

☐ Real images are formed on the opposite side of the lens from the object. For these images, i is positive. Virtual images are formed on the same side of the lens as the object. For a virtual image, i is negative. A surface radius is negative if the center of curvature is on the same side of the surface as the object and is positive if the center of curvature is on the opposite side. The focal length of a lens does not depend on which side faces the object. Lenses with positive focal lengths are said to be converging, while lenses with negative focal lengths are said to be diverging.

☐ A lens has two focal points, located equal distances $|f|$ on opposite sides of the lens. The first focal point, denoted F_1, is on the same side of the lens as the object for a converging lens (f positive) and on the opposite side for a diverging lens (f negative). Light rays that are along lines that pass through F_1 are bent by the lens to become parallel to the axis. For a converging lens, the rays before refraction actually pass through F_1. For a diverging lens, the rays strike the surface, so only their extensions into the other side pass through F_1.

☐ The second focal point, denoted by F_2, is on the side opposite the object for a converging lens (f positive) and on the same side as the object for a diverging lens (f negative). Light rays that are parallel to the central axis are bent by the lens to lie along lines that pass through F_2. For a converging lens, the rays after refraction actually do pass through F_2. For a diverging lens the, backward extensions of the rays pass through F_2.

☐ The position of an off-axis image can be found graphically by tracing two or more rays originating at the same point on the object. One of these might be along a line through F_1 and then parallel to the axis; another might be parallel to the axis and then along a line through F_2. A third ray you might use goes through the center of the lens. It is not refracted. The image is at the point of intersection of these rays or their backward extensions.

☐ In terms of the object and image distances, the lateral magnification associated with a lens is given by $m = -i/p$, an expression that is identical to the expression for a spherical mirror. If m is negative, the image is inverted; if m is positive, the image not inverted. A virtual image formed by a single thin lens is never inverted; a real image is always inverted.

☐ Some optical instruments, such as telescopes and microscopes, consist of a series of lenses. They can be analyzed by tracing a few rays as they pass through each lens in succession or by applying the lens equation to each lens in succession. Consider a system of two lenses, a distance ℓ apart on the same central axis. Let p_1 be the distance from the object to the first lens struck. This lens forms an image at i_1, given by $1/p_1 + 1/i_1 = 1/f_1$, where f_1 is the focal length of the lens. This image is the object for the second lens; the object distance is $p_2 = \ell - i_1$ and the image distance i_2 is given by $1/p_2 + 1/i_2 = 1/f_2$, where f_2 is the focal length of the second lens.

☐ For some systems, the object distance for the second lens may be negative. This occurs if the image formed by the first lens is behind the second lens, so the light exiting the first lens is converging as it strikes the second lens. Such objects are called *virtual* objects. The equation relating object and image distances is still valid — just substitute a negative value for p_2.

☐ The overall lateral magnification of a multi-lens system is the product of the lateral magnifications of the individual lenses.

34–8 Optical Instruments

☐ Because it takes into account the apparent diminishing of the size of an object with distance, the **angular magnification** is often a better measure of the usefulness of a lens used for viewing than is the lateral magnification.

☐ If an object has a lateral dimension h and is a distance d from the eye, its angular size is given in radians by $\theta = h/d$. If the object is viewed through a lens and the image is a distance d from the eye and has a lateral dimension of h', then the angular size of the image is $\theta' = h'/d'$. The angular magnification of the lens is the ratio of these, or $m_\theta = \theta'/\theta = h'd/hd'$.

☐ A single converging lens used as a magnifying glass is usually positioned so the object is just inside the focal point F_1. The image is then virtual and far away. Its lateral size is $h' = mh = -ih/p$, where m is the lateral magnification. Thus $m_\theta = -(i/p)(d/d')$. The eye is close to the lens so $d' \approx |i|$. Furthermore, $p \approx f$. Once these substitutions are made, the result is $m_\theta = d/f$. Usually d is taken to be the distance to the near point of the eye, about 25 cm, so $m_\theta = (25\,\text{cm})/f$.

☐ The near point is used since the object is in focus and has its largest angular size when it is this distance from an unaided eye. The angular magnification then tells us how much better the lens is than the best the unaided eye can do.

☐ A simple compound **microscope** consists of an objective lens and an eyepiece (or ocular). The focal length of the objective lens is small and the lens is positioned so the object lies just outside its first focal point. The image produced by the objective is real, large, inverted, and far from the lens. If the object has a lateral dimension h, then the image has a lateral dimension $h' = |m|h = |i|h/p$. The magnification is great since i is large and p is small.

☐ The distance s between the second focal point of the objective lens and the first focal point of the eyepiece is called the **tube length** of the microscope. The magnification of the objective lens is given by $|m| = h'/h = s/|f_{\text{ob}}|$.

☐ The eyepiece is positioned with its first focal point at the image produced by the objective lens. The eye is placed close to the eyepiece, which then acts as a simple magnifying glass with an angular magnification of $(25\,\text{cm})/f_{\text{ey}}$, where f_{ey} is the focal length of the eyepiece. The overall magnification is given by $M = mm_\theta = -(s/f_{\text{ob}})(25\,\text{cm}/f_{\text{ey}}$. The expression compares the angular size of the image produced by the microscope with the angular size of the object when it is at the near point of an unaided eye.

☐ Telescopes are used to view objects that are far away. Rays entering the objective lens are essentially parallel to the central axis and that lens produces an image that is close to its second focal point. To compute the angular magnification of a telescope, we compare the angular size of the object at its far-away position (not the near point) to the angular size of the image produced by the telescope. The result is $m_\theta = -f_{\text{ob}}/f_{\text{ey}}$. To obtain a large angular magnification, a telescope should have an objective lens with a long focal length and an eyepiece with a short focal length.

Hints for Questions

1 Trace a ray or two. In (a) the fish is the object and in (b) the stalker is the object. Remember that rays are bent toward the normal in going from material with a lower index of refraction into material with a higher index and are bent away from the normal in going from material with a higher index into material with a lower index.

[Ans: (a) a; (b) c]

3 Think about $(1/p) + (1/i) = (1/f)$, where p is the object distance, i is the image distance, and f is the focal length. The solution for i is $i = pf/(p - f)$. In this case f is positive and p is greater than f and increasing. The magnification is $m = -i/p = -f/(p - f)$.

[Ans: (a) from infinity to the focal point; (b) decrease continually]

5 The images are smaller than they would be in a plane mirror. The smallness of the images makes the objects seem further away than they actually are.

[Ans: convex]

7 Consider the lens maker's equation

$$\frac{1}{f} = (n-1)\left(\frac{1}{r_1} - \frac{1}{r_2}\right).$$

The focal length f is large if r_1 and r_2 have the same sign and it is small if they have different signs and neither is infinite. It is intermediate if one of the radii is infinite and the other is not. A plane surface has infinite radius.

[Ans: d (infinite), tie of a and b, then c]

9 If the image formed by lens 1 is virtual and lens 2 is a converging lens then the final image is to the right of lens 2. If the image formed by lens 1 is virtual and lens 2 is a diverging lens then the final image is to the left of lens 2. If the image formed by lens 1 is real we cannot tell since position of the final image is different if the intermediate image is in front of F_2 than if it is behind. Remember that a converging lens does not invert the image if $p < f$ and does invert the image if $p > f$. A diverging lens always inverts the image.

[Ans: (a) all but variation 2; (b) for 1, 3, and 4: right, inverted; for 5 and 6: left, same]

Hints for Problems

3 The intensity of light from an isotropic source falls off in proportion to the reciprocal of the square of the distance from the source. The image acts as a second source and the intensity of light falls off as the reciprocal of the square of the distance from the image. Add the two intensities.

[Ans: 1.11]

15 The focal length of a concave spherical mirror is positive. Solve $(1/p) + (1/i) = (1/f)$ for i. The magnification is given by $m = -i/p$. If i is positive the image is real; otherwise it is virtual. If m is positive the image is no inverted, otherwise it is. A real image is on the same side of the mirror as the object; a virtual image is on the opposite side.

[Ans: (a) -16 cm; (b) -4.4 cm; (c) $+0.44$; (d) V; (e) NI; (f) opposite]

17 Use $(1/p) + (1/i) = (2/r)$. Solve for i, then differentiate the resulting expression with respect to the time t. Substitute v_O for dp/dt and v_I for di/dt.

[Ans: (b) 0.56 cm/s; (c) 11 m/s; (d) 6.7 cm/s]

21 The type of mirror is found from the sign of the focal length. Solve $(1/p) + (1/i) = (1/f)$ for i. If i is positive the image is real and is on the same side of the mirror as the object; otherwise it is virtual and is on the opposite side. The magnification is given by $m = -i/p$. If m is positive the image is not inverted; otherwise it is.

[Ans: (a) concave; (c) $+40$ cm; (e) $+60$ cm; (f) -2.0; (g) R; (h) I; (i) same]

27 The type of mirror is found from the sign of the focal length. The radius is related to the focal length. Solve $(1/p) + (1/i) = (1/f)$ for p. The magnification is given by $m = -i/p$. If m is positive the image is not inverted; otherwise it is. Look at the sign of i to see if the image is real or virtual and to determine if the image is on the same side of the mirror as the object or is on the opposite side.

[Ans: (a) convex; (c) -60 cm; (d) $+30$ cm; (f) $+0.50$; (g) V; (h) NI; (i) opposite]

37 Use $(n_1/p) + (n_2/i) = (n_2 - n_1)/r$. Solve for r. The sign of i tells if the image is real or virtual and the side of the surface where it is formed.

[Ans: (c) +30 cm; (e) V; (f) same]

47 The height of the image on the film is $h' = |m|h$, where h is the height of the person and m is the magnification of the lens. Solve $(1/p) + (1/i) = (1/f)$ for i, then use $m = -i/p$ to calculate the magnification. Here p is the object distance, i is the image distance, and f is the focal length of the lens.

[Ans: 5.0 mm]

55 The type of lens tells the sign of the focal length. Solve $(1/p) + (1/i) = (1/f)$ for i and use $m = -i/p$. The sign of i tells if the image is real or virtual and where its position is relative to the lens. The sign of m tells if the image is inverted or not.

[Ans: (a) −8.6 cm; (b) +0.39 (c) V; (d) NI; (e) same]

61 Solve the lens maker's equation for i. Use $m = -i/p$ to find the magnification. The sign of i indicates if the image is real or virtual and the side of the lens on which the image appears. The sign of m tells if the image is inverted or not.

[Ans: (a) −18 cm; (b) +0.76; (c) V; (d) NI; (e) same]

71 Use $m = -i/p$ to obtain i. Solve $(1/p) + (1/i) = (1/f)$ for f. The sign of f tells the type of lens. The sign of i tells if the image is real or virtual and indicates the side of the lens on which the image is formed. The sign of m tells if the image is inverted or not.

[Ans: (a) D; (b) −5.3 cm; (d) −4.0 cm; (f) V; (g) NI; (h) same]

75 Solve $(1/p) + (1/i) = (1/f)$ for i. The sign of i tells if the image is real or virtual and indicates the side of the lens on which the image is formed. The magnification is given by $m = -i/p$ and its sign tells if the image is inverted or not.

[Ans: (a) C; (d) −10 cm; (f) V; (g) NI; (h) same]

85 The type of lens tells the sign of the focal length. Solve $(1/p_1) + (1/i_1) = (1/f_1)$ for i_1, then calculate $p_2 = d - i_1$ and solve $(1/p_2) + (1/i_2) = (1/f_2)$ for i_2. The sign of i_2 tells if the image is real or virtual and gives the side of lens 2 on which the image is formed. The magnification is the product of the individual magnifications and so is given by $m = (i_1/p_1)(i_2/p_2)$. Its sign tells if the image is inverted or not.

[Ans: (a) −4.6 cm; (b) +0.69; (c) V; (d) NI; (e) same]

93 Solve $1/p + 1/i + 1/f$ for i.

[Ans: (a) 5.3 cm; (b) 3.0 mm]

97 (a) The image produced by the lens must be 4.00 cm in front of the mirror and therefore 6.00 cm behind the lens. Solve $(1/p) + (1/i) = (1/f)$ for p.
(b) The image in the mirror now becomes the object for the lens. The object distance is 14.0 cm. Solve $(1/p) + (1/i) = (1/f)$ for i.

[Ans: (a) 3.00 cm; (b) 2.33 cm]

101 Use the equation for a spherical refracting surface: $(n_1/p) + (n_2/i) = (n_2 - n_1)/r$. Medium 1 is air and medium 2 is glass. The object distance p is the distance from the watcher to the front surface of the glass and the radius of curvature is infinite since the surface is a plane. Solve for the image distance i. The image is the object for the second surface. Calculate the object distance and use $(n_1/p) + (n_2/i) = (n_2 - n_1)/r$ again, but this time medium 1 is glass and medium 2 is water. Part (b) is solved in a similar manner. The fish is now the object.

[Ans: (a) 20 cm; (b) 15 cm]

109 Solve $(1/p) + (1/i) = (1/f)$ for i and explain what happens to i if p is less than f and decreasing. The angle in radians subtended at the eye by the image is $\theta' = h'/|i|$, where h' is the height of the image. Since $h' = |m|h$, and $|m| = |i|/p$, $\theta' = h/p$, where h is the height of the object. The maximum usable angular magnification occurs if the image is at the near point of the eye.

[Ans: (b) P_n]

115 All three lenses are converging lenses, so their focal lengths are all positive. Solve $(1/p_1) + (1/i_1) = (1/f_1)$ for i_1. The object distance for lens 2 is $p_2 = d_{12} - i_1$. Solve $(1/p_2) + (1/i_2) = (1/f_2)$ for i_2. The object distance for lens 3 is $p_3 = d_{23} - i_2$. Solve $(1/p_3) + (1/i_3) = (1/f_3)$ for i_3. This gives the position of the final image relative to lens 3. Its sign tells if the image is real or virtual and indicates the side of lens 3 on which it is formed. The overall magnification is the product of the individual magnifications: $m = m_1 m_2 m_3 = (-i_1/p_1)(-i_2/p_2)(-i_3/p_3)$.

[Ans: (a) +8.6 cm; (b) 2.6; (c) R; (d) NI; (e) opposite]

121 Use $(n_1/p_1) + (n_2/i_1) = (n_2 - n_1)/r$ to find the image distance i_2 for the image formed by the left surface of the sphere. Here medium 1 is air and medium 2 is glass. Set n_1 equal to 1 and replace n_2 with n, the index of refraction for the glass. The incident rays are parallel, so $p_1 = \infty$. The surface is convex so r is positive. The object distance for the right side of the sphere is $p_2 = 2r - i_1$. Solve $(n_1/p_2) + (n_2/i_2) = (n_1 - n_2)/r$ for i_2. Now medium 1 is glass, so $n_1 = n$, and medium 2 is air, so $n_2 = 1$. The surface is concave to the incident light, so r is negative. The sign of i_2 tells if the image is to the right or left of the right side of the sphere.

[Ans: (a) $(0.5)(2 - n)x/(n - 1)$; (b) right]

125 Solve $(1/p_1) + (1/i_1) = (1/f_1)$ for i_1. This is the image distance for the image produced by lens 1. This image is the object for lens 2, so the object distance for that lens is $p_2 = d - i_1$. Solve $(1/p_2) + (1/i_2) = (1/f_2)$ for i_2. The sign of i_2 tells if the image is real or virtual. The overall magnification is the product of the individual magnifications: $m = m_1 m_2 = (-i_1/p_1)(-i_2/p_2)$. The sign of m tells if the final image is inverted or not.

[Ans: (a) −50 cm; (b) 5.0; (c) virtual; (d) inverted]

131 Use $m = -i/p$ to eliminate i from $(1/p) + (1/i) = (1/f)$, then solve for p. Here p is the object distance, i is the image distance, m is the lateral magnification, and f is the focal length. Choose the sign of f so that p is positive. The sign tells if the mirror is concave or convex.

[Ans: (a) convex; (b) 1.60 m]

135 Draw a diagram showing the surface, the point A, and the virtual image of A, on the normal to the surface the same distance behind the surface as A is in front. Also draw a ray from A to some point O on the surface. The reflected ray must be along the line from the image through O. Thus B must be on this line and the distance from the image to B must be the shortest possible distance. A little trigonometry shows that $\theta = \phi$.

Quiz

Some questions might have more than one correct answer.

1. The image produced by a plane mirror of a point object in front of it
   A. appears to be behind the mirror
   B. appears to be twice the distance from the object as the object is from the mirror
   C. appears to be twice the distance from the mirror as the object is from the mirror
   D. is a virtual image
   E. is a real image

2. If light rays diverge as they leave a mirror
   A. the image is a virtual image
   B. the image is a real image
   C. the image appears to be behind the mirror
   D. the image appears to be in front of the mirror
   E. no image is formed

3. The focal point of a spherical mirror is
   A. the place where all images are formed
   B. the place where an image of a far-away point object is formed
   C. behind the mirror if the mirror is convex
   D. behind the mirror if the mirror is concave
   E. virtual for all spherical mirrors

4. Which of the following situations produce virtual images?
   A. a point object in front of a concave spherical mirror a distance less than the focal length
   B. a point object in front of a concave spherical mirror a distance greater than the focal length
   C. a point object in front of a convex spherical mirror a distance less than the focal length
   D. a point object in front of a convex spherical mirror a distance greater than the focal length
   E. a point object at the focal point of a concave mirror

5. Consider light rays that make small angles with the central axis and are incident on a concave spherical mirror. Which of the following statements are true?
   A. a ray that is parallel to the central axis is reflected through the focal point
   B. a ray that is parallel to the central axis is reflected through the center of curvature
   C. a ray that passes through the center of curvature is reflected back through the center of curvature
   D. a ray that passes through the center of curvature is reflected through the focal point
   E. a ray that passes through the focal point is reflected through the center of curvature

6. The image formed by a concave spherical mirror of an object that is perpendicular to the central axis is

   A. larger than the object if the object is anywhere closer to the mirror than the focal point
   B. larger than the object if the object is anywhere farther from the mirror than the focal point
   C. larger than the object if the object is anywhere farther from the mirror than twice the focal point
   D. inverted if the object is anywhere closer to the mirror than the focal point
   E. inverted if the object is anywhere farther from the mirror than the focal point

7. In the mirror equation $\frac{1}{p} + \frac{1}{i} = \frac{1}{f}$, p is the object distance, i is the image distance and f is the focal length. Which of the following statements are true?

   A. f is positive for a concave mirror and negative for a convex mirror
   B. f is positive for a convex mirror and negative for a concave mirror
   C. i is positive for virtual images and negative for real images
   D. i is positive for real images and negative for virtual images
   E. i is positive for images that appear to be behind the mirror and negative for images that appear to be in front of the mirror

8. The focal length of a thin lens depends on

   A. the radii of curvature of the two surfaces
   B. the index of refraction of the lens material
   C. the object distance
   D. which side of the lens faces the object
   E. the index of refraction of the medium surrounding the lens

9. The lensmaker's equation for a thin lens in air is $\frac{1}{f} = (n - 1)\left(\frac{1}{r_1} - \frac{1}{r_2}\right)$, where f is the focal length, r_1 and r_2 are the radii of curvature of the lens surfaces, and n is the index of refraction of the lens material. Which of the following statements are true?

   A. the radii are both positive for all thin lenses
   B. the radii are both negative for all thin lenses
   C. a radius is positive if the surface is concave when viewed from the side of the incoming light
   D. a radius is negative if the surface is concave when viewed from the side of the incoming light
   E. a radius is positive if the surface is concave when viewed from just outside the surface

10. Which of the following statements are true for a thin lens?

    A. all thin lenses with positive focal lengths produce real images of all objects

    B. this lenses with positive focal length produce real images if the object is closer to the lens than the focal point

    C. this lenses with positive focal length produce real images if the object is farther from the lens than the focal point

    D. all thin lenses with negative focal lengths produce real images of all objects

    E. all thin lenses with negative focal lengths produce virtual images of all objects

11. The object distance for the second lens of a two-lens optical system is

    A. never negative

    B. always negative

    C. negative if the image formed by the first lens is behind the second lens

    D. negative if the image formed by the first lens is in front of the second lens

    E. negative if the image formed by the first lens is virtual

Answers: (1) A, D; (2) A, C; (3) B, C; (4) A, C, D; (5) A, C; (6) E; (7) A, D; (8) A, B, E; (9) D; (10) C, E; (11) C

Chapter 35
INTERFERENCE

You studied the fundamentals of interference in Chapter 16. Now the results are specialized to electromagnetic waves and applied to double-slit interference, thin-film interference, and the Michelson interferometer, an important instrument for measuring distances. Pay special attention to the role played by the distances traveled by interfering waves in determining their relative phase. Also remember that there may be a phase change on reflection.

Important Concepts

- ☐ interference
- ☐ Huygens' principle
- ☐ diffraction
- ☐ double-slit interference pattern

- ☐ coherence
- ☐ thin-film interference
- ☐ phase change on reflection
- ☐ Michelson interferometer

Overview

35–2 Light as a Wave

☐ Maxwell's equations lead to Huygens' principle, a geometrical construction that shows how to construct the wavefront for an electromagnetic wave at some time from the wavefront at an earlier time. According to the principle, each point on a wavefront acts as a point source of spherical waves, called Huygens wavelets. After time Δt, the radius of the wavelets will be $v\Delta t$, where v is the wave speed, and the wavefront will be tangent to the wavelets.

☐ The law of refraction follows directly from Huygens' principle. If a wavefront in the incident medium is not parallel to the refracting surface, then part of it enters the second medium before the rest. If the index of refraction of the second medium is greater than the index of refraction of the first, then the wavelets travel slower in the second medium and the wavefront turns to become more nearly parallel to the surface. The argument can be made quantitative and results in the law of refraction.

☐ When light with wavelength λ in vacuum propagates from a medium with index of refraction n_1 into a medium with index of refraction n_2, it speed changes from c/n_1 to c/n_2 and its wavelength changes from λ/n_1 to λ/n_2. Its frequency does not change.

35–3 Diffraction

☐ When a wave is incident on a hole in a barrier, the wave flares out into the region beyond the hole. This is **diffraction**. Huygens' principle explains the phenomenon as the propagation of wavelets into the region, emanating from the portion of the incident wavefront that is in the hole. The more narrow the hole, the greater is the spread of the waves on the other side. If the hole is wide (compared to a wavelength), there is little spreading.

35-4 Young's Interference Experiment

☐ **In Young's experiment**, a monochromatic plane wave is incident normally on a barrier with two slits and the intensity is observed on a viewing screen behind the slits. Wavelets arrive at every point on the screen from both slits and interfere to produce an intensity pattern there. The electric fields of the two waves are vectors and, strictly speaking, the fields should be added vectorially. Assume, however, that the waves are plane polarized and the fields are along the same line. Then scalar addition can be used. This is a good approximation for most of the situations considered in this chapter.

☐ Assume the slits are so narrow that only one wavelet from each slit is required. Suppose the wavelet from one slit travels the distance r_1 to get to some point on the screen and the wavelet from the other slit travels the distance r_2 to get to the same point. At that point on the screen, the phase of the wavelet from the first slit is $\omega t - k r_1$ and the phase of the wavelet from the other slit is $\omega t - k r_2$, where $k = 2\pi/\lambda$ and λ is the wavelength. The phase difference is $\phi = k(r_2 - r_1) = 2\pi(r_2 - r_1)/\lambda$. A maximum in the intensity occurs if ϕ is a multiple of 2π rad; then the wave crests of the two wavelets arrive at the screen at the same time. A minimum in the intensity occurs if ϕ is an odd multiple of π rad; then the crests of one wavelet arrive at the same time as the troughs of the other.

☐ If the screen is far from the slits, then rays from the slits to any point on the screen are nearly parallel to each other and the difference in the distances can easily be written in terms of the angle θ between a ray and a line normal to the barrier. The geometry is shown in the diagram on the right. Since the screen is far away, the distance from the dotted line between the rays to a point on a screen is the same along each ray. The lower ray is longer than the upper by $d \sin \theta$, where d is the slit separation. The phase difference at the screen is $\phi = (2\pi d/\lambda) \sin \theta$.

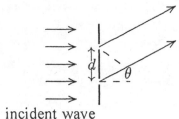

incident wave

☐ The condition for a maximum is $(2\pi d/\lambda) \sin \theta = 2\pi m$, or $d \sin \theta = m\lambda$, where m is an integer. The condition for a minimum is $(2\pi d/\lambda) \sin \theta = (2m + 1)\pi$, or $d \sin \theta = (m + \frac{1}{2})\lambda$, where m is an integer.

☐ Light emanates from each slit in *all* forward directions but only the portions of wavefronts that follow the rays at the angle θ get to the point on the screen being considered. Other portions of the wavefronts get to other places on the screen and for them θ has a different value. For different values of θ, the phase difference is different and, as a result, so are the resultant amplitude and intensity. Alternating bright and dark regions (fringes) are seen on the screen. Centers of bright fringes (maxima of intensity) occur at points for which $d \sin \theta = m\lambda$; centers of dark fringes (minima of intensity) occur at points for which $d \sin \theta = (m + \frac{1}{2})\lambda$.

☐ Constructive interference occurs at the point on the screen directly in back of the slits, for which $\theta = 0$. The angular separation of the first minima on either side of the central maximum is a measure of the extent to which the intensity pattern is spread on the screen. This is given by $2\theta_0$, where $\sin \theta_0 = \lambda/2d$. As d decreases or λ increases, the angular separation increases and the pattern spreads. If $d = \lambda/2$, then $\theta_0 = 90°$ and no bright fringes appear beyond the central maximum.

35–5   Coherence

☐ Two sinusoidal waves are **coherent** if the difference in their phases is constant. Light is emitted from atoms in bursts lasting on the order of nanoseconds and each burst may have a different phase constant associated with it. Thus light from atoms emitting independently of each other is not coherent.

☐ If an extended incoherent source, such as an incandescent lamp, is used to produce light that is incident on a double-slit barrier, the light from each atom goes through each slit and combines on the other side to form an interference pattern. Light from different atoms, however, form patterns that are shifted with respect to each other, the amount of the shift depending on the separation of the atoms in the source. The intensity at any point on the screen fluctuates rapidly between maximum and minimum. Since the eye cannot respond to the rapid fluctuations, a nearly uniform intensity is seen.

☐ A single-slit barrier, placed between the light source and the double-slit barrier, ensures that light from only a small region of the source passes through the double-slit barrier and that the intensity patterns produced at the screen by light from two different atoms nearly coincide.

☐ All the light from a laser is coherent, even though many different atoms are emitting simultaneously. When this light is incident on a double slit, it produces an interference pattern without additional apparatus.

35–6   Intensity in Double-Slit Interference

☐ Phasors can be used to sum waves. A phasor is a rotating arrow with length equal to the amplitude of the wave and angular velocity equal to the angular frequency of the wave. If a phasor has length E_0 and rotates counterclockwise with angular velocity ω, then its projection on the vertical axis is $E(t) = E_0 \sin(\omega t)$.

☐ The phasors for the electric fields of the wavelets from the two slits are drawn to the right. The waves are assumed to have the same amplitude E_0 and their phase difference is ϕ. The phasor for the resultant amplitude E is also drawn. Geometry can be used to show that $\beta = \phi/2$ and that $E = 2E_0 \cos\beta = 2E_0 \cos(\phi/2)$.

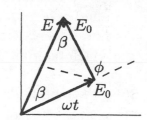

☐ The intensity is proportional to the square of the amplitude, so

$$I = 4I_0 \cos^2(\tfrac{1}{2}\phi),$$

where $\phi = (2\pi d/\lambda)\sin\theta$ and I_0 is the intensity of a single wave. Maximum intensity occurs if $\cos(\phi/2) = \pm 1$; then $\phi = 2m\pi$, $E = 2E_0$, and $I = 4I_0$. Minimum intensity occurs if $\cos(\phi/2) = 0$; then $\phi = (2m + 1)\pi$, $E = 0$, and $I = 0$.

35–7   Interference from Thin films

☐ If light is incident normally on a thin film, some is reflected from the front surface and some from the back. After reflection, the two waves interfere and the resultant intensity may be large or small, depending on their phases and, hence, on the thickness of the film.

☐ For normal incidence on a film of thickness L and index of refraction n, the wave reflected from the back surface travels a distance $2L$ further than the wave reflected from the front surface. The wavelength in the film is $\lambda' = \lambda/n$, so the difference in path length produces a difference of $4\pi nL/\lambda$ in the phase of the two waves.

☐ For one or both of the waves, the medium beyond the reflecting surface might have a higher index of refraction than the medium of incidence. If it does, the wave suffers a phase change of π rad on reflection. Suppose the light is incident in a medium with index of refraction n_1, the film has index of refraction n_2, and the medium beyond the film has index of refraction n_3. Then the phase difference for waves reflected from the two surfaces is $\phi = 4\pi n_2 L/\lambda$ if $n_1 > n_2 > n_3$ or $n_1 < n_2 < n_3$. In the first case neither wave suffers a phase change on reflection. In the second case they both suffer a phase change of π rad. If $n_1 < n_2 > n_3$ or $n_1 > n_2 < n_3$, the phase difference is $\phi = (4\pi n_2 L/\lambda) \pm \pi$. In the first case the wave reflected from the front surface suffers a phase change of π rad but the wave reflected from the back surface does not. In the second case the wave reflected from the back surface suffers a phase change of π rad but the wave reflected from the front surface does not. It is immaterial whether the sign in front of π is plus or minus.

☐ If the phase difference ϕ is a multiple of 2π rad, the waves interfere constructively to produce a bright reflection. If ϕ is an odd multiple of π rad, the interference is completely destructive and the light is transmitted through the film; it is not reflected.

☐ When white light (a combination of all wavelengths of the visible spectrum) is incident on a thin film, the reflected light is colored. It consists chiefly of those wavelengths for which interference of the two reflected waves produces a maximum or nearly a maximum of intensity. Wavelengths for which interference produces a minimum are missing. This phenomena accounts for the colors of oil films and soap bubbles, for example.

☐ When monochromatic light is incident on a thin film with a varying thickness, like a wedge, interference produces bright and dark bands. Bright bands appear in regions for which the film thickness is such that constructive interference occurs; dark bands appear in regions for which the film thickness is such that destructive interference occurs.

35–8 Michelson's Interferometer

☐ A schematic of the instrument is shown below. Light from a source S is incident on a half-silvered mirror M, where the beam is split. One beam travels to mirror M_2, back to M, and then to the eye at E. The other beam travels to mirror M_1, back to M, and then to E. The two beams reaching the eye interfere. To measure a distance, one of the mirrors M_1 or M_2 is moved, thereby changing the interference pattern at the eye. The change in the interference pattern is directly related to the distance moved by the mirror.

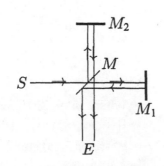

☐ Suppose an intensity maximum is produced at the eye, then either M_1 or M_2 is moved so a neighboring intensity minimum is produced. This means the mirror moved half a wavelength.

Hints for Questions

1 If the length of the nanostructure is a multiple of a wavelength there is a peak at the other end. If the length is an odd multiple of half a wavelength there is a valley.

[Ans: (a) peak; (b) valley]

3 The wave on the left goes a distance of $5d$ and the wave on the right goes a distance of $3d$. For the waves to be exactly out of phase when they emerge the path length difference must be an multiple of half a wavelength.

[Ans: (a) $2d$; (b) (odd number)$\lambda/2$; (c) $\lambda/4$]

5 There is a maximum if the path length difference is a multiple of a wavelength and a minimum if the path length difference is an odd multiple of half a wavelength. See if the path length difference is closer to an integer times a wavelength or to a half-integer times a wavelength.

[Ans: (a) intermediate closer to maximum, $m = 2$; (b) minimum, $m = 3$; (c) intermediate closer to maximum, $m = 2$; (d) maximum, $m = 1$]

7 The angular positions θ of the fringes are given by $\sin\theta = m\lambda/nd$, where λ is the wavelength, n is the index of refraction of the medium in which the apparatus is immersed, and m is an integer. Blue light has a shorter wavelength than red light. Cooking sherry has a larger index of refraction than air.

[Ans: (a) decrease; (b) decrease; (c) decrease; (d) blue]

9 The distances traveled by the two waves from their sources to any point on path 1 are the same. The distances they travel from their sources to any point on path 2 differ by 1.5λ. The distance they travel to points on path 3 vary according to the point on the path.

[Ans: (a) maximum; (b) minimum; (c) alternates]

11 Ray 3 is not reflected. Ray 4 is reflected twice at a boundary with a medium of lower index of refraction than the medium of incidence. Ray 4 has a path of length $3L$ while ray 3 has a path of length L.

[Ans: (a) no; (b) 0; (c) $2L$]

Hints for Problems

5 One wave goes a distance $2L$ further than the other so the phase difference of the waves after they exit the mirror system is $4\pi L/\lambda$. If the waves are exactly out of phase the phase difference must be an odd multiple of π.

[Ans: (a) 155 nm; (b) 310 nm]

11 If a wave with wavelength λ (in air) travels a distance L through a medium with index of refraction n it phase changes by $2\pi Ln/\lambda$. The brightness is great if the phase difference is a multiple of 2π and is low if the phase difference is an odd multiple of π.

[Ans: (a) 1.70; (b) 1.70; (c) 1.30; (d) all tie]

23 The phase difference due to the path-length difference is $2\pi(r_A - r_B)\lambda$. There is also a phase difference of $\pi/2$ because the sources are out of phase.

[Ans: 0]

25 The path-length difference is $\Delta x = \sqrt{y^2 + x^2} - x$, where y is the coordinate of source 2. The phase difference at the detector is $2\pi\,\Delta x/\lambda$, where λ is the wavelength. If the light intensity at the detector is a minimum the phase difference at the detector is an odd multiple of π.

[Ans: $7.88\,\mu\text{m}$]

33 Use the phasor method. Take the phasor associated with wave 1 to be on the horizontal axis. Its horizontal component is $10.0\,\mu\text{V/m}$ and its vertical component is zero. The phasor associated with wave 2 is $45°$ below the horizontal axis. Its horizontal component is $(5.00\,\mu\text{V/m})\cos45°$ and its vertical component is $-(5.00\,\mu\text{V/m})\sin45°$. The phasor associated with wave 3 is $45°$ above the horizontal axis. Its horizontal component is $(5.00\,\mu\text{V/m})\cos45°$ and its vertical component is $+(5.00\,\mu\text{V/m})\sin45°$. Add the horizontal components to find the horizontal component of the resultant and add the vertical components to find the vertical component of the resultant. Divide the vertical component by the horizontal component to find the tangent of the phase angle of the resultant.

[Ans: $(17.1\,\mu\text{V/m})\sin[(2.0\times10^{14}\,\text{rad/s})t]$]

37 Write an expression for the phase difference of the two waves on exit from the thin layer, taking into account the difference in path length and the changes in phase on reflection. For the reflection to be bright the phase difference should be a multiple of 2π. Solve for the thickness of the coating.

[Ans: $70.0\,\text{nm}$]

41 Write an expression for the phase difference of the two waves on exit from the thin layer. One travels further than the other and there may be phase changes on reflection. Since the reflected intensity is a minimum the phase difference is an odd multiple of π. Solve for the values of the wavelength and choose those that are in the visible range.

[Ans: $560\,\text{nm}$]

51 Write an expression for the phase difference of the two waves on exit from the thin layer. One travels further than the other and there may be phase changes on reflection. Since the reflected intensity is a minimum the phase difference is an odd multiple of π. Solve for the values of the thickness and choose the second smallest.

[Ans: $161\,\text{nm}$]

59 Write an expression for the phase difference of the two waves on exit from the thin layer. One travels further than the other and is reflected twice. There may be phase changes at the reflections. Since the transmitted intensity is a minimum the phase difference is an odd multiple of π. Solve for the values of the wavelength and choose those that are in the visible range.

[Ans: $455\,\text{nm}$]

67 Write an expression for the phase difference of the two waves on exit from the thin layer. One travels further than the other and is reflected twice. There may be phase changes at the reflections. Since the transmitted intensity is a minimum the phase difference is an odd multiple of π. Solve for the values of the thickness and choose the second smallest.

[Ans: $161\,\text{nm}$]

73 The tangent of the angle between the slides is given by $\tan\theta = \Delta y/\Delta x$, where Δy is the difference in the thickness of the air film from one dark band to the next and Δx is the distance between adjacent dark bands. Δx is given. To compute Δy consider the reflection at a dark band. One wave travels a distance $2y$ further than the other and both waves are reflected. Write an expression for the difference in phase of the two waves as they exit the air film, taking into account the difference in path length and any changes in phase on reflection. The total phase difference is an odd multiple of π. Now compute the difference in y for two adjacent dark bands.

$\left[\text{Ans: } 0.012° \right]$

<u>77</u> Find an expression for the thickness of the air film so that the phase difference of the waves reflected from the top and bottom surfaces of the film is a multiple of π. Take into account the path length difference and the changes in phase on reflection, if any. The radius r of the bright ring, the radius R of curvature of the lower surface of the lens, and the thickness d of the film are related by $d = R - \sqrt{R^2 - r^2}$, as a little geometry shows. See Fig. 35–46. Write the expression twice, once for the nth ring and once for $(n + 20)$th ring. Use one of the equations to eliminate n from the other, then solve for R.

$\left[\text{Ans: } 1.00 \, \text{m} \right]$

<u>83</u> One wave goes a distance $7d$ and the other goes a distance $2d$. Set the path-length difference equal to an odd multiple of the wavelength. In part (a) the wavelength is the wavelength λ in air and in part (b) it is λ/n, where n is the index of refraction of the protein solution.

$\left[\text{Ans: (a) } 50.0 \, \text{nm; (b) } 36.2 \, \text{nm} \right]$

<u>87</u> Equation 35–14 gives $\sin \theta = m\lambda/d$, where λ is the wavelength and d is the slit separation. The position of the bright fringe on the screen is $y = \tan \theta$ and since the angle is small and $\tan \theta$ is nearly the same as $\sin \theta$, $y = mD\lambda/d$. The exact path-length difference is $\sqrt{(y + d/2)^2 + D^2}$. Set this equal to $m\lambda$ and solve for y. The percent error is the difference between the exact value and the approximate value, divided by the exact value and multiplied by 100 (to convert to per cent).

$\left[\text{Ans: } 0.032\% \right]$

<u>97</u> Before the slab is in place one wave travels the distance $|x_1|$ and the other travels the distance $|x_2|$, where x_1 and x_2 are the coordinates of the sources. Both waves have wavelength λ. After the slab is in place one of the waves travels a distance $x_1 - T$ with wavelength λ and then the distance T with wavelength λ/n, where T is the thickness of the slab and n is the index of refraction of the slab.

$\left[\text{Ans: (a) } 1.6 \, \text{rad; (b) } 0.79 \, \text{rad} \right]$

<u>99</u> Use $d \sin \theta = (m + \frac{1}{2})\lambda$ to find θ, then $\Delta y = D \tan \theta$ to find D. Here λ is the wavelength, D is the distance from the slits to the screen, and Δy is the separation of the dark band from the central maximum on the screen. Assume θ is small, so that $\tan \theta$ can be approximated by $\sin \theta$. For the second dark band $m = 1$.

$\left[\text{Ans: } 6.4 \, \text{m} \right]$

<u>105</u> Since the wavelength in a medium with index of refraction n is λ/n, where λ is the wavelength in air, and since the waves start in phase, the phase difference after passing through the layers is $(2\pi L/\lambda)(n_1 - n_2)$, where L is the thickness of a layer. If the waves start out of phase, π must be added to this. If the phase difference is a multiple of 2π the result is

brightness but if it is an odd multiple of π the result is darkness.

[Ans: (a) 0.87; (b) intermediate, closer to maximum brightness; (c) 0.37; (d) intermediate, closer to complete darkness]

109 The speed of a light wave in a medium with index of refraction n is c/n so the travel time through a layer of thickness T is nT/c.

[Ans: (a) 42.0 ps; (b) 42.3 ps; (c) 43.2 ps; (d) 41.8 ps; (e) 4]

113 One wave travels a distance $2L$ further than the other, where L is the thickness of the coating. The phases of both waves change by π on reflection, so the phase difference is $4\pi L n/\lambda$. This should be an odd multiple of π.

[Ans: 0.20]

121 The electric field of one wave is $E_1 = E_0 \sin(\omega t)$ and the electric field of the other wave is $E_2 = 2E_0 \sin(\omega t + \phi)$, where $\phi = (2\pi d/\lambda) \sin\theta$, where d is the center-to-center separation of the slits and λ is the wavelength. Draw a phasor diagram and use the trigonometric law of cosines to find the square of the resultant electric field magnitude. You should obtain $E^2 = E_0^2(5 + 4\cos\phi)$. Thus the intensity is $I = I_0(5 + 4\cos\phi)$, where I_0 is the intensity of the first wave alone, in the absence of the second wave. Use the trigonometric identity $\cos\phi = 2\cos^2(\phi/2) - 1$ to obtain the final result.

[Ans: $I_0\left[1 + 8\cos^2(\phi/2)\right]$, with $\phi = (2\pi d/\lambda)\sin\theta$]

Quiz

Some questions might have more than one correct answer.

1. Five pipes have different lengths L and different indices of refraction n, given below. A pulse of light travels through which pipe in the shortest time?

    A. $L = L_0$, $n = n_0$
    B. $L = L_0/3$, $n = 1.5n_0$
    C. $L = L_0/2$, $n = n_0 1.5$
    D. $L = 2L_0$, $n = 1.5n_0$
    E. $L = 3L_0$, $n = 1.2n_0$

2. Two sources produce waves that are in phase and have wavelength λ. Wave 1 travels a distance L in air and wave 2 travels the same distance in a medium with index of refraction 1.5. They then meet and their interference is completely destructive. The smallest value L might be is

    A. λ
    B. 0.50λ
    C. 2.0λ
    D. $2.0\pi\lambda$
    E. $0.50\pi\lambda$

3. A monochromatic plane wave is incident normally on an opaque barrier with two narrow slits. The waves from the slits produce total destructive interference at some points on a screen because

    A. they are not in phase at the slits
    B. they travel distances to any of the points that differ by a wavelength
    C. they travel distances to any of the points that are multiples of a wavelength
    D. they travel distances to any of the points that are odd multiples of half a wavelength
    E. they travel distances to any of the points that differ by an odd multiple of half a wavelength

4. At a certain point on a screen waves from two slits in an otherwise opaque barrier produce total destructive interference. Interference at that point can be changed to completely constructive if transparent material is placed over one slit and that material

    A. changes the phase of the wave going through the slit by $\pi/4$ rad
    B. changes the phase of the wave going through the slit by $\pi/2$ rad
    C. changes the phase of the wave going through the slit by π rad
    D. changes the phase of the wave going through the slit by $3\pi/4$ rad
    E. changes the phase of the wave going through the slit by $3\pi/2$ rad

5. Two slits in an otherwise opaque barrier are used to create an interference pattern on a screen. The separation between adjacent bright fringes increases if

    A. the separation between the slits is increased
    B. the separation between the slits is decreased
    C. light with a longer wavelength is used
    D. light with a shorter wavelength is used
    E. the screen is moved closer to the barrier

6. Two slits in an otherwise opaque barrier are used to create an interference pattern on a screen. The angular width of the central fringe increases if

    A. the separation between the slits is increased
    B. the separation between the slits is decreased
    C. light with a longer wavelength is used
    D. light with a shorter wavelength is used
    E. the screen is moved closer to the barrier

7. To obtain an interference pattern using a double-slitted barrier

    A. the phase difference of the waves from the two slits cannot be changing rapidly
    B. the amplitudes of the two waves must be constant
    C. the slits cannot be closer together than half a wavelength
    D. the slits cannot be farther apart than a wavelength
    E. the slits must be illuminated at the same time

8. The bright fringes of a two-slit interference pattern

    A. have equal angular spacing
    B. are more greatly separated in angle at large angles than at small angles
    C. have equal intensities
    D. have intensities that are twice the intensity of the light that is incident on the slits
    E. have intensities that are four times the intensity of the light that is incident on the slits

9. When light is reflected from a surface

    A. its phase changes by π rad
    B. its phase changes by $\pi/2$ rad
    C. its phase changes by π rad only if the index of refraction of the medium beyond the surface is greater than the index of the medium of incidence
    D. its phase changes by $\pi/2$ rad only if the index of refraction of the medium beyond the surface is greater than the index of the medium of incidence
    E. its phase changes by $\pi/2$ rad only if the index of refraction of the medium beyond the surface is less than the index of the medium of incidence

10. White light is incident normally on a thin film with an index of refraction that is greater than the index of the surrounding air. The reflected light does not contain those colors for which the film thickness is

    A. a multiple of the wavelength in air
    B. a multiple of the wavelength in the film medium
    C. an odd multiple of half the wavelength in air
    D. an odd multiple of half the wavelength in the film medium
    E. an odd multiple of a quarter of the wavelength in the film medium

Answers:(1) B; (2) A; (3) D; (4) C; (5) B, C; (6) B, c; (7) A, C; (8) C, E; (9) A, C; (10) B

Chapter 36
DIFFRACTION

Diffraction is the flaring out of light as it passes by the edge of an object or through an opening in a barrier. Pay attention to the description of this phenomenon in terms of Huygens wavelets and learn how the interference of the wavelets produces bright and dark fringes in a diffraction pattern. Single- and double-slit patterns are considered in detail. Pay attention to the slit characteristics that determine the maxima and minima of each pattern. Two important applications are discussed. Diffraction patterns produced by diffraction gratings are used to study spectra and patterns produced by crystals are used to study crystalline structure. Your goals should be to learn what the patterns look like and to understand the details of wave interference that lead to these patterns.

Important Concepts

- ☐ single-slit diffraction
- ☐ double-slit diffraction
- ☐ circular aperture diffraction
- ☐ Rayleigh criterion
- ☐ multiple-slit diffraction

- ☐ diffraction grating
- ☐ dispersion
- ☐ resolving power
- ☐ x-ray diffraction
- ☐ Bragg's law

Overview

36–2 Diffraction and the Wave Theory of Light

☐ Diffraction is discussed in terms of Huygens wavelets. If there is no barrier, the wavelets combine to produce a wave that continues moving in its original direction. If a barrier blocks some of the wavelets, then those that are not blocked move into the geometric shadow. In addition, if the light is coherent, the Huygens wavelets interfere to form a series of bright and dark bands, called the diffraction pattern of the object. Figs. 36–1, 36–2, and 36–3 of the text show some diffraction patterns.

36–3 Diffraction By a Single Slit: Locating the Minima

☐ Consider plane waves of monochromatic light incident normally on a barrier with a single slit of width a, as shown on the right. To find the intensity at a point P on a screen, add the Huygens wavelets emanating from the slit and take the limit as the number of wavelets becomes infinite. Suppose the viewing screen is far away from the slit and consider parallel rays.

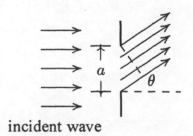

incident wave

☐ The region of screen directly behind the slit is bright. This is the central maximum. The first minimum of the diffraction pattern occurs when every wavelet from the upper half of the slit can be paired with a wavelet from the lower half and the two wavelets of a pair are π rad out of phase. The distance between the two source points for a pair of wavelets is $a/2$, where a is the width of the slit, and the difference in the distance traveled by the wavelets is $(a/2)\sin\theta$. Thus the condition for the first minimum is $(a/2)\sin\theta = \lambda/2$, or $a\sin\theta = \lambda$. Other minima occur for $a\sin\theta = m\lambda$, where m is an integer.

☐ As a decreases, the angle θ for the first minimum increases. The central maximum is broader for narrow slits than for wide slits and diffraction is more pronounced. If a is less than λ, no zeros of intensity occur ($\sin\theta$ cannot be greater than 1) and the entire screen is within the central maximum.

36–4   Intensity in Single-Slit Diffraction, Qualitatively

☐ For $\theta = 0$, all the wavelets are in phase at the screen and they pro-duce the bright central fringe. For other directions, the phases of wavelets from neighboring points in the slit differ and the phasors form the arc of a circle, as shown below. The amplitude of the resultant wave and the intensity are less than for $\theta = 0$. For larger θ, the arc covers a greater portion of a circle, so the amplitude and intensity are still less. At the first minimum, the phasors form a complete circle and the amplitude of the resultant is zero. For θ beyond the first minimum, the phasors wrap around a circle more than once. Since the total arc length is the same, the amplitude and intensity at a secondary maximum are less than at the central maximum.

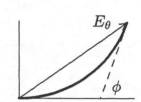

36–5   Intensity in Single-Slit Diffraction, Quantitatively

☐ The angle ϕ in the diagram above is the difference in phase of wavelets from the upper and lower edges of the slit. It is also the angle subtended by the arc at its center. If the arc has radius R, then its length is given by $E_m = R\phi$, for ϕ in radians. E_m is the sum of the amplitudes of all the wavelets and thus is the amplitude if they all have the same phase. E_θ, the amplitude at the screen, is the chord of the arc. A little geometry shows that $E_\theta = 2R\sin(\phi/2)$. Eliminating R between these two expressions yields

$$E_\theta = \frac{2E_m \sin(\phi/2)}{\phi}\,,$$

where $\phi = (2\pi a/\lambda)\sin\theta$. The expression for the amplitude is often written in terms of $\alpha = \phi/2$, rather than in terms of ϕ. Then the amplitude at the screen is

$$E_\theta = \frac{E_m \sin\alpha}{\alpha}$$

and the intensity is

$$I = I_m \left(\frac{\sin\alpha}{\alpha}\right)^2,$$

where I_m is the intensity at the center of the central fringe and $\alpha = (\pi a/\lambda)\sin\theta$.

☐ Carefully study Fig. 36–7 of the text. It shows the intensity as a function of θ for several values of the slit width a. The most prominent feature is a broad central maximum, centered at $\theta = 0$. Since, in the limit as $\alpha \to 0$, $(\sin \alpha)/\alpha \to 1$, the intensity equation predicts the maximum. If the slit width is small, the central maximum spreads to cover the entire screen and no zeros of intensity occur. For a wide slit, the central maximum is narrow and is followed on both sides by secondary maxima. These are narrower and of considerably less intensity than the central maximum. They are roughly midway between zeros of intensity. The number of secondary maxima that appear depends on the slit width.

36–6 Diffraction By a Circular Aperture

☐ When plane waves pass through a circular aperture and onto a screen, a diffraction pattern is formed there. The pattern consists of a bright central disk, followed by a series of alternating dark and bright rings. Fig. 36–3 shows the pattern. The first minimum occurs at an angle θ that is given by $\sin \theta = 1.22\lambda/d$, where d is the diameter of the aperture and θ is measured from the normal to the aperture. The angle for the first dark ring can be used as a measure of the angular size of the central disk.

☐ Stars are effectively point sources of light and lenses act like circular apertures. The image of a star formed by a lens is broadened by diffraction from a point to a disk and rings. Two stars do not form distinct images if the central disks of their diffraction patterns overlap too much. According to the Rayleigh criterion, two far-away point sources are resolved if the centers of their diffraction patterns are no closer than the radius of the first dark ring.

36–7 Diffraction By a Double Slit

☐ Consider two identical slits, each of width a, with a center-to-center separation d. Any point on a screen is reached by a wave from each slit, the resultant of the Huygens wavelets from that slit, and we may think of the pattern as being formed by the interference of these two waves. If light is incident normally on the slits and the observation point is far away, the wave that reaches it from each slit has amplitude $E_m(\sin \alpha)/\alpha$, where $\alpha = (\pi a/\lambda) \sin \theta$, and the two waves differ in phase by $(2\pi d/\lambda) \sin \theta$. When they are combined, the resultant amplitude is

$$E_\theta = E_m (\cos \beta)^2 \left(\frac{\sin \alpha}{\alpha} \right)^2 ,$$

where $\beta = (\pi d/\lambda) \sin \theta$.

☐ Minima of the double-slit interference pattern occur for $\beta = (2m + 1)\pi/2$ or $\sin \theta = (2m + 1)\lambda/2d$. Single-slit diffraction minima occur for $\alpha = m\pi$ or $\sin \theta = m\lambda/a$. In each case, m is an integer but it may have different values in the two expressions. Since d necessarily must be larger than a, the interference minima must be closer together than the diffraction minima. The single-slit diffraction pattern forms an envelope, with the interference pattern inside. Study Fig. 36–14 of the text.

☐ Three parameters are important for the double-slit pattern: the wavelength λ, the slit width a, and the center-to-center slit separation d. The ratio a/λ determines the width of the central diffraction maximum, which extends from $\theta = -\sin^{-1}(\lambda/a)$ to $\theta = +\sin^{-1}(\lambda/a)$. This ratio also determines the positions of the secondary diffraction maxima, if any. The ratio d/λ controls the angular positions of the interference maxima and minima. Finally, the ratio d/a determines how many interference maxima fit within the central diffraction maximum or any of the secondary diffraction maxima.

36–8 Diffraction Gratings

☐ A diffraction grating consists of many thousands of closely spaced rulings on either a transparent or highly reflecting surface. When light is incident on a grating, a multiple-slit diffraction pattern is formed. Because light with different wavelengths produces lines at different angles, diffraction gratings are often used to analyze the spectra of light sources.

☐ Consider a barrier with N parallel slits. Monochromatic plane waves are incident normally on the barrier and the intensity pattern formed by waves passing through the slits is viewed on a screen far away. If the slits are narrow, single-slit diffraction can be ignored. That is, the central maximum of the single-slit diffraction pattern fills the viewing area and what is seen is the multi-slit interference pattern.

☐ The pattern on the screen consists of a series of intense, narrow bands, called **lines**. Secondary maxima lie between but they are much less intense and are usually not important. The pattern is usually described in terms of the angle θ made with the normal by a ray from the slit system to a point on the screen. A diffraction line occurs when the phases at the screen of waves from any two adjacent slits are either the same or differ by a multiple of 2π rad. Since waves from two adjacent slits travel distances that differ by $d\sin\theta$, where d is the slit separation, the condition for a line is $d\sin\theta = m\lambda$, where λ is the wavelength. The integer m in this equation is called the **order** of the line. High order lines occur at greater angles than low order lines.

☐ The locations of the lines are determined by the ratio d/λ and are independent of the number of slits. The lines occur at different angles for different wavelengths of light. For the same order line, the angle for red light is less than that for violet light. If white light is incident on the barrier, the color of an observed band continuously varies from red at one end to violet at the other.

☐ The width of a line is indicated by the angular separation $\Delta\theta$ of the minima on either side. For the line that occurs at angle θ, $\Delta\theta = \lambda/Nd\cos\theta$. It depends on the number of slits. In fact, as the number of slits increases without change in their separation, the width of every line decreases. Maxima near the normal (small θ) are sharper than maxima away from the normal (larger θ).

36–9 Gratings: Dispersion and Resolving Power

☐ Two parameters are used to measure the quality of a grating. The **dispersion** measures the angular separation of lines of the same order for wavelengths differing by $\Delta\lambda$. It is defined by $D = \Delta\theta/\Delta\lambda$ and for order m, occurring at angle θ, it is given by $D = m/d\cos\theta$, where d is the slit separation. Dispersion does not depend on the number of rulings but does depend on their separation. Large dispersion means large angular separation.

☐ The **resolving power** measures the difference in wavelength for lines with an angular separation equal to half their angular width; that is, for two lines that obey the Raleigh criterion for resolution. Mathematically, it is defined by $R = \lambda/\Delta\lambda$, where $\Delta\lambda$ is the difference in wavelength. For an N-ruling grating, the resolving power for order m is given by $R = Nm$.

36–10 X-Ray Diffraction

☐ Atoms in a crystal form a periodic array in three dimensions and x-ray radiation scattered by their electrons produces a diffraction pattern. To obtain a measurable scattering angle, the wavelength of the radiation used should be about the same as the distance between atoms. Thus x ray radiation, with a wavelength on the order of 0.1 nm, is used.

☐ Diffraction occurs only when the x rays are incident at certain angles to **crystal planes**. A crystal plane is a plane that passes through atomic equilibrium positions.

☐ Suppose monochromatic x rays with wavelength λ are incident at an angle θ on a set of crystal planes with separation d. If the angle θ between the beam and the planes satisfies $2d\sin\theta = m\lambda$, where m is an integer, then a high intensity beam is radiated at the angle θ to the planes. This is Bragg's law for x-ray diffraction.

☐ X-ray diffraction is used to find the orientations and separations of a great many sets of crystal planes. These are used to reconstruct the crystal. Crystals with known atomic arrangements are used as filters to separate x rays of a given wavelength from an incident beam containing a mixture of wavelengths.

Hints for Questions

1 For an intensity minimum the path-length difference for rays from the top and center of the slit is half a wavelength. The path-length difference for rays from the center and bottom of the slit is also half a wavelength. For an intensity maximum both these path length differences are approximately halfway between the differences for two adjacent minima.

[Ans: (a) the $m = 5$ minimum; (b) (approximately) the maximum between the $m = 4$ and $m = 5$ minima]

3 The more narrow the slit the greater the spreading. To judge left-right spreading examine the left-right widths of the slits. To judge up-down spreading examine the up-down widths.

[Ans: (a) 1 and 3 tie, then 2 and 4 tie; (b) 1 and 2 tie, then 3 and 4 tie]

5 Recall that $\beta = (\pi d/\lambda)\sin\theta$, where d is the slit separation and λ is the wavelength. The curves are portions of sine curves with multiplying factors that are proportional to the slit separation. The number of interference maxima within the central diffraction envelope is d/a, where a is the slit width.

[Ans: (a) A, B, C; (b) A, B, C]

7 Since the diffraction angle to the first minimum is proportional to λ/d, where λ is the wavelength and d is the diameter of the circular diffracting object, we expect the radius of the ring to increase with increasing wavelength. The wavelength of red light is longer than the wavelength of blue light.

[Ans: (a) larger; (b) red]

9 The line associated with wavelength λ is at the angle θ given by $d\sin\theta = m\lambda$, where d is the ruling separation and m is an integer. The half-width of the line at angle θ is given by $\lambda/Nd\cos\theta$, where N is the number of grating rulings.

$\left[\text{Ans: (a) decrease; (b) same; (c) in place}\right]$

11 The width of the line at angle θ is given by $\lambda/Nd\cos\theta$, where N is the number of grating rulings and λ is the wavelength. A greater number of rulings means a narrower line. A wider line means a larger value of θ and hence a greater value of m.

$\left[\text{Ans: (a) A; (b) left; (c) left; (d) right}\right]$

Hints for Problems

1 For a diffraction minimum $a\sin\theta = m\lambda$, where a is the slit width, θ is the diffraction angle, λ is the wavelength, and m is an integer. The first minimum occurs for $m = 1$ and the angle between two minima of the same order is 2θ.

$\left[\text{Ans: } 60.4\,\mu\text{m}\right]$

9 The path-length difference is $\Delta x = (a/2)\sin\theta$, where a is the slit width and θ is the diffraction angle. The phase difference in radians is $2\pi\,\Delta x/\lambda$.

$\left[\text{Ans: } 2.8\,\text{rad } (160°)\right]$

11 The distance D to the screen, the separation Δy of the point on the screen from the central axis, and the diffraction angle θ are geometrically related by $\tan\theta = \Delta y/D$. The parameter α is given by $\alpha = (\pi a/\lambda)\sin\theta$. The intensity ratio is $(\sin\alpha/\alpha)^2$.

$\left[\text{Ans: (a) } 0.18°; \text{ (b) } 0.46\,\text{rad; (c) } 0.93\right]$

19 The angular separation in radians of the two points is the diffraction angle of the first minimum of the diffraction pattern of a circular aperture. That is, it is $\theta = 1.22\lambda/d$, where λ is the wavelength and d is the diameter of the aperture. In this case d is the diameter of the mirror (5.1 m). The distance between the two points on the Moon's surface is $R\theta$, where R is the Earth-Moon distance.

$\left[\text{Ans: } 50\,\text{m}\right]$

27 The angular separation of the stars in radians is the diffraction angle of the first dark ring and so is $\theta = 1.22\lambda/d$, where λ is the wavelength and d is the diameter of the lens. The distance between the stars is $R\theta$, where R is the distance from Earth to them. Convert light years to kilometers or meters (see Appendix D). Since the stars are far away compared to the focal length f of the lens, the image is at the focal point and the radius of the first dark ring of the image is $f\theta$.

$\left[\text{Ans: (a) } 8.8 \times 10^{-7}\,\text{rad; (b) } 8.4 \times 10^7\,\text{km; (c) } 0.025\,\text{mm}\right]$

29 The minima of the one-slit diffraction pattern are at angles θ such that $a\sin\theta = m\lambda$, where a is the slit width, λ is the wavelength, and m is an integer. You want the number of bright two-slit interference fringes between angles θ_1 and θ_2, where $a\sin\theta_1 = \lambda$ and $a\sin\theta_2 = 2\lambda$. Now there are 5 bright interference fringes on each side of the central bright fringe and within the central diffraction envelope. The sixth bright fringe coincides with the first diffraction minimum. Since the bright interference fringes obey $d\sin\theta = m\lambda$, where d is the slit separation, $d \sin \theta_1 = 6\lambda$, and $d \sin \theta_2 = 12\lambda$. Now you can count the bright interference fringes.

[Ans: 5]

35 The first minimum of the single-slit diffraction pattern occurs at the angle θ such that $a \sin \theta = \lambda$, where a is the slit width and λ is the wavelength. Read θ from the graph and calculate a. The maxima of the two-slit interference pattern occur at angles $]theta$ such that $d \sin \theta = m\lambda$, where d is the slit separation and m is an integer. According to the graph the $m = 4$ maximum occurs at the same angle as the first minimum in the single-slit diffraction problem. Use this angle to compute d. The intensity is given by

$$ I = I_m(\cos^2 \beta) \left(\frac{\sin \alpha}{\alpha} \right)^2 , $$

where $\alpha = (\pi a/\lambda) \sin \theta$ and $\beta = (\pi d/\lambda) \sin \theta$. Use $d \sin \theta = m\lambda$, where m is the order of the interference minimum, to compute $\sin \theta$.

[Ans: (a) 5.0 μm; (b) 20 μm]

37 Spectral lines appear at angles θ such that $d \sin \theta = m\lambda$, where d is the separation between adjacent rulings, λ is the wavelength, and m is an integer. First find the largest integer m_m for which $m_m\lambda/d$ is less than 1 (the maximum value for a sine function). Then find θ for $m = m_m$, $m = m_m - 1$, and $m = m_m - 2$.

[Ans: (a) 62.1°; (b) 45.0°; (c) 32.0°]

49 The resolving power is given by $R = Nm$, where N is the total number of rulings on the grating and m is the order of the spectral line. The dispersion is given by $m/d \cos \theta$, where d is the separation of adjacent rulings, m is the order, and θ is the diffraction angle. Use $d \sin \theta = m\lambda$, where λ is the wavelength, to find the diffraction angle.

[Ans: (a) 0.032°/nm; (b) 4.0 × 10⁴; (c) 0.076°/nm; (d) 8.0 × 10⁴; (e) 0.24°/nm; (f) 1.2 × 10⁵]

55 For an order m reflection from crystal planes with separation d, $2d \sin \theta = m\lambda$, where θ is the Bragg angle and λ is the wavelength. Solve for d.

[Ans: 0.26 nm]

57 Use $2d \sin \theta = m\lambda$, where d is the separation of the crystal planes, λ is the wavelength, and m is an integer. Assume the first peak is the $m = 1$ peak for one wavelength and calculate that wavelength. The second peak is either the $m = 2$ peak for the wavelength you have found or the $m = 1$ peak for the second wavelength. Check the first possibility by calculating θ for that peak and comparing the result with the graph.

[Ans: (a) 25 pm; (b) 38 pm]

59 The reflections obey $2d \sin \theta = m\lambda$. Write the equation once for each reflection. One equation can be solved for d and the other for the wavelength A.

[Ans: (a) 0.17 nm; (b) 0.13 nm]

65 At the first minimum of the diffraction pattern $a \sin \theta = \lambda$, where a is the width of the speaker opening, θ is the diffraction angle, and λ is the wavelength, which is given by $\lambda = v/f$, where v is the speed of sound and f is the frequency. Take the speed of sound to be 343 m/s and calculate θ. The distance along the wall to the first minimum is $y = D \tan \theta$, where D is the distance from the speaker cabinet to the wall.

[Ans: 41.2 m]

<u>77</u> The Bragg reflection angle is the same for both wavelengths. If wavelength λ_1 undergoes reflection of order m_1 and wavelength λ_2 undergoes reflection of order m_2, then $m_1 \lambda_1 = m_2 \lambda_2$. Select the smallest integers m_1 and m_2 for which this equality holds, then solve $2d \sin \theta = m_1 \lambda_1$ for θ. Here d is the separation of the reflecting planes and θ is the Bragg angle. Since θ is the angle between the beams and the planes, the angle between the incident and reflected beams is $180° - 2\theta$.

[Ans: 106°]

<u>81</u> You want the long wavelength end of a spectrum of order m to be at a greater angle than the short wavelength end of the spectrum of the next highest order $(m+1)$. If λ_1 is the long wavelength and λ_2 is the short wavelength you want $m\lambda_1 > (m+1)\lambda_2$. Find the smallest value of m for which the inequality holds. The complete spectrum is present in order m if $m\lambda_1/d$ is less than 1 (the greatest value the sine function can have). Here d is the separation of adjacent rulings on the grating.

[Ans: (a) fourth; (b) seventh]

<u>89</u> Solve $a \sin \theta = m\lambda$ for θ. Here a is the slit width, θ is the diffraction angle, λ is the wavelength, and $m = 2$ for the second diffraction minimum. The distance on the screen between the second diffraction minimum and the center of the pattern is given by $y = D \tan \theta$, where D is the distance from the slit to the screen.

[Ans: 53.4 cm]

<u>95</u> Use $a \sin \theta = m\lambda$, where a is the slit width, θ is the diffraction angle, λ is the wavelength, and m is the order of the spectrum. You want to show that θ for $m = 2$ and $\lambda = 700$ nm is greater than θ for $m = 3$ and $\lambda = 400$ nm for every value of d.

<u>97</u> In the Rayleigh criterion limit the separation d of the flowers is $\Delta x = D \tan \theta$, where D is the distance from you to the flowers. The angle θ is the diffraction angle for the first minimum in the pattern of a circular aperture. That is, $\theta = 1.22\lambda/d$, where d is the diameter of your pupil and λ is the wavelength of yellow light (about 575 nm).

[Ans: 4.9 km]

Quiz

Some questions might have more than one correct answer.

1. Variations in light intensity from place to place in a diffraction pattern can be explained by means of Huygens' principle. According to this principle
   A. light waves always travel faster in a material medium than in a vacuum
   B. each point on a wavefront acts as a point source for secondary spherical waves
   C. the actual wavefront is tangent to the wavefronts of the secondary waves
   D. all light waves have spherical wavefronts
   E. light waves with different frequencies travel with different speeds

2. A uniform beam of monochromatic light is incident on a sphere and the region around it. The light that passes the sphere is viewed on a screen. The intensity pattern on the screen

    A. is dark at its center
    B. is bright at its center
    C. is mottled at its center
    D. shows circular bright and dark fringes near the edge of the shadow of the sphere
    E. shows only the shadow of the sphere

3. Monochromatic light is incident normally on an opaque barrier with a single slit and the intensity pattern is displayed on a screen beyond the barrier. A dark fringe

    A. occurs wherever light from one end of the slit and light from its center travel distances to the screen that differ by a multiple of a wavelength
    B. occurs wherever light from one end of the slit and light from its center travel distances to the screen that differ by an odd multiple of half a wavelength
    C. occurs wherever light from one end of the slit and light from its center travel distances to the screen that differ by an odd multiple of a quarter of a wavelength
    D. occurs wherever light from one end of the slit and light from the other end travel distances to the screen that differ by a multiple of a wavelength
    E. does not occur anywhere

4. Monochromatic light is incident normally on an opaque barrier with a single slit and the intensity pattern is displayed on a screen beyond the barrier. If the slit width is narrowed

    A. the dark fringes move closer together
    B. the dark fringes move further apart
    C. the dark fringes do not move
    D. there are no dark fringes until the slit width becomes less than a wavelength
    E. there are dark fringes until the slit width becomes less than a wavelength

5. The secondary maxima of a single-slit diffraction pattern

    A. have the same intensity as the central maximum
    B. have intensities that decrease with increasing separation from the central maximum
    C. have intensities that increase with increasing separation from the central maximum
    D. become more numerous if the slit width is decreased
    E. become more numerous if the slit width is increased

6. The intensity at the center of a single-slit diffraction pattern

    A. depends on the slit width
    B. depends on the wavelength
    C. is equal to the intensity of the light incident on the slit
    D. is greater than the intensity of the light incident on the slit
    E. is less than the intensity of the light incident on the slit

7. The number of interference maximum within the central diffraction fringe of a double-slit diffraction pattern

   A. depends on the ratio of the slit separation to the slit width
   B. depends on the ratio of the slit separation to the wavelength
   C. depends on the wavelength alone
   D. increases if the slit separation is increased without changing the slit width
   E. increases if the slit separation is decreased without changing the slit width

8. The lines produced by a diffraction grating illuminated by monochromatic light

   A. are essentially the interference pattern of a single light ray from each slit
   B. are essentially the diffraction pattern of a single slit
   C. are all within the central maximum of the single-slit diffraction pattern, which is very wide
   D. all have nearly the same intensity
   E. are uniformly spaced in angle

9. The resolving power of a diffraction grating

   A. measures the separation of two adjacent lines corresponding to the same frequency light
   B. measures the separation of two lines of the same order corresponding to slightly different frequencies of light
   C. is given by the product of the number of slits and the order
   D. is given by the product of the number of slits per unit length of the grating and the order
   E. depends on the slit separation

10. X-ray diffraction patterns

    A. are useful in determining crystal structure because the wavelength is comparable to atomic spacings in crystals
    B. can be used to determine crystal structures because the wavelength is much greater than atomic spacings in crystals
    C. are essentially interference patterns produced by one ray from each atom
    D. show a bright spot where the diffracted ray makes the same angle as the incident ray with some crystal plane

Answers: (1) B, C; (2) B, D; (3) B, D; (4) B, E; (5) B, E; (6) C; (7) A, D; (8) A, C; (9) B, C; (10) A, C, D

Chapter 37
RELATIVITY

When two observers who are moving relative to each other measure the same physical quantity, they may obtain different values. The special theory of relativity tells how the values are related to each other when both observers are at rest in different inertial frames. Although the complete theory deals with all physical quantities, the ones you consider here are the coordinates and time of an event and the velocity, momentum, and energy of a particle.

Important Concepts

- ☐ special theory of relativity
- ☐ event
- ☐ simultaneity
- ☐ time dilation
- ☐ proper time
- ☐ Lorentz factor
- ☐ length contraction

- ☐ proper length
- ☐ Lorentz transformation
- ☐ velocity transformation
- ☐ relativistic Doppler effect
- ☐ relativistic momentum
- ☐ relativistic kinetic energy
- ☐ rest energy

Overview

37–2 The Postulates

- ☐ The **special theory of relativity** is based on two postulates. The first (the relativity postulate) deals with the laws of physics: The laws of physics are the same for observers in all inertial frames. Keep in mind that the laws of physics are relationships between physical quantities, not the quantities themselves. Newton's second law and the conservation principles are examples of laws. The momentum of a system has a different value for different reference frames but if it is conserved for one inertial frame, it is conserved for all inertial frames, according to the postulate.

- ☐ The second postulate (the speed of light postulate) is: The speed of light in free space has the same value in all directions and in all inertial frames. If a light source sends a pulse of light toward you, the speed of the pulse, as observed by you, is the same if you are at rest relative to the source, you are traveling at high speed toward (or away from) the source, or the source is traveling at high speed toward (or away from) you. This postulate is consistent with the notion that electromagnetic radiation does not require a medium for its propagation.

- ☐ Special relativity deals with measurements made in *inertial reference frames*. If the total force on a particle is zero, then its acceleration, as measured in an inertial frame, is zero.

37–3 Measuring an Event

☐ An **event** has four numbers associated with it: three are coordinates that designate the location of the event; the fourth designates the time of the event. To measure the coordinates of an event, meter sticks must be laid out in an inertial reference frame, at rest with respect to the frame. To measure the time of an event, a clock must be present at the location of the event and it must be at rest with respect to the reference frame. In addition, it must be synchronized with other clocks at rest in the frame. Synchronization is accomplished, for example, by sending a light pulse from a master clock to all other clocks and using its arrival time at any clock to set that clock, taking into account the transit time of the pulse.

☐ The coordinates and time of any event may be measured by meter sticks and clocks at rest in any inertial frame. Relativity theory tells how the measurements made in one frame are related to those made in another.

37–4 The Relativity of Simultaneity

☐ Two events, separated in space and simultaneous to one observer, are NOT simultaneous to another observer, moving with respect to the first along the line joining the positions of the events. Carefully study Fig. 37–4 of the text. It shows two events, labeled Red and Blue, that are simultaneous according to Sam. He knows they are simultaneous because the events occurred at the ends of his spaceship and electromagnetic waves from the events meet at the midpoint. Since the waves travel at the same speed and go the same distance, they must have started at the same time.

☐ The events also occur at the ends of Sally's space ship but she is moving away from the position of the Blue event and toward to the position of the Red event. Waves from the Red event reach the midpoint of her spaceship before waves from the Blue event. In her frame, the waves move with the speed of light, just as they do in Sam's frame, so she knows that the Red event occurred before the Blue event.

☐ The situation is exactly symmetric. If two events are simultaneous in Sally's frame, they are not simultaneous in Sam's.

37–5 The Relativity of Time

☐ The interval of time between two events is different when measured with clocks at rest in two inertial frames that are moving relative to each other. The diagram on the right shows a clock. The flash unit F emits a light pulse that travels to mirror M and is reflected back to the flash unit. It is detected there and immediately triggers the next flash. If the flash 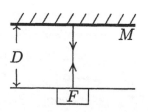 unit and mirror are separated by a distance D, then the time interval between flashes is given by $\Delta t_0 = 2D/c$.

The diagram on the right shows what Sam sees when Sally carries a clock like this at speed v past him. If Δt is the time interval between emission and detection of the pulse, as measured by Sam, then during this interval the flash unit travels a distance $\ell = v\,\Delta t$ and the light pulse moves a distance $2L = 2\sqrt{(\frac{1}{2}v\,\Delta t)^2 + D^2}$. This follows because L is the hypotenuse of a right triangle with sides of length D and $v\,\Delta t/2$. Substitute $2L = c\Delta t$ and solve for Δt:

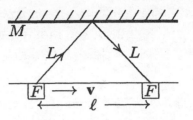

$$\Delta t = \frac{2D}{\sqrt{c^2 - v^2}} = \frac{\Delta t_0}{\sqrt{1 - (v/c)^2}}.$$

Δt is greater than Δt. An observer comparing a moving clock with his clocks concludes that the moving clock ticks at a slower rate. There is a longer time between ticks. This phenomenon is called **time dilation**. It is not important that the clock utilize light as does the clock used to derive the relationship above. Any clock will do.

☐ The time dilation equation is often written

$$\Delta t = \gamma \Delta t_0 \, ,$$

where the **Lorentz factor** γ is $1/\sqrt{1 - (v/c)^2}$. The speed is often given as a fraction of the speed of light. In terms of β ($= v/c$), $\gamma = 1/\sqrt{1 - \beta^2}$. Since the speed of a reference frame is always less than the speed of light, the factor γ is always greater than one.

☐ Time dilation is symmetric with respect to the two frames. If Sally watches a clock at rest with respect to Sam, she sees it tick at a slower rate than her clocks.

☐ If two events, such as the emission and detection of a light pulse, occur at the same coordinate in one frame, then the time interval between them, as measured in that frame, is the **proper time** between the events. The time interval between the same two events, as measured in a frame that is moving relative to the first, is longer; it is multiplied by the factor γ.

37–6 The Relativity of Length

☐ If you measure the length of a rod that is moving past you at high speed, the result is less than if you measure it when it is at rest with respect to you. The length measured with the rod at rest relative to the meter stick is called the **proper length** of the rod.

☐ Suppose the speed of the rod is v. Place a marker on a coordinate axis parallel to the rod's velocity and measure the interval from the time the front end of the rod is at the marker to the time the back end of the rod is at the marker. If that interval is Δt_0, then the length of the moving rod is $L = v\,\Delta t_0$. A single clock at the position of the marker can be used, so Δt_0 is the proper time interval between the two events. L is NOT the proper length because the rod is moving relative to the frame used to measure the length.

Now consider the same events from the point of view of someone moving with the rod. He sees the marker move with speed v from the front to the back of the rod in time Δt and gives the length of rod as $L_0 = v\,\Delta t$. This is the proper length because the rod is at rest relative to the observer. Since Δt_0 is the proper time interval between the events, Δt and Δt_0 are

☐ If the particle is free, then $U = 0$ and Schrödinger's equation becomes

$$\frac{d^2\psi}{dx^2} + \frac{8\pi^2 m}{h^2}E\psi = 0 \, .$$

The wave function for a free particle traveling in the positive x direction is $\psi(x) = Ae^{ikx}$ and the wave function for a free particle traveling in the negative x direction is $\psi(x) = Ae^{-ikx}$, where $k = \sqrt{2mE/h^2}$. These functions are sinusoidal and each has a wavelength λ that is related to k by $k = 2\pi/\lambda$.

38–8 Heisenberg's Uncertainty Principle

☐ If the wave function of a particle does not have the form Ae^{ikx}, then the particle does not have a definite momentum. If we measure the momentum many times, with the particle always in the same state, we obtain a distribution of values. Quantum mechanics can predict the probability that the particle's momentum will be in any given range but it cannot predict the result of any of the momentum measurements. The situation is quite similar to the measurement of position: quantum mechanics can predict the probability that a particle is in a given region of space but it cannot predict its position at any given time.

☐ The position of a particle is localized in space if its wave function is large only over a small region and is zero or nearly zero everywhere else. If the wave function of a particle is changed so the particle's position is more localized, then its position is known with a higher certainty. The particle's momentum then becomes less certain.

☐ In general, if Δx is the uncertainty in the position of a particle and Δp is the uncertainty in its momentum, then $\Delta x \cdot \Delta p \geq h$, where $h = h/2\pi$. Similar relationships hold for each coordinate and the corresponding momentum component.

☐ The Planck constant h appears in the uncertainty relations. If the classical limit is obtained by setting $h = 0$, then the relationship for the x component becomes $\Delta p \cdot \Delta x = 0$. Both the position and momentum can be entirely certain.

38–9 Barrier Tunneling

☐ A matter wave extends into the region beyond a potential energy barrier if the barrier has finite height and width. This means a particle may escape from such a region. Suppose a particle going to the right approaches a potential energy barrier with a height U_b that is greater than its energy E. Classically, the particle cannot travel to the other side of the barrier. Quantum mechanically, it has a non-zero probability of being on either side.

☐ Tunneling through a barrier is described quantitatively by a **transmission coefficient** T and a **reflection coefficient** R. The transmission coefficient is defined so that it gives the probability the particle tunnels through the barrier, while the reflection coefficient gives the probability the particle does not. The sum of the two is one.

☐ If the transmission coefficient is small, it is given by

$$T = e^{-2k_b L} \, ,$$

where

$$k_b = \sqrt{\frac{8\pi^2 m(U_b - E)}{h^2}},$$

E is the energy of the particle (excluding mass energy), and L is the width of the barrier. If the barrier is made wider (L is increased), then T decreases and if the barrier is made higher (U_b is increased), then k_b increases and T decreases. The parameter k_b depends on $U_b - E$. If the energy of the particle is increased (but is still not greater than U_b), then k_b decreases and T increases.

Hints for Questions

1 The greater the work function the less the kinetic energy of the ejected electrons for a given frequency of incident light.

[Ans: potassium]

3 Look at Section 33–2 to find the wavelengths of these waves. The frequency of the wave is proportional to the reciprocal of the wavelength and the photon energy is proportional the frequency.

[Ans: (a) microwave; (b) x ray; (c) x ray]

5 Look at Fig. 38–1 and think about what happens to the plate when electrons are ejected.

[Ans: positive charge builds up on the plate, inhibiting further electron emission]

7 The energy acquired by the electron is equal to the energy lost by the photon. The energy of the photon is proportional to the frequency of its wave and inversely proportional to the wavelength of the its wave.

[Ans: none]

9 Look at Section 33–2 to find the wavelengths of these waves. The frequency of the wave is proportional to the reciprocal of the wavelength and the photon energy is proportional the frequency. The photon momentum is proportional to the reciprocal of the wavelength.

[Ans: (a) B; (b) – (d) A]

11 The scattering is from electrons.

[Ans: no essential change]

13 The wavelength λ is related to the magnitude of the momentum p by $\lambda = h/p$. The kinetic energy K and momentum magnitude are related by $K = p^2/2m$ for nonrelativistic particles.

[Ans: electron]

15 The field either increases or decreases the speed of the electron, depending on the direction of motion. Increasing speed means increasing momentum and the wavelength is proportional to the reciprocal of the momentum.

[Ans: (a) decreasing; (b) increasing; (c) same; (d) same]

17 Look at Eq. 38–22. The change in k_b is the same for a the same change in U_b or E but E is less than U_b.

[Ans: a]

19 The probability of tunneling depends on the product of the barrier thickness and the parameter k, which in turn is proportional to the square root of the difference between the particle kinetic energy and barrier height.

[Ans: all tie]

Hints for Problems

7 The energy emitted by the bulb is the product of the rate with which energy is supplied, the efficiency, and the time interval. This energy is also the product of the number of photons and the energy of each photon, which is the product of Planck constant and the frequency. The frequency is the speed of light divided by the wavelength.

[Ans: 4.7×10^{26} photons]

15 The kinetic energy of the fastest ejected electron is the difference between the photon energy and the work function.

[Ans: 676 km/s]

17 The kinetic energy K of the fastest electron is the difference between the photon energy and the work function. It is also eV_0, where V_0 is the stopping potential and it is $\frac{1}{2}mv^2$, where m is the mass and v is the speed of the electron.

[Ans: (a) 1.3 V; (b) 680 km/s]

21 Write the photoelectric effect equation $K = hf - \Phi$ twice, once for each wavelength and solve the two equations simultaneously for the second wavelength and for the work function.

[Ans: (a) 382 nm; (b) 1.82 eV]

23 The radius of the orbit is given by mv/eB, where v is the speed of the fastest electrons and B is the magnetic field. Solve for v and use the result to calculate the kinetic energy of the fastest electrons. The work done in removing them from the foil (the work function) is the difference between the kinetic energy and the photon energy.

[Ans: (a) 3.1 keV; (b) 14 keV]

31 The fractional change in photon energy is $(f' - f)/f$, where f is the frequency associated with the incident photon and f' is the frequency of the scattered photon. Calculate the wavelength shift and then the wavelength for the scattered photon. Use $f = c/\lambda$ to compute both frequencies.

[Ans: (a) -8.1×10^{-9} %; (b) -4.9×10^{-4} %; (c) -8.8 %; (d) -66 %]

37 You need to know the frequencies associated with the photon before and after scattering. Use $E = hf$ to calculate the frequency of the incident photon and $\lambda = c/f$ to calculate the wavelength before scattering. Use the Compton effect equation to compute the wavelength shift. Then compute the wavelength after scattering. Now you can use $f = c/\lambda$ to compute the frequency and $E = hf$ to compute the photon energy after scattering. The energy of the electron is the change in energy of the photon.

[Ans: (a) 41.8 keV; (b) 8.2 keV]

43 The frequency associated with a photon of energy E is $f = E/h$ and the wavelength is $\lambda = c/f$. The wavelength associated with an electron is $\lambda = h/p$, where p is the magnitude of its momentum. A 1.00 eV electron is nonrelativistic and you can use $K = p^2/2m$ to compute its momentum. A 1.00 GeV electron is relativistic and you should use $(K + mc^2)^2 = (pc)^2 + (mc^2)^2$ but K is so much larger than mc^2 that the expression reduces to $K = pc$. The energies given here are kinetic energies.

$\Big[$Ans: (a) 1.24 μm; (b) 1.22 nm; (c) 1.24 fm; (d) 1.24 fm $\Big]$

51 Having the same resolving power means the waves have the same wavelength. The wavelength of the gamma rays is $\lambda = c/f$, where f is the frequency. The frequency is $f = E/h$, where E is the photon energy. The momentum of an electron is $p = h/\lambda$ and its energy is $K = p^2/2m$. The accelerating potential V is related to the kinetic energy by $K = eV$.

$\Big[$Ans: 9.76 kV $\Big]$

59 Assume the electron is moving along the x axis and use $\Delta x \cdot \Delta p_x \geq h$, where Δx in the uncertainty in position and Δp_x is the uncertainty in momentum. If Δp_x is to have its least possible value $\Delta x \cdot \Delta p_x = h$. Solve for Δp_x.

$\Big[$Ans: 2.1×10^{-24} kg $\cdot$ m/s $\Big]$

63 The transmission coefficient is given by $T = e^{-2k_b L}$, where $k_b = \sqrt{8\pi^2 m(U_b - E)/h^2}$, where L is the width of barrier, U_b is the height of the barrier, E is the energy of the particle, and m is its mass. Convert energy values to joules using $1 \, \text{eV} = 1.60 \times 10^{-19}$. Energy is conserved so the energy of a transmitted particle and the energy of a reflected particle are the same as the energy of an incident particle.

$\Big[$Ans: (a) 9.02×10^{-6}; (b) 3.0 MeV; (c) 3.0 MeV; (d) 7.33×10^{-8}; (e) 3.0 MeV; (e) 3.0 MeV $\Big]$

69 Take room temperature to be $T = 300$ K and calculate the average kinetic energy K. Use $K = p^2/2m$ to calculate the magnitude p of the momentum and $\lambda = h/p$ to calculate the de Broglie wavelength λ. The mass of an atom is M/N_A, where M is the molar mass and N_A is the Avogadro constant. Use the ideal gas law $pV = NkT$, where p is the pressure, V is the volume, N is the number of atoms, and k is the Boltzmann constant, to compute the number of atoms per unit volume N/V. The average distance between atoms is the cube root of the reciprocal of this.

$\Big[$Ans: (a) 73 pm; (b) 3.4 nm; (c) yes, their average de Broglie wavelength is smaller than their average separation $\Big]$

75 The magnitude of the bullet's momentum is $p = mv$ and its de Broglie wavelength is $\lambda = h/p$.

$\Big[$Ans: 1.7×10^{-35} m $\Big]$

79 Assume that the electron is initially at rest and after the collision has the same energy and momentum as the photon. That is, the energy is hf and the momentum is $h/\lambda = hc/f$, where f is the frequency of the electromagnetic wave. Substitute into the relativistic expression for the relationship between the kinetic energy and momentum of a relativistic particle $((pc)^2 = K^2 + 2Kmc^2)$ and show that the result predicts that there is no photon initially.

Quiz

Some questions might have more than one correct answer.

1. Which of the following statements are true for monochromatic electromagnetic radiation?
   A. its energy might have any value
   B. its energy must be a multiple of hf, where f is the frequency and h is the Planck constant
   C. the magnitude of its momentum might have any value
   D. the magnitude of its momentum must be a multiple of h/λ, where λ is the wavelength and h is the Planck constant
   E. its frequency must be a multiple of the fundamental frequency f_0

2. In a photoelectric effect experiment
   A. the number of electrons emitted per unit time depends on the intensity of the incident light
   B. the energy of the most energetic electron emitted is proportional to the wavelength of the incident light
   C. the energies of the emitted electrons depend on the frequency of the incident light but not on its intensity
   D. there is a minimum light frequency for which electrons are emitted
   E. none of the above are true

3. In analyzing the energetics of the photoelectric effect the work function of the material must be taken into account because
   A. the incident photons do work on electrons of the material
   B. an electric potential difference is applied between the material and the collector cup
   C. electrons can repel photons
   D. electrons must overcome the attraction of ions in the material to escape the material,
   E. electrons collide with atoms on their way out of the material

4. In an interaction with an essentially stationary free electron
   A. a photon must either be absorbed or not interact at all
   B. some of the photon's energy might be transferred to the electron
   C. some of the photon's momentum might be transferred to the electron
   D. the wave associated with the photon might be reduced in frequency
   E. the wave associated with the photon might be increased in frequency

5. The change in wavelength of the wave associated with a photon brought about by scattering by a free electron
   A. is greater for greater scattering angle
   B. is less for greater scattering angles
   C. cannot be more than about 2.43×10^{-12} m
   D. cannot be less than about 2.43×10^{-12} m
   E. is always equal to about 2.43×10^{-12} m

6. The intensity of an electromagnetic wave in any given small region

   A. has no physical meaning since electromagnetic radiation consists of photons
   B. has a physical meaning that is completely unrelated to the photon theory
   C. is proportional to the square of the probability that a photon is in that region
   D. is proportional to the probability that a photon is in that region
   E. is proportional to the photon speed

7. Which of the following statements are true about the sinusoidal matter wave that is associated with a free particle?

   A. its speed is equal to the speed of the particle
   B. its frequency is proportional to the energy of the particle
   C. its frequency is proportional to the momentum of the particle
   D. its wavelength is proportional to the energy of the particle
   E. its wavelength is proportional to the momentum of the particle

8. Which of the following statements are true for the wave function $\Psi(x, t)$ for a particle that is constrained to the x axis?

   A. $|\Psi(x, t)|\, dx$ gives the probability that the particle is between x and $x + dx$ at time t
   B. $|\Psi(x, t)|^2$ gives the probability that the particle has coordinate x at time t
   C. $|\Psi(x, t)|^2\, dx$ gives the probability that the particle is between x and $x + dx$ at time t
   D. Ψ obeys Maxwell's equations
   E. Ψ obeys Schrödinger's equation

9. Quantum theory correctly predicts

   A. that a particle placed on one side of a potential energy barrier might be found on the other side even though it does not have sufficient energy to surmount the barrier
   B. that the position and momentum of particle at any instant cannot both be known with certainty
   C. that if the position of a particle at some instant is known with small uncertainty, then its momentum at that instant can be known only with great uncertainty
   D. that electrons incident on a barrier with two slits display an interference pattern
   E. that the masses of all particles are multiples of a fundamental mass

Answers: (1) B, D; (2) A, C, D; (3) D; (4) B, C, D; (5) A, C; (6) D; (7) B; (8) B, E; (9) A, B, C, D

Chapter 39
MORE ABOUT MATTER WAVES

In this chapter you learn some details about matter waves that will prove useful when you study atoms. Learn that the energy of a particle is quantized when the particle is confined in space, as electrons in atoms are. Also learn the mathematical forms of matter waves for an electron in a one-dimensional trap and in a hydrogen atom. Pay attention to the quantum numbers used to designate hydrogen atom states.

Important Concepts

- ☐ trapped electron
- ☐ energy quantization
- ☐ normalized wave function
- ☐ zero-point energy
- ☐ ground state
- ☐ excited state
- ☐ radiative transitions
- ☐ hydrogen atom energies

- ☐ hydrogen atom wave functions
- ☐ Bohr radius
- ☐ radial probability function
- ☐ principal quantum number
- ☐ orbital quantum number
- ☐ orbital magnetic quantum number

Overview

39–2 String Waves and Matter Waves

☐ If a particle is confined to a limited region of space, a finite segment of the x axis, say, then its energy is quantized. That is, the energy can have any of a set of definite discrete values but no values between. This result is closely related to a similar result for a wave on a string that is clamped at both ends. In the absence of an external driving force, the string can vibrate with any of a set of discrete frequencies, the normal modes or harmonics.

39–3 Energies of a Trapped Electron

☐ An energy is associated with each possible state of the electron. For a particle of mass m, in a one-dimensional trap of width L, with infinite potential energy outside, possible values of the energy are given by

$$E_n = \frac{n^2 h^2}{8mL^2},$$

where n is an integer. The energy is said to be quantized. It can have any of the values given above but it can have no others.

□ Notice that the particle has non-zero energy when it is in its lowest (or ground) state. This energy, which is given by $h^2/8mL^2$, is called its **zero-point energy**. The particle cannot be at rest under any conditions.

□ A trapped particle may absorb a photon and thus increase its energy but this happens only if the photon energy equals the difference in energy between the initial state of the particle and some higher state. Suppose the electron is initially in the state with quantum number n_i when it absorbs a photon and goes to the state with quantum number n_f. Then, the photon energy is

$$E_{\text{photon}} = E_f - E_i = \frac{(n_f^2 - n_i^2)h^2}{8mL^2}.$$

Similarly a particle in an excited state can make the transition to a state with lower energy by emitting a photon. The energy of the photon equals the difference in energy between the initial and final states.

39–4 Wave Functions of a Trapped electron

□ Suppose an electron is trapped on the x axis by infinite potential energy barriers at $x = 0$ and $x = L$. No force acts on it when it is in the interior but when it reaches either end of the trap, an infinite force pushes it toward the interior. Inside the trap ($0 \leq x \leq L$) the coordinate-dependent part of the wave function ψ satisfies Schrödinger's equation

$$\frac{\mathrm{d}^2\psi}{\mathrm{d}x^2} + \frac{8\pi^2 m}{h^2}E\psi = 0.$$

Outside the trap, $\psi = 0$. ψ obeys the boundary conditions $\psi(0) = 0$ and $\psi(L) = 0$.

□ There are many possible wave functions for a particle in a trap. They are quite similar to the functions that describe the displacement of a vibrating string that is fixed at both ends. Inside the trap (from $x = 0$ to $x = L$),

$$\psi_n = \sqrt{\frac{2}{L}} \sin\left(\frac{n\pi x}{L}\right),$$

where n is a positive integer. Outside the trap, $\psi_n = 0$. The integer n distinguishes one function from another and is used to label the functions. It is called a *quantum number*. Notice that $\psi_n = 0$ at $x = 0$ and at $x = L$.

□ If the wave function is ψ_n, then the probability density is

$$|\psi_n|^2 = \left(\frac{2}{L}\right)\sin^2\left(\frac{n\pi x}{L}\right).$$

Fig. 39–6 shows graphs of the probability densities for four states.

□ The functions have been **normalized**. That is, the constant in front ($\sqrt{2/L}$) is chosen so that

$$\int_{-\infty}^{+\infty} |\psi|^2\,\mathrm{d}x = 1.$$

This condition is derived from the statement that $|\psi|^2\,\mathrm{d}x$ is the probability that the particle is between x and $cx + \mathrm{d}x$. Since the particle must be somewhere on the x axis, the sum of probabilities for all segments of the axis must be 1.

☐ Notice that the wave functions are quite similar to the functions that describe the displacement of a vibrating string fixed at both ends. Both the matter wave and the string displacement are the sum of sinusoidal waves with the same frequency and wavelength, traveling in opposite directions. For the standing wave to have zero amplitude at $x = 0$ and $x = L$, the wavelength λ and the width of the trap must be related by $\lambda = 2L/n$. If the traveling matter waves extend along the entire x axis, they would be associated with a free particle that has momentum with magnitude $p = h/\lambda = nh/2L$. The energy is $E = p^2/2m = n^2h^2/8mL^2$.

39–5 An Electron in a Finite Well

☐ If the barriers that form the walls of the trap are finite, then particle energies are quantized if they are less than the barrier height and are not quantized if they are greater. If $E > U_0$, where U_0 is the barrier height, then the particle is not confined to the trap. Fig. 39–9 shows an example of some energy levels.

☐ The wave functions are sinusoidal inside the trap and tend exponentially to zero in the barriers. Fig. 39–8 shows some probability densities. There is some chance that the particle will be outside the trap, in the region forbidden to it by classical mechanics.

39–6 More Electron Traps

☐ Atoms are electron traps. The positive charge of the nucleus confines the electron and causes its energy to be quantized.

☐ Crystals with dimensions on the order of nanometers can trap electrons. Such crystals absorb all incident light with wavelength below a certain threshold value that depends on the crystal size and transmit all light with wavelength above the threshold. The frequency of the long wavelength light, and, hence, the energy of its photons, is so small that it cannot excite an electron from a low-lying energy level to a higher level. Thus the crystal size controls the color of the reflected light.

☐ Quantum dots are constructed by sandwiching a tiny piece of semiconductor between insulating materials. The semiconductor becomes an electron trap. Such devices are used in the electronic and optical communications industry.

☐ Quantum corrals are formed by manipulating atoms on the surface of a solid so they trap electrons.

39–7 Two- and Three-Dimensional electron Traps

☐ Suppose an electron is trapped by infinite potential energy walls in a rectangle of length L_x along the x axis and L_y along the y axis. Then possible values of its energy are given by

$$E_{nx,ny} = \frac{h^2}{8m}\left(\frac{n_x^2}{L_x^2} + \frac{n_y^2}{L_y^2}\right).$$

where n_x and n_y are positive integers. For some values of the ratio L_x^2/L_y^2 two states might have the same energy. For example, if $L_x = L_y$, then the state with $n_x = 1$ and $n_y = 2$ has the same energy as the state with $n_x = 2$ and $n_y = 1$. Such states are said to be degenerate.

☐ If an electron is trapped by infinite potential energy walls in a box of length L_x along the x axis, L_y along the y axis, and L_z along the z axis, then the allowed values of its energy are given by

$$E_{nx,ny,nz} = \frac{h^2}{8m} \left(\frac{n_x^2}{L_x^2} + \frac{n_y^2}{L_y^2} + \frac{n_z^2}{L_z^2} \right).$$

where n_x, n_y, n_z are positive integers. Degenerate states are again possible.

39–8   The Bohr Model of the Hydrogen Atom

☐ In a hydrogen atom a single electron orbits a single proton. The proton exerts a force on the electron that is given by Coulomb's law. The magnitude is $F = (1/4\pi\epsilon_0)e^2/r^2$ and it is radially inward, toward the proton.

☐ Bohr assumed that classical physics could be applied to this situation with the additional hypothesis that the magnitude of the angular momentum of the electron is a multiple of $\hbar$ (the Planck constant divided by 2π). As a result the model predicts that the electron moves in an orbit of radius $r = an^2$, where $a = h^2\epsilon_0/\pi me^2 = 52.292$ pm and n is a positive integer. The radius a of the smallest orbit is called the **Bohr radius**.

☐ In the Bohr model the orbit radius is quantized. It cannot have any value but only one of a set of discrete values, distinguished by the integer n.

☐ The Bohr model also predicts that the orbital energy of a hydrogen atom is quantized. Its possible values are given by

$$E_n = \frac{me^4}{8\epsilon_0^2 h^2} \frac{1}{n^2} = -\frac{13.60 \text{ eV}}{n^2}.$$

☐ A hydrogen atom can absorb an incident photon if the photon energy is equal to the difference in energy of two of the hydrogen levels: $hf = E_f - E_i$, where f is the frequency of the wave associated with the photon, E_i is the initial energy of the atom, and E_f is its final energy.

☐ A hydrogen atom in an excited state (above the $n = 1$ state) and go to a lower energy state by emitting a photon with an energy that is equal to the difference in energy of the two states. Possible wavelengths of emitted light are given by

$$\frac{1}{\lambda} = R \left(\frac{1}{n_{low}^2} - \frac{1}{n_{high}^2} \right),$$

where $R = me^4/8\epsilon_0^2 h^3 c = 1.097 \times 10^7 \text{ m}^{-1}$.

39–9   Schrödinger's Equation and the Hydrogen Atom

☐ Although the Bohr model correctly predicts the allowed values of the energy of a hydrogen atom it cannot be quantum mechanically correct because it also predicts definite orbits for the electron. In reality the electron does not have a definite position. All we can do is calculate the probability it is in some given volume at some time. We solve Schrödinger's equation for the wave functions.

☐ The set of wave functions appropriate to an electron in a hydrogen atom differs from the set appropriate to an electron in another situation (trapped in a well with zero potential energy inside, for example) because the force acting on the electron is different for the two cases. The important quantity for a quantum mechanical calculation of a particle wave function is the potential energy function and this is different if different forces act. For hydrogen, which consists of an electron and proton, this function is

$$U(r) = -\frac{e^2}{4\pi\epsilon_0 r},$$

where r is the separation of the two particles.

☐ The electron is trapped by the proton and quantum mechanics predicts a discrete set of energy levels. They are given by

$$E_n = -\frac{me^4}{8\epsilon_0^2 h^2 n^2} = -\frac{13.6\,\text{eV}}{n^2},$$

where n is a positive integer. These are the same values as are predicted by the Bohr model. Notice that even in the lowest energy state ($n = 1$), the electron has energy and is not at rest.

☐ Possible transitions are grouped into series, with all members of a series having the same value n_f of the quantum number for the final state. If $n_f = 1$, the transition is a member of the Lyman series; if $n_f = 2$, the transition is a member of the Balmer series; and if $n_f = 3$, the transition is a member of the Paschen series. The series limit of any series is obtained by setting the quantum number n_i for the initial state equal to ∞. For any series, this radiation has the greatest frequency and the shortest wavelength of any radiation in the series.

☐ Wave functions for electrons in atoms are identified by a set of three quantum numbers: the principal quantum number n, the orbital quantum number ℓ, and the orbital magnetic quantum number m_ℓ. The first is closely associated with the energy, the second with the magnitude of the orbital angular momentum, and the third with one component of the orbital angular momentum. The values of ℓ are restricted by the value of n. For a given value of n, ℓ can have any integer value from 0 to $n - 1$. The values of m_ℓ are restricted by the value for ℓ. For a given value of ℓ, m_ℓ can have any integer value from $-\ell$ to $+\ell$. For the ground state of hydrogen, $n = 1$, $\ell = 0$, and $m_\ell = 0$. The allowed values of the energy for a hydrogen atom depend only on n, not the other quantum numbers. For other atoms, they also depend on ℓ.

☐ The ground state ($n = 1$) probability density for the electron in a hydrogen atom is

$$|\psi|^2(r) = \frac{1}{\pi a^3}e^{-2r/a},$$

where a is the **Bohr radius** (5.292×10^{-11} m). This is the orbit radius predicted by the Bohr model.

☐ If dV is an infinitesimal volume a distance r from the proton, then the probability the electron is in that volume is given by $|\psi|^2\, dV$. The **radial probability density** $P(r)$ is defined so that $P(r)\, dr$ gives the probability of finding the electron in a spherical shell of width dr

a distance r from the proton. Since the volume of the shell is $dV = 4\pi r^2\,dr$, the radial probability density is $P(r) = 4\pi r^2|\psi|^2$. The radial probability density for the ground state wave function of hydrogen is

$$P(r) = \frac{4}{a^3}\,r^2\,e^{-2r/a}.$$

Fig. 39–20 shows a graph of this function. The maximum occurs at $r = a$. The fraction of the time the electron is inside a sphere of radius a is about 0.32; the fraction of the time it is outside the sphere is about 0.68.

☐ The $n = 1$ state is spherically symmetric; the wave function depends only on the distance r from the proton and not on any angular variables. The same is true of the $n = 2$, $\ell = 0$ state and, in fact, of any state for which $\ell = 0$. Wave functions for the three $n = 2$, $\ell = 1$ states, however, depend on the angular coordinates θ and ϕ as well as on r. But the sum of the probability densities for these three states is spherically symmetric. Thus the probability distribution for a completely filled subshell is spherically symmetric.

☐ A hydrogen atom in an excited state can emit a photon and make a transition to a lower energy state. If n_i is the principal quantum number associated with the initial state and n_f is the principal quantum number associated with the final state, then the energy of the photon is given by

$$E_{\text{photon}} = \left(\frac{me^4}{8\epsilon_0^2 h^2}\right)\left(\frac{1}{n_i^2} - \frac{1}{n_f^2}\right)$$

and the frequency of the associated wave is given by $f = E_{\text{photon}}/h$.

39–7 Quantum Weirdness: An Example

☐ The text describes what is called the Einstein, Podolsky, Rosen thought experiment. An atom sends out two photons in opposite directions, each having a property called X. When X is measured, two values are possible, X_1 and X_2. The system is such that if photon A has the value X_1, then photon B will have the property X_2 and vice versa. We may think of the system as having two possible states, which can be denoted by (AX_1, BX_2) and (AX_2, BX_1). The actual wave function is a combination of these two states, with equal probability of either one occurring. When the photons are far apart, X is measured for photon A. Either result can be obtained. If X_1 is obtained, then the wave function collapses to (AX_1, BX_2). If X is now measured for B, the result is X_2, without exception. The question is: How did photon B know that the state changed to (AX_1, BX_2)?

Hints for Questions

1 The ground state energy is proportional to the reciprocal of the square of the well width.

   [Ans: a, c, b]

3 The ends of the trap must strongly repel the particle.

   [Ans: c]

5 Count the minima and maxima for the probability density function $|\psi|^2 = (2/L)\sin^2(n\pi x/L)$.

[Ans: (a) 18; (b) 17]

7 The probability density is $|\psi|^2 = (2/L)\sin^2(n\pi x/L)$ for both particles.

[Ans: equal]

9 The quantum number ℓ is limited to 0 through $n-1$ and the quantum number m_ℓ is limited to $-\ell$, $+\ell$ and all integers between.

[Ans: b, c, and d]

11 The wave function for a higher energy state has more nodes and antinodes than the wave function for a lower energy state. Image squeezing more peaks uniformly into the same width L. Since the wave function is normalized it must decrease somewhere if it increases elsewhere.

[Ans: (a) decrease; (b) increase]

13 The de Broglie wavelength is proportional to the reciprocal of the wavelength of the wave.

[Ans: $n = 1$, $n = 2$, $n = 3$]

15 Look at Schrödinger's equation to see the parameters that determine the wave functions and the allowed values of the energy.

[Ans: same]

17 The third excited state is the $n = 4$ state. The bigger the jump the greater the frequency and the shorter the wavelength of the light. If light is absorbed the electron goes to a higher n state and if light is emitted it goes to a lower n state.

[Ans: (a) $n = 3$; (b) $n = 1$; (c) $n = 5$]

Hints for Problems

5 Solve $E_n = n^2h^2/8mL^2$, with $n = 3$, for L. E_n should be in joules. Use $1\,\text{eV} = 1.60 \times 10^{-19}\,\text{J}$.

[Ans: 0.85 nm]

11 Allowed values of the energy are given by $E_n = n^2h^2/8mL^2$ and you want the difference between the energies of the $n = 4$ and $n = 2$ states. Use $1\,\text{eV} = 1.60 \times 10^{-19}\,\text{J}$ to convert from joules to electron volts. When the electron is de-excited the energy of the photon emitted is equal to the energy difference ΔE of these states. The wavelength of the light is $\lambda = ch/\Delta E$. The electron might jump directly to the $n = 1$ state, or it might jump to the $n = 2$ state and then to the $n = 1$ state, or it might make various other jumps.

[Ans: (a) 72.2 eV; (b) 13.7 nm; (c) 17.2 nm; (d) 68.7 nm; (e) 41.2 nm; (f) 68.7 nm]

15 Read the energy E_2 of the $n = 2$ state from the graph. Add 400 eV and subtract U_0. The value for U_0 can be found in Sample Problem 39–4 of the text.

[Ans: 59 eV]

19 The allowed values of the energy are

$$E_{nx,ny,nz} = \frac{h^2}{8m}\left(\frac{n_x^2}{L_x^2} + \frac{n_y^2}{L_y^2} + \frac{n_z^2}{L_z^2}\right).$$

Since none of the integers can be zero, the ground state has $n_x = 1$, $n_y = 1$ and $n_z = 1$. Use $1\,\text{eV} = 1.60 \times 10^{-19}\,\text{J}$ to convert the result to electron volts.

[Ans: 3.08 eV]

23 The allowed values of the energy are

$$E_{nx,ny,nz} = \frac{h^2}{8m}\left(\frac{n_x^2}{L_x^2} + \frac{n_y^2}{L_y^2} + \frac{n_z^2}{L_z^2}\right).$$

Use trial and error to find the values of n_x, n_y, and n_z that produce the five lowest values. List all the energy differences ΔE. The frequency of the light is $\Delta E / h$.

[Ans: (a) 7; (b) 1.00; (c) 2.00; (d) 3.00; (e) 9.00; (f) 8.00; (g) 6.00]

27 The energy levels for hydrogen are given by $E_n = -(13.6\,\text{eV})/n^2$. The energy of the photon is the energy difference ΔE for the $n = 3$ and $n = 1$ states. Its momentum is $p = \Delta E / c$ and its wavelength is $\lambda = h/p$.

[Ans: (a) 12.1 eV; (b) 6.45×10^{-27} kg $\cdot$ m/s; (c) 102 nm]

31 The ground-state wave function is $\psi = (1/\sqrt{\pi}a^{3/2})e^{-r/a}$ and the probability density is $|\psi|^2$. The radial probability density is $P = 4\pi r^2|\psi|^2$. Substitute $r = a = 5.29 \times 10^{-11}$ m.

[Ans: (a) 291 nm^{-3}; (b) 10.2 nm^{-1}]

35 The total energy is the sum of the kinetic and potential energies. The potential energy is $U = -e^2/4\pi\epsilon_0 r$.

[Ans: (a) 13.6 eV; (b) -27.2 eV]

39 The probability is the integral $\int P(r)\,dr$ of the radial probability function from $r = a$ to $r = \infty$. Find an expression for the radial probability function in Section 39–9 of the text. The integral can be found in Appendix E.

[Ans: 0.68]

51 The orbital quantum number ℓ may have the value 0 and any positive integer value up to and including $n - 1$. The orbital magnetic quantum number m_ℓ may have the values $-\ell$, $+\ell$ and all integer values between, including 0. You should find it useful to know that the sum of an arithmetic series is the average of the first and last terms multiplied by the number of terms.

[Ans: (a) n; (b) $2\ell + 1$; (c) n^2]

55 The square of a one-dimensional wave function has a unit of a reciprocal meter since its product with a length is a probability, which is dimensionless. The Planck constant has a unit of a joule·second. A joule is a kilogram·meter squared per second squared. The unit of energy is the joule and the unit of mass is the kilogram.

[Ans: (b) meter$^{-2.5}$]

Quiz

Some questions might have more than one correct answer.

1. Quantum theory correctly predicts that
    A. the energy of a confined particle might have one of a set of certain discrete values
    B. a confined particle's energy might have any value
    C. the linear momentum of a confined particle with a given wave function has a distribution of values
    D. the linear momentum of a confined particle with a given wave function has a definite value
    E. an atom may make the transition from an initial state to a state with lower energy, emitting a photon with energy equal to the energy difference of the electron states

2. The allowed values of the energy of a particle trapped in a one-dimensional infinite potential well are
    A. proportional to the square of the well width
    B. proportional to the reciprocal of the square of the well width
    C. proportional to integers: 1, 2, 3, . . .
    D. proportional to the squares of integers: 1, 4, 9, . . .
    E. proportional to the mass of the particle

3. The wave functions for a particle trapped in a one-dimensional infinite potential well are sinusoidal within the well and zero outside. Which of the following statements are true?
    A. there is a greater chance of finding the particle in a small region where the wave function is large and positive than in a region of the same width where it is large and negative
    B. there is a greater chance of finding the particle in a small region where the wave function is large and positive than in a region of the same width where it is nearly zero
    C. the wave function is zero at the boundaries of the well
    D. the greater the energy of the particle the more peaks the wave function has
    E. the greater the energy of the particle the fewer peaks the wave function has

4. Quantum theory correctly predicts that one of the allowed energy values
    A. for a confined particle is always zero
    B. for a confined particle is zero for some situations
    C. for a confined particle is never zero
    D. for a free particle is always zero
    E. for a free particle is never zero

5. If a particle is in a well with a finite potential beyond the well boundaries and its quantum number is n,

   A. its wave function is zero beyond the well boundaries
   B. its wave function is not zero beyond the well boundaries
   C. its energy is greater than the energy it would have if it were in a infinite well and had the same quantum number
   D. its energy is less than the energy it would have if it were in a infinite well and had the same quantum number
   E. its energy is the same as the energy it would have if it were in a infinite well and had the same quantum number

6. For a particle in a cubic trap, with infinite potential outside,

   A. three quantum numbers are required to designate a quantum mechanical state
   B. only one quantum number is required to designate a quantum mechanical state since all the sides of the trap have the same length
   C. no quantum mechanical states have the same energy
   D. some quantum mechanical states have the same energy
   E. the ground state energy is the same as the ground state energy of the same particle in a one-dimensional trap with width equal to the length of a cube side

7. The allowed values of the energy of the electron in a hydrogen atom are proportional to

   A. integers: 1, 2, 3, ...
   B. the squares of integers: 1, 4, 9, ...
   C. the reciprocals of integers: 1, 1/2, 1/3, ...
   D. the reciprocals of the squares of integers: 1, 1/4, 1/9, ...
   E. the cubes of integers: 1, 8, 27, ...

8. The restrictions on the quantum numbers for the electron in a hydrogen atom are

   A. the principal quantum number may be any (positive or negative) integer or zero
   B. the principal quantum number may be any positive integer (greater than zero)
   C. the orbital quantum number may be any positive integer or zero
   D. the orbital quantum number may be any positive integer, including zero, that is less than the principal quantum number
   E. the orbital magnetic quantum number may be any (positive or negative) integer or zero with magnitude less than or equal to the orbital quantum number

Answers: (1) A, C, E; (2) B, D; (3) B, C, D; (4) C; (5)B, D; (6) A, D; (7) D; (8) B, D, E

Chapter 40
ALL ABOUT ATOMS

Here you learn about the structure of atoms and, in particular, about the states of multi-electron atoms. Atomic structure is reflected in the periodic table of chemistry and is fundamental to our understanding of the properties of materials. An important new idea is the Pauli exclusion principle, which greatly influences the structures of atoms and their chemical interactions and determines many important details of the periodic table. The chapter closes with an explanation of how a laser works.

Important Concepts

- ☐ spin angular momentum
- ☐ spin quantum number
- ☐ spin magnetic quantum number
- ☐ magnetic dipole moment
- ☐ Pauli exclusion principle
- ☐ subshell
- ☐ periodic table

- ☐ continuous x-ray spectrum
- ☐ cut-off wavelength
- ☐ characteristic x-ray spectrum
- ☐ Moseley plot
- ☐ laser
- ☐ stimulated emission
- ☐ metastable state
- ☐ population inversion

Overview

40-2   Some Properties of Atoms

☐ Some of the properties of atoms that quantum physics is used to explain are:

1.   Atomic properties are periodic. The periodic table of chemistry is arranged in rows, with 2, 8, 18, 18, and 32 chemical elements in successive rows. Atomic properties change in a regular way along a row and are quite similar for atoms in the same column. See Appendix G for the periodic Table and Fig. 40-2 for a plot of ionization energies as a function of atomic number. Carefully note the regularities.

2.   Atoms emit and absorb electromagnetic radiation. An atom has a set of discrete allowable energy values and a photon is emitted when an atom goes from a higher energy state to a lower energy state. When a photon is absorbed, the atom goes from a lower state to a higher state. Conservation of energy yields $hf = |E_i - E_f|$, where f is the frequency of the electromagnetic wave, E_i is the initial energy of the atom, and E_f is the final energy.

3.   Atoms have angular momentum and magnetic dipole moments. Angular momentum is associated with the orbital motion of an electron and with its spin. Atomic nuclei also have spin angular momentum. A magnetic dipole moment is associated with each of these angular momenta.

40–3 Electron Spin

☐ Electrons (and other particles) have intrinsic angular momenta, which have nothing to do with their motions. Although there is no evidence that an electron is actually spinning, its intrinsic angular momentum is called its **spin angular momentum** or just its spin.

40–4 Angular Momenta and Magnetic Dipole Moments

☐ The magnitude of an electron's orbital angular momentum is quantized. The allowed values are given by

$$L = \sqrt{\ell(\ell+1)}(h/2\pi),$$

where ℓ is zero or a positive integer and is called the **orbital quantum number**. The constant $(h/2\pi)$ is the Planck constant divided by 2π. When the atom is in a state with principal quantum number n, the quantum number ℓ can take on the values 0, 1, 2, ..., $n-1$.

☐ Each component of the angular momentum is also quantized. The z component, for example, may have only the values

$$L_z = m_\ell(h/2\pi),$$

where m_ℓ is an integer (positive, negative, or zero) and is called the **magnetic orbital quantum number**. If the orbital quantum number is ℓ, the possible values of m_ℓ are $-\ell$, $+\ell$. and all integer values between.

☐ Since the z component of the angular momentum is quantized, the angle between the angular momentum vector and the z axis can have only certain values, given by $\cos\theta = L_z/L = m_\ell/\sqrt{\ell(\ell+1)}$. For a given value of ℓ, the smallest angle occurs when $m_\ell = \ell$.

☐ A magnetic dipole moment is associated with the orbital motion of an electron. At the atomic level, magnetic dipole moments are conveniently measured in units of the Bohr magneton μ_B ($= eh/4\pi m = 9.274 \times 10^{-24}$ J/T $= 5.788 \times 10^{-5}$ eV/T).

☐ If an electron is in a state with magnetic quantum number m_ℓ, then in terms of m_ℓ and μ_B, the z component of its orbital magnetic dipole moment is $\mu_{\text{orb},z} = -m_\ell\mu_B$. The direction of the dipole moment is opposite to the direction of the angular momentum because the electron is negatively charged.

☐ The magnitude of the spin angular momentum is given by $S = \sqrt{s(s+1)}(h/2\pi)$, where s is called the **spin quantum number** and has the value $s = 1/2$ for electrons. The value of the z component is given by $S_z = m_s(h/2\pi)$, where the **spin magnetic quantum number** m_s is either $-1/2$ or $+1/2$.

☐ A magnetic dipole moment is associated with spin angular momentum. If an electron has spin magnetic quantum number m_s then the z component of its spin dipole moment is $\mu_{s,z} = -2m_s\mu_B$. Except for a factor 2, this is the same as the relationship between the z components of the orbital dipole moment and the orbital angular momentum.

☐ Spin angular momentum and its quantum number are not predicted by Schrödinger's equation but they are predicted by relativistic quantum theory.

☐ The total angular momentum of an atom is denoted by $\vec{J}$ and is the vector sum of the orbital angular momenta and spin angular momenta of all its electrons. Similarly, the magnetic dipole moment of an atom is the vector sum of the orbital and spin dipole moments of all the electrons. The dipole moment may not be parallel to $\vec{J}$.

40–5 The Stern-Gerlach Experiment

☐ This experiment provides an important experimental verification of space quantization. A beam of atoms passes through a non-uniform magnetic field to a screen. If the field is in the z direction and varies with z, then it exerts a force on the atoms in either the positive or negative z direction, depending on whether μ_z is positive or negative. If the angle between the dipole moment $\vec{\mu}$ and the magnetic field $\vec{B}$ is θ, then the force is given by $F_z = \mu(dB/dz)\cos\theta$.

☐ Atoms with different orientations of their magnetic moments are deflected in different directions. The number of directions can be counted and used to calculate ℓ.

40–6 Magnetic Resonance

☐ **Nuclear magnetic resonance** measurements detect changes in the spin directions of protons in materials when the protons in a constant magnetic field $\vec{B}$ are subjected to a sinusoidal electromagnetic field. If the frequency f and total magnetic field B are related by $hf = 2\mu B$, the electromagnetic wave causes the proton spins to flip and energy is preferentially absorbed from the wave. Here μ is the magnitude of the proton magnetic dipole moment. Resonance data is used in analytical chemistry to identify compounds and in medicine to make images of the interior of a body.

40–7 The Pauli Exclusion Principle

☐ Electrons (and many other particles) obey the **Pauli exclusion principle**, which states that only a single electron in a system can be assigned to any quantum state. For an atom, at least one of the quantum numbers n, ℓ, m_ℓ, and m_s must be different for each electron.

☐ The principle has important and far-reaching consequences. For example, all the electrons in a multi-electron atom cannot be in the lowest energy single-particle state.

40–8 Multiple Electrons in Rectangular Traps

☐ Suppose several electrons are in the one-dimensional trap discussed in Chapter 39. There are two quantum numbers, n and m_s, and m_s can have two values ($+1/2$ and $-1/2$). The energy of an electron depends only on its value of n. Thus two electrons may have the energy associated with $n = 1$, two may have the energy associated with $n = 2$, and similarly for other values of n.

☐ There are three quantum numbers for electrons in a two-dimensional trap: n_x, n_y, and m_s. The electron energy depends only on n_x and n_y. If the only degeneracies are those that arise from the two possible values of m_s ($+1/2$ and $-1/2$) only two electrons may have the energy associated with $n_x = 1$, $n_y = 1$ or any other values of these quantum numbers. If the same energy value is associated with two different sets of values of n_x and n_y, then four electrons may have that energy.

☐ Similarly, there are four quantum numbers for electrons in a three-dimensional trap: n_x, n_y, n_z, and m_s. The electron energy depends only on n_x, n_y, and n_z. If the only degeneracies are those that arise from the two possible values of m_s only two electrons may have the energy associated with $n_x = 1$, $n_y = 1$, $n_z = 1$ or any other values of these quantum numbers.

☐ To find the ground state energy of one of these systems, first list the single-electron energies in order from the lowest and write the quantum numbers of the states associated with each energy. If there are no degeneracies there will be only one set of quantum numbers for each energy but if there are degeneracies there will be more than one set. Now place the electrons in the states one at a time, starting with the lowest energy state and proceeding upward in energy until all the electrons have been assigned different states. The total energy of the system is the sum of the single-electron energies.

40–9   Building the Periodic Table

☐ The same quantum numbers n, ℓ, m_ℓ, and m_s can be used to label electron states in all atoms, even though the corresponding wave functions and energies are different for electrons in different atoms.

☐ When an atom is in its ground state, the electrons fill the various states in such a way that the total energy is the least possible. You may start with an atom with atomic number $Z - 1$ and construct an atom with atomic number Z by adding a single proton to the nucleus and a single electron outside. If the atom is in its ground state, the electron goes into the state that gives the atom the lowest possible energy without violating the Pauli exclusion principle.

☐ All electrons with the same values of n and ℓ are said to belong to the same **subshell**. The various values of ℓ are designated by lower case letters when labeling a subshell: $\ell = 0$ is designated s, $\ell = 1$ is designated p, $\ell = 2$ is designated d, $\ell = 3$ is designated f, $\ell = 4$ is designated f, and $\ell = 5$ is designated h. A particular subshell is designated by a symbol such as 3d, which indicates that $n = 3$ and $\ell = 2$. The number of electrons in the subshell is written as a superscript. Thus $3d^2$ indicates that there are two electrons in the 3d subshell.

☐ The number of states in a subshell is $2(2\ell + 1)$. The orbital and spin angular momentum vectors point in all possible directions for electrons in a filled subshell, making the total orbital angular momentum and the total spin angular momentum both zero. The contribution of a filled orbital to the magnetic dipole moment of the atom is zero.

☐ A neon atom has ten electrons. Two of these occupy the $n = 1$, $\ell = 0$ subshell, two occupy the $n = 2$, $\ell = 0$ subshell, and six occupy the $n = 2$, $\ell = 1$ subshell. The ten electrons completely fill three subshells, so a neon atom in its ground state has a total angular momentum of zero and a magnetic moment of zero. Electrons in filled subshells are strongly bound and interact only weakly with neighboring atoms. Neon is chemically inert.

☐ A sodium atom has eleven electrons. In the ground state, ten of them are in the same subshells as the ten electrons of a neon atom. The other is in the $n = 3$, $\ell = 0$ subshell. The angular momentum and magnetic dipole moment of a sodium atom are due entirely to the spin of this electron. A single electron outside completely filled subshells is weakly bound and readily interacts with neighboring atoms. Sodium is chemically active.

☐ A chlorine atom has seventeen electrons. In the ground state, ten have the same configuration as neon, two fill the 3s subshell and five are in the 3p shell. Chlorine is very active chemically.

It combines easily with atoms that have an additional electron outside a closed shell. This electron can occupy the sixth state in the 3p subshell.

☐ An iron atom has twenty-six electrons. Eighteen of them occupy closed subshells. These are the 1s, 2s, 2p, 3s, and 3p subshells. Six are in the 3d subshell and two are in the 4s subshell. Note that the 4s subshell starts filling before the 3d subshell is full. The $3d^6 4s^2$ configuration has a lower energy than the $3d^8$ configuration.

☐ The periodic table of the chemical elements groups atoms with similar chemical properties in the same column, with atoms in neighboring columns of the same row differing by one in the number of electrons. Application of quantum mechanics and the Pauli exclusion principle explains the similarities of the properties of atoms in the same column and the variation of properties from column to column.

40–10   X Rays and the Numbering of the Elements

☐ When highly energetic electrons strike nearly any target, x-ray radiation is produced. The radiation spectrum has two parts: a broad, low-level, **continuous spectrum** and a series of sharp peaks. The sharp peaks can be used to investigate electrons in deep-lying energy states of the atoms and to assign atomic numbers to the chemical elements.

☐ Energetic electrons are produced by accelerating electrons from a hot filament through a potential difference. If the potential difference is V, the kinetic energy of an electron is $K = eV$.

☐ An important feature of the continuous spectrum is the **cut-off wavelength**; for any fixed incident electron energy, no x rays with wavelengths less than λ_{min} are emitted.

☐ After an electron enters the target it may lose some or all of its kinetic energy in a decelerating encounter with a nucleus. An x-ray photon is emitted in one of these events. Radiation with the greatest possible frequency (and shortest possible wavelength) is emitted by those electrons that lose *all* their kinetic energy in a single event. The maximum frequency is, therefore, given by $hf_{max} = eV$, so the shortest possible wavelength is given by $\lambda_{min} = hc/eV$. The value of λ_{min} is independent of the target material.

☐ Radiation of the characteristic spectrum (the sharp peaks) is emitted when an incoming electron knocks another electron from a low-energy state, an $n = 1$ state, for example. An electron from a state with higher energy makes the transition to the empty state and a photon is emitted. For most atoms, the difference in energy of the two states is sufficient to produce a photon in the x-ray region of the electromagnetic spectrum. The K_α x-ray line is produced when electrons fall from an $n = 2$ state (labeled an L state) to an $n = 1$ state (labeled a K state). The K_β line is produced when electrons fall from an $n = 3$ state (labeled an M state) to a K state. Radiation with other frequencies is produced when electrons with still higher energies fall into vacated states.

☐ X-ray emissions can be used to determine the position of a chemical element in the periodic table. Moseley showed that the frequency f of the K_α line, for example, changes in a regular way from element to element in the table: a plot of $\sqrt{f}$ as a function of position in the table is a nearly straight line. Moseley concluded that the order of atoms in the table depends on the number of protons in the nucleus.

☐ You should be able to use the results of quantum theory to understand this result. If there are Z protons in the nucleus, then the effective charge that acts on an electron in a K state is closely approximated by $(Z - 1)e$. According to quantum theory, the energy associated with a deep lying state with principal quantum number n is

$$E_n = -\frac{m(Z - 1)^2 e^4}{8\epsilon_0^2 h^2 n^2}.$$

This is the same as the expression for hydrogen atom energies except that e^4 in the numerator has been replaced by $(Z - 1)^2 e^4$. The frequency associated with K_α-photons is given by $(10.2\,\text{eV})(Z-1)^2$. This shows that $\sqrt{f}$ is proportional to $Z-1$. A Moseley plot is essentially a graph of $\sqrt{f}$ versus Z and is nearly a straight line. Z is called the atomic number of the chemical element. The expression given above for E_n gives a very poor approximation to the energies of outer electrons in many-electron atoms but it is quite good for the innermost electrons responsible for x-ray emission.

40–11 Lasers and Laser Light

☐ Laser light has the following important properties:
1. It is highly monochromatic.
2. It is highly coherent.
3. It is highly directional.
4. It can be sharply focused.

40–12 How Lasers Work

☐ Laser light is produced in the following process. An electron in a state with high energy is stimulated by an incoming photon to drop to a state with lower energy and emit a photon. The energy of the incoming photon must match the difference in energy of the two states ($hf = E_2 - E_1$) and the energy of the stimulated photon is exactly the same. After the emission, there are two photons, identical in every way. If there are sufficient electrons in the higher energy state, the process continues and the number of identical photons quickly multiplies.

☐ Laser light is monochromatic because all the photons produced by stimulation have the same energy; all the waves associated with them have the same frequency. Laser light is highly coherent because all the waves associated with the stimulated photons have the same phase. Laser light does not spread significantly because all the stimulated photons travel in the same direction. Actually some spreading does occur because the waves are diffracted as they pass through the window of the laser.

☐ To produce laser light, the upper of the two states must be **metastable**. Atoms must remain in those states until the transition to a lower state is stimulated by a photon. In addition, the number of atoms in the upper states must be greater than the number in the lower states, so there are more downward than upward transitions. Since there are normally more atoms in lower energy states than in higher energy states, the desired condition is called **population inversion**.

In a helium-neon laser, two neon energy levels, both above the ground state, are responsible for laser emission. An electric current causes, through collisions, a helium level near the upper neon level to become occupied. When an excited helium atom collides with a neon atom in its ground state, energy is transferred to the neon atom and its upper level becomes occupied. The lower level, being above the ground state, is essentially unoccupied and so collisions with excited helium atoms bring about population inversion. The excited neon atoms are stimulated to emit photons by photons that are already present.

Hints for Questions

<u>1</u> A silver atom has 47 electrons. An s subshell can hold 2, a p subshell can hold 6, and a d subshell can hold 10.

$\Big[$Ans: same number (10)$\Big]$

<u>3</u> For $n = 2$ ℓ can be 0 or 1; for $n = 5$ ℓ can be 0, 1, 2, 3, or 4. For each value of ℓ there are $2(2\ell + 1)$ two states.

$\Big[$Ans: (a) 2; (b) 8; (c) 5; (d) 50$\Big]$

<u>5</u> The quantum number m_ℓ can have the values $-\ell$, $+\ell$ and all integers between.

$\Big[$Ans: -1, 0, 1, and 2$\Big]$

<u>7</u> For a single electron atom the energy depends only on the principal quantum number. The radial probability density depends on the angular momentum hence electrons in different states of a multi-electron atom may sample different parts of the distribution of the other electrons.

$\Big[$Ans: (a) n; (b) n and $\ell$$\Big]$

<u>9</u> The largest possible value of ℓ is $n - 1$. m_ℓ can have the values $-\ell$, $+\ell$ and all integers between. The number of subshells is equal to the number of possible values of ℓ. The number of subshells for a given value of n is equal to the number of possible values of ℓ

$\Big[$Ans: all true$\Big]$

<u>11</u> Electrons in the L shell can have $\ell = 0$ or 1. Electrons in the M shell (which give rise to the K_β line) can have $\ell = 0$, 1, or 2.

$\Big[$Ans: (a) 2; (b) 3$\Big]$

<u>13</u> Internal energy is not the only kind of energy a helium atom can have.

$\Big[$Ans: In addition to the quantized energy, a helium atom has kinetic energy; its total energy can equal 20.66 eV$\Big]$

Hints for Problems

<u>3</u> The magnitude of the orbital angular momentum is given by $L = \sqrt{\ell(\ell + 1)}(h/2\pi)$ and the projection of the angular momentum vector on the z axis is $m_\ell(h/2\pi)$, where m_ℓ can be $-\ell$, $+\ell$ or any integer between.

$\Big[$Ans: (a) 3.65×10^{-34} J · s; (b) 3.16×10^{-34} J · s$\Big]$

7 The magnitude of m_ℓ must be less than or equal to ℓ and n must be greater than ℓ.

[Ans:]

13 Since the orbital angular momentum is zero the cosine of the angle is given by S_z/S, where S is the magnitude of the spin angular momentum and S_z is the z component. Use $S = \sqrt{s(s+1)}(h/2\pi)$ and $S_z = m_s(h/2\pi)$, where $s = 1/2$ and $m_s = -1/2$ or $+1/2$.

[Ans: (a) 54.7°; (b) 125°]

19 The allowed values of the single-electron energies are given by $h^2 n^2/8mL^2$, where n is a positive integer. Two electrons can have each value of n. Find the value of n for each of the electrons so the total energy is the least possible.

[Ans: 44]

21 Consider the various possibilities for the distribution of the seven electrons among the lowest four single-electron states, with no more than two electrons with the same value of n. Pick out the three states with the lowest energy above the ground state energy.

[Ans: (a) 51; (b) 53; (c) 56]

25 Add the electrons one at time, with each electron going into the lowest-energy unfilled state. Remember that an s subshell holds 2 electrons, a p subshell holds 6, and a d subshell holds 10.

[Ans: (a) 4p; (b) 4; (c) 4p; (d) 5; (e) 4p; (f) 6]

31 The energy of the photon produced in the first collision is $K_0 - K_1$ and this must be hc/λ, where λ is the wavelength of the electromagnetic wave associated with the photon. The energy of the second photon is $K_2 - K_1$.

[Ans: 49.6 pm, 99.2 pm]

39 The kinetic energy of the electron must be sufficient to knock a K electron out of the atom and this energy is eV, where V is the accelerating potential. A photon associated with the minimum wavelength is emitted if the electron gives all its kinetic energy to the photon. The wavelength is $\lambda_{min} = hc/K$, where K is the electron's kinetic energy. The energy of a K$_\alpha$ photon is the difference in the L and K energy levels and the energy of a K$_\beta$ photon is the difference in the M and K energy levels.

[Ans: (a) 69.5 kV; (b) 17.8 pm; (c) 21.3 pm; (d) 18.5 pm]

43 Use the expression for the hydrogen energy levels to find an expression for the energy difference of the $n = 1$ and $n = 2$ levels in terms of the fundamental constants. Divide by h to find the frequency associated with the emitted photon and multiply by $Z - 1$ to take into account the atomic number of the emitting atom. Compare the result with Eq. 40–27 to find the expression for the constant C.

[Ans: (a) −24%; (b) −15%; (c) −11%; (d) −7.9%; (e) −6.4%; (f) −4.7%; (g) −3.5%; (h) −2.6%; (i) −2.0%; (j) −1.5%]

45 The length of the pulse is given by $c \Delta t$, where Δt is the duration. The energy in the pulse is Nhf, where N is the number of photons and f is the frequency. Use $c = \lambda f$ to substitute for the frequency.

[Ans: (a) 3.60 mm; (b) 5.25×10^{17}]

53 There is a node at each end of the laser, so the length of the laser is a multiple of half a wavelength. The wavelength inside the laser is λ/n, where n is the index of refraction of the lasing material and λ is the wavelength of the light after emission. The frequency, both inside and outside the laser, is c/λ. Since the wave speed in the lasing material is c/n the travel time is Ln/c, where L is the length of the laser.

$\left[\text{Ans: (a) } 3.03 \times 10^5; \text{ (b) } 1.43\,\text{GHz; (d) } 3.31 \times 10^{-6}\right]$

57 The photon energies are $\Delta E_1 = E_1 - E_0$ and $\Delta E_2 = E_2 - E_0$, where E_1 and E_2 are the upper energies and E_0 is the lower energy that participate in the transitions. ΔE_1 and ΔE_2 are also the photon energies. Use $\Delta E = hc/\lambda$, where λ is the wavelength, to compute their values. The difference in ΔE_1 and ΔE_2 is the same as the difference in the upper energies. The energy difference is $2\mu B$, where μ is the magnitude of the electron spin magnetic dipole moment and B is the magnitude of the internal magnetic field.

$\left[\text{Ans: (a) } 2.13\,\text{meV; (b) } 18\,\text{T}\right]$

67 If the electron had no spin, an s state could hold only one electron, a p state could hold 3, a d state could hold 5, and an f state could hold 7. Look at Appendix G to find the atomic numbers of the noble gases and which have just the right number to leave no subshell partially filled.

$\left[\text{Ans: argon}\right]$

Quiz

Some questions might have more than one correct answer.

1. Which of the following statements are true for the spin angular momentum of an electron?

   A. it depends on the motion of the electron
   B. it is an intrinsic property of an electron
   C. its magnitude is given by $s(h/2\pi)$, where $s = 1/2$
   D. its magnitude is given by $\sqrt{s(s+1)}(h/2\pi)$, where $s = 1/2$
   E. its component along any axis is given by $m_s(h/2\pi)$, where m_s is either $+1/2$ or $-1/2$

2. The spin magnetic dipole moment $\vec{\mu}_s$ of an electron is

   A. related to its spin angular momentum $\vec{S}$ by $\vec{\mu}_s = (e/2m)\vec{S}$
   B. related to its spin angular momentum $\vec{S}$ by $\vec{\mu}_s = (e/m)\vec{S}$
   C. related to its spin angular momentum $\vec{S}$ by $\vec{\mu}_s = -(e/2m)\vec{S}$
   D. related to its spin angular momentum $\vec{S}$ by $\vec{\mu}_s = -(e/m)\vec{S}$
   E. has a magnitude of $-\sqrt{3}e(h/2\pi)/2m$

3. The orbital magnetic dipole moment $\vec{\mu}_{orb}$ of an electron in an atom

    A. is related to its orbital angular momentum $\vec{L}$ by $\vec{\mu}_{orb} = (e/2m)\vec{L}$

    B. is related to its orbital angular momentum $\vec{L}$ by $\vec{\mu}_{orb} = -(e/2m)\vec{L}$

    C. has a magnitude that is given by $\mu_{orb} = (e/2m)\sqrt{\ell(\ell+1)}(h/2\pi)$, where ℓ is the orbital quantum number

    D. has a component along an arbitrary z axis that is given by $\mu_{orb,\, z} = -e(h/2\pi)/2m)m_{\ell}$, where m is the mass of an electron and m_{ℓ} is its orbital magnetic quantum number

    E. has a component along an arbitrary z axis that is given by $\mu_{orb,\, z} = -m_{\ell}\mu_B$, where m_{ℓ} is its orbital magnetic quantum number and μ_B is the Bohr magneton

4. The total magnetic dipole moment of an atom

    A. is in the direction of its total angular momentum

    B. is in the direction opposite to that of its total angular momentum

    C. is not parallel to its total angular momentum but its effective magnetic moment is in the direction of its total angular momentum

    D. is not parallel to its total angular momentum but its effective magnetic moment is in the direction opposite that of its total angular momentum

    E. is not parallel to its total angular momentum but its effective magnetic moment is perpendicular to the direction of its total angular momentum

5. The energies E and degeneracies d of the three lowest energy levels of a certain electron trap are given below.

level 1 $E = E_0$, $d = 2$
level 1 $E = 2E_0$, $d = 2$
level 3 $E = 3E_0$, $d = 4$

There are 5 electrons in the trap. The ground state energy of this system is

    A. $5E_0$

    B. $9E_0$

    C. $10E_0$

    D. $15E_0$

    E. $18E_0$

6. A certain energy level that is occupied by an electron separates into two levels in a magnetic field $\vec{B}$. Which of the following statements are true?

   A. when the electron is in the higher-energy state its spin magnetic moment is parallel to the magnetic field

   B. when the electron is in the higher-energy state its spin magnetic moment is antiparallel to the magnetic field

   C. when the electron is in the higher-energy state its spin magnetic moment is perpendicular to the magnetic field

   D. if the magnetic field is in the positive z direction the difference in energy between the two states is $\Delta E = 2\mu_{s\,z}B$

   E. the frequency of the light wave associated with the photon that is emitted when the spin of the electron flips is $f = 2\mu_{s\,z}B/h$

7. Because electrons obey the Pauli exclusion principle

   A. no two electrons in an atom have the same set of values for the four quantum numbers that characterize an electron in an atom (principal, orbital, orbital magnetic, and spin magnetic quantum numbers)

   B. no more than two electrons in an atom have the lowest single-electron energy

   C. no more than $2(2\ell + 1)$ electrons in an atom can have the same value of the principal quantum number n and the same value of the orbital quantum number ℓ

   D. no more than $2(2n + 1)$ electrons in an atom can have the same value of the principal quantum number n

   E. the chemical elements can be arranged in the periodic table

8. The characteristic x-ray spectrum of a chemical element

   A. is continuous in frequency

   B. consists of several discrete frequencies

   C. is emitted when an incident electron knocks an electron in a low energy state from the atom and a higher-energy electron fills the vacant state

   D. can be used to find the molar mass of the element

   E. can be used to find the atomic number of the element

9. For a two energy level laser to work

   A. more atoms must be in the higher-energy state than in the lower-energy state

   B. atoms in the higher-energy state must not spontaneously drop to the lower-energy state over relatively long times

   C. when atoms are stimulated to drop from the higher-energy state to the lower-energy state they must be replaced by some means

   D. the temperature must be kept moderate

   E. the pressure must be kept moderate

10. Laser light is

    A. coherent because the wave associated with a photon emitted by stimulation has the same phase as the wave associated with the stimulating photon

    B. monochromatic because the wave associated with a photon emitted by stimulation has the same frequency as the wave associated with the stimulating photon

    C. highly directional because the photon emitted by stimulation travels in the same direction as the stimulating photon

    D. can be intense because it can be focused to a small region

    E. can be intense because sometimes many photons are emitted simultaneously from the same atom

Answers: (1) B, D, E; (2) D, E; (3) B, C, D, E; (4) D; (5) B; (6) B, D, E; (7) A, B, C, E; (8) B, C, E; (9) A, B, C; (10) A, B, C, D

Chapter 41
CONDUCTION OF ELECTRICITY IN SOLIDS

The ideas of modern physics are used to understand one of the important properties of solids, their electrical conductivity. Pay attention to distinctions between metals, insulators, and semiconductors. These materials differ greatly in the fraction of their electrons that participate in electrical conduction and in the changes that occur in their resistivities when the temperature changes. You should understand how the differences come about. Later sections of the chapter are devoted to solid-state devices, so pervasive in modern technology.

Important Concepts

□ insulator

□ metal

□ semiconductor

□ energy band

□ energy gap

□ Fermi energy

□ density of states

□ occupation probability

□ density of occupied states

□ conduction band

□ valence band

□ hole

□ doping

□ *p-n* junction

□ junction rectifier

□ light-emitting diode

□ solid state laser

□ integrated circuit

Overview

41–2 The Electrical Properties of Solids

□ Electrical **insulators** have very few electrons that are free to contribute to an electrical current; their resistivities are large. **Metals**, with large conduction electron concentrations, are good conductors. They have low resistivities. In between are **semiconductors**. Metals have positive temperature coefficients of resistivity; their resistivities increase with increasing temperature. The temperature coefficient of resistivity of a typical semiconductor is larger than that of a metal and is negative.

41–3 Energy Levels in a Crystalline Solid

□ Allowed energies for electrons in a crystalline solid form **bands**, large groups of closely spaced levels separated by gaps. Typically, bands are a few electron volts wide; gaps may range from somewhat less than an electron volt to several electron volts. The number of quantum mechanical states in a band is a small multiple of the number of atoms in the crystal. Energy bands arise as atoms are brought close together and the wave functions originally associated with different atoms overlap; several neighboring atoms exert electrical forces on any electron.

☐ The distribution of electrons among the quantum states determines if a substance is a metal or not.

41–4 Insulators

☐ Electrons obey the Pauli exclusion principle: there is at most one electron in each state. At temperature $T = 0\,\mathrm{K}$, the N electrons of a solid fill the N states with the lowest energy. For an insulator, the number of electrons is exactly right to completely fill an integer number of bands, so there are no partially filled bands.

☐ A completely filled band makes no contribution to an electrical current because every electron can be paired with another that is traveling with the same speed but in the opposite direction. An electric field cannot change this situation because no empty states are nearby for the electron to occupy. The gap between the highest filled band and the empty band above is so large that essentially no electrons are excited across it by thermal agitation or by an applied electric field.

41–5 Metals

☐ For a metal, the states in one band are partially occupied. States in lower energy bands are essentially all occupied and states in higher energy bands are essentially all unoccupied. Electrons in the partially filled band are the conduction electrons and are responsible for the current when an electric field is turned on.

☐ The energy of the highest occupied state at $T = 0\,\mathrm{K}$ is called the **Fermi energy** and is denoted by E_F. For an electron in a metal, with energy equal to the Fermi energy, nearly all its energy is kinetic energy, so the speed of an electron with the Fermi energy is $v_F = \sqrt{2E_F/m}$. This is called the **Fermi speed** and typically has a value of about $10^6\,\mathrm{m/s}$.

☐ In the absence of an electric field in the metal, the average electron velocity is zero because, for every electron traveling in any direction, another is traveling with the same speed in the opposite direction. Because the band is partially filled, there are unoccupied states with energy near the Fermi energy and with velocity opposite the direction of the field. An electric field causes a small fraction of the conduction electrons to occupy these states. A current results.

☐ If the electrons did not suffer collisions with atoms of the solid, their velocities would continue to increase in the direction opposite the field. Electrons with energies near the Fermi energy, however, do suffer collisions. Electrons with less energy do not because there are no nearby empty states for them to occupy.

☐ The resistivity of a metal is determined by the number n of electrons per unit volume in the partially filled band and by the average time τ between collisions (the **relaxation time**): $\rho = m/ne^2\tau$, where m is the mass of an electron and e is the magnitude of its charge. See Chapter 26.

- [ ] The **density of states** $N(E)$ is defined so $N(E)\,dE$ gives the number of quantum mechanical states per unit volume of sample that have energies between E and $E+dE$. For a free-electron metal, it is given by

$$N(E) = \frac{8\sqrt{2}\pi m^{3/2}}{h^3} E^{1/2}\,.$$

- [ ] Not all states are filled. The probability that at absolute temperature T a state with energy E contains an electron is given by the occupancy probability

$$P(E) = \frac{1}{e^{(E-E_F)/kT} + 1}\,,$$

where k is the Boltzmann constant and T is the absolute temperature. This function takes the Pauli exclusion principle into account: no state may be occupied by more than one electron. The **density of occupied states** N_0, which is the number of electrons per unit volume of sample per unit energy interval, is given by the product of $N(E)$ and $P(E)$: $N_0(E) = N(E)P(E)$.

- [ ] At the absolute zero of temperature, the probability that a state with energy less than E_F is occupied is one; these states are guaranteed to be filled. The probability that a state with energy greater than E_F is occupied is zero; these states are guaranteed to be empty. The Fermi energy at $T = 0\,K$ is determined by the condition that the number of occupied states per unit volume in the partially filled band is the same as the number of conduction electrons per unit volume in the metal. The determining condition is written mathematically as

$$n = \int_0^{E_F} N_0(E)\,dE\,,$$

where the energy for the bottom of the band was taken to be zero. Once the integral is evaluated, the result can be solved for the Fermi energy:

$$E_F = \left(\frac{3}{16\sqrt{2}\pi}\right)^{2/3} \frac{h^2}{m} n^{2/3} = \frac{0.121 h^2}{m} n^{2/3}\,.$$

41–6 Semiconductors

- [ ] There are precisely enough electrons in a pure semiconductor to completely fill all the states in an integer number of bands, with none left over. At $T = 0\,K$, the most energetic electron is in the highest energy state of one of the bands and is separated in energy from the empty state above it by a gap. The highest filled band is called the **valence band** and the lowest empty band is called the **conduction band**.

- [ ] Since a completely filled band does not contribute to an electrical current, a pure semiconductor is an insulator at $T = 0\,K$. Electrons must be promoted across the gap before an electric field will generate a current.

- [ ] At higher temperatures, electrons are thermally promoted across the gap from the valence to the conduction band. Since the gap for a semiconductor is much less than the gap for an insulator, the conduction electron concentration is much greater for a semiconductor than for an insulator at the same temperature. On the other hand, it is much less for a semiconductor than for a metal. Even at high temperatures, the resistivity of a typical semiconductor is much greater than that of a metal.

☐ When electrons have been excited to the conduction band of a semiconductor, both the conduction and valence bands are partially filled and both contribute to the current when an electric field is turned on. Rather than deal with the contributions of the vast number of electrons in the valence band, the band is thought to consist of a much smaller number of fictitious particles, called **holes**, one for each empty state. These particles have positive charge and are accelerated in the direction of an applied electric field.

☐ The relaxation time decreases somewhat as the temperature increases, but both the number of conduction band electrons and the number of valence band holes increase dramatically with temperature. As a result, the resistivity of a semiconductor decreases as the temperature increases.

41–7 Doped Semiconductors

☐ The number of conduction band electrons or the number of valence band holes can be greatly increased by **doping** a semiconductor: that is, by adding certain impurity atoms.

☐ To increase the number of electrons in its conduction band, a semiconductor is doped with **donor** impurity atoms. Each such atom contributes an electron in a hydrogen atom-like state around the impurity, with energy in the gap between the valence and conduction bands. It is easily promoted to the conduction band by thermal agitation. Semiconductors that are doped with donors are said to be n-type semiconductors.

☐ To increase the number of holes in its valence band, a semiconductor is doped with **acceptor** atoms. Each of these contributes an empty hydrogen atom-like state with energy in the gap. Electrons can easily be promoted from the valence band to impurity states, thereby creating a hole in that band. Semiconductors that are doped with acceptors are said to be p-type semiconductors.

41–8 The p-n Junction

☐ The basic building block of nearly all semiconductor devices is the p-n junction, consisting of p-type and n-type semiconducting materials in contact. Electrons from the n-type material diffuse into the p-type material, where they fall into holes. Holes from the p-type material diffuse into the n-type material, where electrons combine with them. The electron and hole currents, called **diffusion currents**, are in the same direction, from the p side toward the n side.

☐ Electron and hole diffusion leaves a narrow region near the boundary on the n side with positively charged ions, the donors that have lost their electrons, and a narrow region near the boundary on the p side with negatively charged ions, the acceptors that have gained electrons. As a result, an electric field exists near the boundary. It pushes electrons toward the n side and holes toward the p side. This is the **drift current**. If the circuit is not completed, the drift and diffusion currents cancel and the net current is zero. The difference in the electric potential of the two sides is called the **contact potential difference**. The region in which the electric field exists is called the **depletion region** because it has few charge carriers.

41–9   The Junction Rectifier

☐ *p-n* junctions are often used as **rectifiers**. An *ideal* rectifier has infinite resistance for current in one direction and zero resistance for current in the other direction. Rectifiers are used to change an alternating current into a direct current.

☐ When a source of emf is connected to a *p-n* junction, with the positive terminal at the *p* side, the junction is said to be *forward biased*. The electrical resistance of the junction is small and the current is large. If the positive terminal of the emf is connected to the *n* side, the junction is said to be *back biased*. The electrical resistance is large and the current is small.

☐ A forward bias on a *p-n* junction lowers the potential barrier and increases the diffusion current dramatically. A back bias on the junction raises the potential barrier and decreases the diffusion current. When the junction is back biased, holes flow from the *n* to the *p* side and electrons flow from the *p* to the *n* side. Because there are so few holes on the *n* side and so few electrons on the *p* side, the current is severely limited.

41–10   The Light-Emitting Diode (LED)

☐ If conditions are right, light is emitted from a semiconductor when an electron falls from the conduction band to the valence band. In many cases, the energy lost by the electron equals the energy gap between the two bands. If the gap is E_g, then the frequency of the light is $f = E_g/h$ and its wavelength is $\lambda = hc/E_g$. For most semiconductors, this is infrared or red light. Forward-biased *p-n* junctions are used to construct light-emitting diodes because then the number of photons emitted is much greater than the number absorbed. The emission of light by recombination of electrons and holes is also the basis of **solid state lasers**.

41–11   The Transistor

☐ Field-effect transistors make use of the dependence of the depletion region width on an applied bias. Two *n*-type regions, for example, are imbedded in a *p*-type semiconductor and are connected by a channel of *n* type material. A potential difference applied to the regions generates a current in the channel. The width of the channel, and hence its resistance, is controlled by a back bias applied to the junction formed by the channel and the substrate. Large changes in the current are produced by small changes in the bias (or gate) potential, so the device can be used for amplification.

☐ Modern electronic devices employ thousands (or in some cases millions) of transistors, all fabricated into a single small chip. The result is called an **integrated circuit**.

Hints for Questions

<u>1</u> Each of the corner atoms is shared by 8 adjoining cells and there are 8 corners. Each of the atoms at the centers of the faces is shared by two cells and there are 6 faces.

[Ans: 4]

3 Remember that drift speed is the additional speed acquired by an electron because it is accelerated by an applied electric field.

[Ans: much less than]

5 The electrons in the highest occupied band of a metal are essentially free electrons. The energy limits of the gaps do not depend on the sample size but the number of levels in a band does. Note that the density of states depends on the energy and on the volume of the sample.

[Ans: b, c, and d]

7 The probability that an electron will jump the gap from the valence to the conduction band increases dramatically as the gap decreases. For a pure semiconductor a hole is created in the valence band for every electron that leaves that band to enter the conduction band.

[Ans: b and d]

9 A phosphorus atom has five electrons in its outer shell and is neutral. Both electrons and holes contribute to the electrical current in an applied electric field.

[Ans: none]

11 Each ion bonds with 4 others.

[Ans: $+4e$]

13 The electric field is produced by donor replacement atoms on the n side that have lost electrons and acceptor replacement atoms on the p side that have gained electrons. The field increases if the field produced by the bias is in the same direction as the contact field and decreases if it is in the opposite direction.

[Ans: (a) right to left; (b) back bias]

15 The color of the light depends on the gap between the valence and conduction bands. A photon of blue light has more energy than a photon of red light.

[Ans: blue]

Hints for Problems

1 The mass of a copper atom is its molar mass divided by the Avogadro constant. The number of copper atoms per unit volume is the density of copper divided by the mass of a copper atom. Since copper is monovalent the number of conduction electrons per unit volume is the same as the number of copper atoms per unit volume.

[Ans: 8.49×10^{28} m^{-3}]

11 Solve $P = 1/[e^{(E-E_F)/kT} + 1]$ for E by first solving for $e^{(E-E_F)/kT}$, then taking the natural logarithm of both sides. Evaluate $N(E) = (8\sqrt{2}\pi m^{3/2}/h^3)E^{1/2}$ for the density of states and $N_0 = N(E)P(E)$ for the density of occupied states.

[Ans: (a) 6.81 eV; (b) 1.77×10^{28} m$^{-3} \cdot$ eV^{-1}; (c) 1.59×10^{28} m$^{-3} \cdot$ eV^{-1}]

13 The density of occupied states is given by $N_0(E) = P(E)N(E)$, where $P(E)$ is the probability that a state with energy E is occupied and $N(E)$ is the density of states at energy E. The probability of occupation is given by $P = 1/[e^{(E-E_F)/kT} + 1]$, where E_F is the Fermi energy. The density of states is given by $N(E) = (8\sqrt{2}\pi m^{3/2}/h^3)E^{1/2}$, where m is the electron mass.

[Ans: (a) $1.36 \times 10^{28}\,\text{m}^{-3} \cdot \text{eV}^{-1}$; (b) $1.67 \times 10^{28}\,\text{m}^{-3} \cdot \text{eV}^{-1}$; (c) $9.0 \times 10^{27}\,\text{m}^{-3} \cdot \text{eV}^{-1}$; (d) $9.5 \times 10^{26}\,\text{m}^{-3} \cdot \text{eV}^{-1}$; (e) $1.7 \times 10^{18}\,\text{m}^{-3} \cdot \text{eV}^{-1}$]

<u>17</u> The Fermi energy of a metal is given by $E_F = (3/16\sqrt{2}\pi)(h^2/m)n^{2/3}$, where m is the electron mass and n is the number of conduction electrons per unit volume. Solve for n. Now you need the number of atoms per unit volume. This is the density divided by the mass of an atom. The mass of an atom is the molar mass divided by the Avogadro constant. Be sure to use consistent units.

[Ans: 3]

<u>23</u> According to the result of Problem 21 the average kinetic energy of the conduction electrons of a metal is $(3/5)E_F$, where E_F is the Fermi energy. Sample Problem 41–5 gives 7.0 eV for the Fermi energy of copper. The total kinetic energy of the conduction electron system is the average energy multiplied by the conduction electron number density and the volume of the sample. The conduction electron number density is calculated in Problem 1 and found to be $8.49 \times 10^{28}\,\text{m}^{-3}$.

[Ans: 57.1 kJ]

<u>25</u> According to Problem 24 the fraction of the conduction electrons in a metal that have energies greater than the Fermi energy is given by fract $= 3kT/2E_F$, where T is the temperature on the Kelvin scale, k is the Boltzmann constant, and E_F is the Fermi energy.

[Ans: 4.7×10^2 K]

<u>29</u> Use $P(E) = 1/[e^{(E-E_F)/kT} + 1]$ to calculate the occupation probability. For a state at the bottom of the conduction band $E - E_F = E_g/2$ and for a state at the top of the valence band $E - E_F = -E_g/2$, where E_g is the gap between those bands. The probability that a state is not occupied is $1 - P$.

[Ans: (a) 1.5×10^{-6}; (b) 1.5×10^{-6}]

<u>33</u> Use $P(E) = 1/[e^{(E-E_F)/kT} + 1]$ to calculate the occupation probability. For a state at the bottom of the conduction band $E - E_F$ is half the gap for the pure semiconductor and is 0.11 eV for the doped semiconductor. For the donor state $E - E_F$ is $0.11\,\text{eV} - 0.15\,\text{eV} = -0.04\,\text{eV}$.

[Ans: (a) 4.79×10^{-10}; (b) 0.0140; (c) 0.824]

<u>41</u> Start with Eq. 41–5, $N(E) = (8\sqrt{2}\pi m^{3/2}/h^4)E^{1/2}$, for the density of states in a metal and replace E with E_F. Use Eq. 41–9, $E_F = (3/16\sqrt{2}\pi)^{2/3}(h^2/m)n^{2/3}$, to substitute for E_F. Since copper is monovalent the number density of conduction electrons is the same as the number of copper atoms per unit volume and this is the density of copper divided by the mass of a copper atom. The mass of a copper atom is the molar mass of copper divided by the Avogadro constant.

[Ans: $1.8 \times 10^{28}\,\text{m}^{-3} \cdot \text{eV}^{-1}$]

<u>47</u> The number density of conduction electrons in copper can be found as the result of Problem 1. Use the ideal gas law $pV = NkT$, where p is the pressure, V is the volume, and T is the

temperature on the Kelvin scale. Set N/V equal to the conduction electron number density and solve for p.

[Ans: 3.49×10^3 atm]

Quiz

Some questions might have more than one correct answer.

1. Unlike a metal, a semiconductor

   A. has a small charge carrier number density
   B. has a large charge carrier number density
   C. has a large resistivity
   D. has a positive temperature coefficient of resistivity
   E. has a negative temperature coefficient of resistivity

2. Allowed energy values for conduction electrons in crystalline solids

   A. are very nearly the same as the allowed energy values for an electron in an isolated atom of the same material
   B. form a continuous spectrum
   C. are closely spaced in a series of bands, separated by gaps in which there are no allowed energies
   D. differ from the allowed values for an electron in an isolated atom because conduction electrons in a solid are influenced by the electric fields of many atoms
   E. are above the potential energy barriers between atoms

3. The Fermi energy

   A. is an allowed energy level in either a metal or a semiconductor
   B. is not necessarily an allowed energy level in either a metal or a semiconductor
   C. is the energy of the most energetic electron at all temperatures
   D. is the energy of the least energetic electron at temperature $T = 0$
   E. is an energy that is greater than or equal to the energy of the highest occupied state at temperature $T = 0$ and below the energy of the lowest unoccupied state at that temperature

4. The Fermi energy of a crystalline solid

   A. cannot be calculated but is experimentally measured
   B. is calculated using the condition that it is the energy for which the probability of occupation is exactly 0.5 for temperatures above absolute zero
   C. is calculated as the energy of the highest energy quantum mechanical state
   D. is calculated using the condition that the number of quantum mechanical states with energy less than the Fermi energy is equal to the number of electrons
   E. must be calculated for each temperature of interest

5. The probability of occupation of a quantum mechanical state with energy E depends on

   A. the number of states with that energy

   B. the number of electrons with that energy

   C. the temperature

   D. the number of atoms in the solid

   E. the Fermi energy

6. The number of quantum mechanical states with energy in a narrow range around some energy E in the conduction band of a metal

   A. depends strongly on the temperature

   B. depends only weakly on the temperature

   C. depends on the Fermi energy

   D. is proportional to the square of E

   E. is proportional to the square root of E

7. The resistivity of a crystalline metal is given by $\rho = m/e^2 n\tau$, where n is the number density of conduction electrons, τ is the means free time for conduction electrons, and m is the mass of an electron. Which of the following statements are true?

   A. n depends on the number of quantum mechanical states with energy near the Fermi energy

   B. n depends on the probability of occupation of a state with energy near the Fermi energy

   C. n is proportional to the number of collisions per unit time involving conduction electrons

   D. τ is proportional to the number of collisions per unit time involving conduction electrons

   E. τ is proportional to the reciprocal of the number of collisions per unit time involving electrons

8. Which of the following statements are true?

   A. the temperature coefficient of resistivity of a pure semiconductor is negative because the mean time between collisions increases strongly with an increase in temperature

   B. the temperature coefficient of resistivity of a pure semiconductor is negative because the number density of conduction electrons increases strongly with an increase in temperature

   C. the temperature coefficient of resistivity of a metal is positive because the mean time between collisions increases with an increase in temperature while the number density of conduction electrons remains constant

   D. the temperature coefficient of resistivity of a metal is positive because the mean time between collisions decreases with an increase in temperature while the number density of conduction electrons remains constant

   E. the temperature coefficient of resistivity of a metal is positive because the number density of conduction electrons decreases strongly with an increase in temperature

9. A doped semiconductor is

    A. n type if the replacement atoms have more than four electrons in their outer shells
    B. n type if the replacement atoms have fewer than four electrons in their outer shells
    C. p type if the replacement atoms have more than four electrons in their outer shells
    D. p type if the replacement atoms have fewer than four electrons in their outer shells
    E. n type if the Fermi energy is small and p type of it is great

10. Which of the following statements are true?

    A. a pn junction is formed when a semiconductor that is doped with donors is joined to a semiconductor that is doped with acceptors
    B. if a forward bias is applied to a pn junction the current increases in proportion to the applied potential difference
    C. if a forward bias is applied to a pn junction the electrical resistance decreases as the applied potential difference increases
    D. if a reverse bias is applied to a pn junction the current is nearly independent of the applied potential difference over a wide range
    E. the electrical resistance is smaller if a pn junction is reverse biased than if it is forward biased with the same applied potential difference

Answers: (1) A, C, E; (2) C, D; (3) B, E; B, D, E; (4) B, D, E; (5) C, E; (6) B, E; (7) A, B, D; (8) B, D; (9) A, D; (10) A, D

Chapter 42
NUCLEAR PHYSICS

A nucleus takes up only an extremely small fraction of the atomic volume but accounts for most of the mass of an atom. You will learn about the constituents of nuclei and about nuclear decay. Pay attention to the energy considerations that tell if a nucleus is stable or not and, if it is not, learn to find the energy of the decay products. Also learn about the mathematics of random decay.

Important Concepts

- ☐ atomic nucleus
- ☐ nuclide
- ☐ isotope
- ☐ atomic number
- ☐ neutron number
- ☐ mass number
- ☐ strong nuclear force

- ☐ disintegration constant
- ☐ binding energy
- ☐ half life
- ☐ disintegration energy
- ☐ alpha decay
- ☐ beta decay
- ☐ nuclear models

Overview

42–2   Discovering the Nucleus

☐ Rutherford's experiment consists of firing a beam of alpha particles at a gold foil. Most are deflected through small angles, but a few are deflected by nearly 180°. This is possible only if the nucleus is extremely small and positively charged.

42–3   Some Nuclear Properties

☐ Nuclei consist of **nucleons**, a term that encompasses both protons and neutrons. The **atomic number** (Z) is the number of protons; the **neutron number** (N) is the number of neutrons, and the **mass number** (A) is the total number of nucleons. The word **nuclide** is used to designate a nuclear species. Two nuclides are different if they differ in either their atomic or neutron numbers. Two nuclides that have the same atomic number but differ in neutron number are called **isotopes**.

☐ Nucleons attract each other via the strong nuclear force, an extremely strong force with an extremely short range, about 10^{-15} m. In addition, protons repel each other via electrical forces. As a result, stable nuclei with small mass numbers have the same number of neutrons as protons but stable nuclei with large mass numbers have more neutrons than protons. The extra neutrons participate in the strong nuclear interactions, which hold the nucleus together, but not in the electrical interactions, which tend to tear it apart.

☐ The surface of a nucleus is somewhat ill defined but an average radius R can be measured. It is given by $R = R_0 A^{1/3}$, where A is the mass number and R_0 is about 1.2 fm. A femtometer is 10^{-15} m.

☐ Nuclear masses are commonly measured in **atomic mass units** (abbreviated u): 1 u is about 1.661×10^{-27} kg. The mass of a nucleus in u, rounded to the nearest integer, is its mass number. Because the volume of a nucleus and its mass are both proportional to A, the densities of all nuclei are nearly the same.

☐ The mass of a stable nucleus is less than the sum of the masses of its constituents. The difference accounts for the binding together of the nucleons. If Δm is the mass difference, then the **binding energy** is given by $E = \Delta m\, c^2$, where c is the speed of light. This is the energy that must be given a nucleus to separate it into its constituent particles, well-separated and at rest. Atomic, not nuclear, masses are usually tabulated. The mass of an appropriate number of electrons must be subtracted from the atomic mass to obtain the nuclear mass, or else the mass of an atom with Z protons and N neutrons must be compared to the sum of the masses of Z hydrogen atoms and N neutrons.

☐ Nuclides in the vicinity of iron have the greatest binding energy per nucleon; the binding energy per nucleon drops as A becomes either larger and smaller. See Fig. 42–6. Energy is released in a nuclear fusion event, when two nuclei with low mass numbers combine to form a single nucleus. Energy is also released in a nuclear fission event, when a nucleus with high mass number breaks into smaller fragments.

☐ The internal energy of a nucleus has a discrete set of allowed values, a different set for each nuclide. The difference in energy of adjacent low-lying states is on the order of MeV and when a nucleus changes state from a higher to a lower energy, the photon emitted is in the gamma-ray portion of the electromagnetic spectrum.

☐ Nuclei also have intrinsic angular momenta and magnetic dipole moments. Values of the angular momentum are the same as for electrons in atoms, but nuclear magnetic dipole moments are smaller by a factor of about 1000 because nuclear masses are that much greater than the electron mass.

42–4 Radioactive Decay

☐ An unstable nucleus may spontaneously turn into another nucleus with the emission of one or more particles. Alpha decay involves the emission of a helium nucleus and beta decay involves the emission of either an electron or a positron, along with a neutrino or antineutrino.

☐ Although any nucleus in a collection of identical unstable nuclei might decay in any given time interval, which will actually decay cannot be predicted. The number that decay in any small time interval Δt is proportional to the number N of undecayed nuclei present and to the interval itself: $\Delta N = -\lambda N\, \Delta t$, where the constant of proportionality λ is called the **disintegration constant**. The negative sign appears because ΔN is negative (N decreases). In the limit as $\Delta t \to 0$, this equation becomes

$$\frac{dN}{dt} = -\lambda N$$

and its solution is $N(t) = N_0 e^{-\lambda t}$, where N_0 is the number of undecayed nuclei at time $t = 0$.

☐ The decay rate or activity R also follows an exponential law:

$$R = -\frac{dN}{dt} = R_0 e^{-\lambda t},$$

where R_0 is the decay rate at $t = 0$. The exponent is the same as the exponent in the law for N. Activities are often measured in becquerels (1 Bq = 1 disintegration per second) or curies (1 Ci = 3.7×10^{10} Bq).

☐ A radioactive decay is often characterized by its **half-life** $T_{1/2}$, the time for half the undecayed nuclei initially present to decay. It is related to the disintegration constant by $T_{1/2} = (\ln 2)/\lambda$.

☐ The mass of the parent nucleus is greater than the sum of the masses of the decay products. The reduction in mass that occurs with a decay, multiplied by c^2, gives the **disintegration energy**, denoted by Q. This energy appears as the kinetic energy of the decay products and as the excitation energy of the daughter nucleus, if it is left in an excited state.

42–5 Alpha Decay

☐ When a nucleus decays by emission of an alpha particle, the daughter nucleus has an atomic number that is two less than that of parent nucleus, a neutron number that is two less and a mass number that is four less.

☐ The half-lives of alpha emitters range from less than a second to times that are longer than the age of universe. The combination of the strong nuclear force of attraction and the electrical repulsion of the daughter nucleus for the alpha particle leads to a high potential energy barrier that tends to hold the alpha inside. Classically, it can never escape because its energy is less than the barrier height, but quantum mechanical tunneling is possible. The half-life depends sensitively on the height and width of the barrier.

42–6 Beta Decay

☐ In β^- decay, a neutron is converted to a proton, with the emission of an electron and a neutrino (more precisely an antineutrino). The daughter nucleus has an atomic number that is one greater than that of the parent nucleus, a neutron number that is one less, and a mass number that is the same. In β^+ decay, a proton is converted to a neutron with the emission of a positron and a neutrino. The daughter nucleus has an atomic number that is one less than that of parent nucleus, a neutron number that is one greater, and a mass number that is the same. Neutron-rich nuclides tend to decay by β^- emission while proton-rich nuclides tend to decay by β^+ emission. The β particle and neutrino are created in the decay process; they do not exist before the decay.

☐ The disintegration energy is shared by the β particle, neutrino, and recoiling daughter nucleus, with nearly all of it going to the first two. For identical parent nuclei, the energies of the emitted β particles span a wide range, but the sum of the β and neutrino energies is always the same.

☐ The mass of a neutrino is difficult to measure and may be zero. Neutrinos interact only extremely weakly with matter. Enormous numbers pass through Earth and our bodies every second.

42–7   Radioactive Dating

☐ Radioactive decay is used to date the deaths of ancient organisms. The atmosphere contains a small amount of radioactive carbon, which enters all living things. When an organism dies, the nuclei lost to decay are not replenished and the time of death can be fixed by measuring the amount of undecayed radioactive carbon left in the organism.

42–8   Measuring Radiation Dosage

☐ The gray (Gy) is used to measure the energy per unit mass of target actually delivered to a target: 1 Gy is equivalent to 1 J/kg delivered. The **sievert** (Sv) is used to measure the dose equivalent and takes into the account the biological effect of the radiation. To find the dose equivalent in sieverts, the dose in grays is multiplied by the *relative biological effectiveness factor* (RBE), a factor that is tabulated for various radiations in handbooks.

42–9   Nuclear Models

☐ The **collective model** treats a nucleus as a drop of liquid. It is useful for comparisons of the masses and binding energies of various nuclei and in discussions of nuclear fission. The **independent particle model** treats the nucleons in a nucleus as independent particles. Protons and neutrons have their own sets of energy levels and quantum numbers, like electrons in atoms. The model predicts the existence of shells, which are particularly stable when completely filled. The **combined model** treats closed shells as a liquid drop and nucleons outside as independent particles.

Hints for Questions

1 The incident particle stops when all its initial kinetic energy is converted to electrostatic potential energy and this is proportional to the product of the charges on the incident particle and target nucleus and to the reciprocal of the distance between them. The charge on the proton is half the charge on the alpha.

[Ans: less]

3 The greater the binding energy the more stable the nucleus.

[Ans: above]

5 A neutron has changed into a proton. The atomic number increases by 1 and the mass number stays the same.

[Ans: (a) ^{196}Pt; (b) no]

7 The nucleus is well above the region of stable nuclei. It does not have enough neutrons to be stable. We expect some of the protons to change into neutrons.

[Ans: (a) on the $N = Z$ line; (b) positrons; (c) about 120]

9 The decay rate is proportional to $e^{-\lambda t}$. The greater the value of λ the steeper the decline in the decay rate.

[Ans: yes]

11 Exponential decay is a direct result of the randomness of the decay process.

[Ans: no]

13 The process is random. Every nucleus has the same chance of decaying as any other (in the same time interval).

[Ans: no effect]

15 The greater the energy of the alpha the greater are its chances of tunneling through the potential energy barrier that holds it in the nucleus.

[Ans: ^{209}Po]

17 Protons and neutrons separately exist in closed shells and are particularly stable when their numbers are magic numbers.

[Ans: d]

Hints for Problems

3 Both kinetic energy and total linear momentum are conserved in the collision. The particles are nonrelativistic, so you can use $\frac{1}{2}mv^2$ for the kinetic energy and $m\vec{v}$ for the momentum of a particle. The collision is one dimensional. See Section 9–10 of the text.

[Ans:]

13 Let f_{24} be the abundance of ^{24}Mg, f_{25} be the abundance of ^{25}Mg, and f_{26} be the abundance of ^{26}Mg. Let M_{24}, M_{25}, and M_{26} be the masses of the nuclei. Then the average atomic mass is $f_{24}M_{24} + f_{25}M_{25} + f_{26}M_{26}$. Furthermore, the abundances must sum to 1: $f_{24} + f_{25} + f_{26} = 1$. Solve these two equation simultaneously for f_{25} and f_{26}.

[Ans:]

23 The initial decay rate is $R_0 = \lambda N_0$, where λ is the disintegration constant and N_0 is the number of nuclei present. The disintegration constant is related to the half-life by $\lambda = (\ln 2)/T_{1/2}$. The number of nuclei is given by M/m, where M is the mass of the sample and m is the mass of a nucleus (67 u). The decay rate at time t is $R = R_0 e^{-\lambda t}$.

[Ans: (a) 7.5×10^{16} s^{-1}; (b) 4.9×10^{16} s^{-1}]

33 Calculate the mass of the undecayed nuclei at $t = 14.0$ h and at $t = 16.0$ h. The difference is the mass of the nuclei that decayed between those two times. The number of undecayed nuclei at time t is given by $N = N_0 e^{-\lambda t}$, where N_0 is the number at $t = 0$ and λ is the disintegration constant, which is related to the half-life by $\lambda = (\ln 2)/T_{1/2}$. Multiply by the mass of a nucleus to obtain $m = Me^{-\lambda t}$ for the mass of undecayed nuclei at time t. Here M is the mass of the sample.

[Ans: 265 mg]

39 When the activity reaches its final value the number of undecayed nuclei is constant and the rate of production must equal the activity. Suppose that ^{56}Mn is being produced at the rate R. It is decaying at the rate λN, where N is the number of undecayed ^{56}Mn nuclei present and λ is the disintegration constant, which is related to the half-life by $\lambda = (\ln 2)/T_{1/2}$. The rate of change of N is $dN/dt = R - \lambda N$. After a long time N approaches a constant value and dN/dt approaches zero. Then $R = \lambda N$. Solve for N. To find the total mass of undecayed nuclei multiply N by the mass of a nucleus, which is 56 u.

[Ans: (a) 8.88×10^{10} s^{-1}; (b) 1.19×10^{15}; (c) 0.111 μg]

43 Assume the ^{238}U nucleus is initially at rest and the ^{234}Th nucleus is in its ground state. Then the disintegration energy is given by $Q = K_{Th} + K_\alpha$, where K_{Th} is the kinetic energy of the recoiling ^{234}Th nucleus and K_α is the kinetic energy of the alpha particle (4.196 MeV). Linear momentum is conserved, so $0 = p_{Th} + p_\alpha$, where p_{Th} is the momentum of the ^{234}Th nucleus and p_α is the momentum of the alpha particle. Use $K_{Th} = p_{Th}^2/2m_{Th}$ and $K_\alpha = p_\alpha^2/2m_\alpha$ along with $p_{Th} = -p_\alpha$, where m_{Th} and m_α are the masses, to show that $K_{Th} = (m_\alpha/m_{Th})K_\alpha$.
$\left[\text{Ans: } 4.269\,\text{MeV}\right]$

49 The disintegration energy is the change in the nuclear masses, multiplied by c^2. Since a carbon atom has 6 electrons and a boron atom has 5, the mass of a carbon nucleus is $m_C - 6m_e$ and the mass of a boron nucleus is $m_B - 5m_e$, where m_C and m_B are the atomic masses.
$\left[\text{Ans: (b) } 0.961\,\text{MeV}\right]$

53 The number of ^{238}U nuclei in the rock is given by M/m, where M is the total mass of that isotope and m is the mass of a single nucleus (238 u). A similar expression holds for ^{206}Pb. A ^{206}Pb nucleus is created for every ^{238}U nucleus lost so the number of ^{238}U nuclei in the rock is the sum of the number of nuclei of both types now present. Use $N = N_0 e^{-\lambda t}$ to find the age t of the rock. The disintegration constant λ is related to the half-life by $\lambda = (\ln 2)/T_{1/2}$.
$\left[\text{Ans: (a) } 1.06 \times 10^{19}; \text{ (b) } 0.624 \times 10^{19}; \text{ (c) } 1.68 \times 10^{19}; \text{ (d) } 2.97 \times 10^9\,\text{y}\right]$

61 The number of plutonium atoms ingested is the mass of plutonium ingested divided by the mass of a plutonium atom (239 u). Since 12 h is much less than the half-life you can use $\Delta N = R\,\Delta t = \lambda N\,\Delta t$ to compute the number of alpha particles produced in the time interval Δt. Here R is the decay rate (which is λN), N is the number of ^{239}Pu atoms present, and λ is the disintegration constant, which is related to the half-life by $\lambda = (\ln 2)/T_{1/2}$. Multiply by 0.95 to find the number of alpha stopped by the body. The energy absorbed is the product of the number of alphas absorbed and the energy of each alpha. The physical dose in grays is the energy absorbed divided by the mass of the worker and the dose equivalent in sieverts is the product of this and the RBE factor.
$\left[\text{Ans: (a) } 6.3 \times 10^{18}; \text{ (b) } 2.5 \times 10^{11}; \text{ (c) } 0.20\,\text{J}; \text{ (d) } 2.3\,\text{mGy}; \text{ (e) } 30\,\text{mSv}\right]$

67 The counting rate at time t is $R = R_0 e^{-\lambda t}$, where R_0 is the rate at $t = 0$ and λ is the disintegration constant, which is related to the half-life by $\lambda = (\ln 2)/T_{1/2}$. In part (a) solve for t. In part (b) evaluate R_0/R.
$\left[\text{Ans: (a) } 59.5\,\text{d}; \text{ (b) } 1.18\right]$

79 The atomic number of Neptunium (Mp) is 93. Find the atomic number and the mass number of the final isotope, then look up the isotope in Appendix F.
$\left[\text{Ans: } ^{225}\text{Ac}\right]$

85 If M_H is the actual mass of ^{1}H and M is the actual mass of some isotope, then the mass of the isotope in the new unit is $(M/M_H)(1.000\,000\,\text{u})$. See Appendix B for M_H.
$\left[\text{Ans: (a) } 11.906\,83\,\text{u}; \text{ (b) } 236.2025\,\text{u}\right]$

87 Use $r = r_0 A^{1/3}$, where $r_0 = 1.2$ fm and A is the mass number. Solve for A.
$\left[\text{Ans: } 27\right]$

Quiz

Some questions might have more than one correct answer.

1. The atomic number of a nuclide

   A. uniquely identifies the nuclide
   B. uniquely identifies the chemical element
   C. is equal to the number of nucleons in the nucleus
   D. is equal to the number of electrons in the nucleus
   E. is equal to the number of protons in the nucleus

2. Two isotopes of each other

   A. have the same number of protons in their nuclei
   B. have the same number of neutrons in their nuclei
   C. have the same number of nucleons in their nuclei
   D. have the same number of electrons in their nuclei
   E. can be distinguished from each other by their mass numbers

3. The radius of a nucleus is roughly

   A. proportional to the cube root of its atomic number
   B. proportional to the cube of its atomic number
   C. proportional to the cube root of its neutron number
   D. proportional to the cube root of its mass number
   E. proportional to the cube of its mass number

4. The most stable nuclei

   A. with small mass numbers have an equal number of neutrons and protons
   B. with small mass numbers have more neutrons than protons
   C. with large mass numbers have an equal number of neutrons and protons
   D. with large mass numbers have more neutrons than protons
   E. with large mass numbers have more protons than neutrons

5. The binding energy per nucleon of a nucleus

   A. can be computed from the mass of the nucleus and the masses of its nucleons when they are isolated
   B. is greater for less stable nuclei than for more stable nuclei
   C. is greatest for nuclides with mass numbers of about 57
   D. is least for nuclides with mass numbers of about 57
   E. is reduced when a heavy nucleus splits apart

6. The probability that a given nucleus in a collection of identical radioactive nuclei will decay in a given time interval is proportional to

    A. the number of nuclei in the collection
    B. the half life of the nuclei
    C. the disintegration constant of the nuclei
    D. the number of nuclei initially present in the collection
    E. none of the above

7. At the end of two half lives the rate of decay of identical nuclei in a collection of radioactive nuclei is

    A. zero
    B. half the initial rate of decay
    C. one-fourth the initial rate of decay
    D. twice the initial rate of decay
    E. four times the initial rate of decay

8. In an alpha decay

    A. the mass number of the decaying nucleus decreases by two
    B. the mass number of the decaying nucleus decreases by four
    C. the atomic number of the decaying nucleus decreases by two
    D. the atomic number of the decaying nucleus decreases by four
    E. the mass number and the atomic number do not change

9. In a beta decay

    A. a hydrogen nucleus might be emitted
    B. an alpha particle might be emitted
    C. an electron and a neutrino might be emitted
    D. a positron and a neutrino might be emitted
    E. an electron and a positron might be emitted

10. In a β^- decay

    A. the mass number of the emitting nucleus increases by one
    B. the mass number of the emitting nucleus decreases by one
    C. the atomic number of the emitting nucleus increases by one
    D. the atomic number of the emitting nucleus decreases by one
    E. the mass of the daughter nucleus is greater than the mass of the parent

answers: (1) B, E; (2) A; (3) D; (4) A, D; (5) A, C; (6) C; (7) C; (8) B,C; (9) C, D; (10) C

Chapter 43
ENERGY FROM THE NUCLEUS

Both the fission of a heavy nucleus into two lighter nuclei and the fusion of two light nuclei into a heavier nucleus convert internal energy associated with nucleon-nucleon interactions into kinetic energy of the products. In each case, your chief goal should be to understand the basic process. Learn what the products are, what energy is released, and what inhibits the process. In addition, you will learn about occurrences of the phenomena in nature and about human attempts to produce and control nuclear energy in a sustained fashion.

Important Concepts

- ☐ nuclear fission
- ☐ fission fragment
- ☐ prompt neutron

- ☐ delayed neutron
- ☐ nuclear fusion
- ☐ proton-proton cycle

Overview

43–2   Nuclear Fission: The Basic Process

☐ In the basic **fission** process, a thermal neutron is absorbed by a heavy nucleus (^{235}U, say) to form a compound nucleus (^{236}U) and the compound nucleus breaks into two medium-mass nuclei (fission fragments), with the emission of one or more neutrons. The fission fragments may also emit neutrons and undergo beta decay.

☐ Thermal neutrons, with energies of about 0.04 eV, are used because the chance of capture is much better than for fast neutrons.

☐ Starting with the same nucleus, the mass numbers of the fragments may be different for different fission events and experiments give the relative probabilities for the various possible outcomes. Look at Fig. 43–1 of the text to see the distribution of fragments for the fission of ^{236}U. The fragments are not usually identical. The most likely fragments have mass numbers around 95 and 137.

☐ Fission takes place in several steps. First, the compound nucleus breaks into what are called the primary fragments and these fragments immediately emit one or more neutrons, called the **prompt neutrons**. The resulting fragments are usually still neutron-rich and they decay by β^- emission. This decay often leaves a fragment nucleus in an excited state and if the excitation energy is sufficient, it may decay via the emission of more neutrons, although usually the decay is via gamma emission. Neutrons, if they are emitted, are called **delayed neutrons**.

☐ The disintegration energy is given by $Q = \Delta m\, c^2$, where Δm is the change in the total mass. Consider the fission of the heavy compound nucleus F into the fragments X and Y, with the production of b neutrons: $F \rightarrow X + Y + bn$. Let m_F be the mass of F, m_X be the mass of X, m_Y be the mass of Y, and m_n be the mass of a neutron. Then the disintegration energy is given by $Q = (m_F - m_X - m_Y - bm_n)c^2$. Q is positive, indicating the release of energy. Most appears as the kinetic energy of the fragments but some appears as the kinetic energy of the neutrons, and if X and Y are the final stable nuclei, some appears as the kinetic energies of electrons and neutrinos produced in the β decays.

43–3 A Model for Nuclear Fission

☐ The collective model is used to discuss fission. After the thermal neutron is captured, the fragments are formed and pull apart from each other, much as a single drop splitting into two.

☐ Not all neutron-rich heavy nuclides fission. The absorbed neutron must supply enough energy for the separating fragments to overcome the energy barrier produced by their strong mutual attraction. Once the fragment separation is beyond the peak in the potential energy curve, the potential energy decreases with separation because the fragments are both positively charged and repel each other electrically.

☐ The most energy that can be supplied by a thermal neutron is equal to its binding energy E_n. If this energy is not about the same as the barrier height E_b or greater, fission does not occur and the compound nucleus loses the excitation energy by means of gamma emission. Fission is favored if $E_n \geq E_b$.

43–4 The Nuclear Reactor

☐ To produce a large number of fission events in a reactor, the fissionable material is made to undergo a **chain reaction**: neutrons produced in one fission event are used to initiate other fission events.

☐ If fission takes place in a solid or liquid, most of the released energy appears as thermal energy. In a reactor power generator, the energy is used to raise the temperature of water and the resulting steam is used to drive a turbine.

☐ Neutron production is proportional to the volume of the reactor; the number of neutrons that leak out is proportional to the surface area. Neutron leakage from a reactor is controlled by making the reactor large.

☐ Neutrons produced in fission events are energetic and must be slowed before they can trigger other events. The fuel is mixed with a **moderator** (usually water), so fast neutrons are slowed by collisions with protons in the moderator.

☐ As a neutron slows, it passes a critical energy range in which it is particularly susceptible to nonfissionable capture by a ^{238}U nucleus. The fuel and moderator are clumped so neutrons pass this critical energy range while they are in the moderator and they are slowed to thermal energies before entering the fuel.

☐ If the number of neutrons present in a reactor remains constant with time, the reactor is said to be **critical**; if the number decreases, it is said to be **subcritical**; and if the number increases, it is said to be **supercritical**. The value of the **multiplication factor** k indicates the tendency. If N neutrons that will participate in fission events are present at some time, then after the fission events occur, there will be kN neutrons present to participate in the next round of fission events. The value of k is 1 if the reactor is critical, less than 1 if it is subcritical, and greater than 1 if it is supercritical. For steady power generation, $k = 1$.

☐ The multiplication factor is varied by inserting and withdrawing **control rods**, which readily absorb neutrons. To increase the number of fission events per unit time, some rods are withdrawn for a short time, during which k is greater than one. The rods are then inserted so k becomes one again. Because some neutron emissions are delayed, the rods are effective even though there is a response time for adjusting them.

43–6 Thermonuclear Fusion: The Basic Process

☐ Nuclear fusion occurs when two light nuclei fuse together to form a single heavier nucleus. For low mass numbers, the binding energy per nucleon is much greater for heavier nuclei than for lighter. The heavy nucleus has much less internal energy and the excess appears as the kinetic energy of the products. Products include the final nucleus and perhaps neutrons, β particles, and photons.

☐ The light nuclei are positively charged and repel each other electrically. They must overcome (or tunnel through) a potential energy barrier before they get close enough for the strong force of attraction to be effective. In a hot plasma of protons or other light nuclei, there are many nuclei with sufficient kinetic energy to fuse, even if the mean kinetic energy is much less than the barrier height. Look at Fig. 19–7 to see the distribution of speeds for molecules in a gas.

43–7 Thermonuclear Fusion in the Sun and Other Stars

☐ The proton-proton cycle of fusion processes are responsible for the internal energy of the Sun and other stars. It proceeds by the following steps:

^{1}H + ^{1}H → ^{2}H + e^+ + ν Two protons fuse to form a deuteron. A positron (e^+) and a neutrino (ν) are emitted. The positron annihilates with a free electron, producing two photons.

^{2}H + ^{1}H → ^{3}He + γ A deuteron and a proton fuse to form a light helium nucleus (with only one neutron) and another photon is emitted.

^{3}He + ^{3}He → ^{4}He + ^{1}H + ^{1}H   Two light helium nuclei fuse to form an alpha particle and two protons.

The overall process can be written 4^1H + $2e^-$ → ^{4}He + 2ν + 6γ.

☐ The disintegration energy for the entire process is about 26.7 MeV. About 0.5 MeV is carried away by the neutrinos, but most is available to maintain the internal energy and keep the Sun shining. It is estimated that the proton-proton cycle can continue for about 5×10^9 y before the Sun's hydrogen is gone.

☐ When all the hydrogen has been fused into helium, the Sun's core will collapse and as gravitational potential energy is converted to kinetic energy, it will get hotter. In the hot dense core, helium can fuse into heavier elements. Chemical elements with mass numbers beyond $A = 56$ cannot be formed in fusion events because the binding energy per nucleon decreases beyond this mass number.

43–8 Controlled Thermonuclear Fusion

☐ Currently, attempts are being made to produce sustained controlled fusion for purposes of generating electric power. The goal is to maintain a high-temperature, high-density gas of particles for sufficient time that a significant number of fusion events take place. If n is the particle concentration and τ is the confinement time, then **Lawson's criterion** for a successful fusion reactor is $n\tau \geq 10^{20}$ s $\cdot$ m^{-3}. In addition, the temperature must be high enough to allow particles to overcome the Coulomb barrier to their fusion.

☐ A tokamak makes use of a magnetic field to maintain high particle concentrations and to keep the hot plasma away from walls. Magnetic confinement methods seek to obtain a long confinement time, although particle concentrations are low.

☐ The inertial confinement method uses a high-power laser beam to ionize and compress small pellets of deuterium and tritium. Fusion takes place in the core of the pellet. This method uses a short confinement time but a high particle concentration in an attempt to reach the critical value of the Lawson number $n\tau$.

Hints for Questions

<u>1</u> The fragments have mass numbers of about 115.

[Ans: a]

<u>3</u> Rank the nuclides according to the per cent yield, as read from the graph.

[Ans: b, e, a, c, d]

<u>5</u> Before the decay the nucleus has 92 protons and 147 neutrons. After the decay it has 93 protons and 146 neutrons. Look up the atomic numbers of the nuclides in Appendix F.

[Ans: c]

<u>7</u> Write expressions for the volume and surface area of each solid, then divide the expression for the surface area by the expression for the volume. You should get a result that depends on a and can be used to rank the solids.

[Ans: c, a, d, b]

<u>9</u> The binding energy per nucleon of the result of fusion must be greater than that for the original nuclei. Look at Fig. 42–6. Find the mass numbers of the given elements in Appendix F.

[Ans: d]

<u>11</u> An ^{3}H is larger than an ^{2}H nucleus so the nuclei must get closer together in a d-d reaction than in a d-t reaction before the strong nuclear force dominates.

[Ans: c]

Hints for Problems

5 The disintegration energy is given by $Q = (M_{Cr} - 2M_{Mn})c^2$. Use $c^2 = 931.5\,\text{MeV/u}$.

[Ans: $-23.0\,\text{MeV}$]

7 The rate of spontaneous fission decays is given by λN, where λ is the disintegration constant for that type decay and N is the number of ^{235}U nuclei in the sample. The disintegration constant is related to the half-life by $\lambda = (\ln 2)/T_{1/2}$ and the number of nuclei in the sample is M/m, where M is the mass of the sample and m is the mass of a nucleus (235 u). The ratio of the decay rates is equal to the ratio of the disintegration constants and also to the reciprocal of the ratio of the half-lives.

[Ans: (a) 16 fissions/day; (b) 4.3×10^8]

13 Calculate the total energy output over the $3.00\,\text{y}$ as the product of the power and the time interval. Assume each fission event produces $200\,\text{MeV}$ and calculate the number of fission events. The original number of nuclei is double this. Multiply by the mass of a single nucleus to obtain the total mass of the original fuel.

[Ans: $462\,\text{kg}$]

17 Consider $1.00\,\text{g}$ of ^{90}Sr, which produces thermal energy at a rate of $0.93\,\text{W}$. The rate of thermal energy generation is given by $P = Q_{\text{eff}}R$, where R is the fission rate. The fission rate is $R = \lambda N$, where N is the number of λ is the disintegration constant and is related to the half-life by $\lambda = (\ln 2)/T_{1/2}$. The number of nuclei is given by M/m, where M is $1.00\,\text{g}$ and m is the mass of a ^{90}Sr nucleus.

[Ans: (a) $1.2\,\text{MeV}$; (b) $3.2\,\text{kg}$]

23 If the power at time 0 is P_0, then, according to the result of Problem 18, the power at time t is $P_0 k^{t/t_{\text{gen}}}$, where k is the multiplication factor and t_{gen} is the neutron generation time. Solve for k.

[Ans: 0.99938]

31 The barrier height is $q^2/4\pi\epsilon_0 d$, where q is the charge on a nucleus and d is the center-to-center separation of the nuclei. That is, $d = 2r$, where r is the radius of a nucleus. Use $r = r_0 A^{1/3}$, where A is the mass number and $r_0 = 1.2\,\text{fm}$, to compute the radius.

[Ans: $1.41\,\text{MeV}$]

35 The energy E and the mass m of the Sun are related by $E = mc^2$. The rate of energy loss is $P = dE/dt$. Solve for dm/dt. The energy loss during an interval Δt is $\Delta E = P\,\Delta t$ and the mass loss is $\Delta m = (P/c^2)\,\Delta t$.

[Ans: (a) $4.3 \times 10^9\,\text{kg/s}$; (b) 3.1×10^{-4}]

37 The number of fusion events per unit time is P/Q, where P is the rate of energy radiation and Q is the energy produced per event. P is given in Problem 35 and Q is given in Section 43–7. Each event produces 2 neutrinos. The fraction of the neutrinos that reach Earth is equal to the ratio of the cross-sectional area of Earth to the surface area of a sphere with radius equal to the Earth-Sun distance.

[Ans: (a) $1.8 \times 10^{38}\,\text{s}^{-1}$; (b) $8.2 \times 10^{28}\,\text{s}^{-1}$]

43 Use $Q = -\Delta mc^2$, where Δm is the change in total mass. Calculate the number of deuterium atoms in $500\,\text{kg}$ by dividing the total mass by the mass of a deuterium atom. Multiply by

0.300 to obtain the number of nuclei that undergo fusion. Multiply by Q to obtain the energy generated. Use 1 megaton TNT $= 2.6 \times 10^{28}$ MeV to compute the rating.

[Ans: (a) 24.9 MeV; (b) 8.65 megaton TNT]

47 Multiply the density by 0.35 to obtain the hydrogen density, then divide by the mass of a hydrogen atom to obtain the number density of protons contained in hydrogen. Carry out a similar calculation for helium, remembering that each helium nucleus contains two protons. Add the two number densities. Use the ideal gas law to find the number density of particles in an ideal gas.

[Ans: (a) 3.1×10^{31} protons/m^3; (b) 1.2×10^6 times]

51 According to Sample Problem 2 the rate of consumption by neutron capture is one-quarter of the rate of fission events, so the mass of ^{235}U should be 25% higher than the value computed in Problem 13.

[Ans: 6×10^2 kg]

53 According to the kinetic theory of gases the average kinetic energy of a particle in a gas at temperature T on the Kelvin scale is given by $K_{avg} = \frac{3}{2}kT$, where k is the Boltzmann constant.

Quiz

Some questions might have more than one correct answer.

1. Which of the following immediately result from a typical fission event?

   A. two nuclei that are both heavier than the original nucleus
   B. two nuclei that are lighter than the original nucleus
   C. an alpha particle
   D. protons
   E. neutrons

2. A typical fission event is initiated by the absorption of

   A. a thermal alpha particle
   B. a thermal electron
   C. a thermal proton
   D. a thermal neutron
   E. none of these

3. Fission fragments

   A. are usually not radioactive
   B. usually decay by alpha emission
   C. usually decay by emission of an electron and a neutrino
   D. usually decay by emission of a positron and a neutrino
   E. usually undergo fission

4. Typically the total kinetic energy of the fission fragments is about
   A. 20 eV
   B. 20 keV
   C. 200 keV
   D. 20 MeV
   E. 200 MeV

5. In a chain reaction
   A. the fissioning nuclei are arranged on straight lines
   B. fission events produce fissionable fragments
   C. all fission events occur simultaneously
   D. neutrons from one fission event trigger other fission events
   E. beta particles from one fission event trigger other fission events

6. Which of the following statements are true for a nuclear reactor?
   A. the moderator slows neutrons so they have a greater probability of triggering a fission event
   B. the moderator is adjusted so the reactor is critical
   C. control rods are adjusted so the reactor is supercritical
   D. control rods are adjusted so the reactor is critical
   E. the positions of the control rods determine the power output

7. In a fusion event
   A. two heavy nuclei merge to form a nucleus that fissions
   B. two light nuclei merge to form a nucleus that fissions
   C. two light nuclei merge to form a nucleus with a mass that is greater than the sum of the masses of the combining nuclei
   D. two light nuclei merge to form a nucleus with a mass that is less than the sum of the masses of the combining nuclei
   E. two heavy nuclei merge to form a nucleus with a mass that is greater than the sum of the masses of the combining nuclei

8. For a fusion event
   A. the mass of the final nucleus is greater than the combined masses of the original nuclei
   B. the mass of the final nucleus is less than the combined masses of the original nuclei
   C. the binding energy per nucleon for the final nucleus is greater than the binding energy per nucleon of the original nuclei
   D. the binding energy per nucleon for the final nucleus is less than the binding energy per nucleon of the original nuclei
   E. the binding energy per nucleon for the final nucleus is the same as the binding energy per nucleon of the original nuclei

9. For a fusion event to occur
   A. the combining nuclei must be close enough to each other for the strong nuclear force to dominate
   B. the combining nuclei must be close enough to each other for the electrostatic force to dominate
   C. the initial total kinetic energy of the combining nuclei must be greater than the peak of the potential energy barrier
   D. the total kinetic energy of the combining nuclei can be slightly less than the peak of the potential energy barrier
   E. the combining nuclei must attract each other electrostatically

10. the total energy released in a proton-proton fusion reaction ($4^1H + 2e^- \longrightarrow {}^4He + 2\nu + 6\gamma$) is about
    A. 20 eV
    B. 20 keV
    C. 200 keV
    D. 20 MeV
    E. 200 MeV

answers: (1) B, E; (2) D; (3) C; (4) E; (5) D; (6) A, D; (7) D; (8) B, C; (9) A, D; (10) D

Chapter 44
QUARKS, LEPTONS, AND THE BIG BANG

This chapter deals with two closely related topics: the most fundamental constituents of matter and the evolution of the universe from its start at the Big Bang. Pay attention to the properties of particles and to the mechanisms by which they interact. This information is used to unravel the history of the early universe. Also learn of the evidence for the Big Bang and for dark matter, on which the future of the universe depends.

Important Concepts

- [ ] fermion
- [ ] boson
- [ ] hadron
- [ ] meson
- [ ] baryon
- [ ] lepton
- [ ] antiparticle
- [ ] lepton numbers
- [ ] baryon number
- [ ] strangeness

- [ ] eightfold way
- [ ] quark
- [ ] messenger particle
- [ ] W and Z particles
- [ ] gluon
- [ ] expansion of the universe
- [ ] Hubble's law
- [ ] microwave background radiation
- [ ] dark matter
- [ ] Big Bang

Overview

44–2   Particles, Particles, Particles

- [ ] Intrinsic angular momentum (spin) is used to classify particles. The maximum value of any cartesian component of the spin is $S_z = m_s(h/2\pi)$, where m_s is either an integer or half integer and $(h/2\pi)$ is the Planck constant divided by 2π. Particles that have m_s equal to 0 or an integer are called **bosons** while particles that have m_s equal to half an odd integer are called **fermions**. Fermions obey the Pauli exclusion principle; bosons do not. Electrons, protons, and neutrons are fermions; photons and pions are bosons.

- [ ] Particles are also classified according to the types of interactions they have with other particles. Particles that interact via the strong nuclear force are called **hadrons**; particles that do not interact via the strong nuclear force but do interact via the weak nuclear force are called **leptons**. Protons, neutrons, and pions are hadrons; electrons, positrons, and neutrinos are leptons.

☐ Hadrons are further categorized according to their spin. Particles that interact strongly *and* have integer (or zero) spin, such as the pion, are called **mesons**. Particles that interact strongly *and* have half-integer spin, such as the neutron and proton, are called **baryons**.

☐ An **antiparticle** is associated with each particle. A particle and its antiparticle have the same mass and spin but charges of opposite sign (if charged). Other quantum numbers also have different signs for a particle and its antiparticle. Except for the positron (e^+), an antiparticle is denoted by a bar over the symbol for the particle: $\bar{p}$ represents an antiproton, for example.

44–3   An Interlude

☐ Energy, charge, momentum, and angular momentum are conserved in all high energy reactions and decays. Mass is not necessarily conserved.

44–4   The Leptons

☐ Counting the neutrinos but not the antiparticles, there are six members of the lepton family. They are: the electron (e), the muon (μ), the tauon (τ), the electron neutrino (ν_e), the muon neutrino (ν_μ), and the tauon neutrino (ν_τ). All are fermions. The electron, muon, and tauon are charged and participate in electromagnetic interactions; neutrinos are not charged. No evidence of any internal structure has been observed for any lepton. They are thought to be truly fundamental.

☐ Electrons and electron neutrinos are assigned an electron lepton number of +1; positrons and electron antineutrinos are assigned an electron lepton number of −1; all other particles have electron lepton numbers of 0. In a similar manner, muon lepton numbers and taon lepton numbers are also assigned. In every decay or reaction, each of the lepton numbers is conserved: the sum for all particles before an event is the same as the sum for all particles after the event, even if the particles change identities.

44–5   The Hadrons

☐ **Baryon number** $B = +1$ is assigned to each baryon, baryon number $B = -1$ is assigned to each antibaryon, and baryon number $B = 0$ is assigned to every other particle. The sum of the baryon numbers before an event is always equal to the sum of the baryon numbers after the event.

☐ No similar conservation law holds for mesons. Mesons can be created or destroyed in any number without violating a conservation law.

44–6   Still Another New Conservation Law

☐ **Strangeness** S is a particle quantum number that is conserved in strong and electromagnetic interactions but not in weak interactions. Particles are assigned strangeness S by observing the results of decays and interactions. K^+ and K^0, for example, have $S = 1$, K^- and $\bar{K}^0$ have $S = -1$.

☐ When a decay or reaction takes place via either the strong or electromagnetic interaction, the total strangeness of the products must be the same as the total strangeness of the original particles. If the decay or reaction takes place via the weak interaction, the total strangeness may change.

44–7 The Eightfold Way

☐ When the strangeness and charge quantum numbers of the eight baryons with spin quantum number $\frac{1}{2}$ are plotted, with the charge on a sloping axis, the result is a hexagonal pattern. The same pattern is formed by the nine mesons with spin zero. See Fig. 44–4. All particles can be arranged in this pattern or a triangular pattern of ten particles. These **eightfold way patterns** are related to the particles in much the same way as the periodic table of chemistry is related to atoms.

44–8 The Quark Model

☐ Baryons and mesons are thought to be made of smaller, more fundamental particles called **quarks**. Every meson is a combination of a quark and an antiquark, whereas every baryon is a combination of three quarks and every antibaryon is a combination of three antiquarks.

☐ The baryon number of a quark is $+1/3$ and that of an antiquark is $-1/3$, so the model predicts that the baryon number of a meson is 0, the baryon number of a baryon is 1, and the baryon number of an antibaryon is -1, in agreement with observation.

☐ Six quarks (and associated antiquarks) have been discovered, in the sense that they are required for the construction of observed mesons or baryons. Many physicists believe no others exist. The lowest mass mesons and baryons consist of u (up), d (down), and s (strange) quarks, along with the associated antiquarks. The quark content of the π^- meson, for example, is $u\bar{d}$; the quark content of the π^+ meson is $d\bar{u}$. Kaons, which are strange mesons, have an s or $\bar{s}$ quark. The quark content of a proton is uud and the quark content of a neutron is ddu. Σ baryons are strange and have s or $\bar{s}$ quarks.

☐ Quarks are charged and so participate in electromagnetic interactions, but the charge of a quark is a fraction of the fundamental charge e, not a multiple of it. A u quark has a charge of $+2e/3$, a d quark has a charge of $-e/3$, and an s quark has a charge of $-e/3$. The charge on an antiquark is the negative of the charge on the associated quark. The charge of a hadron is simply the sum of the charges on its constituent quarks.

☐ Quark number is conserved in strong interactions. The strong interaction can create and destroy quark-antiquark pairs (such as $d\bar{d}$ or $u\bar{u}$) but it cannot change one type quark into another. The weak interaction, however, can change one type quark into another. In a beta decay of a proton ($p \rightarrow n + e^+ + \nu$), a u quark is changed into a d quark. In a beta decay of a neutron ($n \rightarrow p + e^- + \bar{\nu}$), a d quark is changed into a u quark.

44–9 The Basic Forces and Messenger Particles

☐ In modern quantum theory, the basic interactions are pictured as exchanges of **messenger particles**. The messenger particles associated with the weak interaction are called W and Z, the messenger particle associated with the electromagnetic interaction is called the photon, and the messenger particles associated with the strong interaction are called **gluons**.

☐ When a particle emits a messenger particle, its rest energy is reduced. Nevertheless, it retains its character as long as it absorbs a messenger particle a short time later. Energy is not conserved during the interval between emission and absorption. For a given loss in energy ΔE, the time interval between emission and absorption must be less than the minimum predicted by the Heisenberg uncertainty principle. Thus $\Delta t < h/\Delta E$. This limits the time for events, such as decays, and also limits the range of an interaction.

☐ Quarks interact with each other via the strong interaction, which involves the exchange of gluons. Inside a baryon or meson, the interaction force of two quarks is very weak but the interaction becomes extremely strong if the separation is larger than the particle size. Free quarks have never been observed. When a high-energy bombarding particle strikes a baryon or meson, other quarks are created (in quark-antiquark pairs) and these form combinations with existing quarks and are emitted as other mesons or baryons.

☐ The property of quarks that causes them to emit and absorb gluons is called **color**. It is somewhat similar to charge, the property that causes particles to absorb and emit photons and thus participate in the electromagnetic interaction. The chief difference is that a gluon carries color away from the quark that emits it and thereby changes the color of the quark. On the other hand, photons do not carry charge. The three types of color are called red, yellow, and blue.

☐ The condition that the net color of quarks in a hadron be neutral explains the limited quark combinations observed. A particle is color neutral only if it contains three quarks of different color, three antiquarks of different color, or a quark and antiquark of the same color.

☐ The basic interactions may be different manifestations of a single interaction. The electromagnetic and weak interactions have already been successfully unified into what is known as the **electroweak** interaction. A theory that unifies the strong and electroweak interactions is called a **grand unification theory** (GUT) and a theory that unifies all four interactions is called a **theory of everything** (TOE).

44–10   A Pause for Reflection

☐ At the present time all phenomena that have been observed in nature can be explained in terms of the up quark, the down quark, the electron, the electron neutrino, and the messenger particles of their interactions. The other quarks and leptons are observed only in high energy experiments. However, they were present in the early universe and greatly influenced its evolution.

44–11   The Universe Is Expanding

☐ Distant astronomical objects are seen as they were when the light left them, perhaps more than 10^9 y ago.

☐ Most physicists believe that the universe started in a state of extremely high density and temperature, perhaps a highly concentrated mixture of quarks, leptons, and messenger particles. A "big bang" initiated an expansion (and cooling) that will continue for some time into the future, perhaps forever.

☐ Experimental evidence exists for the expansion of the universe and for the Big Bang. Doppler shift measurements are used to determine the speeds and directions of travel of distant galaxies. The galaxies are found to be moving away from each other. Hubble's law relates the speed with which a galaxy is receding from us to the distance between us and the galaxy. Mathematically it is $v = Hr$, where H is the Hubble parameter, about $80 \pm 17 \mathrm{km}/(\mathrm{s} \cdot \mathrm{Mpc})$.

Since a parsec is 3.256 ly, this is the same as 19.3 mm/(s · ly). An observer in *any* part of the universe sees galaxies at the same distance receding with the same speed.

☐ If the rate of expansion of the universe has been constant, the reciprocal of the Hubble constant gives the age of the universe: about 1.5×10^{10} y.

44–12 The Cosmic Background Radiation

☐ The cosmic microwave background radiation fills the entire universe with an intensity that is nearly the same in every direction. There is no distinguishable source. It is believed to have originated shortly after the Big Bang. Although the background radiation started as a hot gas of highly energetic photons, in the gamma ray region of the electromagnetic spectrum, today it has a thermal spectrum corresponding to a temperature of about 2.7 K and is chiefly in the microwave region of the spectrum. The decrease in frequency accompanied the cooling of the universe.

44–13 Dark Matter

☐ Whether the universe will continue to expand indefinitely or will eventually collapse (perhaps in preparation for another big bang) depends on the amount of mass it contains. The universe seems to contain much more mass than is visible and the postulated invisible matter is called **dark matter**. The chief evidence for dark matter comes from a study of the rotation rates of distant galaxies.

44–14 The Big Bang

☐ The text describes five eras in the history of the universe as it evolved from an extremely hot gas of quarks, leptons, and messenger particles to its present state.

Hints for Questions

1 The magnetic force is in the direction of the vector product $\vec{v} \times \vec{B}$ for positively charged particles and in the opposite direction for negatively charged particles. Here $\vec{v}$ is the particle velocity and $\vec{B}$ is the magnetic field. The force points from the particle toward the center of its circular orbit.

[Ans: into]

3 The radius of curvature of the track is proportional to the momentum of the particle.

[Ans: the π^+ pion whose track terminates at point 2]

5 The mass of a quark and its antiquark are the same. The spin quantum number of all quarks and antiquarks is 1/2. Quarks and their antiquarks have charge of opposite sign. All quarks and antiquarks have lepton number 0. Quarks have baryon number +1/3 and antiquarks have baryon number −1/3.

[Ans: c, f]

7 A proton has electron lepton number 0, muon lepton number 0, baryon number +1, and spin quantum number 1/2. Electrons and neutrinos have electron lepton number +1 while their anti particles have electron lepton number −1. All have muon lepton number 0, baryon

number 0, and spin quantum number 1/2. Electrons have charge quantum number -1, positrons have charge quantum number $+1$, and neutrinos have charge quantum number 0.

[Ans: baryon number]

9 Look at Table 44–5 to obtain the quark properties. Since the spin quantum number of the particle is 1/2 it must consist of three quarks or three antiquarks. Since the strangeness quantum number is -1 it must contain one s quark. The other two quarks are chosen so the total charge quantum number is $+1$.

[Ans: c]

11 Negative leptons have lepton number $+1$, positive leptons have lepton number -1, neutrinos have lepton number $+1$, antineutrinos have lepton number -1, and all other particles have lepton number 0.

[Ans: (a) 0; (b) $+1$; (c) -1; (d) $+1$; (e) -1]

13 Look at Tables 44–2, 44–3, 44–4, and 44-5.

[Ans: 1b, 2c, 3d, 4e, 5a]

Hints for Problems

1 Since momentum is conserved the gammas have momenta of equal magnitude. They therefore have equal energies. The sum of their energies equals the rest energy of the pion. The wavelength is $\lambda = c/f = hc/E$, where f is the frequency and $E \ (= hf)$ is the photon energy.

[Ans: 18.4 fm]

5 The kinetic energy K of a relativistic particle is related to it speed v by $K = (\gamma - 1)mc^2$, where m is its mass and $\gamma = 1/\sqrt{1 - (v/c)^2}$. Solve for v. Since the energy is so much greater than the rest energy you should find the algebraic solution and then make a binomial expansion in powers of the ratio of the rest energy to the total energy.

[Ans: 0.0266 m/s.]

13 For each quantum number the total for the particles entering the reaction must equal the total for the particles leaving the reaction. The pion has a charge quantum number of $+1$ and the proton has a charge quantum number of $+1$. The pion has a baryon number of 0 and the proton has a baryon number of $+1$. The pion and the proton have strangeness quantum numbers of zero.

[Ans: (a) 0; (b) -1; (c) 0]

17 Use conservation of charge, baryon number, and angular momentum to find the values of these quantities for the unknown particle. Conservation of angular momentum requires that if an odd number of fermions enter the reaction then an odd number must leave and if an even number (including 0) enter then an even number must leave.

[Ans: (a) K^+; (b) $\bar{n}$; (c) K^0]

27 Use $v = Hr$, where r is the distance to the galaxy and H is the Hubble constant, to find the speed v of the galaxy, then use the Doppler shift equation $\Delta\lambda/\lambda = v/c$ to find the shift $\Delta\lambda$ in the wavelength. The observed wavelength is $\lambda + \Delta\lambda$.

[Ans: 666 nm]

31 The number of molecules in an excited state with energy E above the ground state is given by $N = N_0 e^{-E/kT}$, where N_0 is the number in the ground state, k is the Boltzmann constant, and T is the temperature on the Kelvin scale. Put N/N_0 equal to 0.25 and solve for E.

$\left[\text{Ans: (a) } 256 \,\mu\text{eV; (b) } 4.84\,\text{mm}\right]$

35 The energy of the photon is given by $E = hf = hc/\lambda$, where f is the frequency of its wave. To create an electron-positron pair the energy must be at least twice the rest energy of the electron, or $1.022\,\text{MeV}$.

$\left[\text{Ans: (b) } 2.39 \times 10^9\,\text{K}\right]$

41 The relativistic expression for the momentum is $p = mv/\sqrt{1 - (v/c)^2}$, where m is the mass and v is the speed of the particle. For the antiproton $mc^2 = 938.3\,\text{MeV}$ and for the pion $mc^2 = 193.6\,\text{MeV}$. For both particles $pc = 1.19\,\text{GeV}$. The time interval between the signals from S_1 and S_2 is $\Delta t = d/v$, where $d \ (= 12\,\text{m})$ is the distance between the scintillation counters.

$\left[\text{Ans: (a) } 0.785c\text{; (b) } 0.993c\text{; (c) C2; (d) C1; (e) } 51\,\text{ns; (f) } 40\,\text{ns}\right]$

Quiz

Some questions might have more than one correct answer.

1. Which of the following statements are true?

   A. particles with a spin angular momentum quantum number of $(2n + 1)(h/2\pi)/2$, where n is a positive integer or zero, obey the Pauli exclusion principle

   B. particles with a spin angular momentum quantum number of $n(h/2\pi)$, where n is a positive integer or zero, obey the Pauli exclusion principle

   C. particles with a spin angular momentum quantum number of $(2n + 1)(h/2\pi)/2$, where n is a positive integer or zero, are called bosons

   D. particles with a spin angular momentum quantum number of $(2n + 1)(h/2\pi)/2$, where n is a positive integer or zero, are called fermions

   E. particles with a spin angular momentum quantum number of $n(h/2\pi)$, where n is a positive integer or zero, are called bosons

2. The strong nuclear force acts on

   A. all hadrons
   B. all bosons
   C. all fermions
   D. all leptons
   E. all photons

3. The weak nuclear force acts on
    A. all hadrons
    B. all bosons
    C. all fermions
    D. all leptons
    E. all photons

4. The electromagnetic force acts on
    A. all hadrons
    B. all bosons
    C. all fermions
    D. all leptons
    E. all electrically charged particles

5. Which of the following statements are true?
    A. all hadrons are fermions
    B. some hadrons are fermions and some are bosons
    C. all leptons are fermions
    D. some leptons are fermions and some are bosons
    E. some leptons are hadrons

6. A particle and the corresponding antiparticle
    A. have charge of the opposite sign, if they are charged
    B. have mass of the opposite sign
    C. have the same mass
    D. have the same spin angular momentum
    E. can annihilate each other

7. There are three lepton families, each consisting of a particle and antiparticle with mass, a neutrino, and an antineutrino (perhaps without mass). Which of the following statements are true?
    A. the particles of a lepton family, but not the antiparticles, are assigned a lepton family number of +1
    B. the antiparticles of a lepton family, but not the particles, are assigned a lepton family number of +1
    C. particles that are not leptons are assigned a lepton number of +1
    D. all three lepton family numbers are conserved in all interactions
    E. all three lepton family number are conserved in all interactions except weak interactions

8. If a negative muon decays, one of the products must be

    A. a muon neutrino
    B. a muon antineutrino
    C. a negatively charged particle
    D. a positively charged particle
    E. a fermion
    F. a boson

9. A baryon is

    A. a hadron that is also a boson
    B. a hadron that is also a fermion
    C. a lepton that is also a boson
    D. a lepton that is also a fermion
    E. a meson that is also a fermion

10. Baryon numbers are assigned as follows:

    A. baryon particles have baryon number +1
    B. baryon antiparticles has baryon number +1
    C. mesons have baryon number +1
    D. leptons have baryon number +1
    E. photons have baryon number +1

11. Net baryon number is conserved in

    A. all interactions
    B. all interactions except the weak interaction
    C. all interactions except the electromagnetic interaction
    D. only the strong interaction
    E. only the strong and electromagnetic interactions

12. Which of the following statements are true?

    A. all particles have a strangeness number of either +1 or −1
    B. some baryons have a strangeness number of either +1 or −1
    C. some mesons have a strangeness number of either +1 or −1
    D. all hadrons have a strangeness number of either +1 or −1
    E. some leptons have a strangeness number of either +1 or −1

13. Which of the following statements are true?

    A. all baryons are made of three quarks
    B. all mesons are made of three quarks
    C. all mesons are made of a quark and an antiquark
    D. all mesons are made of two quarks
    E. all leptons are made of two quarks

14. The quark content of a hadron determines its

    A. baryon number
    B. charge
    C. spin angular momentum qunatum number
    D. strangeness
    E. none of the above

15. The following are messenger particles:

    I. photons
    II. W and Z particles
    III. gluons

Associate the messenger particles with the forces they mediate.

    A. weak – I and II, electromagnetic – I, strong – I and III
    B. weak – II, electromagnetic – II, strong – I and III
    C. weak – III, electromagnetic – I and III, strong – II
    D. weak – I, electromagnetic – I, strong – II and III
    E. weak – II, electromagnetic – I, strong – III

Answers: (a) A, D, E; (2) A, B; (3) A, B, C, D; (4) E; (5) B, C; (6) A, C, D, E; (7) A, D; (8) A, C; (9) B; (10) A; (11) A; (12) B, C; (13) A, C; (14) A, B, C, D; (15) E

NOTES